AF564415

2nd Fully Revised and Enlarged Edition

Irrigation and Drainage Engineering

NIPA® GENX ELECTRONIC RESOURCES & SOLUTIONS P. LTD.
New Delhi-110 034

About the Author

Prof. R.K. Biswas is an M.Tech and Ph.D in Agricultural Engineering with specialization in Soil & Water Conservation Engineering. He has been actively engaged in teaching and research since he joins in Bidhan Chandra Krishi Viswavidyalaya in 1984. During this period Prof. Biswas has uninterruptedly undertaken the courses of irrigation, drainage & water management in PG & UG level and conducted research works funded by ICAR, MoWR, Govt. of India and other financial institutions. The outcome of the research works has been appreciated by the funding agencies and the professionals.

Prof. Biswas has served the State Govt. Departments (West Bengal) as the technical expert, Ex-officio director & TSG in the area of drought & water resources management, agro-industries and micro-irrigation development respectively. Prof. Biswas has been all along worked for the academic development of the institution and betterment of the students. The effort of Prof. Biswas in introducing UG & PG programs in Bidhan Chandra Krishi Viswavidyalaya is well recognized. Prof. Biswas has served the university at various capacities, viz. Dean, Post Graduate Studies; Dean, Faculty of Agricultural Engineering and Head, Dept. of Soil & Water Engineering.

2nd Fully Revised and Enlarged Edition

Irrigation and Drainage Engineering

Ranajit Kumar Biswas

Dean, Faculty of Agricultural Engineering
Bidhan Chandra Krishi Vishwavidayalaya
District Nadia-741252, West Bengal
Member, Water Resource and Drought Mission (GoWB)

NIPA® GENX ELECTRONIC RESOURCES & SOLUTIONS P. LTD.
New Delhi-110 034

NIPA® GENX ELECTRONIC RESOURCES & SOLUTIONS P. LTD.

101,103, Vikas Surya Plaza, CU Block
L.S.C. Market, Pitam Pura, New Delhi-110 034
Ph : +91 11 27341616, 27341717, 27341718
E-mail:newindiapublishingagency@gmail.com
www: www.nipabooks.com

For customer assistance, please contact
Phone: +91-11-27 34 17 17
Fax: +91-11-27 34 16 16
E-Mail: feedbacks@nipabooks.com

Print ISBN: 978-93-95319-70-6

ebook ISBN: 978-93-95319-71-3

NIPA also publishes books in a variety of electronic formats. Some content that appears in print may not be available in electronic books, and vice versa.

Composed and Designed by NIPA.

Preface to the First Edition

This book **Irrigation & Agricultural Drainage Engineering** is intended as a text book in the area of irrigation and drainage for the students of agricultural engineering in particular and agricultural science in general. However, this book also may be useful for agricultural extension workers and the professional working in this area. Irrigation and drainage is a vast area. It is not possible to include everything of it in a book. Drip and sprinkler are the most important and advanced methods of irrigation. It deserves more detail discussion. These are being tried in a separate book. The contents of the book may enable one to acquire some basic requirements which an irrigation and drainage manager must have. The contents include basics along with some information toward research achievements, importance and usefulness so that the students get interested to the subject and at the same time help them to attend the institutional and competitive examinations. The book contains good numbers of numerical as example and task to get the students familiar to the requirements, complicacies, and possible remedies in actual working condition. Excepting the traditional broad and short questions multiple choice questions are also set in every chapter to assist the students in successful preparation for the entrance examinations in PG programs and the competitive examinations like State and Union PSC, etc.

The author is thankful to Indian Council of Agricultural Research (ICAR), Ministry of Agriculture, Govt. of India and Indian National Committee for Irrigation and Drainage (INCID), Ministry of Water Resources, Govt. of India for financial support in conducting research works, the outcome which are found relevant have been included in this book. The author gratefully acknowledges the support of Bidhan Chandra Krishi Viswavidyalaya and my esteemed colleagues in this university who are the basis of everything in preparation of this book.

I would acknowledge the great contribution of Dr.A.K.Bhattacharya, erstwhile eminent Scientist and Teacher, IARI, New Delhi. The touch of his personality and knowledge and recognition I enjoyed during the research work conducted under his guidance funded by ICAR and the complementary offer of his book 'LAND DRAINAGE-Principles, Methods and Applications' inspired me a lot to put the best effort to academics and research. In this opportunity I would like to mention my students over the years that I have come across; their respect and thankfulness, their curiosity and originality in ideas has encouraged me to go ahead. I will feel rewarded if this endeavor can help them in anyway. Every effort is made to acknowledge the sources of information in the text. If any omission remains, it is inadvertent, and will be corrected if noted or pointed out.

I would like to thank NIPA for the support of publication of this book.

R K. Biswas

Preface to the First Edition

Preface to the Second Edition

The second edition of this book is renamed **as Irrigation & Drainage Engineering,** looking after the interest of the students and other users good numbers of numerical analysis particularly the GATE questions of Agricultural Engineering, a new chapter by the name **Groundwater Hydraulics & Wells** and the topic **Grassed Waterways** have been included. Hopefully, these will help to get the book more useful.

I would like to thanks and acknowledge **NIPA,** New Delhi for the consistent persuasion in publishing this second edition.

R.K. Biswas

Contents

Preface to the First Edition .... *v*
Preface to the Second Edition vii

1. Methods of Irrigation 1
1.1 Introduction 1
1.2 Border Irrigation 4
1.3 Check Basin 27
1.4 Furrow Irrigation 36
1.5 Sub-surface Irrigation 46
Questions and Problems *47*
References 53

2. Measurement of Irrigation Water 55
2.1 Volumetric Method 55
2.2 Velocity-Area Method 57
2.3 Measuring Structures 69
2.4 Tracer Method 84
Questions and Problems *88*
References 91

3. Irrigation Efficiencies 93
3.1 Introduction 93
3.2 Various Irrigation Efficiencies 93
Questions and Problems *106*
References 108

4. Scheduling Irrigation 109
4.1 When to Irrigate 109
4.2 How Much to Irrigate 118
Questions and Problems *130*
References 134

5. Design of Irrigation Channels 135

5.1 Irrigation Channels 135

5.2 Design Considerations 141

5.3 Regime Approach 163

Questions and Problems *185*

References 189

6. Soil-Water-Plant Relationship 191

6.1 Soils 191

6.2 Soil-Water Relationship 199

6.3 Classes of Soil Water 206

6.4 Infiltration 212

6.5 Soil Moisture Characteristics 228

6.6 Water Movement in Soils 236

6.7 Soil Water Uptake by Plants 261

Questions and Problems *262*

References 266

7. Consumptive Use of Water 269

7.1 General 269

7.2 Energy Balance 275

7.3 Radiation 283

7.4 Measurement of Evapotranspiration 289

Questions and Problems *332*

References 335

8. Groundwater Hydraulics & Wells 337

8.1 water Bearing Formation 337

8.2 Aquifer Properties 340

8.3 Equation of Motion 355

8.4 Flow in Wells 365

Questions and Problems *378*

References 382

9. Lining Irrigation Channels 383

9.1 Introduction 383

9.2 Types of Lining 384

9.3 Selection of Lining Materials 403

9.4 Economics of Canal Lining 405

9.5 Measurement of Seepage 416

Questions and Problems *427*

References 431

10. Salt Problem and Irrigation Water Quality 433

10.1 Causes of Salt Problems 433

10.2 Terminology 434

10.3 Classification of Salt Affected Soils 436

10.4 Irrigation Water Quality and Effects on Soil and Crop 439

10.5 Quality and Classification of Irrigation Water 441

10.6 Control of Salinity Problem 446

10.7 Infiltration Problem 461

10.8 Toxicity 464

Questions and Problems *465*

References *468*

11. Water Resources and Irrigation Development in India 469

11.1 Land and Water Resources 469

11.2 History of Irrigation in India 473

11.3 Irrigation Development Under the Five Year Plans 477

11.4 Water Resources of the World 480

11.5 Quality of Ground Water Resources of India 481

Questions and Problems *486*

References 488

12. Surface Drainage 489

12.1 Introduction 489

12.2 Surface Drainage System 491

12.3 Design of Surface Drainage System 497

12.4 Grassed Waterways 514
12.5 Curve Number Method 520
Questions and Problems 530
References 534

13. Sub-surface Drainage 535
13.1 Design Considerations 535
13.2 Replacement of the Ditch by Tile Drain and Concept of Equivalent Depth 544
13.3 Ernst Equation 568
13.4 Non-steady State Sub-surface Drainage 572
13.5 Layout of the Tile Drains 580
13.6 Drainage Material 584
13.7 Mole Drain 587
Questions and Problems 588
References 593

Index* *595

1

Methods of Irrigation

1.1 Introduction

Irrigation is the artificial application of water needed to the crop field. It is the process of uniformly wetting the root zone depth. The wetting may be done over the land surface or below of it. The application of water in excess to the requirement or accumulation of water in crop field due to rain or unplanned seepage is not considered as irrigation. However, irrigation is supplemental to rainfall. Irrigation water is considered as single most important input for crop production.

Functions of irrigation water

i) It acts as the solvent for the nutrients in soil derived by the plants.

ii) It provides essential moisture for the germination of seeds, promotes chemical process, bacterial growth and favorable environment for the plants.

iii) It helps in controlling soil temperature from summer and winter variation in temperature.

iv) It softens soil clods and helps in easy tillage operation.

v) It leaches out the salts in soil for reclaiming it under controlled condition.

vi) It reduces hazard of soil piping.

vii) It enables application of fertilizers.

viii) It helps in maintaining the turgidity of plants. In absence of water wilting to death of plants will occur.

Classification of irrigation methods

The irrigation methods may be classified as below.

1. Surface irrigation
2. Sub-surface or sub-irrigation
3. Overhead irrigation
4. Drip irrigation

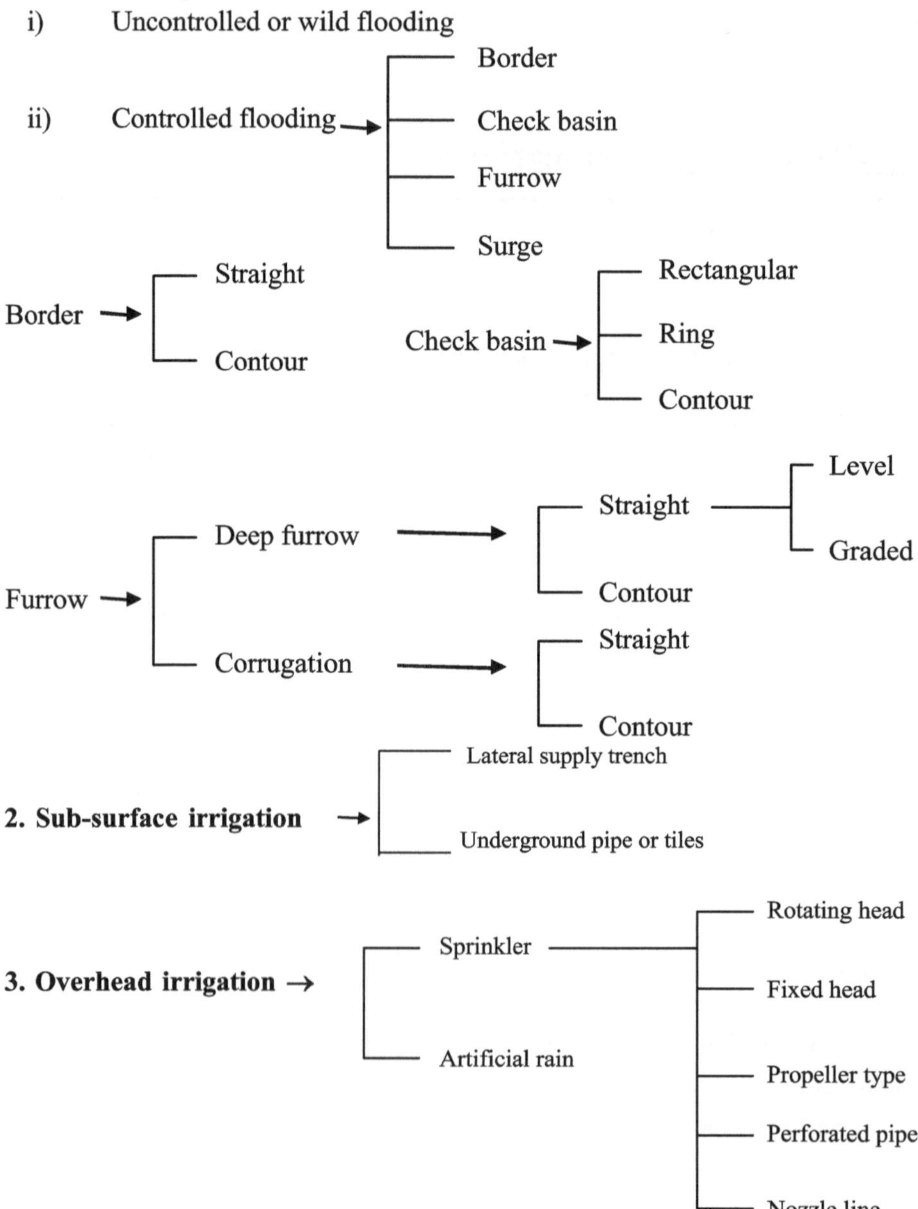

4. Drip or trickle irrigation

Surface irrigation method

In surface irrigation method water is allowed to flow over the land surface towards the down slope after receiving it at the upper reach of the field from the supply channel. In general the lands are divided in to so many pieces to facilitate uniform distribution of water. Land slope, crop grown, size of the stream, extent of land level, etc., determine the selection of particular method of surface irrigation. Surface method is most conventional method of irrigation throughout the world.

Uncontrolled or wild flooding

In this method water is applied relatively to a vast tract of fairly smooth or flat field without much control over the flow by the channels and bunds. The stream size is usually large.

Advantages

1. Suitable for pasture and forage crops.
2. Where land leveling is difficult or expensive.
3. Skill of labour is not required.

Disadvantages

1. Uniformity of water distribution is poor.
2. Wasteful use of water.
3. Suitable only where water is in plenty and inexpensive.
4. It is used in deep soil, which does not crust badly.

Controlled flooding

In controlled flooding the lands are suitably divided into plots of suitable size depending on the porosity of soil. The irrigation is done by applying water to the higher elevation point of the plot from where it proceeds to lower end. The plots are comparatively small in flat field. Flat fields are provided in inundation irrigation. Spacing of the water courses are designed depending on type of soil, size of stream and topography of the field.

Advantages

1. It helps to store the required amount of water in capillary zone of the soil.
2. Little loss of water in the form of runoff.
3. Uniformity of application of water is good.
4. Soil erosion problem is less.
5. Machineries can be used for land preparation, cultivation, etc.

Disadvantages

1. Efficiency is lower than the sub-surface, sprinkler or drip irrigation.
2. Considerable quantity of water is lost through evaporation or deep percolation particularly when stands for a long time.
3. High skill is required for getting higher uniformity of application of water.
4. 7 to 10% land is lost due to watercourses and ditches.

Factors involved in hydraulics of controlled flooding

1. Slope of the field.
2. Length of run and time required.
3. Roughness of the field.
4. Size of the stream
5. Depth of water to be applied.

1.2 Border Irrigation

In border irrigation the area to be irrigated is divided into a number of strips separated by low height border ridges or bunds (Fig 1.1). The water is delivered to the upper end of the border from where it flows towards tail end in the form of thin sheet guided by the borders. Borders are having some slopes (<5%) towards the direction of its length. There is practically no slope across the border. The efficiency of irrigation in a border largely depends on the uniformity of land leveling both length and crosswise. The border method of irrigation is suitable for close growing, non-sown or drilled crops and orchards excepting in any soil condition where infiltration rate is very high or low or cost involvement in land shaping is economically not viable for border irrigation. This method is best suited up to slope of less than 0.5%.

Cross border: Poor land leveling leads too much uneven distribution of water. The water applied to the border use to advance to the tail end covering entire width of the border in a uniformly leveled border. In an uneven border the watercourse follow the low levels or fill up the depressions by avoiding the areas at comparatively higher levels. This causes to reach the water to the tail end much earlier or later than a uniformly leveled border. To divert the water from lower level to higher-level cross bunds or borders are used (Fig. 1.3). This is the practice, which

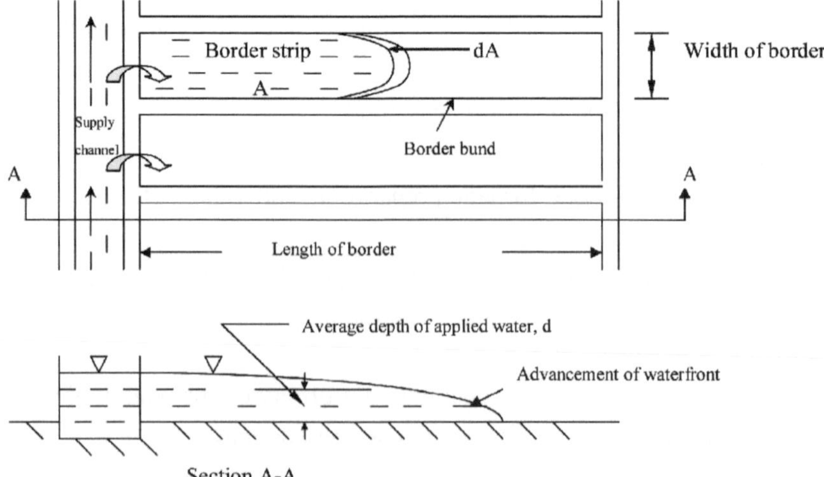

Fig. 1.1 Diagram of border strip

provides scope to cover entire width of border by the waterfront. The percentage of uniformity of application of water can be increased up to 10-15% following this practice (Biswas, 2001). The number, length and angle of the cross border depend on the width, length, stream size and degree of land leveling. Different level of undulations develops in the field during the laddering operation is almost unavoidable unless much extra care is taken involving additional cost. The farmers' practice of cross-border requires little cost but provides significant water saving (Table 1.1).

Fig. 1.2 Border irrigation in wheat field

The farmers' practice of using cross-bunds for better distribution of water in the borders of wheat were studied in the Haringhata area of Nadia district in West Bengal for wheat crop in consecutive three years. Similar to the farmers the lands were prepared, borders were made and seeds were sown with the usual leveling operation. All the borders were provided with the required irrigations and scheduled agronomic treatment throughout the growing period. There were 10 numbers of borders of which 5 were provided cross-bund and the other 5 without cross-bund to have the comparison of performance of these two groups of borders. The details of the borders, cross-bunds, water application, water storage, application and distribution efficiencies of water in both borders with cross-bunds and without cross-bunds are described in Table 1.1 as on one year result of study.

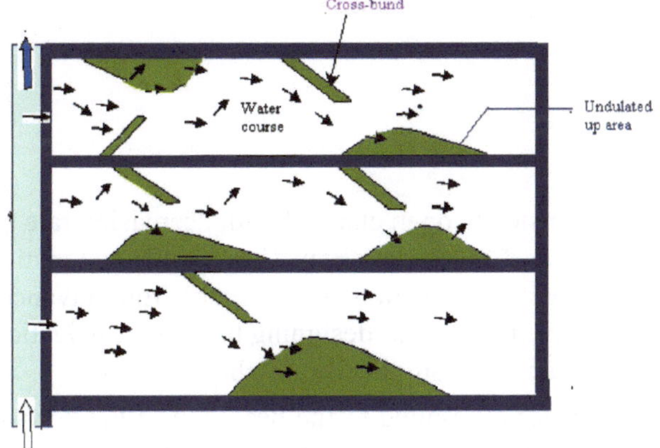

Fig. 1.3 Cross-border in border strip

Straight and Contour border

In straight border the fields are divided into straight strips and borders are taken straight when the field is of gentle slope and can be leveled without affecting the soil productivity by removing the fertile topsoil. It is found that straight border is not possible due to excess land slope or extreme undulation which can not be properly leveled economically then borders may be constructed along the general slope following the contour which are called as contour borders. The design criteria are same for straight or contour border. There should have longitudinal slope and level cross-slope in each type of borders. The direction of the contour border strip may change to approximately right angle to the direction of the general land slope. In straight border general land slope is being used for border slope. Contour border may adjust the border slope by adjusting the direction of the strip (Fig. 1.4). Irrigation channels in contour border may follow the general slope instead of following the direction perpendicular to the general land slope followed in case of straight border. The contour borders are constructed independently and therefore after the construction these may have the look of steps. The step shape irrigation channel requires some measures against scouring effect of falling water.

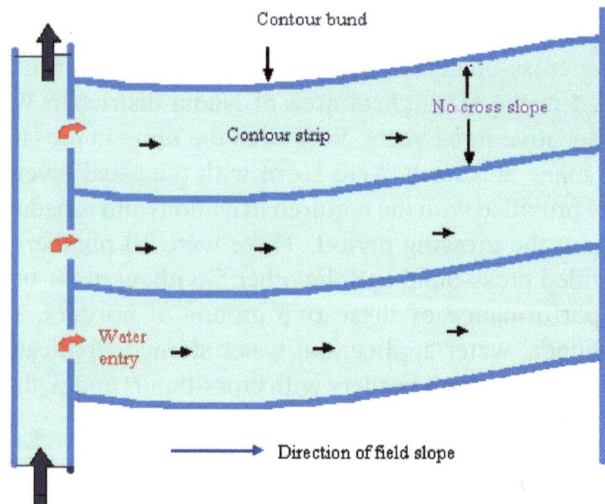

Fig. 1.4. Contour border

Hydraulics of border irrigation

The flow of water in a border is of unsteady open channel with decreasing rate of discharge. The discharge rate decreases due to infiltration. The flow characteristics in a border greatly govern by the hydraulic phenomena of water front advance, ponding storage and recession. For the purpose of designing the border irrigation system these hydraulic phenomena are to be studied thoroughly.

Water front advance: Irrigation water as applied to the head end of the border, the major portion of it flows downward and remaining portion infiltrates in to soil.

As the time elapse the water front advance further and cover larger area of the border and thereby allowing further area for infiltration. It is desirable to advance the waterfront in the border at such a rate that will provide a uniform infiltration opportunity time to every section of the borders. This is possible by adjusting the stream size with width, length and slope of the border. Sometime, in contrary, width and length of the border are adjusted depending on the availability of stream size.

Ponding of water: The uniform infiltration at every section in a border is possible only when there is uniform opportunity of ponding. The delivery of water in a border usually stops when it reaches to the tail end or at certain predetermined section or length. The ponding of water starts at head end as soon as it receives supply of water but at tail end it starts much later as the water takes time to reach at there. However, once the irrigation is stop the standing water at head end disappears earlier than the tail end. The disappearing of water approaches slowly towards the tail end. The time in between the receiving and disappearing of ponded water at the head and tail end should be same in an efficient irrigation practice.

Recession of flow: The disappearance of water from the head end and recedes to the downstream in a border is the recession of flow. The recession and advancement of water occurs in similar way. Therefore, curves can be drawn taking the distance as X-axis and the elapsed time as Y-axis in same graph paper (Fig. 1.5). The parallelism of the advance and recession curves expresses the uniform distribution of water.

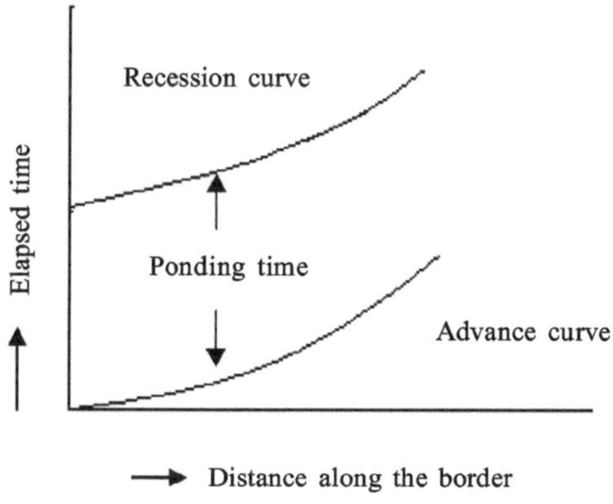

Fig. 1.5. Advance and recession curve

Infiltration opportunity time or time of ponding: This is the time period in between the water reaches to any particular point in a border and disappears from there to recedes to tail end. The infiltration opportunity time of any point in the border is the vertical distance between the advance and recession curves at that point.

Table 1.1. Performance of borders for cross-bunds and without cross-bunds in wheat crop

No. of border	Width of border (m)	Length of border (m)	Border Slope (%)	No. of cross bunds	Total length of cross-bunds in The borders (m)	Length of cross-bunds/ 100m border length (m)	Angle of cross-bunds (degree)	Water applied (cm)	Water stored in root zone depth (cm)	Application efficiency (%)	Distri-bution efficiency (%)	Discharge/ meter border width(lit)
1	3	18.50	0.26	-	-	-	-	5.21	4.28	82.15	82.32	2.10
2	3	19.00	0.26	4	5.0	26.31	95.16	5.18	4.86	93.82	94.15	2.03
3	3	20.70	0.26	-	-	-	-	5.37	4.33	80.63	81.22	1.94
4	5	25.70	0.15	4	5.8	22.57	115.26	4.90	4.64	94.69	95.58	2.14
5	3.5	15.80	0.10	-	-	-	-	4.75	4.00	84.21	83.23	1.75
6	3.5	15.80	0.10	4	3.6	22.78	105.87	4.83	4.60	95.24	93.97	1.88
7	5	20.60	0.15	-	-	-	-	5.13	4.29	83.62	81.82	1.79
8	5	20.60	0.15	4	4.1	19.90	93.14	4.68	4.40	94.02	93.89	2.25
9	5	25.30	0.15	-	-	-	-	4.72	3.91	82.84	83.51	2.04
10	5	25.60	0.15	4	5.75	22.40	110.86	4.93	4.62	93.71	94.05	1.84
Average						22.80		4.97	4.39	82.69,92.28	82.42,94.49	

Border specification: In any irrigation it is desirable to wet the soil root zone depth uniformly throughout the field. To ensure such irrigation the width and lengths of the borders are determined depending on the size of the stream, land slope, soil type and degree of land leveling. The following are some assumptions in general in regard to these.

Width of border: The width of a border depends on size of the stream and degree of land leveling practicable. As the stream is small the border width is reduced. Usually border widths vary from 3 to 15m. Sometime the farmers use the border width as low as 1.5m when the degree of land leveling is poor and further leveling proved to be uneconomic. When the width of border in a field decreases the number of border increases accordingly involving some additional cost and wastage of land for making the bunds. However, the skilled farmers' cultivation practice are such that the crops are grown so close to these bunds that it can not be said misuse of land.

Border slope: Usually a border is having the gentle slope towards the tail end. It is obvious that higher slope will cause more quickly the water to reach at tail end and provides more infiltration opportunity time at there. In such a condition to satisfy the head end with required depth of water, there is the possibility of deep percolation at an incremental rate towards the tail end. It will be opposite if the border slope is excessively low. Higher slope in border also may cause soil erosion and breach of bund at tail end. The recommended safe limit of land slope at different type of soil is tabulated in Table 1.2.

Table 1.2. Recommended safe limit of border slope

Type of soil	Percent slope
Clay to clay loam (Heavy soil)	0.05-0.20
Medium loam (Medium soil)	0.20-0.40
Sandy loam to sandy (Light soil)	0.25-0.65

Stream size: The appropriate size of irrigation stream depends on the rate of infiltration of the soil and width of the border. The coarse soil having higher infiltration requires large stream so that in a short time the water may spread over the strip avoiding the excessive deep percolation loss. The size of stream if wisely used can compensate the inadequacies of selected width and length of the border. In fact, the size of the stream can be selected at some particular rate with per unit width of the border. The total discharge is the multiplication of rate of discharge per unit width to the width of the border. Table 1.3 gives some typical values of stream sizes at different soil and slopes.

Table 1.3. Some typical values of stream sizes at different soil sizes and slopes

Soil type		Border slope, %		Stream size, l/m/s	
Sandy soil, infiltration rate 2.5cm/h		0.20-0.40	0.40-0.65	10-15	7-10
Loamy sand, ,,	,, 1.8 to 2.5 ,,	0.20-0.40	0.40-0.60	7-10	5-8
Sandy loam, ,,	,, 0.6 to 0.8 ,,	0.20-0.40	0.40-0.60	5-7	4-6
Clay loam, ,,	,, 0.0 to 0.8 ,,	0.15-0.3	0.30-0.4	3-4	2-3
Clay, ,,	,, 0.2 to 0.6 ,,	0.1-0.2		2-4	

Source: Michael (1985)

Time required irrigating a given area of border strip

Let us consider an area A irrigated at any time t and the differential area dA flooded during the time dt (Fig. 1.1). The total quantity of water received in the border in small interval, dt, will be the sum of the infiltration and the quantity of surface flow during the interval.

Therefore, $Qdt = fAdt + d.dA$ (1.1)

Where

Q = Discharge of water, ha-m/h

f = Rate of infiltration, m/h

A = Area irrigated at any time t, ha

d = Average depth of sheet of water, m.

From **Eq.1.1,** we get

$$dt = \frac{d.dA}{Q - fA}$$

$$= \frac{d.dA}{Q\left(1 - \frac{fA}{Q}\right)}$$

$$\text{Let, } Z = 1 - \frac{fA}{Q} \therefore dZ = \frac{-fdA}{Q}$$

$$\text{or, } dA = \frac{-QdZ}{f}$$

$$\therefore dt = \frac{-d.QdZ}{fQZ} = \frac{-d.dZ}{fZ}$$

Integrating, t $= \frac{-d}{f} ln Z + ln C$

$$= \frac{-d}{f} ln\left(1 - \frac{fA}{Q}\right) + C$$

When, t = 0, A or d = 0

$\therefore$ C = 0

$$\therefore t = \frac{-d}{f} ln\left(1 - \frac{fA}{Q}\right)$$

$$= \frac{d}{f} ln\left(\frac{Q}{Q - fA}\right) \quad (1.2)$$

Maximum area of border strip at certain Q and f

Rearranging the **Eq.1.2**

$$\ln\left(\frac{Q}{Q - fA}\right) = \frac{ft}{d}$$

$$\text{or, } \log_{10}\left(\frac{Q}{Q - fA}\right) = \frac{1}{2.313}\frac{ft}{d} = y(\text{say})$$

$$\therefore \frac{Q}{Q - fA} = 10^y$$

or , $Q = 10^y (Q - fA)$

$= 10^y Q - 10^y fA$

or, $10^y fA = Q (10^y - 1)$

$$\therefore A = \frac{Q(10^y - 1)}{10^y f} = \frac{Q}{f} \quad (1.3)$$

Advantages of border irrigation

1. Large stream can be safely used
2. Requires less labour and time
3. Efficiency of water application is high
4. Borders (bunds) can be utilized for growing crops
5. Supply channel can be used as drainage channel

6. Borders can be constructed by using simple farm implements viz., moldboard plough, ridger or bund former.

Disadvantages

1. Precise land leveling is required
2. Initial cost of land leveling and grading is high
3. Skill is required for application of water
4. Light irrigation of less than 5cm is difficult to apply efficiently.

Experimental approach to design borders

Border strip irrigation method has good application efficiency and suitable in all type of soils for close growing crops. Appropriate design and efficient management of this necessitates a thorough idea of optimum size of the stream, required depth of infiltration, length and breadth of the border, slope of the land and roughness factor. However, in actual practices maintaining a constant stream size is hardly possible. Roughness factor also varies with the development of crops. Even the infiltration characteristics change with the antecedent moisture content and other surface changes. Therefore, a mathematically perfect designed border may not necessarily yield better irrigation or water use efficiency. Under such variations, the appropriate cut-off ratio and width of border (length is generally fixed in Indian field condition (Grampurahit & Ingle, 1983) may be considered for attaining an application and distribution efficiency of the irrigation method.

The following details of a research work conducted at BCKV Research Station, Memari, Bardhaman (WB) during the consecutive winter of 1992-93 and 1993-94 may enable one to understand the field practices and practicality on design of borders.

The surface soil of the experimental site was sandy clay loam in texture. Before sowing of wheat, the land (55m x 60m) was leveled nearly to flat using a tractor drawn cultivator. The land was then divided in to 3 sets of 4m (W_1), 5m (W_2) and 6m (W_3) width having 60m lengths with a buffer passage of 1m in between any two borders. Each border plot was properly bunded to avoid seepage. Wheat (Sonalika) seed @ 100 kg/Ha was sown on 2nd December and 29th November in 1992-93 and 1993-94 respectively. Recommended doses of N: P: K fertilizers @ 60:40:40 were applied to the field and proper agronomic measures were followed during the crop growth period. Three cut-off ratios namely, 70% (C_1), 80% (C_2) and 90% (C_3) were followed at any irrigation. The irrigation was applied when IW/CPE reached 1.2.

Gravimetric soil moisture at 0-15, 15-30, 30-45, 45-60 cm depths at 0, 15, 39, 45 and 60 cm length along the borders was determined before and after any irrigation. These data were used to calculate the application and distribution efficiency. The infiltration characteristics of soil before and after the irrigations were measured by using double ring infiltrometer. The relationship of the functions were developed by analyzing the sample data using the procedure suggested by Davis (1943) as

$Y=at^b+c$ (1.4)

Where

Y = Depth of water infiltration, cm

a, b, c = Constants

During irrigation, the advance time of water through the borders recorded by putting stakes at 10m interval along the borders. The rectangular shape cement-blocks were also placed at 10m intervals along the borders to measure the depth of flowing water. The surfaces of the cement blocks were in the same level of the respective point. The elevations of the different cement block surfaces were recorded before the first irrigation by using a contour marker. The average of the depths of the flowing water at different cement block points were used to determine the hydraulic slopes.

The water advance time and distance along the border was analyzed and represented by the general equation of the form

$X= at^b$ (1.5)

Where

X = Advance distance in meter

t = Advance time in minutes

a & b = Constants

The designed time of water application in a border was calculated following the equation suggested by Parker (1915) and latter modified by Israelson (1935) and derived as Eq. 1.2,

$$t = \frac{d}{f} ln \frac{Q}{Q - fA}$$

Where

t = Time required in hours in irrigating a border strip

d = Average depth of the sheet of flowing water over the border in meter

f = Infiltration capacity of the soil in meter/hour

A = Area of border to be irrigated in hectare.

The average of distribution and application efficiencies of different irrigation for different borders showed that the distribution efficiencies of various treatment combination were ranged between 73.41 to 93.21% with the highest distribution efficiency in W_1C_3, the application efficiencies were ranged in between 85.17 to 96.51% where the highest efficiency was observed in W_2C_3 in comparison to distribution efficiencies (73.41 to 93.21%), the variation in application are much less (85.19 to 97.90%) and the overall average is 93.55% which is considerably better.

Considering the cut-off ratio and the width of border independently, it is found that 90% cut-off and 6m-border width had the highest distribution efficiency (92.61 and 87.33%) as well as application efficiency (96.45 and 94.48%) respectively. However, these efficiency values for 80% cut-off and 5m-border width were found close to 90% cut-off and 6m-border. On the other hand the efficiencies were poor in the borders of 70% cut-off and 4m-border width.

The depth of water stored at different distances along the borders (Table 1.4) showed that more water was stored at the up-stream of the borders with an average of 6.51cm at '0' distance and 4.01cm at 60m. However, this variation is much less in 90% cut-off in comparison to other two cut-off treatments.

The application rates were varied from 0.99 to 1.41l/m/s in different borders. However, this variation caused a little change in depth of water applied which was in between 5.42 and 6.29cm. The designed depth of water application calculated following the Eq. 1.2 was also in between 3.94 to 4.45cm among the treatments (Table 1.5).

Time of inflow and the designed time of inflow are shown in Table 1.5 which showed that the average designed time of inflow were less than the actual time of inflow for all the cases and were less by average 26.61%. The actual time of inflow showed better application efficiency. The lesser time required in lower cut-off ratio could not provide better application efficiency. The average actual time is 26.61% more than the average design time (Table 1.5). Therefore, irrigating the borders in a crop field like wheat the designed time of inflow calculation following the Eq. 1.2 should have multiplied by 1.26 for favourable application rate in present situation.

There were little variation in between land and hydraulic slopes in the treatments (Table 1.5). The average depths of flowing water over the borders also varied less (3.017 to 3.775 cm).

The infiltration equations of the experimental site for before and after the first irrigation for both the year 1993 and 1994 are shown in Table 1.6. The accumulated infiltration was more in case of before irrigation compared to after irrigation. There was a little variation in infiltration rate of the beginning but at the later stage it is almost same. The average of before and after irrigation may be taken as the representative infiltration characteristics for any case.

The values of a and b in Eq. 1.5 and slopes are given in Table 1.7. It is found that the value of 'b' did not vary much in comparison to 'a' under different treatments. The value of 'a' increased in most of the cases when slopes were increased. However, in some cases this assumption did not stand good.

The changes in inflow rates during the season did not show much variation over the average depth of water application for both the cases of time of water application fixed on the basis of percentage of cut-off and the designed time followed by Israelsons' (1935) equation.

Table 1.4 Distribution and application efficiencies of different treatment of the borders (average of 5 irrigation).

Border	Depth of water stored, cm Distance along the border, m					Distribution efficiency %	Application efficiency %	Efficiency based on cut-off ratio & widths of borders	
	0	15	30	45	60			Distribution efficiency %	Application efficiency %
W_1C_1	7.14	6.07	5.67	5.72	1.94	73.41	85.19	C_1=77.89	C_1=89.50
W_2C_1	6.68	5.4	5.16	5.41	2.6	77.1	90.21	C_2=87.24	C_2=94.60
W_3C_1	6.03	5.37	5.51	5.15	3.29	83.41	93.12	C_3=92.61	C_3=96.45
W_1C_2	6.29	5.68	5.36	5.2	3.98	89.03	95.25	W_1=85.17	W_1=91.76
W_2C_2	6.89	6.31	5.71	5.5	4.98	86.19	95.13	W_2=85.28	W_2=94.41
W_3C_2	6.6	5.84	5.42	5.03	4.08	86.61	93.71	W_3=94.48	W_3=94.48
W_1C_3	6.3	5.96	5.68	5.46	5.12	93.31	94.83		
W_2C_3	6.31	5.75	5.62	5.39	5.24	92.56	97.9		
W_3C_3	6.37	5.52	5.54	5.34	4.88	91.98	96.61		
Average	6.51	5.76	5.51	5.35	4.01	85.95	93.55	C=85.91 W=88.31	C=93.51 W=93.55

Table 1.5 Hydraulic parameters in the borders

Border	Average depth of flowing water, cm	Land slope,%	Hydraulic slope, %	Inflow rate l/s	Inflow rate l/s/m	Time of inflow, min	Designed time of inflow, min	Depth of water applied, cm	Designed depth of water, cm	Variation in between time of inflow and designed time of inflow,%
W_1C_1	3.775	0.242	0.254	5.337	1.334	48.50	34.46	6.29	4.27	28.94
W_2C_1	3.305	0.207	0.239	5.580	1.116	52.50	39.15	5.68	4.10	25.42
W_3C_1	3.030	0.172	0.214	5.967	0.994	57.25	41.51	5.42	3.94	27.49
W_1C_2	3.742	0.184	0.216	5.632	1.408	41.25	33.78	5.53	4.45	18.11
W_2C_2	3.215	0.245	0.265	5.597	1.119	58.00	39.21	5.70	4.05	32.39
W_3C_2	3.017	0.223	0.259	6.157	1.025	61.00	45.04	6.02	4.09	26.16
W_1C_3	3.775	0.054	0.127	5.535	1.393	48.75	34.92	5.80	4.37	28.36
W_2C_3	3.385	0.054	0.105	5.887	1.177	53.00	38.48	6.77	4.28	27.39
W_3C_3	3.225	0.064	0.128	6.102	1.017	63.25	47.25	5.79	4.28	25.29
Average	3.385	0.160	0.200	5.75	1.174	53.72	39.31	5.79	4.20	26.61

Table 1.6 The infiltration characteristics of the experimental site

Year	Before or after irrigation	Infiltration equation
1993	Before	$Y=0.857t^{0.29}+0.032$
	After	$Y=0.376t^{0.39}+0.045$
1994	Before	$Y=3.962t^{0.088}-2.49$
	After	$Y=0.431t^{0.275}-0.009$

Table 1.7 Values of constants 'a' and 'b' in different irrigation for different hydraulic slopes.

Year	Irrigation No.	Border	Inflow rate, l/s	Average hydraulic slope	Values of 'a'	Values of 'b'
1993	I	W_1C_1	4.32	0.253	0.65	3.00
		W_2C_1	4.32	0.263	0.80	1.39
		W_3C_1	4.32	0.175	0.78	1.27
		W_1C_2	4.32	0.258	1.20	0.40
		W_2C_2	4.32	0.235	0.72	2.10
		W_3C_2	4.32	0.173	0.66	2.20
		W_1C_3	3.6	0.216	0.72	2.70
		W_2C_3	3.8	0.071	0.72	2.20
		W_3C_3	3.8	0.116	0.85	1.11
	II	W_1C_1	3.75	0.293	0.79	2.13
		W_2C_1	4.32	0.246	0.78	1.68
		W_3C_1	5.23	0.163	0.75	1.83
		W_1C_2	4.09	0.190	1.10	0.89
		W_2C_2	3.95	0.248	0.74	2.00
		W_3C_2	5.00	0.278	0.86	1.46
		W_1C_3	4.59	0.113	0.88	1.59
		W_2C_3	5.00	0.083	0.95	1.10
		W_3C_3	4.78	0.130	0.96	0.99
1994	I	W_1C_1	6.62	0.196	1.49	0.16
		W_2C_1	6.63	0.166	1.43	0.23
		W_3C_1	6.82	0.200	1.33	0.36
		W_1C_2	6.52	0.156	1.33	0.22
		W_2C_2	6.62	0.290	1.21	0.36
		W_3C_2	8.33	0.256	1.16	0.52
		W_1C_3	6.25	0.086	1.18	0.40
		W_2C_3	7.25	0.126	1.07	0.65
		W_3C_3	8.33	0.106	1.09	0.71
	II	W_1C_1	6.82	0.276	1.29	0.27
		W_2C_1	7.25	0.240	1.26	0.36
		W_3C_1	7.75	0.258	1.24	0.40
		W_1C_2	7.50	0.260	1.29	0.23
		W_2C_2	7.50	0.300	1.24	0.32
		W_3C_2	7.50	0.300	1.23	0.36
		W_1C_3	7.50	0.096	1.31	0.18
		W_2C_3	7.50	0.143	1.29	0.22
		W_3C_3	7.50	0.163	1.28	0.27

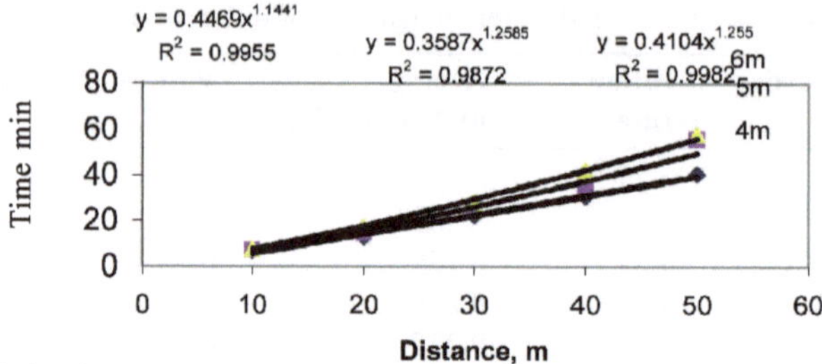

a) First irrigation, 70% cut

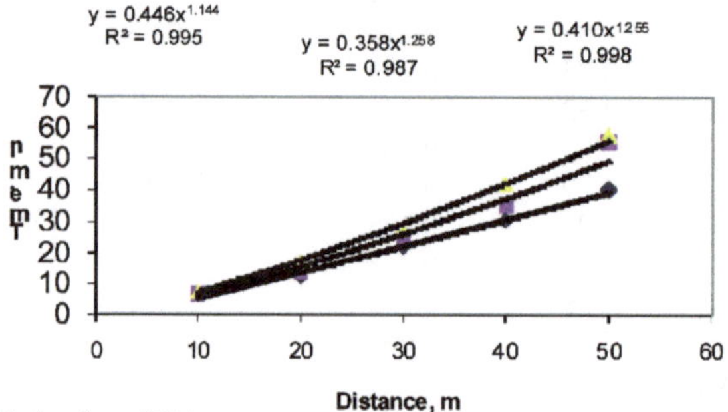

b) Second irrigation, 70% cut

Fig. 1.6 Advance curves in different cut-off ratio and border width

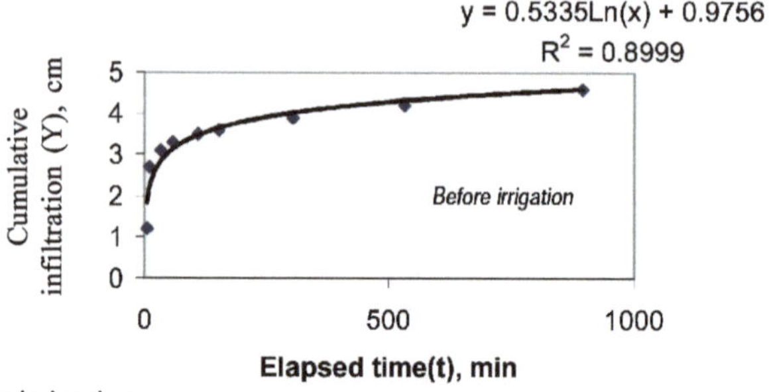

a) Before irrigation

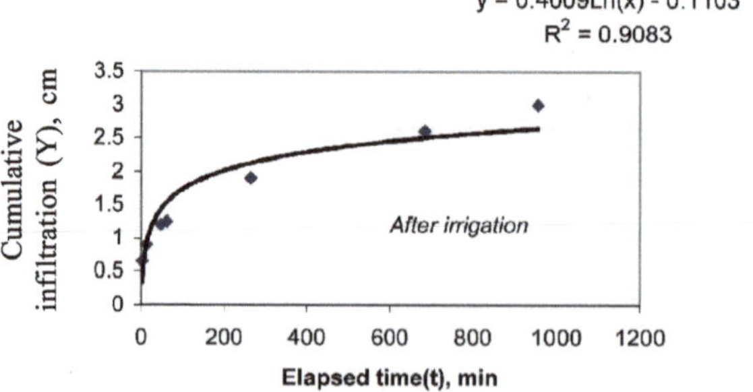

b) After irrigation

Fig 1.7 Infiltration characteristics of soil

Example 1.1 Determine the time required to irrigate a border of 0.025ha by a stream size of 10l/s. Assume rate of infiltration 4.5cm/h and average depth of water flow 7.5cm. What is the maximum area that can be irrigated by this stream?

Given: Area of the border, A =0.025ha

Stream, Q =10l/s

Rate of infiltration, f =4.5cm/s

Average depth of water flow =7.5cm

Required: Time of irrigation, t =?

Maximum possible border strip, A_{max}=?

Solution: We have, $t = \frac{d}{f} ln\left(\frac{Q}{Q - fA}\right)$

$$= \frac{7.5cm}{4.5cm/h} ln\left(\frac{10l/s}{10l/s\text{-}4.5cm/hx0.025ha}\right)$$

$$= \frac{7.5\text{h}}{4.5} \ln\left(\frac{10l/s}{10l/s - \frac{4.5\times0.025\times10{,}000\times10{,}000}{3600\times1000} l/s}\right)$$

$$= 1.67hx\,ln\left(\frac{10}{10\text{-}3.125}\right)$$

=1.67ln 1.45h

= 0.62h = 37.23 min.

Maximum area, A_{max}= Q/f

$$= \frac{10l/s}{4.5cm/h}$$

$$= \frac{10\times1000cm^3/s}{4.5cm/3600s} = 8000000 \text{ cm}^2$$

= 800 m^2 = 0.08 ha

Example 1.2 Water is applied at a rate of 30l/s in a border strip of 6m wide. The estimated depth of water flow is 7.0cm and the rate of infiltration is 3.5cm/h. Determine (i) the time required for the water to reach a distance of 350m, (ii) the average depth of water applied, and (iii) the maximum area and length of border can be irrigated.

Given:

Width of border, w	=	6m
Discharge, Q	=	30l/s
Depth of water flow, d	=	7.0cm
Rate of infiltration, f	=	3.5cm/h
Distance in border strip, L	=	350m

Required:

i) Time, t = ?

ii) Average depth, d_a = ?

iii) Maximum area of strip, A_{max} = ?

iv) Maximum length of strip, L_m = ?

We have, $t = \frac{d}{f} ln\left(\frac{Q}{Q - fA}\right)$

$$= \frac{7cm}{3.5cm/h} ln\left(\frac{30l/s}{30l/s - 3.5cm/h \times 350m \times 6m}\right)$$

$$= \frac{7cm}{3.5cm/hr} ln\left(\frac{30l/s}{30l/s - \frac{3.5 \times 350 \times 100 \times 600}{3600s}}\right)$$

$$= 2h\, ln\left(\frac{30l/s}{30l/s - \frac{3.5 \times 350 \times 100 \times 600}{3600 \times 1000l/s}}\right)$$

$$= 2h\, ln\left(\frac{30}{30\text{-}20.416}\right) = 2\text{ln } 3.13\text{h}$$

= 2 x 1.14h = 2.28h

Volume of water applied during the period t

= 30l/sx 2.28h

$$= \frac{30}{1000} \times 2.28 \times 3600 m^3 = 246.24\ m^3$$

Average depth of water applied,

$$D = \frac{246.24\text{m}^3}{350 \times 6\text{m}} = 0.11725\text{m} \cong 11.73\text{cm}$$

For maximum area, $A_{max} = Q/f$

$$= \frac{30l/s}{3.5\text{cm/h}} = \frac{30 \times 1000\text{cm}^3/s}{3.5\text{cm}/3600\text{s}} = 30857142.86\text{cm}^2 = 0.308\text{ha}$$

$$\text{Maximum length} = \frac{\textit{Maximum area}}{\textit{Width of border}} = \frac{0.308\text{ha}}{6\text{m}}$$

$$= \frac{0.308 \times 10000\text{m}^2}{6\text{m}} = 514.28\text{m}$$

Example 1.3 An irrigation stream size of $1x10^{-3} m^3 s^{-1}$ per meter width is supplied for 5 hours to a vegetated border strip of 250m long and 5m wide. The average longitudinal slope of the border is 0.2 percent. The volume of water collected as runoff loss at the end of the border strip is 36m^3. Determine (a) depth of flow and (b) water application efficiency. Assume Manning's n as 0.1. (GATE, 2004)

Solution: Stream size Q = $0.001\text{m}^3/\text{s/m}$

Area of border strip = 5m x 250m = 1250m^2

Volume of water applied in 5 hours = 0.001m^3 / s / m x 5 m x 5 h

= 0.001 x 5 x 3600 x 5 = 90m^3

Volume of water retained in the border strip = 90-36=54m^3

Depth of water infiltrated during the time of irrigation (5h)

$$= \frac{54\text{m}^3}{5\text{m} \times 250\text{m}} = 0.0432\text{m}$$

Rate of infiltration, $f = \dfrac{0.0432\text{m}}{5\text{h}} = 8.64 \times 10^{-3}\text{m/h}$

Water application efficiency =

$$\frac{\textit{Water retained in the root zone depth}}{\textit{water applied}} = \frac{54\text{m}^3}{90\text{m}^3} = 60\%$$

Manning's equation,

$$V = \frac{1}{n} R^{2/3} S^{1/2} = \frac{1}{0.1}\left(\frac{\textit{wetted x-sectional area of flow, A}}{\textit{Wetted perimeter, P}}\right)^{2/3} (0.002)^{1/2}$$

$$=\frac{1}{0.1}\left(\frac{5d}{5+2d}\right)^{2/3}(0.002)^{1/2}$$

Rate of application of water, $Q = AV = 5dV = 5 \, x \, 0.001m^3/s = 0.005m^3/s$

$$Or, V = \frac{0.005}{5d} = \frac{0.001}{d}$$

$$\therefore \frac{0.001}{d} = \frac{1}{0.1}\left(\frac{5d}{5+2d}\right)^{2/3}(0.002)^{1/2} = \left(\frac{5d}{5+2d}\right)^{2/3} \times 0.447$$

$$Or,\ 2.235 \times 10^{-3} = \left(\frac{5d}{5+2d}\right)^{2/3} \times d$$

Assuming, d = 0.0257m,

$$\text{RHS} = \left(\frac{5\times0.0257}{5+2\times0.0257}\right)^{2/3}\times0.0257 = (0.0254)^{2/3}\times0.0257 = 2.22\times10^{-3}$$

$\therefore$ LHS = RHS

Thus, the normal depth of flow = 2.57cm

Example 1.4 A wheat border of 6mx6m was irrigated by the stream of 15l/s for 45 and 50 minutes in two consecutive irrigations respectively. The soil sampling for moisture contents at different depth and length of the border before and after the irrigation was as follows.

Depth, cm	**Length along the border**					
	10m		**30m**		**50m**	
	Moisture content, %		**Moisture content, %**		**Moisture content, %**	
	1st Irrigation	**2nd Irrigation**	**1st Irrigation**	**2nd Irrigation**	**1st Irrigation**	**2nd Irrigation**
	Before & after irrigation	**Before & after Irrigation**	**Before & after Irrigation**	**Before & after Irrigation**	**Before & after irrigation**	**Before & after Irrigation**
0-15	14 - 22	13.5 –23	15-22	14-23	13-21	12.5-21.5
15-30	15 –22	13 –23	15-21	14.5-22	13-22	12.5-21
30-45	16 –23	15 –22.5	15.5-21	16-23	14.5-22	13-22
45-60	17- 22.5	16.5 -22	16-22	16-22.5	15.5-21	14.5-22.5

Determine the depth of water penetrated at different lengths and application efficiencies of irrigation. The apparent specific gravity of soil is 1.45.

Solution

The depth of water retained at 10m lengths in first irrigation

$$=\{(22\text{-}14) + (22\text{-}15) + (23\text{-}16) + (22.5\text{-}17)\}\ \frac{1}{4}\times 1.45$$

= 27.5 x 0.25 x 1.45 = 9.97cm

Similarly, the water retained at 30 & 50m lengths of border are 8.88 & 10.88cm respectively.

The depth of water retained at 10 m lengths in second irrigation

$$=\{(23\text{-}13.5) + (23\text{-}13) + (22.5\text{-}15) + (22\text{-}16.5)\}\ \frac{1}{4}\times 1.45$$

= 32.5 x 0.24 x 1.45 = 11.78cm

Similarly, the water retained at 30 & 50m lengths of border are 10.88 & 12.52cm respectively.

Average depth of water applied in first irrigation,

$$d_1 = \frac{9.97+8.88+10.88}{3} = \frac{29.73}{3} = 9.91cm$$

Volume of water retained = 9.91cm x 6m x 60m = 35.68 m^3

Average depth of water applied in second irrigation

$$d_2 = \frac{11.78+10.88+12.51}{3} = \frac{35.17}{3} = 11.72cm$$

Volume of water retained = 11.72cm x 6m x 60m = 14.19m^3

Volume of water applied in first irrigation,

$$V_1 = 15l/s \times 45\text{min} = \frac{15}{1000} m^3/s \times 45 \times 60s = 40.5m^3$$

Volume of water applied in second irrigation,

$$V_2 = 15l/s \times 50\text{min} = \frac{15}{1000} m^3/s \times 50 \times 60s = 45m^3$$

Application efficiency in first irrigation

$$= \frac{\text{Volume of water retained}}{\text{Volume of water applied}} = \frac{35.68}{40.5} = 0.88\, or\, 88\%$$

Application efficiency in second irrigation

$$= \frac{\text{Volume of water retained}}{\text{Volume of water applied}} = \frac{42.19cm}{45cm} = 0.9376\, or\, 93.76\%$$

Example 1.5 The following data were recorded in 6mX60m wheat border where irrigation was applied following 90% cut off ratio. The infiltration characteristics of soil are represented as $I = 0.86t^{0.29} + 0.03$ where, I is the cumulative infiltration in cm and t is the time in minute.

Distance along the border, m	Inflow rate, l/s	Advance time, min	Recession time, min
0	6.8	0	1200
10		7	1220
15		10	1230
20		15	1240
25		21	1250
30		30	1280
45		50	1320
54		68	1350
60		85	1400

i) Develop the equations and curves for advance and recession of water flow.

ii) Find out the infiltration opportunity time at different length of the border from the fitted curves.

iii) Determine the depth of water penetrated at different length of the border and calculate for efficiency of distribution of water.

iv) If the average depth of water flow was 3.3 cm, what is the theoretical time required to irrigate the border?

Solution:

i) The data are plotted using the advancement and recession of water front along the border as X-axis and the time required for these along the Y-axis as shown in Fig 1.8.

x	y1	y2
0	0	1200
10	7	1220
15	10	1230
20	15	1240
25	21	1250
30	30	1280
45	50	1320
54	68	1350
60	85	1400

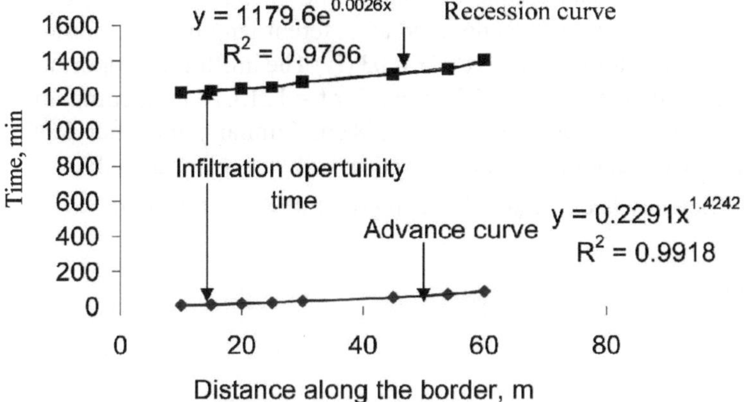

Fig. 1.8 Advance and recession curves in a border

The available fitted equations are:

i) Advance of water front, $t = 0.2291x^{1.4242}$ (i)

ii) Recession of water front, $t = 1179.6e^{0.0026x}$ (ii)

Where,

t = Advance time, min.

x = Advancement or recession along the border, m.

The infiltration opportunity time at different border length as recorded and calculated from the fitted advance and recession curves of waterfront are given below.

Distance along the border, m	Advance time, min	Recession time, min	Infiltration opportunity time, min (Col.(3)-Col.(2)	Calculated infiltration opportunity time, min	Depth of water penetration, cm
1	2	3	4	5	6
0	0	1200	1200	1179.6	6.71
10	7	1220	1213	1204.59	6.75
15	10	1230	1220	1215.67	6.77
20	15	1240	1225	1226.23	6.79
25	21	1250	1229	1236.38	6.8
30	30	1280	1250	1246.2	6.83
45	50	1320	1270	1274.19	6.87
54	68	1350	1282	1290.22	6.89
60	85	1400	1315	1300.68	6.91

From Eq. (ii) putting the value of x = 10m, we may get, t = 1210.67 min. Similarly, from Eq. (i) putting x = 10, we get t = 6.08 min. Thus, the infiltration opportunity time at 10m length is 1210.67-6.08 = 1204.59 min. Similarly, the infiltration opportunity time for other distances are calculated and entered in Col (5) of the above table (iii) The depth of water penetrated at different length of the border is calculated by using the equation, $I = 0.86\, t^{0.29} + 0.03$ (iv) The infiltration opportunity time at 15m length is calculated as 1215.78 min. For t = 1215.78 min, cumulative infiltration i.e., the depth of water penetrated = 6.78 cm. Similarly, the other depths of water penetration were calculated and entered in Column (6) of the table.

The distribution efficiency is calculated by using the following formula,

$$E_d = \left(1 - \frac{\bar{y}}{\bar{d}}\right) \times 100$$

Where,

E_d = Distribution efficiency, percent

$\bar{y}$ = Average of sum of the numerical deviation from the mean, cm

$\bar{d}$ = Mean of the depth of penetrations, cm.

$\bar{d}$ = (6.71+6.75+6.77+6.79+6.80+6.83+6.87+6.89+6.91)/9=61.32/9=6.81

$\bar{y}$ = [(6.81-6.71)+(6.81-6.75)+(6.81-6.77)+(6.81-6.79)+(6.81-6.80)+ (6.83-6.81)+(6.87-6.81)+(6.89-6.81)+(6.91-6.81]/9=0.49/9=0.0544

So, E_d = (1-0.0544/6.81) 100=99.2%

iv) The theoretical time required to irrigate the border is calculated by using the equation,

$$t = \frac{d}{f} ln\left(\frac{Q}{Q - fA}\right)$$

For t =60 min, I = 2.843 cm. So, f = 2.843 cm/h

$$= \frac{3.3cm}{2.843cm/h} ln\left(\frac{6.8l/s}{6.8l/s - 2.843cm/h \times 360m^2 \times 0.9}\right)$$

$$= \frac{3.3cm}{2.843cm/h} ln\left(\frac{6.8 \times \frac{3600}{1000} m^3/h}{6.8 \times \frac{3600}{1000} m^3/h - 2.843 \times \frac{1}{100} \times 360m^3/h \times 0.9}\right)$$

$$= 1.16 ln\left(\frac{24.48m^3/h}{24.48m^3/h - 9.21m^3/h}\right)$$

= 1.16ln (1.60) = 0.55h = 33 min

Example 1.6 A border strip of 8x250m is being irrigated by a border stream of 50lps. The infiltration capacity of the soil is 25mm/h (assumed to be constant throughout the period of irrigation). The average depth of advancing sheet of water over the land is 70mm. The time required to irrigate the border strip in minutes will be

a) 16.6 b) 25.7

c) 13.06 d) 16.07

(GATE, 2012)

Solution

Area of the border, A = 8x 250 = 2000m^2 = 0.2ha

Border stream size, Q = 50lps

Infiltration rate of soil, f = 25 mm/h = 2.5cm/h

Depth of advancing sheet, d = 70 mm = 7cm

Time required to irrigate the field, $t = \frac{d}{f} ln \frac{Q}{Q - fA}$

$$= \frac{7}{2.5} \ln \frac{50}{50 - 2.5 \times 0.2}$$

$$= 2.8 \ln \frac{50}{50 - 0.5}$$

$$= 2.8 \ln \frac{50}{49.5}$$

= 0.22h

= 13.20 min

1.3 Check Basin

In check basin method of irrigation the field is divided into small plots called checks surrounded by bunds. It is also called as check basin. The smalls are level but the larger checks may have gentle slope. Water from supply channel is supplied to field channel from where water is applied one after another basin. Supply channel is constructed at the upper reach of the field and field channel follow the slope.

Rectangular basin: These are basin of rectangular or sometime square shape used in level or of gentle slope land (Fig. 1.9). Check basin is very common method of irrigation in India and many other countries. The size of the basin may be a few square meters for vegetable crops to as large as about a hectare or more for wet land rice cultivation. The area in a check depends on infiltration rate of soil, land slope and stream size. The area of check is smaller in light soil and larger in heavy soil. The height of the bunds around the check may be 25 to 30 cm for small area check depending on the depth of ponded water. Bunds of checks are made for temporary or semi permanent use. The wide semi permanent bunds may suitably be able to be used for growing crops or movement of farm machineries.

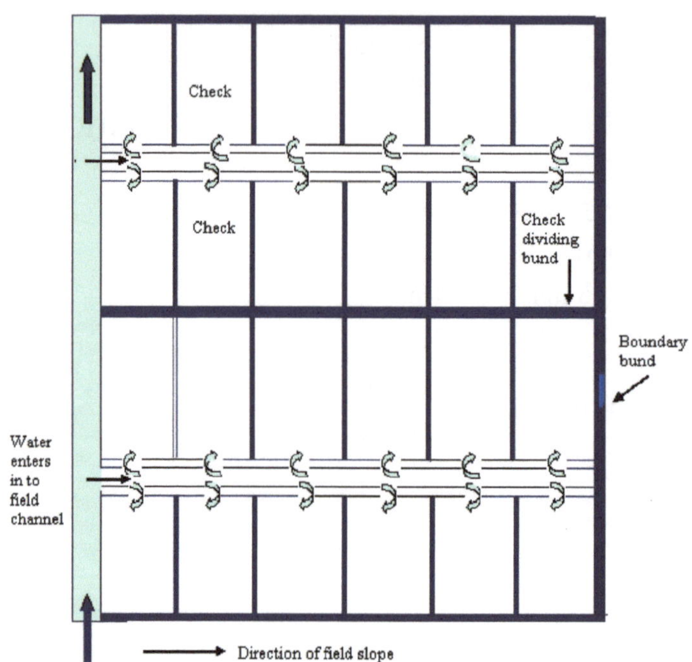

a) Schematic diagram of rectangular check basin

b) Coriander cultivated rectangular check basin

Fig. 1.9 Rectangular check basin

The check method is suitable for row crops (rice, etc.) as well as close growing crops (wheat, finger millet, pearl millet, paragrass, etc.), fodder and vegetables (cabbage, cauliflower, ladies finger, radish, etc,) in soils of moderate to low infiltration rate. The check basin irrigation may be practiced in soils of high infiltration rate by reducing the area of basin and increasing the size of stream so that in a short time the water get spreaded over the basin without causing much loss in deep percolation at the entrance of the stream. Check basin method is suited for waterlogged condition for rice cultivation, leaching in salt affected soil and harvesting almost entire rainwater. The check basin method allows supplying water precisely by measuring the flow at intake of the check and duration of flow. Usually the time required to irrigate a check is one-fourth the time requires infiltrating the entire depth of water applied in it (Dakshinamurti et al, 1973,). The knowledge of infiltration characteristics of the soil and available discharge are very useful information for decide upon most approximate size of the plot. A rough estimation of check size for different soil and stream size are given in Table 1.8.

Table 1.8 Recommended dimension of check basin for different soil and stream size

Place	**Soil**	**Stream size (l/s)**	**Check size m^2**
Hissar	Sandy loam	14	225 (15mx15m)
Jobner	Sand	3.5	25 (5mx5m)
Kharagpur	Sandy loam	7	100 (10mx10m)
Kota	Clay loam	10	50 (5mx5m)
Patnagar	Clay loam	15	150 (10mx15m)

Source: Majumder (2000)

Contour check basin: Contour check basin is constructed by raising bunds following the contour in sloping and rolling land (Fig. 1.10). The design and application of water in similar to rectangular check basin. The checks are usually small, level or gently sloping. The spacing of field channel varies between 15 to 50m depending on (i) grade of land between two field channels, (ii) uniformity of slope and (iii) type of slope. Contour check basins are suitable for vegetables, forage and grain crops including the rice. However, this method involves high labour cost and become unsuitable and leads to adopt other method of irrigation.

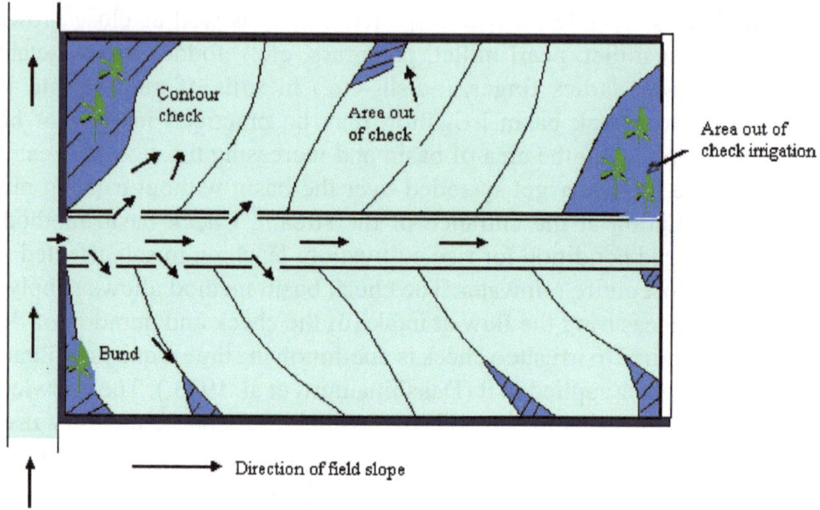

Fig. 1.10 Contour check basin

Advantages of rectangular check basin

1. High distribution and application efficiencies are easily obtained in a well-designed system.
2. Maximum use of rainfall.
3. Method of operation is easy.
4. Surface drainage facilities can be had at low cost.
5. Suitable for wide range of soil.
6. No loss of water through run-off.
7. Leaching down of salts is easily possible.

Disadvantages

1. High labour requirement
2. Large irrigation stream is required.
3. Bunds create hindrance to movement of farm equipments.
4. Unsuitable for soils of moderate to high permeability.
5. Considerable land is wasted for bunds and channels.
6. Use is restricted to relatively smooth lands where land-leveling cost is not much.
7. Method is not suitable to irrigated crops sensitive to wet soil condition.

Advantages and disadvantages of contour check basin

Advantages

1. Suitable for sloping and rolling land.
2. Degree of required land leveling is less than other method of irrigation.
3. Chance of soil erosion is reduced.
4. Suitable for vegetable, forage and grain crops including rice.

Disadvantages

1. High labour cost of irrigation.
2. Not suggested if other method of irrigation is possible.
3. Check gates, turnouts, drops, etc., measures are required to prevent channel bed erosion.
4. Application efficiency is not high.

Ring check basin

Ring check basin is essentially meant for orchards. Sometime basin is excavated for the individual tree or plant (Fig. 1.11). The basins are delivered water through small ditches or laterals. The laterals are joining to main ditch. One or more plants are irrigated in a basin depending on the type of crop and soil it is adopted in flat land. Water is applied rapidly to the basin made of low embankments or dikes, which retain the water till it is infiltrated into the soil.

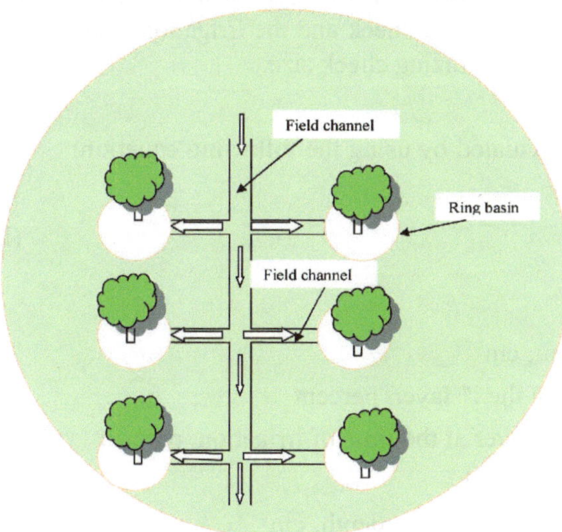

Fig 1.11 Ring check basin

Advantages

1. Especially suitable for orchard cultivation.
2. Entire field need not be flooded.
3. No chance of soil erosion.
4. Rapid irrigation is possible.
5. No irrigation water is lost by run-off.
6. Maximum use of rainwater.
7. Water application efficiency is very high.
8. Land leveling is not required except the basin area.

Disadvantages

1. High labour cost to retain the water.
2. Accurate land leveling is generally required.
3. Adopted for flat land only.

Design of check basin

A thorough study of hydraulics of flow relating size of the stream, size of the check, soil type, level and grade of the land, infiltration characteristics, depth of flow to be applied, etc., are required to have best efficiency in designing check basin. However, as stated earlier, as a thumb rule, the time requires to irrigate a check equals to one-fourth the time requires in infiltrating the entire depth of water applied in soil, the design of check may proceed. The design involves two important parameters, namely, the optimum size of the check and the irrigation stream. The following steps may be followed in optimizing check size.

i) Depth of water to be applied

The net depth of water may be calculated by using the following equation:

$$IR_n = \sum_{i=1}^{n} \frac{\left(M_{fci} - M_{bi}\right) A_s D_i}{100} \tag{i}$$

Where,

IR_n = Net irrigation requirement, cm

M_{fci} = Field capacity of the soil in the ith layer, percent

M_{bi} = Moisture content in the ith layer at the time of irrigation, percent

A_s = Apparent specific gravity of soil

D_i = Depth of the i th soil layer in root zone depth, cm

n = Number of soil layer in root zone depth.

If the soil moisture at field capacity and wilting point and the percentage of soil moisture depletion at the time of irrigation are known, the irrigation requirement may be calculated by the following equation:

$$IR_n = \sum_{i=1}^{n} \frac{\left(M_{fci} - M_{wi}\right) A_s D_i}{100} \text{ x soil moisture depletion (fraction)} \quad \text{(ii)}$$

Where, M_{wi} = soil moisture at wilting point.

The field capacity of the soil and the moisture content at the time of irrigation requires field experimentation.

ii) Infiltration

The following empirical equations may be used for determining infiltration characteristics of soil.

$I = at^{b}$ (iii)

or, $I = at^{b} + c$ (iv)

Where,

I = Cumulative infiltration or depth of water to be applied

t = Infiltration opportunity time i.e., the time required to absorb the applied water by the soil.

a, b, c = Constants

iii) Size of the stream

The available stream size to be known. It may be used to irrigate one check at a time. If the size of the stream is large that may be divided to irrigate several checks at a time.

iv) Length of the check

In consideration to geometry of the plot, a practical value of width of the check is to be selected. The length of the check may be selected by using the equation,

$$L = mt_a^n \quad \text{(v)}$$

Where,

L = Plot length

t_a = One-fourth of the time to infiltrate the applied water and wet the root zone depth

m & n = Constants.

The values of m and n depends on stream per unit width of check, soil type, soil texture and quality of land preparation. The exact values of m and n can be determined only by conducting the experiments. However, workable values can be had from Table 1.9.

Table 1.9 Values of constants 'm' and 'n'

Soil type & land slope	Discharge, l/s/m	m	n
Loamy sand soil (0.32% slope)	3.46	6.1	0.85
	2.37	4.5	0.86
	0.98	1.87	0.88
Loamy sand soil (0% slope)	1.7	5.6	0.63
	2.5	7.2	0.63
	3.4	7.2	0.68
Sandy loam soil (0.1% slope)	2.6	7	0.8

Source: Mal, 1995

It has been observed that the size of the check have significant effect on water requirement, grain yield and water use efficiency. Lower size of check gives better water use efficiency than the larger. Checks preferably are within 120m^2 area in light textured soil (Mal, 1995).

Example 1.7 Design a check basin from the following data:

Size of the plot = 50m x 100m

Field capacity of soil = 19.5%

Permanent wilting point = 9.0%

Time of irrigation = At 50% depletion of available soil moisture

Root zone depth = 90 cm

Crop = Wheat

Soil = Loamy sand

Land slope = 0.3%

Apparent specific gravity = 1.48

Size of the stream = 10l/s

Infiltration characteristics of soil, $I = 0.7t^{0.45}$

Solution

i) Depth of the water applied,

$$IR_n = \sum_{i=1}^{n} \frac{\left(M_{fci} - M_{wi}\right) A_s D_i}{100} \text{ x soil moisture depletion (fraction)}$$

$$= \frac{(19.5\text{-}9.0)\times 1.48\times 90\times 0.5)}{100} = 7.0\,cm$$

ii) Cumulative infiltration, $I = 0.7t^{0.45}$

Or, $7.0 = 0.7t^{0.45}$

Or, $t^{0.45} = 10$

$\therefore$ t = 166.81min

iii) The stream size 10l/s is not large enough. It may be used to irrigate the individual check.

iv) In consideration to geometry of the plot, the width of the check may be taken as 10m.

So, discharge in l/s/m width $=\frac{10}{10}=1$

Using the Table 1.9, the m and n are selected 1.87 and 0.88 respectively.

$$L = mt_a^n = m(t/4)^n$$

$$= 1.87\left(\frac{166.81}{4}\right)^{0.88} = 1.87 \text{ x } 26.53 = 49.84 = 50\text{m } \textit{(say)}$$

Therefore, the plot can be divided into 10 numbers of 10m x 50m checks.

$$\text{Time of irrigation to each basin} = \frac{7.0\text{cm} \times 10\text{m} \times 50\text{m}}{10l/s}$$

$$= \frac{7.0\times10\times50\times1000}{100\times10\times60}\text{min} = 58.33 \text{ minutes}$$

Example 1.8 A stream of 25l/s irrigates a check basin of size 15m x 10m. The maximum water holding capacity of the soil is 15% and the average moisture content in the soil prior to irrigation is 7.5%. How long the irrigation stream be applied to replenish the 1.2m root zone depth assuming no loss of water through deep percolation. The apparent specific gravity of the soil is 1.45.

Given: Check basin size, A = 15m x 10m = 150m^2

Stream size, Q = 25l/s

Maximum water holding capacity, M_m = 15%

Soil moisture content in soil, M_w = 7.5%

Root zone depth, D = 1.2m

Apparent specific gravity, A_s = 1.45gm/cc.

Required: Time to irrigate, T = ?

Solution: Net irrigation requirement = 15-7.5 = 7.5%

= 7.5 x A_s x D m/m

= 7.5 x 1.45 x 1.2 = 13.05cm

$$\text{Volume of water required to check} = \frac{13.05\times150\text{m}^3}{100} = 19.575\text{m}^3$$

Time required to irrigate, $T = \frac{19.575s}{25/1000} = 783s = 13.05min.$

Example 1.9 Design the size of the check in a soil where infiltration characteristics is such that it infiltrates 5cm of water in 1 hour. The size of the available stream is 15l/s and depth of water to be applied is 7 cm.

Solution: Assuming the irrigation time one-fourth of the time required for infiltrating the depth of water applied, time of application of water,

$$t = \frac{7cm}{5cm \times 4} = 0.35h$$

Volume of water applied = $15l / s \text{ x } 0.35h = 18900l$

$$\text{Area of the check} = \frac{18900l}{7cm} = \frac{18900 \times 1000cm^3}{7cm}$$

$$= \frac{18900 \times 1000}{7 \times 100 \times 100} = 270m^2$$

1.4 Furrow Irrigation

The principle of furrow irrigation is to wet some part of the surface area (one-half to one-fifth). Water flows through the furrows soaked into soils laterally and irrigate the areas between the furrows. Furrow method of irrigation is not suitable for light soil where infiltration rate is high. A few furrows are irrigated at a time from the open ditch through the siphon tuber or spile or opening the channel and directly connect the furrows. The slope of the furrows usually varies between 0.1 to 0.5% depending on soil and stream size. The length of furrows efficiently can be irrigated range from 45m on light soil to as high as 800m on heavy soil.

The usual furrow size for row crops is about 25cm wide and 10 to 15cm deep. Spacing of furrow may be 1 to2m in orchards and 3 to 4m in impervious soil. In the furrows large stream is received at the upstream and reduce the size as it advances to lower end at a rate almost equal to the rates the soil absorb the water. In general water stream reaches to the end of the furrow within $1/4^{th}$ of the total time of irrigation. Velocity of flow is important in a furrow as it influences on percolation of water and erosion of soil. The relation gives the empirical formula used for maximum non-erosive stream size in a furrow:

$$Q = \frac{0.6}{S} \quad (1.6)$$

Where

Q = Maximum furrow stream size, l/s

a) Schematic diagram of furrow irrigation

b) Furrow irrigation in potato field

Fig. 1.12 Furrow irrigation (straight)

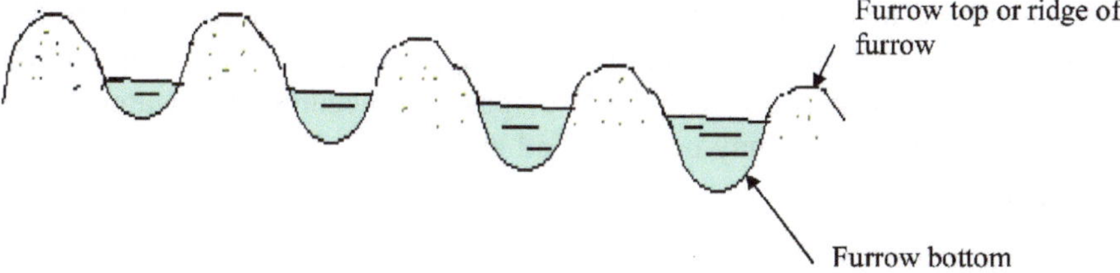

Fig. 1.13 Sectional view of furrows

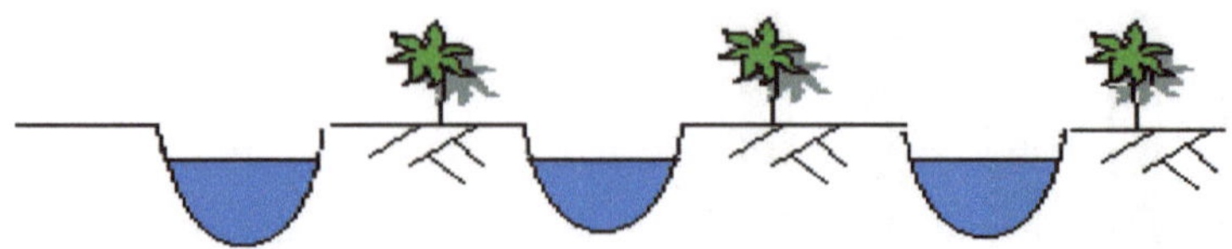

Fig. 1.14 One furrow for one row

Fig 1.15 One furrow for two rows

S = Slope of furrow expressed in percent

The average depth of water applied during irrigation may be calculated as below.

$$D = \frac{Q \times 3600}{S \times L}$$

Where

D = Average depth of water applied (cm) in one hour

S = Furrow spacing, m

L = Length of furrow irrigated in one hour, m

Furrow irrigation is suitable for crops like potatoes, maize, cotton, lettuce, carrots, onions, pointed gourd, sorghum, tobacco, brinjal, tomato, napier grass, sugarcane, etc.

Furrow irrigation is also sometime followed in combination to border irrigation in black cotton soil. The border bunds guide the irrigation water in the strip and the furrows also serve the purpose of quickly dispose the accumulated water during heavy rainfall. The furrows may be classified as below.

i) Based on alignment
   - Straight furrow
   - Contour furrow

ii) Based on size and spacing
   - Deep furrow
   - Corrugation

Straight furrow: Straight furrows are constructed following the existing land slope or by changing the slope to a non-erosive grade called as straight graded furrows. If it is done on level ground then called as straight level furrows. Straight furrows in land slope up to 0.7% are best. However, in the area of high rainfall the suitable land slope to be considered as 0.5%. For irrigating furrows if possible large stream are delivered at the beginning of irrigation to quickly fill up the furrows with water to its full capacity and then at a reduced rate equal to absorbing capacity of the furrows. This practice gives better water application efficiency. A poor water application practice may cause water logging at the tail end of the furrow. Some time a small drainage channel is provided at the end of the furrows to drain out excess water due to carelessness in irrigation or high rainfall. Straight level furrows are suitable for the soils of low infiltration rate and moderate to high water infiltration rate. Level furrows are short in length. The constant stream size is maintained during the irrigation and irrigation stop only when the desired depth of water is applied. Level furrows have certain advantages over the graded furrows. Level furrows provide the less deep percolation and runoff loss, little erosion hazard and more uniformity in application of water. However, it requires large stream size furrows and uniform leveling of land.

Advantages of straight furrows

1. Large and small streams can be used by adjusting the number of furrows irrigated at a time.
2. Low evaporation loss due to wetting of only one-half to one-fifth surface area.
3. Furrows serve the purpose of drainage in heavy rainfall areas.
4. It provides scope for early cultivation in heavy soil.
5. Fairly efficient use of water.

Disadvantages

1. Not suitable for light soil of high infiltration rate.
2. It is expensive in consideration to time and labour.
3. Much attention and skill are required for uniform distribution of water and minimum waste.
4. Facilities are required to dispose of surface runoff.

Contour furrows

The contour furrow method of irrigation is similar to straight furrow excepting that the furrows carry water across the field rather than down the slope. Contour furrows are curved to fit the topography of the land. When the land slopes exceeds the safe limit of straight furrows, furrows are considered along the contour. Contour furrows can be safely made in the land up to 5% slope. However, where soil is stable and cultivation is not frequent up to 8-10% slope may be suitable for contour furrow. Contour furrows may be graded if a gentle slope is allowed along its direction of run. Fields are required to smooth enough to have the uniformly spaced furrow and the field supply channel at the head of the furrows. The field channel is either grassed or provided with structure, which prevent erosion of channel bed and sides. It should have sufficient capacity to feed the entire furrows. Furrows should be provided with waste ditch to dispose off the excess irrigation water and rainfall.

Contour furrows are suitable to all type of soil excepting light sandy soils and soils that crack. In heavy rainfall areas the furrow length should be short enough to dispose of the excess runoff water safely without causing damage to the furrows. Contour furrows are mostly used for row crops and installed annually. However, for orchard or permanent planting the contour furrows may be used for years by properly maintaining the furrows. The waste ditch needs to be cleaned up before the rainy season.

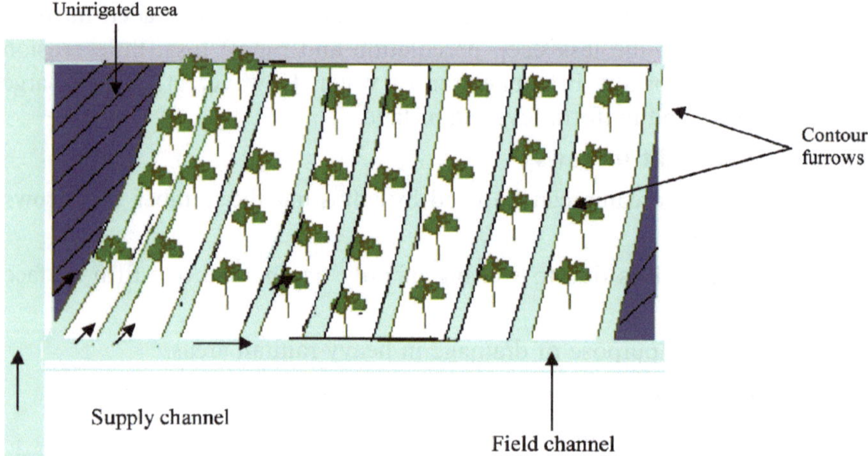

Fig. 1.16 Contour furrows

Advantages

1. Suitable to steep lands where straight furrows are not possible.
2. Reduces the earthwork compared to straight furrow.
3. Can be adopted in sloping and rolling field.
4. Suitable to almost all type of soils.
5. Large stream can be used.

Disadvantages

1. Cannot be used in coarse textured soils and that develop cracks.
2. Furrow breaches can develop erosion hazard.
3. Length of furrow is usually short.
4. Grassed waterways or structures are required to carry the water in field channel.

. Requires constant supervision to possible breaches and repair the same.

Furrow length

The length of the furrows mainly depends on land slope, soil type and stream size. In India, length of furrow usually restricted to short length due to small plot size. Some of the recommended lengths of furrows based on experimental data are described in Table 1.10 & 1.11.

Table 1.10 Maximum suggested furrow lengths (Marr, 1967)

Infiltration rate mm/h	Probable soil texture	Maximum furrow length, m
25-75	Sandy, sandy clay loam	200
13-25	Loam, silt loam	400
6-13	Clay, clay loam, silty clay loam	800
<6	Clay	800

Source: Mal, 1995

Table 1.11 Length of furrow and stream size for different soils, land slopes and water application depth (Reddi, 1967).

Slope, %	Length of furrow for different soil & depth of irrigation												Maximum flow rate, l/s
	Clay				Clay loam & sandy loam				Loamy sand & sand				
	Irrigation depth, mm												
	75	150	225	300	50	100	150	200	50	75	100	125	
0.05	300	400	400	400	120	270	400	400	60	90	150	190	12
0.1	340	440	470	500	180	340	440	470	90	120	190	220	6
0.2	370	470	530	620	220	370	470	530	120	190	250	300	3
0.3	400	500	620	800	280	400	500	600	150	220	280	400	2
0.5	400	500	560	750	280	370	470	530	120	190	250	300	1.2
1	280	400	500	600	250	300	370	470	90	150	220	250	0.6
1.5	250	340	430	500	220	280	340	400	80	120	190	220	0.4
2	220	270	340	400	180	250	300	340	60	90	150	190	0.3

Source: Mal, 1995

Corrugation Irrigation

Corrugation irrigation is the method of application of water through miniature furrows in close growing crops, forage and pasture crops. The method is suitable for fine to moderately coarse soil susceptible to bake and crust. The chance of crust formation greatly reduced due to practice of partial wetting of soil in corrugation irrigation. The corrugations are not stable in sandy soils. The ridges of the corrugation use to collapse due to impact of rainfall or flowing water in moderate to high rainfall areas.

Corrugations are of V-shaped or U-shaped channels of about 10 cm depth spaced 40 to 75 cm depending on the soil type, crops grown and availability of water. Bamboo corrugator or tractor drawn corrugator using several parallel beams or pipes of 8 to 10 cm diameter makes corrugations (Fig. 1.17). Small furrowers or other similar equipments can be used in making the corrugation. The direction of corrugation may not coincide to the direction of line of crops. These are made after the seeds are sown in soil but before the germination. Corrugations use to follow the steepest slope. The spacing is closer in light textured soil. The maximum permissible length may be 50m in light textured soil to 150m in heavy textured soil. As the corrugation irrigation partially wet the soil, the method is some time preferred in early irrigation of the checks or border strips or where seeds are sown through drilling or broadcasting. Partial wetting and lateral movement of water from the corrugations help in germination of the seeds by minimizing the formation of soil crust.

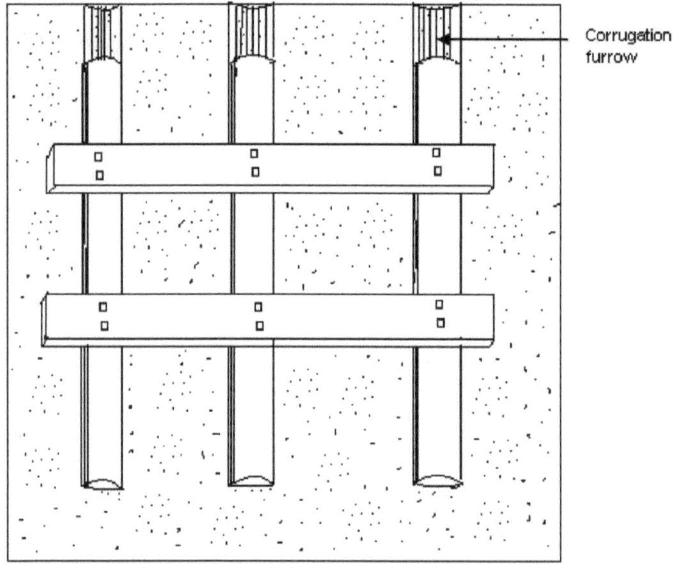

Fig. 1.17 Corrugator

Advantages

1. Small streams can be used.
2. It prevents the formation of soil crust.
3. Water application efficiency is high.
4. No precise land leveling is required.
5. It can be made easily.

Disadvantages

1. Not useful for high rainfall areas and where corrugation ridges are smoothed by the impact of rain.
2. Not suitable for the level land.
3. Not suitable in sandy soil.
4. Corrugations are to form in every crop season.

Deep furrow: The furrows more than 10cm depth and 45cm spacing may be regarded as deep furrow. Usually the furrows mean the deep furrows. Deep furrows are straight and contour.

Surge Irrigation

Surge irrigation is the intermittent application of water to the field surface under gravity flow in the mode of 'on' and 'off' for constant or variable time periods. Surge irrigation is done to reduce the deep percolation in furrows or border strips.

Infiltration rate is higher at initial stage of water application. After the initial infiltration more or less a stable infiltration rate persists. At a smaller stream more time is required for waterfront to reach at tail end of the border strips or furrows. This causes variable infiltration opportunity time and more time of infiltration at the upper end of the furrows. Faster delivery of water to the furrows and borders and cut-off the supply satisfy the initial requirement of water for infiltration. After the first application the next application rate and time are fixed depending on the changing soil infiltration rate. The practice of surge irrigation increases the infiltration uniformly and reduces the deep percolation compared to constant water application.

Example 1.10 The non-erosive stream is applied for a period of 15 minutes in a furrow of 80m long spaced 65cm apart and having a slope of 0.15%. Determine the average depth of water applied.

Solution

Length of furrow, L = 80m

Spacing of furrow, W = 0.65m

Slope of furrow, S = 0.15%

Average depth of water applied, d = ?

We have, $Q = \frac{0.6}{S}$

Where

Q = Non-erosive stream, l/s

S = Slope, percent

$$\therefore Q = \frac{0.6}{0.45} = 4l / s$$

Volume of water applied during the period

= 4 x 15 x 60 = 36001

$$\text{Average depth of water applied} = = \frac{3600l}{80\text{m} \times 65\text{cm}}$$

$$= \frac{3600/1000}{80 \times 65/100} = 0.0692\text{m} \cong 6.92\,\text{cm}$$

Example 1.11 Furrows of 120m length with 0.5% slope are made at 90 cm spacing. The maximum non-erosive stream flow rate is applied in furrow that takes 1.0 hour to reach the lower end. This flow rate is reduced to half of its size and, subsequently continued for another 1.0 hour. The average depth of water is ……cm.

(GATE, 2018)

Solution

$$\text{Maximum non-erosive discharge, } Q = \frac{0.6}{S} = \frac{0.6}{0.5} = 1.2l / s$$

Water applied in first 1 hour =1.2 x 3600 = 43201

Water applied in next 1 hour =1.2/2 x 3600 = 21601

Total water applied = 4320 + 2160 = 64801 = 6.48m^3

$$\text{Depth of water applied} = \frac{6.48}{\text{L} \times \text{S}} = \frac{6.49}{120 \times 0.9} = 0.06\text{m} = 6\,\text{cm}$$

Example 1.12 The following data were obtained from an irrigation.

Stream size, l/s	Distance from u/s end of furrow, m	Time of advance, min	Wetted perimeter, cm	Wetted x-sectional area of furrow corresponding to depth of flow, cm^2
3	50	5.5	23.5	97.42
	60	6.2	23.65	98.37

Compute the depth of infiltration at the different elapsed times.

Solution: For distance 50m:

Accumulated inflow = 3 x 5.5 x 60 = 990 liter

Accumulated storage = 50 mx 97.42 cm^2 = 0.487m^3

Accumulated infiltration (volume) = 990-487.1 = 502.9 liter

Area of wetted furrow = 50m x 23.50 cm^2 = 117500 cm^2

Accumulated infiltration depth = $\frac{502.9 \times 100}{117500} cm$ = 4.28cm

For distance 60m:

Accumulated inflow = 3 x 6.2 x 60 = 1116 liter

Accumulated storage = 60m x 98.37 cm^2 = 0.59m^3 = 590.22 liter

Accumulated infiltration = 1116 - 590.22 = 525.781

Area of wetted furrow = 60m x 23.65 cm = 141900 cm^2

∴ Accumulated infiltration depth = $\frac{525.78 \times 1000}{141900} cm$ = 3.705 cm

Example 1.13 Design the furrow system in growing potato in sandy loam soil. The field capacity and wilting point of the soils are 18.0 % and 8.0% respectively. Irrigation is suggested at 50% depletion of available moisture. Furrows are made along the slope of 0.2%. Apparent specific gravity and infiltration rate of the soil are 1.48 and 1.5cm/h respectively.

Solution: The root zone depth of the potato may be assumed 60cm.

Depth of irrigation water

$$= \sum_{i=1}^{n} \frac{(M_{fci} - M_{wi})A_s D_i}{100} \text{ x soil moisture depletion (fraction)}$$

$$= \frac{(18\text{-}8) \times 1.48 \times 60 \times 0.5}{100} = 4.44 \text{ cm}$$

From **Table 1.10** maximum furrow length is 400m and soil is loam or silt loam. From **Table 1.11** for 0.2% slope and 4.44 cm depth of irrigation water, furrow length for loam soil can be 220m. Finally, the length may be adjusted to 200m.

The non-erosive stream in furrows, $Q = \frac{0.6}{\text{S}} = \frac{0.6}{0.2} = 3l/s$

The final size of the stream in a furrow depends on the discharge from the source. Say, the source is supplying at a rate of 10l/s. therefore, final stream in furrow will be 3.33l/s.

Approximate furrow cross-section = spacing of furrow x depth of irrigation

= 60 cm x 4.44 cm = 266.4 cm^2

Assuming a bottom width of 10 cm and side slope 1:1 the x-sectional area of furrow = $bd + zd^2$

or, 10d + d2 = 266.4

or, d = 12.1 cm

Taking a free board of 20%, d = 12.0 x 1.2 = 14.52 = 15 cm (say)

1.5 Sub-surface Irrigation

Sub-surface irrigation or sub-irrigation is the practice of application of water below the soil surface by constructing trenches or through perforated pipes or tiles. Water applied to the trenches contributes to the root zone area of crops in between the trenches through lateral and vertical movement under capillary action. The perforated pipes or tiles remain full of water during the irrigation. The water is forced; trickle out through the perforation or gaps between the tiles. Due to capillary rise of water the upper soil layer remains unsaturated owing to evaporation loss of water.

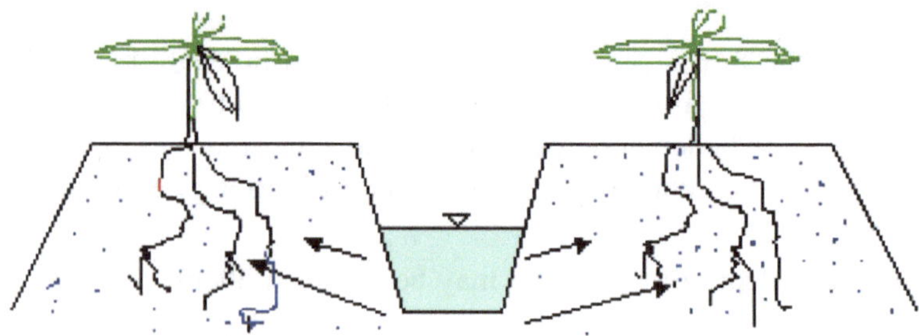

Fig 1.18 Sub-surface irrigation

In sub-irrigation through trenches a constant ground water level is maintained through controlled application of water. High water table to semi-impervious or impervious sub-soil is required to maintain the artificial ground level. Sub-irrigation requires permeable root zone soil, soils of free from salt problems. The depth of open ditch or trenches are created 30 to 100cm and spaced 15 to30m depending on the soil type and lateral movement of water in soils. The topography of the soil should be smooth and nearly level to uniform gentle slope. Various crops can be grown by sub-irrigation. Sometime high price vegetables especially shallow rooted crops cultivated by installing sub-surface perforated pipe. In India sub-surface irrigation is practiced in limited area for vegetable cultivation. However, the crops like wheat, jowar, bajra, potato, beet, peas, etc. can be cultivated through sub-irrigation. In Kerala sub-irrigation is used in coconut palms cultivation.

Example 1.14 For irrigation of sugarcane crop furrows are 75m length is laid 1.0m apart at a slope of 0.2%. Maximum size of the non-erosive stream is used for irrigation. Area covered by each furrow in 30 minutes and size of the stream was reduced to half and continued for another 30 minutes. Calculate the average depth of irrigation. (GATE, 1997)

Solution: Maximum non-erosive stream, $Q = \frac{0.6}{S} = \frac{0.6}{0.2} = 3l/s$

The volume of water applied = 3l/s x 30min. + 1.5l/s x 30min.

= 3 x 30 x 60+1.5 x 30 x 60 = 8100l

Average depth of water applied $= \frac{8100l}{75m^2} = \frac{8100 \times 1000}{75 \times 10000} = 10.8cm$

Questions and Problems

1.1 Define irrigation. What is the function of irrigation water?

1.2 Classify the methods of irrigation.

1.3 What do you mean by uncontrolled and controlled irrigation? Discuss the advantages and disadvantages of these?

1.4 What are the important factors influence to select the area of plots and spacing of supply channels.

1.5 What is border irrigation? Write a short note on border irrigation.

1.6 Discuss the infiltration opportunity time, advance and recession of flow related to border irrigation.

1.7 Discuss the idea of selecting the approximate width, length and stream size in a border.

1.8 Derive the equation for the time required to irrigate a given area in a border strip and the maximum area that can be irrigated for certain discharge and infiltration rate.

1.9 What is cross-border? Explain the benefit of cross-border with diagram.

1.10 What is check basin irrigation? How they are classified?

1.11 What are the advantages and disadvantages of check basin irrigation?

1.12 Discuss the scope of rectangular, contour and ring basin irrigation methods.

1.13 What is furrow irrigation? Describe the different type of furrow irrigation.

1.14 What are the advantages and disadvantages of furrow irrigation?

1.15 What is corrugation irrigation? How does it differ to furrow irrigation?

1.16 The cumulative infiltration is expressed as $I = at^b$. The irrigation of 3.0 & 3.5 cm depths of water got infiltrated in 85 and 140 minutes respectively. What are values of the constants a and b? Ans. a = 0.76, b = 0.31

1.17 The non-erosive streams are applied for 1.0h to two furrows of slope 0.2 & 0.3% respectively. What is the difference in volume of water application between these two furrows? Ans. 3.6m^3

1.18 Derive the equation for the time required to irrigate a given area in border strip and the maximum area that can be irrigated for certain discharge and infiltration rate.

1.19 Advance and recession curves are expressed as $t = 0.3x^{1.45}$ and $t = 950e^{0.0025x}$ respectively. What is the infiltration opportunity time at 25m length? Ans. 979.34 min

1.20 The cumulative infiltration is expressed as $I = 0.85t^{0.35}$. The porosity of soil is 0.3. If irrigation is applied at 60% depletion of available soil moisture, what is the time required to wet 60 cm root zone depth? Ans. 1428.36 min

1.21 Water is applied at a rate of 10l/s for 1.0h to a field of 5mx50m and 0.02% slope. What are the depths of water at head and tail ends just after the irrigation? Assume no infiltration during the irrigation. Ans. 13.9cm, 14.9cm.

1.22 Furrows of 100m length and spaced 1m apart applied with 0.12m depth of water is 50 minutes. What is the required size of irrigation stream? Ans. $4x10^{-3}m^3/s$ (GATE, 2004)

1.23 In an 80m long and 5m wide border, a stream of 8l/s was delivered from the upper end to irrigate wheat crop. The stream was cut off as soon as it had reached the lower end of the border in 70 minutes. Layer-wise soil moisture contents before and after are given below.

Depth, cm	Moisture content, %	
	Before irrigation	After irrigation
0-30	8.7	15.4
30-60	11.8	15.8
60-90	14.6	16.2

Determine the application efficiency if the root zone depth is 90 cm and the apparent specific gravity of soil is 1.4. Ans. 61.5% (GATE, 2002)

1.24 A border of size 0.02ha is irrigated by a stream of 12l/s for a period of 50 minutes. What was the average depth of water flow if soil characteristics permit infiltration rate 3.5cm/h? Ans. 16.5cm

1.25 A wheat border of 6m x 90m was irrigated by the stream of 15l/s for 45 minutes. The soil sampling for moisture contents at different depth and length of the border before and after the irrigation was as follows.

Depth, cm	Length along the border					
	22.5m		45m		67.5m	
	Moisture content, %		Moisture content, %		Moisture content, %	
	Before irrigation	After irrigation	Before irrigation	After irrigation	Before irrigation	After irrigation
0-15	14	22	15	22	13	21
15-30	15	22	15	21	13	22
30-45	16	23	15.5	21	14.5	22
45-60	17	22.5	16	22	15.5	21

Determine the depth of water penetrated at different lengths and application efficiencies of irrigation. The apparent specific gravity of soil is 1.45.

1.26 Design the check basins for a wheat field of 40mx80m in sandy loam soil. The field is almost flat and source of water is a shallow tube well discharges at a rate of 15l/s. The field capacity of the soil is 20%. Irrigation is suggested when soil moisture reaches to 15%. The apparent specific gravity of soil is 1.45 and root zone depth 90cm. The cumulative infiltration in soil is represented by $I = 0.75t^{0.47}$. Ans. Check = 8 # 10m x 40m, t = 29 min

1.27 Select the appropriate answer from the following:

1. Border irrigation is not suitable in the field of
   - a) Very high infiltration b) High infiltration
   - c) Medium infiltration d) Low infiltration
2. Wild flooding is suggested where
   - a) Field is exactly flat
   - b) Field is appropriately leveled and water is scarce
   - c) Field is divided in to small plots and water is in abundance
   - d) Fairly smooth field and water is in plenty and inexpensive.
3. Border irrigation is more suitable to
   - a) Paddy b) Close growing crops
   - c) Water loving plants d) Creepers
4. The cross-borders or cross-bunds are used in borders of
   - a) Large stream b) Large width and length
   - c) Poorly leveled and cloddy d) Shallow soil depth.
5. The size of the check basin does not depend on
   - a) Infiltration rate of soil b) Land slope
   - c) Crop grown d) Size of stream.
6. Ring basin method is suitable for cultivation of
   - a) Rice b) Close growing crops
   - c) Orchard d) Forage
7. Entire field need not to be flooded in
   - a) Rectangular check basin b) Ring check basin
   - c) Border d) Wild flooding.
8. Furrow method of irrigation is not suitable to soil where infiltration rate is
   - a) High b) Medium
   - c) Low d) Very low.
9. Contour furrow is suitable in
   - a) Land slope less than 0.15%
   - b) Steep land where straight furrows are not possible

c) Soils that crack

d) Soils of high infiltration.

10. Corrugation are made to

a) Steepest slope b) Across the slope

c) Direction of line of the crops d) Fields after germinating the seeds.

11. Checks are usually not

a) Large b) Small

c) Level d) Gently sloping.

12. The check basin irrigation method is mostly practiced in

a) Soil of high infiltration rate

b) Soil of moderate infiltration rate

c) Soil of moderate to low infiltration rate

d) Soil of low infiltration rate

13. The size of check basin mainly depends on

a) Type of crop cultivated b) Soil type and size of stream

c) Depth of water applied d) Slope of the check basin

14. The size of the check may be as high as

a) 1ha b) 5ha

c) 7.5ha d) 10ha

15. Ring check is essentially meant for

a) Orchard b) Vegetables

c) Field crop d) Citrus

16. Ring check is adapted to

a) Sloping land b) Sloping and rolling land

c) Flat land d) Land of gentle slope

17. Maximum rain water harvesting is possible following

a) Border irrigation method b) Contour border

c) Check basin d) Ring basin

18. Light soil suggests

a) Smaller check b) Larger check

c) Medium check d) No use of check

19. The principle of furrow irrigation is to wet

a) Half of the surface area

b) One-fourth to one-fifth surface area

c) One-half to one-fifth surface area

d) Entire surface area

20. Furrow irrigation be best adopted to land slope

a) High slope land b) Soils of high infiltration rate

c) Soils of low infiltration rate d) High value crops

21. Furrow method of irrigation is not suitable to

a) Arid zone soil b) Humid zone soil

c) Light soil d) Heavy soil

22. Rice field is usually irrigated by

a) Wild flooding b) Border irrigation

c) Check basin d) Furrow method

23. Surge irrigation follows

a) Continuously increasing of discharge rate

b) Continuously decreasing of discharge rate

c) Intermittent application

d) Corrugation

24. A border field has the discharge rate of 5l/s and infiltration rate 1.0cm/h. Maximum border area is

a) 0.12ha b) 0.14ha

c) 0.16ha d) 0.18ha

25. The moisture content of a field was found 10 & 15% respectively before and after irrigation. If the apparent specific gravity of the soil was 1.5, the depth of water applied to 50 cm root zone depth was

a) 1.87cm b) 3.75cm

c) 7.50cm d) 10.00cm

26. The recession of water in border is expressed by $t = 1200e^{0.002x}$. Water recedes in border up to the length after 30h is approximately

a) 50m b) 100m

c) 175m d) 200m

27. The non-erosive stream is applied for a period of 10 minutes in a furrow of 72m long spaced 50cm apart and having a slope of 0.1%. The average depth of water applied is

a) 5.0cm b) 7.5cm

c) 10.0cm d) 12.5cm

28. The length of check basin is determined by $L = mt_a^n$. The time of infiltration of applied water in a check basin is 2h. If m & n are unity, the length of the basin is

a) 20m b) 30m
c) 40m d) 50m

Ans.

1. a)	2. d)	3. b)	4. c)	5. c)	6. c)	7. b)	8. a)
9. b)	10. a)	11. a)	12. c)	13. b)	14. a)	15. a)	16. c)
17. d)	18. a)	19. c)	20. c)	21. c)	22. c)	23. c)	24. d)
25. b)	26. d)	27. c)	28. b)				

1.28 Write **True** or **False** of the following:

1. In contour border there may be elevation difference between two adjacent borders.
2. Design criteria for straight and contour border are different.
3. Sometime the width and length of the borders are adjusted depending on the stream size.
4. There is cross slope in border strip.
5. The area of check is smaller in light soil.
6. Checks are suitable to soils of moderate to high infiltration.
7. Corrugation irrigation is suitable in level field.
8. Corrugation irrigation is suitable for sandy soil
9. Corrugation irrigation is unsuitable beyond one percent slope.
10. There is a relation between line of crop sowing and direction of corrugation.
11. Corrugation irrigation is suitable for orchard cultivation.
12. Corrugation irrigation is suitable for close growing crops.
13. Small stream can be used in corrugation irrigation.
14. Corrugation irrigation minimizes the function of soil crust.
15. Corrugations are the permanent water conveyance structure.
16. Intermittent water application is done in surge irrigation.
17. Surge irrigation provides more deep percolation.
18. Time period in any irrigation of surge irrigation is constant.
19. Surge irrigation may be done in furrows.
0. Corrugation is suitable in sandy soil.

Ans.

1. True	2. False	3. True	4. False	5. True	6. False	7. False
8. False	9. False	10. False	11. False	12. True	13. True	14. True
15. False	16. True	17. False	18. False	19. True	20. False	

References

Biswas, R.K (2001). Project completion report of Ministry of Water Resources (MoWR), Govt. of India sponsored ad-hoc research scheme entitled "Application and evaluation of some interventions on irrigation system of deep tube well command area of new alluvial agro-climatic zone of West Bengal". BCKV, Mohanpur, West Bengal.

Biswas, R.K. Roy, R. & Pramanik, M. (1994). Effect of cut-off ratio on yield and water use efficiency of wheat in DVC command area. J. of Water Management, ISWM, Rahuri, Maharashtra. 3 (1&2) : 35-36.

Davis, D.S. (1943). Empirical equation and nonmography. McGraw Hill Book Co. New York. p.200.

Dakshinamurti, C., Michael, A.M. & Mohan, S. (1973). Water resources and their optimum utilization in Agriculture. IARI Monograph No.3. Water Technology Centre, IARI, New Delhi. p.392.

Majumdar, D.K. (2000). Irrigation Water Management Principle & Practices. Printice-Hall of India Pvt. Ltd, New Delhi.

Mal,B.C.(1995).Introduction to Soil and Water Engineering. Kalyani Publishers, 1/1, Rajinder Nagar, Ludhiana-141008.

Michael, A.M. (1978). Irrigation Principle and Practices. Vikas Publishing House Pvt. Ltd., New Delhi.

2

Measurement of Irrigation Water

Measurement of water used in irrigation ensures the application of required amount of water in crops. Knowing the rate of flow the adjustment to period of control at gates, etc. can be done to regulate the water at different branches of canals. To get the total control over the supply of water there should have the provision of water measurement at the critical points in irrigation systems.

Units of Measurement

Water is measured at rest or in motion. The water of reservoir, ponds, tanks, etc. at rest is measured in volumetric unit as liter, cubic meter, hectare-meter and hectare-centimeter. Water flowing in rivers, canals, channel, etc. is measured in litre/sec^2, liter per hour, meter per sec and hectare-meter per hour or day.

1 liter	=	1000cm^3
1 hectare	=	10,000m^2
Cumec	=	Cubic-meter per second
1 hectare-centimeter	=	100m^3 = 100,000 liters

One hectare-centimeter (ha-cm) is the volume of water when applied to one hectare land develops a depth of 1cm.

Methods of Measurement

The principles of measuring irrigation water may me grouped as: (i) volumetric measurement method, (ii) velocity-area method, (iii) measuring structures, and (iv) tracer method.

2.1 Volumetric method

This is a simple method of measuring the small discharge to crop fields. The water is collected in any container of known volume by a siphon or spiles for a specified period (Fig. 2.1). The simple bucket, kerosene jar or diesel barrel are the suitable containers. The time of filling the container may be recorded by stopwatch or seconds' wristwatch.

$$\text{Discharge, l/s} = \frac{Volume\ of\ the\ container, litres}{Time\ required\ to\ fill\ the\ container, seeonds}$$

This method is suitable to measure the discharges through the shallow tube wells or low capacity river lift pumps. The diesel barrel of about 180 liter capacity may be used as the container. Completely filling the barrel is difficult without losses of water. The volume of water in the partially filled barrel is measured by knowing the diameter and height of water in it. The method is also suitable to measure the discharge to big tank or reservoir. It is determined by knowing the additional volume of water due to flow in the reservoir for the noted time.

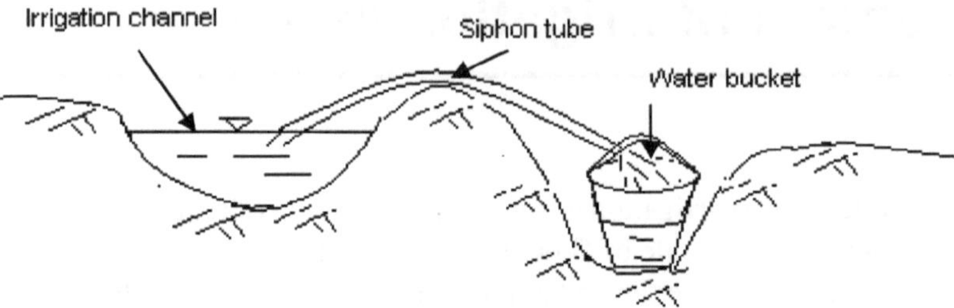

Fig. 2.1 Volumetric method of discharge measurement

Example 2.1 A barrel of diameter 50cm was used to measure the discharge of a shallow tube well. Determine the average discharge from the following.

Trial No.	Time, sec	Height of water in barrel, cm
1	20	75
2	25	92
3	27	100

Solution: Cross-sectional area of the barrel = $\frac{\pi D^2}{4} = \frac{\pi(0.5)^2}{4} = 0.196\text{m}^2$

Volume of water at 75 cm height = 0.147m^3

Volume of water at 92 cm height = 0.181m^3

Volume of water at 100 cm height = 0.196m^3

Rate of discharge at depth 75 cm & time 20 sec = $\frac{0.147m^3}{20\,\text{seconds}} = 7.53 l/s$

Rate of discharge at depth 92 cm & time 25 sec = 7.24 l/s

Rate of discharge at depth 100 cm & time27 cm = 7.26 l/s

Average discharge = 7.28 l/s

Example 2.2 A river lift pump discharges at a rate of 7.5 l/s and operated for 15 hours everyday. How much area can be cultivated if the crop requires 30 cm of water for 120 days irrigation period of the crop.

Solution: Total volume of water pumped during the 120 days irrigation period

= 7.5 l/s x 15 hrs/day x 120 days

= 7.5 x 15 x 3600 x 120 = 48600m^3

Irrigation water required by 1ha area

$$=1ha\times 30\text{cm}=10{,}000\times\frac{30}{100}=3000m^3$$

$$\therefore \text{Area irrigated}=\frac{48600}{3000}=16.2\text{ha}$$

Hose pipe: The flexible rubber pipe is now a days a very common practice to convey the water within 100-150m from the tube well or RLI (River Lift Irrigation) water source (Fig. 2.2). This pipe is made connected to the end of the delivery of the pump and carried up to point of application of water. It reduces the conveyance loss to the extent of 4-5% of loss in earthen channel. Some losses take place hose pipe due to the faulty fittings of it. However, the hose pipe used in soils having murram, plants with spines and snails or other objects having sharp edges cause leakage to the pipes and thereby considerable loss of water. It is cumbersome to shift the pipe from one plot to another when water is flowing through it. Doing so, there is every chance to damage the crops particularly the vegetable crops. The discharge from the pipe can be measured by volumetric method.

Fig. 2.2 Flexible hose pipe for conveying the irrigation water

2.2 Velocity-area Method

The discharge through a section of a pipe or open channel is estimated by multiplying the cross-sectional area of flow at right angle to the direction of flow by the average velocity of flow. This is expressed as,

$$Q=AV$$

Where

Q = Discharge rate, m^3/s

A = Cross-sectional area, m^2

The cross-sectional area is directly measured and velocity of water by using floats, current meter, etc.

Float method: Float method gives a rough estimate of the flow rate in an open channel. The velocity of water is measured by using a float. The partially filled water bottle, dry wood piece, part of plant stem or any thing, which do not submerge in water during the experimentation, can be used as float. A channel section of about 30m long as far as possible straight and uniform in cross-sections are selected (Fig. 2.3). At least three cross-sections preferably one in the middle and two at the ends of the channel section are considered for determining the average width and depth of flow. A string is stretched at each of this section perpendicular to the direction of flow. Several depths are taken at small distances perpendicular to flow i.e., along the string to get the average depth of flow in each selected cross-section of the channel. The multiplication of average width of the wetted part of the channel and depth gives the average cross-sectional area of the channel. The float is placed in the middle of the channel at short distance upstream from the trial section. The time the float takes to cross the trial section each time to be noted. The average of few such readings is the average time of travel. Care should be taken to see that the float does not come in touch to the sides of the channel and its movement get obstructed. Since the velocity of flow on the surface of the water is greater than average velocity of the stream, the float velocity is multiplied by the factor 0.85 to arrive at the correct value. The velocity of flow with respect to depth in a stream cross-section is logarithmic in nature. The following is a depth velocity distribution formula for a rough and turbulent flow (Subramanya, 1994).

$$V = 5.75 V_* \log_{10} \left(\frac{30Y}{K_g} \right) \tag{2.1}$$

Where,

V = Velocity at a point Y above the field

V_* = Shear velocity

K_g = Equivalent sand grain roughness

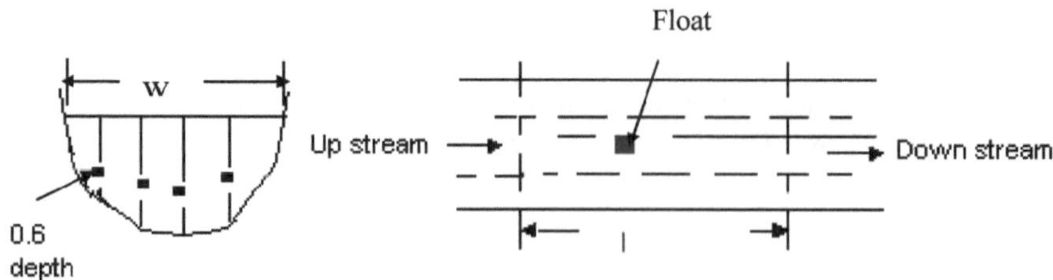

Fig. 2.3 Water measurement by float method

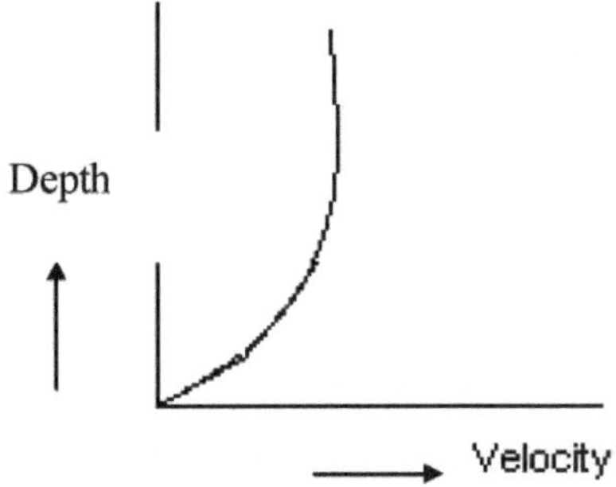

Fig. 2.4 Approximate depth-velocity curve

When there is no suitable equipments or devices to measure the flow in channel, the float method is the way to get the best approximation.

Example 2.3 A channel section of length 20m was selected for determining discharges through it by float method. The widths at upstream, downstream and middle of the section were 45, 50 & 47 cm respectively. The floats took 92, 90 and 89 seconds to pass the section in three consecutive trial runs. The depths of flow at different sections were as below.

At upstream, cm	At middle section, cm	At downstream, cm
0.0, 15.2, 25.8, 26.0, 12.3, 12.9, 0.0	0.0, 16.1, 24.6, 25.7, 21.2, 13.6, 0.0	0.0, 12.9, 21.7, 22.3, 20.8, 16.2, 0.0

Determine the velocity and discharge through the channel.

Solution:

Average depth of the sections:

$$\text{At upstream} = \frac{0.0+15.2+25.8+26.0+22.3+12.9+0.0}{7} = \frac{102.2}{7} = 14.6\text{cm}$$

$$\text{At middle section} = \frac{0.0+16.1+24.6+25.7+21.2+13.6+0.0}{7} = \frac{101.0}{7} 14.45\text{cm}$$

$$\text{At down stream} = \frac{0.0+12.9+21.7+20.8+22.3+16.2+0.0}{7} = \frac{93.9}{7} = 13.42\text{cm}$$

Area of the sections:

At upstream = 14.6 cm x 45cm = 657cm^2

At middle = 14.45cm x 50cm = 722.5cm^2

At down stream = 13.42cm x 47cm = 630.74cm^2

$$\text{Average x-sectional area of the channel section} = \frac{657 + 722.5 + 630.74^2}{3}$$

= 670.08cm^2

Average time taken by the float to pass the channel section $= \dfrac{92 + 90 + 89}{3}$

$= \dfrac{271}{3} = 90.33\text{s}$

Velocity of water on the surface $= \dfrac{20\text{m}}{90.33\text{s}} = 0.22\text{m/s}$

Average velocity through the section = 0.22 x 0.85 = 0.187 m/s

Discharge through the section = Average $\times$-sectional area x average velocity

= 670.08 cm2 x 0.187 m/s

= 13530cm^3/s = 13.53 *l/s*

Current meter method: Current meter is used to measure the velocity of water in a stream or river and discharge is estimated by multiplying the mean velocity and the wetted cross-sectional area of the stream. Current meter is the small instrument containing the conical cups or propeller, which rotates by the movement of water. It is suspended in deep stream by a cable or attached to a rod in shallow stream. The number of revolution of the vane made by the flowing water in a specified time is counted by the number of *bip* sound in a headphone or directly to a digital meter. Headphone or digital meter is connected to the axle of the rotating element by electric wires to actuate the revolution mechanism. The velocity of the point where the current meter is placed in the stream section is determined from the calibration chart of velocity vs. number of revolutions per second. For determining the mean velocity of the channel section, the section is divided into number of sub-areas depending on the size of the stream and the precision of measurement desired. The average of the velocities at 0.2 and 0.8 depths from the surface are taken as the velocity of the channel section. For a shallow depth channel the velocity is taken at 0.6 depths from the surface.

The meters are two types depending on the rotating elements are fitted on the vertical axis or horizontal axis. Vertical axis current meter has the series of conical cups mounted around a vertical axis and rotates in a horizontal plane (Fig. 2.5). The horizontal axis current meter is fitted with propeller at the end of the horizontal shaft, which rotates in vertical plane (Fig. 2.6). Vertical axis current meter is not recommended where vertical component of velocities exists. Horizontal axis current meters are not affected by oblique flows of even 15^0.

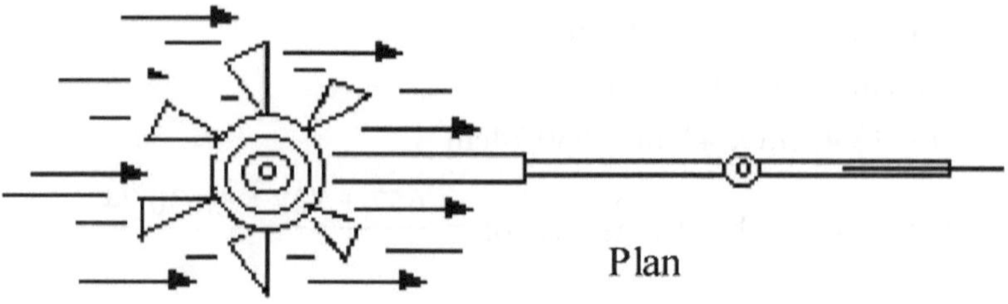

Fig. 2.5 Vertical axis current meter

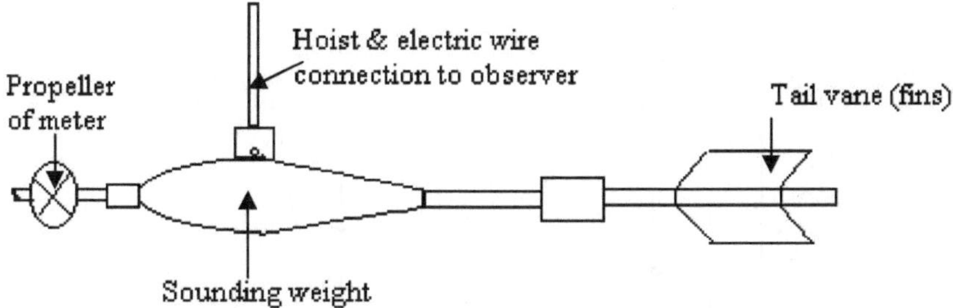

Fig. 2.6 Horizontal axis current meter

Current meter is usually set up at metering bridges in canals or other structures giving good access to the stream. The measuring section is selected where the channel is straight and fairly regular cross-section. The structures with piers divert the direction of flow. Such structures should be avoided as far as possible. The current meter has the limited use in irrigation channels. It is normally recommended for a velocity range of 0.15 to 4.0m/s.

For using the current meter in estimating discharge in a channel section, the section is suitably marked and divided in to large number of subsections (Fig. 2.7). The average velocity of these subsections is determined by using the current meter at desirable depths.

The following formula is used to calculate the velocity of flow measured by the current meter.

$$V = a + bN \tag{2.2}$$

Where

V = Velocity of flow, m/s

N = Number of revolutions per second

a & b = Constants

Each instrument has a threshold velocity below which instrument can not be used. The manufacturers provide the value of constants.

Example 2.4 Determine the flow of a channel from the following observed readings taken at 0.6-stream depth by a current meter. Take a = 0.03 & b = 0.50.

Distance at which readings are taken, m	**No. of revolution**	**Channel depth, m**	**Duration of observation, s**
0	0	0.0	0
1	35	0.4	60
3	70	1.0	60
5	100	1.4	60
7	90	1.3	60
9	40	0.6	60
10	0	0.0	0

Solution:

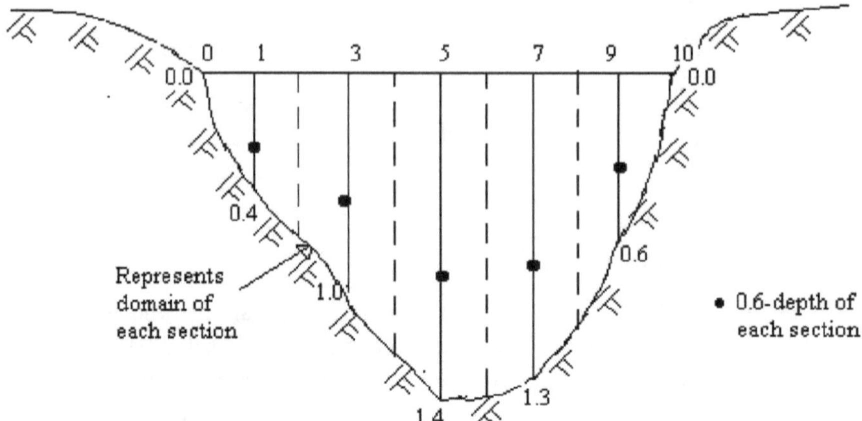

Fig. 2.7 Current meter used in a channel

Distance from the left edge, m	Average width of the segment, m	Stream depth, m	Revolutions/ second	Average velocity, m/s (V=0.03+ 0.50(Col.4)	Discharge rate at each segment, m^3/s (Col.2x Col.3xCol.5)
1	2	3	4	5	6
0	0	0	0	0	0
1	2	0.4	0.58	0.320	0.257
3	2	1.0	1.167	0.613	1.226
5	2	1.4	1.667	0.863	2.417
7	2	1.3	1.500	0.780	2.028
9	2	0.6	0.667	0.363	0.436
10	0	0	0	0	0
				Total = 6.364 m^3/s	

Dethridge method: The Dethridge method is named after its inventor, J.S.Dethridge of Australia. It consists of undershot water, which rotates by the moving water through a concrete pipe fitted at the outlet. The lower half of the wheel is provided only the practicable minimum clearance at its sides and around circumference (Fig. 2.8). The water driven by the wheel axle records number of revolution and gives the direct measurement of the discharge in volumetric units. The meter gives accuracy of measurement more than 95% in normal working condition. The meter is mainly used in Australia.

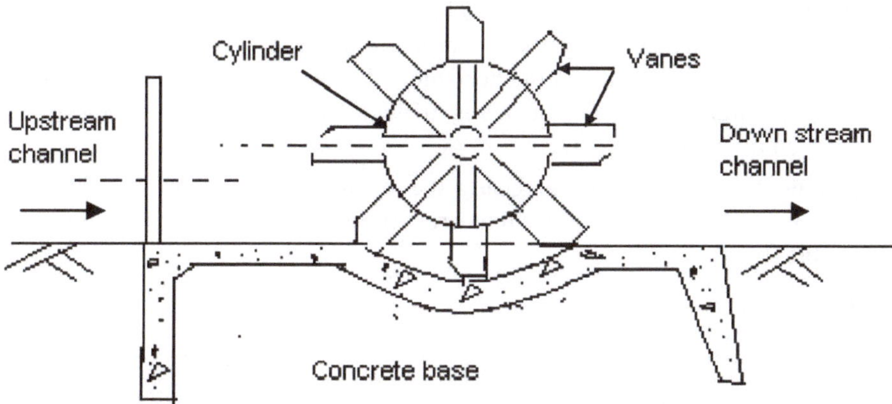

Fig. 2.8 Dethridge meter

Water meter: The water meter consists of multiblade propeller that rotates in a vertical plane and geared to a totalizer with a numerical counter that totals the flow in volumetric unit as desired (Fig. 2.9). The meter is used in water pipes. For using the meter in an open channel, the meter is made to flow through a pipe and meter is fitted in it at the outlet. For perfect operation of the water meter it should flow full and exceed the minimum rated range of flow. Water meter of different capacity are available in the market. The manufacturers do the calibrations. Water meters are costly equipments and usually not used to measure irrigation water.

Fig. 2.9 Water meter

Co-ordinate method: Co-ordinate method is used to measure the discharge through pipes either vertically or horizontally. Vertical discharge is used to estimate the discharge from flowing well and the horizontal discharge for small pumping unit. Accurate measurement of co-ordinates is difficult by this method. Therefore, co-ordinate method is of limited use.

Measurement by vertical pipe: Using the vertical pipe and forcing the flow through it in the form of jet measures the water from a flowing well (Fig. 2.10(a)).

$$Q = CA\sqrt{2gh} \tag{2.3}$$

Q = Discharge, m^3/s

A = Cross-sectional area of pipe, m^2

G = Acceleration due to gravity, $9.81 m/s^2$

C = Coefficient of discharge

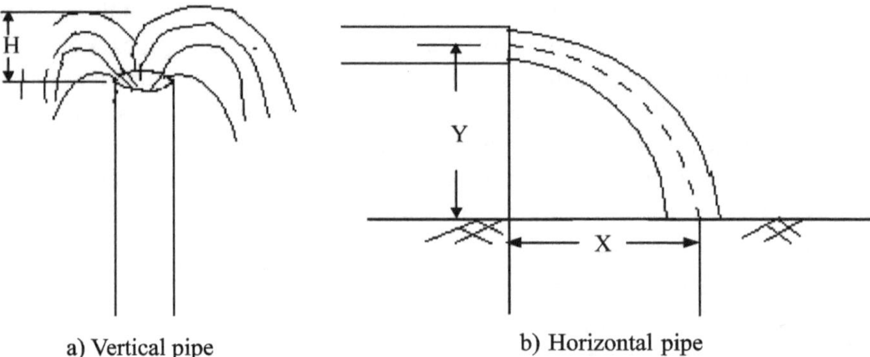

Fig. 2.10 Schematic diagram illustrating the coordinate method of measuring flow through full flowing pipes

Measurement by horizontal pipe: The discharge from a pump may be measured by using horizontal pipe. The water is allowed to flow freely through the pipe and the water jet thus formed drop at certain distance. The horizontal distance X is measured from the end of the pipe to the middle point of the jet drops on the ground. The vertical distance Y is measured from the center of the pipes to the ground. The horizontal and vertical distances are called the horizontal and vertical coordinates (Fig. 2.10(b)).

$$Y = \frac{1}{2}gt^2 \therefore t = \sqrt{\frac{2Y}{g}}, X = Vt$$

$$Thus, \frac{X}{V} = \sqrt{\frac{2Y}{g}}$$

Or, $V = \frac{X\sqrt{g}}{\sqrt{2Y}}$

The discharge through any pipe is

Q = CAV

Putting the value of $V = \frac{X\sqrt{g}}{\sqrt{2Y}}$

$$Q = \frac{CAX\sqrt{g}}{\sqrt{2Y}} \tag{2.4}$$

Example 2.5 The height of water jet forms in a 10cm diameter flowing well is 30cm. Determine the discharge of the well. Assume 0.8 as discharge coefficient.

Solution:

Discharge, $Q = CA\sqrt{2gh}$

$= 0.8 \times \pi \frac{(0.1)^2}{4} \times \sqrt{2 \times 9.81 \times 0.3}$

$= 0.8 \text{ x } 7.85 \text{ x } 10^{-3} \text{ x } 2.426$

$= 0.01523 \text{ m}^3 \text{/s} = 15.23 \; l/s$

Example 2.6 The water jet from a shallow tube well delivery pipe at 0.75m height falls at horizontal distance of 1.55, 1.45 & 1.4m during different time in a day. If the diameter of the delivery pipe 6.5cm, determine the discharge of the shallow tube well. Assume the discharge coefficient as 0.95.

Solution: The average distance of water jet $= \frac{1.55+1.45+1.4}{3} = 1.47\text{m}$

Velocity of water jet $= \frac{X\sqrt{g}}{\sqrt{2Y}}$

Where,

X = Horizontal distance of water jet

g = Acceleration due to gravity

Y = Vertical height of the water jet

Y = Vertical height of the water jet

$V = \frac{1.47\sqrt{9.81}}{\sqrt{2 \times 0.75}} = \frac{4.593}{1.224} = 3.75\text{m/s}$

$Q = \text{CAV}$

$$=0.95\times\frac{\pi(0.065)^2}{4}\times 3.75$$

$$= 0.95 \text{ x } 3.32 \text{ x } 10^{-3} \text{ x } .375$$

$$= 0.0118\text{m/s} = 11.82\ l/s$$

Pitot tube: Pitot tube can be used to measure the flow through open channel as well as flow through pipes. It is an L shaped tubes as shown in Fig. 2.11, which can be placed at any depth in an open channel to measure the velocity of flow. The velocity of water converts to pressure head when enters in to pipes and represented by h in Fig. 2.11. The pitot tube when used in a pipe takes the shape of inverted U (Fig. 2.12). This is done to bring the static head in to consideration. When a pitot tube is positioned properly it develops zero velocity just in front of the open end of the tube. This states that the velocity head is converted to pressure head. The velocity head converted to pressure head is represented in the U tube by the difference of water level h and represented by the equation,

$$V = C\sqrt{2gh}$$

Where C is the pitot tube coefficient whose value varies from 0.95 to 1.0. The discharge through the pipe may be calculated by the multiplication of velocity (V) and cross-sectional area of flow.

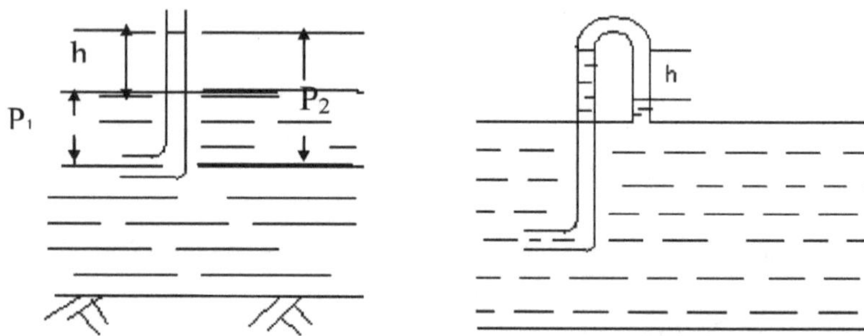

Fig 2.11 Pitot tube

Fig. 2.12 Inverted U tube

Example 2.7 A pitot tube having a coefficient of 0.96 is used to measure the velocity of water at certain depth in an open channel of wetted x-sectional area 2.5m^2 recorded 0.2m water head. Determine the velocity and discharge of water at that depth of channel.

Solution: Velocity of water, $V = C\sqrt{2gh}$

$$= 0.96\times\sqrt{2\times 9.81\times 0.2} = 1.90\text{m/s}$$

Discharge, $Q = A \text{ x } V = 2.5 \text{ x } 1.9 = 4.75\text{m}^3/\text{s}$

Venturi meter: Venturi meter is used to measure the discharge through pipe flowing full. It consists of converging and diverging section in the pipe within a short length. When flow occurs through the converging section the velocity of flow increases and the pressure drops. The pressure drops is measured by using a manometer at the point of constriction as shown in Fig. 2.13. Considering the point 1 and 2 in Fig. 2.13, the cross-sectional areas are A_1 and A_2 respectively. Assuming the pipe in horizontal, applying Bernoulli's theorem and neglecting friction losses,

$$\frac{P_1}{\rho g}+\frac{V_1^2}{2g}=\frac{P_2}{\rho g}+\frac{V_2^2}{2g} \tag{2.5}$$

Or, $\frac{V_2^2}{2g}-\frac{V_1^2}{2g}-\frac{P_1}{\rho g}-\frac{P_2}{\rho g}$

Or, $V_2^2-V_1^2=2g\left(\frac{P_1-P_2}{\rho g}\right)=2g\left(\frac{\rho_m gh-\rho gh}{\rho g}\right)$

$$=2gh\left(\frac{\rho_m}{\rho}-1\right) \tag{2.6}$$

Where,

h = The difference of head indicated in manometer

ρ_m = Density of grease fluid in manometer

The discharge through the pipe,

$Q=A_1V_1=A_2V_2$

$\therefore V_2=\frac{A_1}{A_2}V_1$

Substituting V_2 in Eq. 2.6

$V_1^2\left(\frac{A_1}{A_2}\right)^2-V_1^2=2gh\left(\frac{\rho_m}{\rho}-1\right)$

Or, $V_1^2\left[\left(\frac{A_1}{A_2}\right)^2-1\right]=2gh\left(\frac{\rho_m}{\rho}-1\right)$

Or, $V_1\frac{1}{\sqrt{\left(\frac{A_1}{A_2}\right)^2-1}}.\sqrt{2gh\left(\frac{\rho_m}{\rho}-1\right)}$

As an ideal case, $Q_{ideal}=A_1V_1=\frac{A_1}{\sqrt{\left(\frac{A_1}{A_2}\right)^2-1}}.\sqrt{2gh\left(\frac{\rho_m}{\rho}-1\right)}$

As actual case, $Q_{actual} = C_d.Q_{ideal}$

Where C_d is the coefficient of discharge. The value of C_d varies between 0.97 to 0.99.

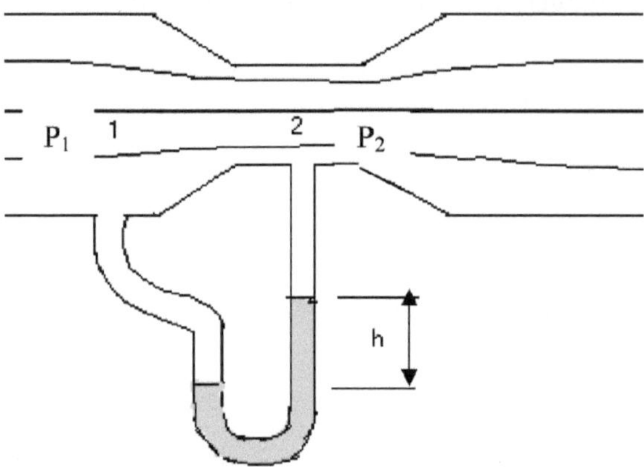

Fig. 2.13 Venturi meter

Example 2.8 A mercury manometer is used in the venturimeter installed in a pipe of diameter 25mm to measure the discharge through it. The diameter of constriction of the venturi is 10mm. At full flow through the pipe it gives a manometer reading of 5cm. Determine the discharge through the pipe.

Solution: Using Eq. 2.5 the velocity of flow through the pipe,

$$V_1 \frac{1}{\sqrt{\left(\frac{A_1}{A_2}\right)^2 -1}} . \sqrt{2gh\left(\frac{\rho_m}{\rho}-1\right)}$$

$$Or,\ V_1 = \frac{1}{\sqrt{\left(\frac{\pi(0.025)^2}{4} / \frac{\pi(0.01)^2}{4}\right)^2 -1}} . \sqrt{2\times 9.81\times \frac{5}{100}\left(\frac{13.6}{1}-1\right)}$$

$$= \frac{1}{\sqrt{38.06}} . \sqrt{12.36}$$

$$= \frac{1}{\sqrt{6.17}} \times 3.516 = 1.41\text{m/s}$$

$$Q_{ideal} = C_d Q_{ideal} = C_d V_1 A_1$$

Assuming $C_d = 0.98$

$$Q_{actual} = 0.98 \times 1.41 \times \pi . \frac{(0.025)^2}{4}$$

$= 6.78 \times 10^{-4}\ m^3/s$

Example 2.9 Two ends of a differential mercury manometer are connected at two points of pipe carrying oil. The manometer shows difference in mercury level of 20 cm. the specific gravity of oil and mercury are 0.8 and 13.6. The density of water is 1000kg/m^3 at 4^{0}C and acceleration due to gravity (g) is 9.81m/s^2. At the same two points in pipe, the difference in pressure in N/m^2 is 25.11, 251.14, 25113.60, and 251136.60. (GATE 2020)

Solution

The difference in pressure at two ends of mercury, $P = \rho_m gh - \rho_0 gh$, ρ_m & ρ_0 are density of mercury and oil respectively and difference of mercury level, h=0.3m.

$$p = \rho \times 13.6gh - \rho \times 0.8gh$$

$$= \rho gh(13.6 - 0.8)$$

$$= 1000kg/m^3 \times 9.81m/s^2 \times 0.2m \times 12.8$$

$$= 25113.6N/m^2$$

2.3 Measuring Structures

The most common devices used in measuring the irrigation water are orifices, weirs, flumes and meter gates. The devices provide measurement of water by the flow reading used in the standard formulae or calibration chart. These devices are very simple, can be locally made and give accurate result if properly constructed and operated.

Orifice: An orifice is a circular or rectangular opening in a metallic plate wall through which water flows when it is placed across the irrigation channel (Fig 2.14). The edges of the opening are sharp. The area of the opening is much smaller than the channel cross-section, which ensures complete contraction of stream flow. An orifice may be free flowing or submerged. A free flowing orifice entirely discharges into air whereas in the submerged orifice water level at the downstream is above the top of the opening. In case of partially submerged orifice water level at downstream is in between the top and bottom of the opening. Partially submergence of orifice is avoided for the sake of correct measurement of flow.

Free flows orifice: Free flow orifice is used for small stream where a sufficient fall is available such as in border, strip furrow or check basin. Free flow condition cannot be achieved in a level channel when loss of head is little. The orifices are of much smaller dimension to channel cross section and may range from 2.5 to 7.5cm diameter. The sheet iron, steel, aluminium, etc., which can be machined for exact round opening are used.

The discharge from an orifice may be measured by using the formula,

$$Q = C \times 10^{-3}\, a\sqrt{2gH} \tag{2.7}$$

Where,

Q = Discharge, l/s

a = Cross-sectional area of the orifice, cm^2

C = Coefficient of discharge, usually found to be 0.61

g = Acceleration due to gravity, 981cm/s^2

H = Depth of water over the center of orifice in case of free flowing and or difference in elevation between upstream and downstream faces in case of submerged orifice, cm.

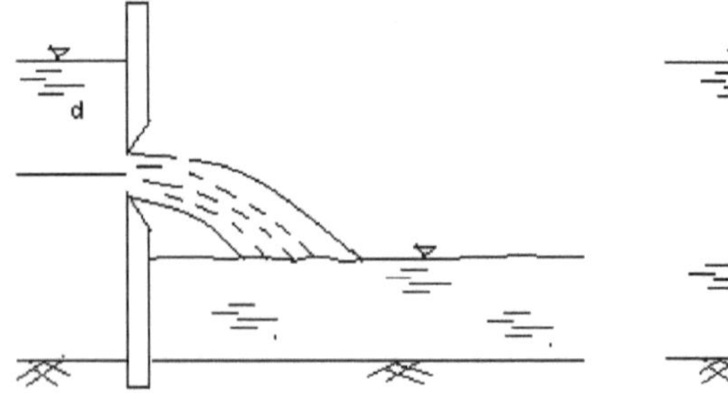

a) Free flows orifice

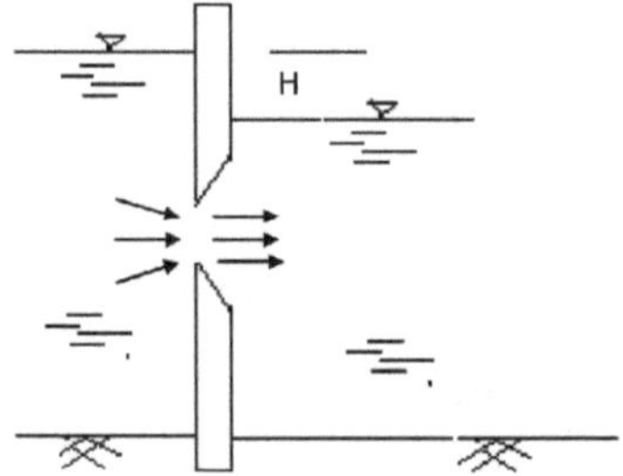

b) Submerge flows orifice

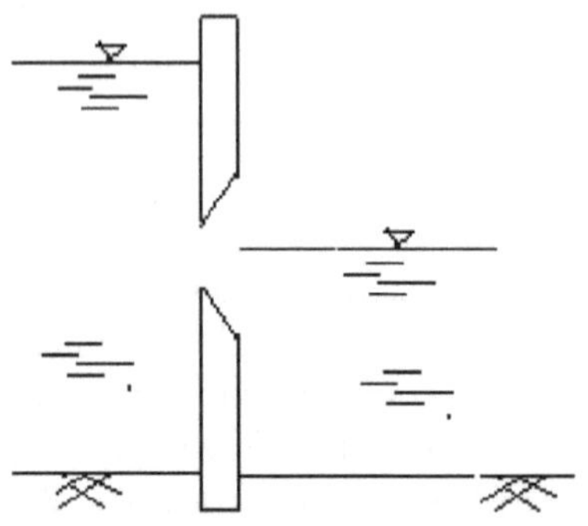

c) Partially submerged orifice flows

Fig. 2.14 Flow through orifices

Submerged orifice: The opening of the submerged orifice may be round or rectangular. The coefficient C varies from 0.61 to 0.80 or more depending on the position of the orifice to the sides and bottom of the water channel and perfections in roundness of edges of the orifice. The opening of the submerged orifice may be increased or decreased to suit the requirement. The width of the rectangular orifice is about 2 to 6 times the height of the orifice. The advantages of using orifices in measuring water are: accuracy of measurement is quite satisfactory, no much head loss and device is so simple in construction and easily installed in a channel. However, the floating materials sometime obstruct flows in the orifices.

Example 2.10 The head of water over a 15cm x 20 cm rectangular orifice is 30 cm. Determine the discharge (a) at free fall, (b) at 15 cm submerged flow and (c) at partially submergence of 3 cm over the bottom of the orifice.

Solution: Let us assume coefficient of discharge, $C = 0.61$

We have, $Q = CAV$

a) At free fall, Q = 0.61 x 15cm x 20cm x $\sqrt{2 \times 981 \times 30}$ = 44397.77

   = 44.39*l/s*

b) At submerged flow, Q = 0.61 x 15cm x 20cm x $\sqrt{2 \times 981 \times 15}$

   = 0.61 x 300 x 171.55 = 31393.97 = 31.39*l/s*

c) Effective depth for back flow = 3 - 3/2 = 1.5cm

   Net depth of water causing flow = 30 - 1.5 = 28.5cm

   Q= 0.61 x 15cm x 20cm x $\sqrt{2 \times 981 \times 28.5}$

   = 0.61 x 300 x 236.47 = 43273.6cm^3 = 43.27*l/s*

Coefficient of velocity (C_v): This is the ratio of actual mean velocity in the cross section of stream (jet) to the theoretical mean velocity suppose to occur without friction loss.

$$C_v = \frac{Actual\ mean\ velocity}{Theoretical\ mean\ velocity} \quad (2.8)$$

Coefficient of contraction (C_c): This is the ratio of the contracted section of the stream (jet) to the area of the opening through which the fluid flows:

$$C_c = \frac{area\ of\ stream\ (jet)}{area\ of\ opening} = \frac{A_{jet}}{A_0} \quad (2.9)$$

Coefficient of discharge (C_d): It is the ratio of actual discharge to the theoretical discharge at any device.

$$C_d = \frac{actual\ flow}{theoretical\ flow} = \frac{Q_a}{Q_t} \quad (2.10)$$

The coefficient of discharge is related to coefficient of velocity and coefficient of contraction as below:

$$C_d = C_v \times C_c \quad (2.11)$$

The coefficient of discharge is determined experimentally as

$$Q = C_d AV = C_d A\sqrt{2gH}$$

Where,

A = Cross sectional of device

H = Total head causing flow.

Example 2.11 The actual velocity of water in the contracted section of a jet flowing from a 25mm diameter orifice was found 10.5m/s at head of 7.5m. The measured discharge was 3.6 l/s. Find the coefficient of velocity, contraction and discharge.

Solution: Actual velocity,

$$V = C_v\sqrt{2gh}$$

$$Or,\ 10.5 = C_v\sqrt{2\times9.18\times7.5} = C_v\ 12.13$$

$$\therefore C_v = \frac{10.5}{12.13} = 0.87$$

Again, $Q = C_d A\sqrt{2gH}$

$$\text{Or, } \frac{3.6}{1000} = C_d \frac{\pi}{4}.\left(\frac{25}{1000}\right)^2 .\sqrt{2\times9.81\times7.5}$$

$= C_d$ x 0.000491 x 12.13

$$\therefore C_d = \frac{0.0036}{0.000491\times12.13} = 0.6$$

Since $C_d = C_v\ x\ C_c$

$$\therefore C_c = \frac{C_d}{C_v} = \frac{0.6}{0.87} = 0.69$$

Example 2.12 Water is being delivered at a rate of 20 l/s from the 65mm diameter delivery pipe of a tube well under pressure head of 5m and at a height of 1.2m. The jet of water drops on the ground at 3.95m from the delivery pipe end. Compute the coefficients.

Solution:

$$Q = C_d A\sqrt{2gH}$$

$$Or, \frac{20}{1000} m^3/s = C_d \frac{\pi}{4}\left(\frac{65}{1000}\right)^2 \sqrt{2\times9.81\times5}$$

$0.02 = C_d$ x 0.0033 x 9.9

$$\therefore C_d = \frac{0.02}{0.033} = 0.6$$

Horizontal drop of the jet, $x = Vt$

Vertical drop of the jet, $y = \frac{1}{2} g t^2$

Eliminating t, $x^2 = \frac{2V^2 y}{g}$

$$\therefore (3.95)^2 = \frac{V^2 \times 2 \times 1.2}{9.81}$$

$$\therefore \quad V^2 = \frac{153.06}{2.4} = 63.77$$

$$\therefore V = 7.99 \text{ m/s}$$

So, $7.99 = C_v \sqrt{2gH}$

$$= C_v \sqrt{2 \times 9.81 \times 5}$$

$$\therefore Cv = \frac{7.99}{9.9} = 0.81$$

Meter gate: A meter gate is modified submerged orifice whose dimension is adjustable. These are used to monitor the water flowing from one channel to another. Meter gates are largely used in USA at the canal outlets to charge water on volumetric basis. The openings of the gates are generally circular. In a partially submerged gate it is difficult to measure the area of a partially flow gate. Therefore, calibration charts are required for determining discharges at different head of water and opening of the gate.

Weirs

A weir may be defined as a notch in a wall built across a stream. The notch may be rectangular, trapezoidal or triangular (Fig. 2.15). Water is allowed to flow over it for its measurement in an irrigation channel. Weirs are very simple devices, which can be used to measure any small supply of water. It is also constructed and installed in a channel easily. The floating materials cannot disturb much in measurement of flow. Weirs are quite durable. Weirs do not work well in the channel carrying silt which deposits on approach channel and destroy the optimum condition for accurate measurement of flow.

Term used

Weir pond: The channel portion immediately upstream of the weir.

Head: The depth of water over the weir crest measured at some distance in the upstream from the weir.

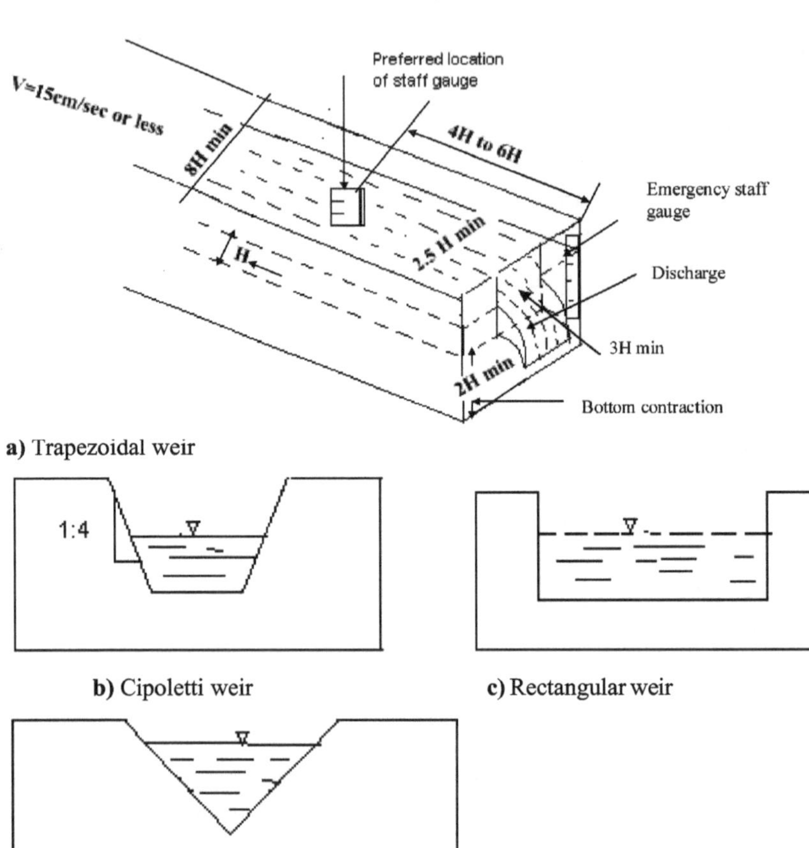

a) Trapezoidal weir

b) Cipoletti weir

c) Rectangular weir

d) V-notch weir

Fig. 2.15 Different type of weirs

Weir crest: The bottom of the weir notch.

Sharp crested weir: The weir having very thin edged crest, which provides maximum surface contact to flowing water.

End contraction: The horizontal distance between the ends of the crest to the sides of the weir ponds.

Bottom contraction: The vertical distance between weir crest to the bottom of the weir pond.

Weir scale or gauge: The scale fastened on the sides of the weir or with the stake in the weir pond to measure the head of water over the weir crest.

Nappe: The free falling sheet of water from a weir is called nappe.

The weir may be sharp crested or broad crested. These are divided to weirs with end contractions and no end contractions. End contraction develops when the notch of the weir is less than the channel width.

The basic formula for calculating the discharge through a weir is

$$Q = CLH^n \quad (2.12)$$

Where

Q = Discharge

C = Coefficient depends on nature of the crest and approach condition

L = Length of crest

H = Head of water on the crest

n = An exponent, depends on the geometry of weir.

Rectangular weir

As shown in Fig. 2.15 (a) if L is the crest length and H is the head of water over the crest, the discharge through the weir is,

$$Q = CAV$$

$$= C'LH\sqrt{2g\frac{H}{2}}$$

$$= C'\sqrt{g}\,LH^{3/2}$$

$$= CLH^{3/2} \quad (2.13)$$

The discharge equation can also be derived mathematically.

As shown in Fig 2.16 a small discharge through the infinitesimal small strip dy at y distance from the water surface on the weir is

$$dQ = Ldy\sqrt{2gy}dy$$

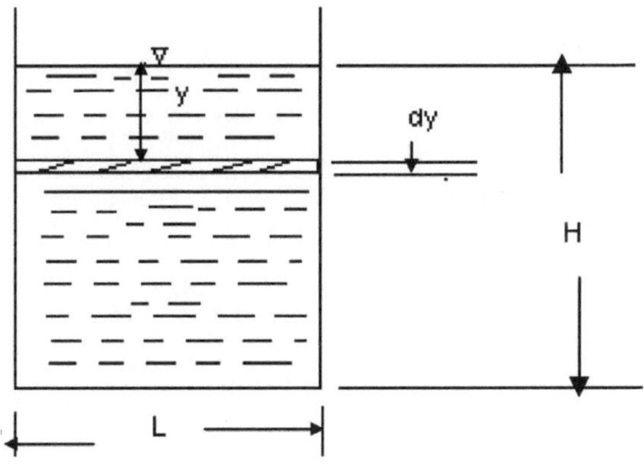

Fig. 2.16 Wetted cross-section of a rectangular weir

On integration,

$$Q = L\sqrt{2g}\left|\frac{y^{3/2}}{\frac{3}{2}}\right|_0^H = \frac{2}{3}\sqrt{2g}\,\mathrm{L\,H}^{3/2}$$

Adding the discharge coefficient

$$Q = C'\frac{2}{3}\sqrt{2g}LH^{3/2} = \mathrm{CLH}^{3/2}$$

The discharge through rectangular weir may be computed by the formula given by Francis as below.

i) Suppressed rectangular weir or the rectangular weir with no end contraction

$$Q = 0.0184\mathrm{LH}^{3/2} \qquad (2.14)$$

Where,

Q = Discharge, l/s

L = Length of crest, cm

H = Head of water over the crest, cm

ii) Rectangular weir with end contraction

$Q = 0.0184\ (\mathrm{L} - 0.2\mathrm{H})\ \mathrm{H}^{3/2}$ for both end contraction

and $Q = 0.0184\ (\mathrm{L} - 0.1\mathrm{H})\ \mathrm{H}^{3/2}$ for one end contraction

Example 2.13 Compute the discharge of a rectangular weir of 50cm width and 15cm head of water for no end contraction, one end contraction and two-end contraction. Use Francis' formula.

Solution: For no end contraction

$Q = 0.0184\ \mathrm{LH}^{3/2} = 0.0184 \times 50 \times 15^{3/2}$

= 53.45 *l/s*

For one end contraction

$Q = 0.0184\ (\mathrm{L}-0.1\mathrm{H})\mathrm{H}^{3/2} = 0.0184 \times 48.5 \times 15^{3/2}$

= 51.84 *l/s*

For two end contraction

$Q = 0.0184\ (\mathrm{L}-0.2\mathrm{H})\mathrm{H}^{3/2}$

$= 0.0184 \times 47 \times 15^{3/2}$

= 50.24*l/s*

Cipoletti weir

Cesare Cipoletti, an Italian engineer developed the trapezoidal weir in which the side of the notch has a slope of 1 horizontal to 4 vertical. Such a rectangular weir is believed to be proportional to the length of the weir crest. Therefore, no end correction is required for the crest length. The weir is sharp crested and beveled to the downstream side only. The discharge through the Cipoletti weir is

$$Q = 0.0186\ LH^{3/2} \quad (2.15)$$

Where

Q = Discharge, l/s

L = Crest length, cm

H = Head of water, cm

V-notch weir

V-shaped notched weir is V-notch weir. The 90^0 V-notch weir is most commonly used. However, the notches of other angle are also in use. The V-notch weir has great practical use than other weirs that it can be used to measure any small stream.

The discharge through a 90^0 V-notch is measured by the formula

$$Q = 0.0138H^{5/2} \quad (2.16)$$

Where

Q = Discharge, l/s

H = Head of water, cm

Derivation for discharge of V-notch of any angle

Referring the Fig. 2.17,

dy = Infinitesimal thickness of water at depth y from the water surface in the weir

x = Width of the strip at depth y

q = V-notch angle

H = Head of water

The discharge through the strip of thickness dy is

$$dQ = xdy\sqrt{2gy}$$

$$= 2\tan\theta/_2\,(H\text{-}y)\sqrt{2gy}dy$$

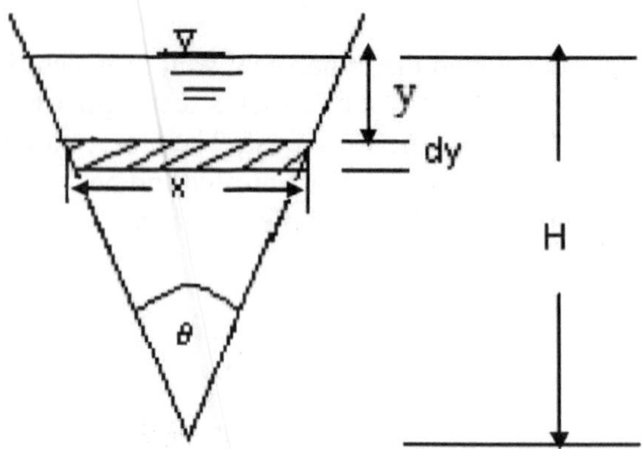

Fig. 2.17 A V-notch weir

On integration,

$$Q = 2\tan\frac{\theta}{2}\sqrt{2g}\left|\frac{Hy^{3/2}}{3/2} - \frac{y^{5/2}}{5/2}\right|_0^H$$

$$= 2\tan\frac{\theta}{2}\sqrt{2g}\left(\frac{H^{3/2}}{3/2} - \frac{H^{5/2}}{5/2}\right)$$

$$= \frac{8}{15}\tan\frac{\theta}{2}\sqrt{2g}.H^{5/2} \tag{2.17}$$

Introducing the coefficient of discharge, C_d

$$Q = C_d\frac{8}{15}\tan\frac{\theta}{2}\sqrt{2g}H^{5/2}$$

Example 2.14 Compute the discharge for 30cm and 35cm head of water measured by 90^0 and 60^0 V-notch respectively. Find the coefficient of discharge in 90^0 V-notch.

For 90^0 V-notch

Without considering the discharge coefficient,

$$Q = \frac{8}{15}\tan\frac{\theta}{2}\sqrt{2g}H^{5/2}$$

$$= \frac{8}{15} \times 1 \times 44.29 \times 4923.5$$

$= 116.44$ *l/s*

Considering the discharge coefficient,

$Q = 0.0138\ H^{5/2}$

$= 0.0138\ (30)^{5/2}$

$= 68.03\ l/s$

Coefficient of discharge, $C = \dfrac{68.03}{116.44} = 0.58$

For 60^0 V-notch:

Without including the discharge coefficient,

$$Q = \frac{8}{15} tan\frac{\theta}{2}\sqrt{2g}H^{5/2}$$

$$= \frac{8}{15} \times 0.577 \times 44.29 \times 7247.192 = 98.77\ l/s$$

Example 2.15 In an irrigation channel of uniform section, water passes through a 90^0 triangular weir measuring 36cm head over the crest. After traveling certain distance in the same channel, water passes through 1.0m long rectangular weir. There is no loss of water in between two weirs. Using Francis's formula, the head over the crest of rectangular weir in cm is 22.1, 18.4, 15.0, 11.8. (GATE 2020)

Solution:

Francis's formula for triangular weir,

$$Q = 0.0138H^{\frac{5}{2}} = 0.138(36)^{\frac{5}{2}} = 107.31l/s$$

Francis's formula for rectangular weir,

$$Q = 0.0186LH^{\frac{3}{2}} = 0.01.86 \times 100H^{\frac{3}{2}} = 1.86H^{\frac{3}{2}} = 107.31l/s$$

The discharge in both the weir is same.

Therefore, head in rectangular weir

$$H = \left(\frac{107.31}{1.86}\right)^{\frac{3}{2}} = (57.69)^{0.667} = 14.95 \approx 15.0cm$$

Example 2.16 A Cipolletti weir has the crest length of 1.2m and a crest level of 0.5m. The average approach velocity of water in m/s on the crest is

a) 0.49 b) 1.18

c) 1.19 d) 1.32 (GATE, 2015)

Solution:

$Q = 0.0186LH^{3/2}$

$= 0.0186 \text{ x } 120\text{cm x } (50)^{3/2}$

$= 789/131/s$

$$V=\frac{Q}{A}=\frac{789.13l/s}{120\text{cm}\times 0.5\text{m}}=\frac{0.789m^3/s}{1.2\text{m}\times 0.5\text{m}}=1.32m/s$$

Installation of a weir

A properly constructed and installed weir gives 95 percent or more accurate measurement. Improper installation may give large error. The following points are to be borne in mind while installing a weir.

i) The weir should be set in straight section whose length in the upstream is not less than ten times the length of the weir.

ii) The weir should be placed perpendicular to the direction of flow and crest should be level.

iii) The thickness of the crest and side of the weir notch should not be more than 3mm and beveled to the downstream side. Notch should be of proper shape, sharp, rigid and straight.

iv) The velocity of flow in the upstream of the weir should be uniform and should not exceed 15cm/s. A pond is created at the upstream close to weir to reduce the turbulence and getting correct measurement of velocity. Baffles may be used in the pond if found necessary.

v) There should not be any obstruction of flow upstream of weir.

vi) The weir crest should be at height from bottom of the approach channel at least twice the depth of water on the crest.

vii) The distance between the end of weir crest and bank of the upstream should be at least twice the depth of water on the crest.

viii) The crest of weir should be high enough to have the free flow. The depth of water over the rectangular weir should not be less than 5cm and more than two-thirds of the crest length.

Flumes

A flume is a channel section of special shape, which is usually constructed and installed, in open channel for flow measurement by stable stage-discharge relationship. Flume may be portable type and are mainly used for measuring small stream in irrigation channels. Potable flumes are two types: Parshall flume and Cut throat flume.

Parshall flume is an open channel type device for measuring the discharge and operates with small drop of head. A Parshall flume consists of three section: (i) upstream or converging section, (ii) throat or middle section, and (iii) downstream or diverging section. The floor of the converging section is level with verticals converging to the throat. The wall of the throat section is parallel and the floor inclined downward at a slope of 9 vertical to 24 horizontal. The floor of the diverging section is inclined upward at a slope of 1 vertical to 6 horizontal and the walls are diverging towards the outlet. There are two-graduated scale in the stilling wells to record the head of water; one at the floor of the convergence section (h_a) and another at the beginning of diverging section (h_b). The discharges through the flume may be free flow or submerged flow. If the water level in the downstream of the flume is high enough to retard the discharge, the flume is to be said submerged. Free flow or submerged flow condition is decided by the ratio h_b/h_a as given below.

Width of throat (cm)	Free flow limits, h_b/h_a
2.5 to 7.5	0.5
15 to 22.5	0.6
30 to 240	0.7

The size of the flume is decided by the size of the throat. The flume may be constructed by using wood, steel sheet, concrete or concrete and bricks. Sheet metal is preferred for portable Parshall flumes (standard dimension of the Parshall flumes and discharge values are given in Fig.2.18 and Table 2.1 respectively). It is better to use the calibration charts to avoid the risk of inaccuracies in dimensions, which use to occur during the construction of it.

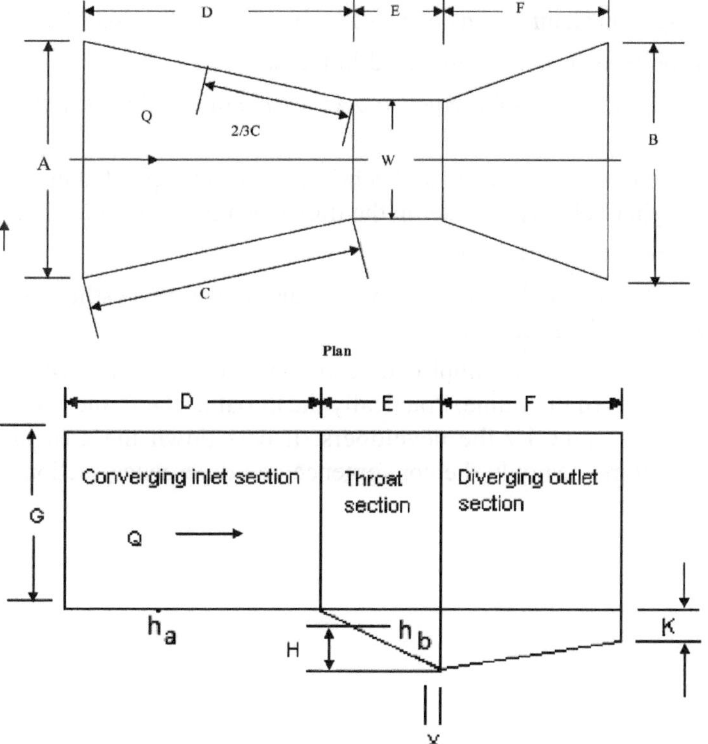

Fig 2.18 Plan and side view of Parshall flume

Table 2.3 Dimensions and capacities of various sizes Parshall flumes

Throat width (cm)	**7.5**	**15**	**23**	**30**
		Dimensions,	**cm**	
A	31	41.4	58.8	91.5
B	45.5	61	86.4	134
C	17.8	39.4	38.1	61
C	26	39.7	57.5	84.5
E	45.5	61	76	91.5
F	15	30.5	30.5	61
G	30.5	61	45.5	91.5
K	2.5	7.6	7.6	7.6
N	5.7	11.5	11.5	23
X	2.5	5.1	5.1	5.1
Y	3.8	7.6	7.6	7.6
	Free flow capacity, l/s			
Minimum	0.85	1.4	2.5	3.13
Maximum	28.4	110.8	23	456.6

Source: Majumdar, 2000

The advantages of Parshall flume are:

i) Stream of small irrigation channel can be measured by portable Parshall flume,

ii) Large stream can be measured if constructed in the canal,

iii) Can withstand a high degree of submergence over a wide range of back water condition,

iv) It is a self cleaning device and sand, silt and debris can not get deposited in the flume because of higher velocity of water in the flume than that in the channel,

v) It gives a accurate result than weir, and

vi) It works well at low head than the weirs. However, the flumes are costlier and difficult in construction than the weirs.

Cutthroat flume: The cutthroat is the simplified version of Parshall flume, which rejects the presence of throat of the flume. Practically the throat of the flume is cut and renamed as cutthroat flume by the developers. It cuts down the cost of construction. Cutthroat flume has only the convergence and divergence section.

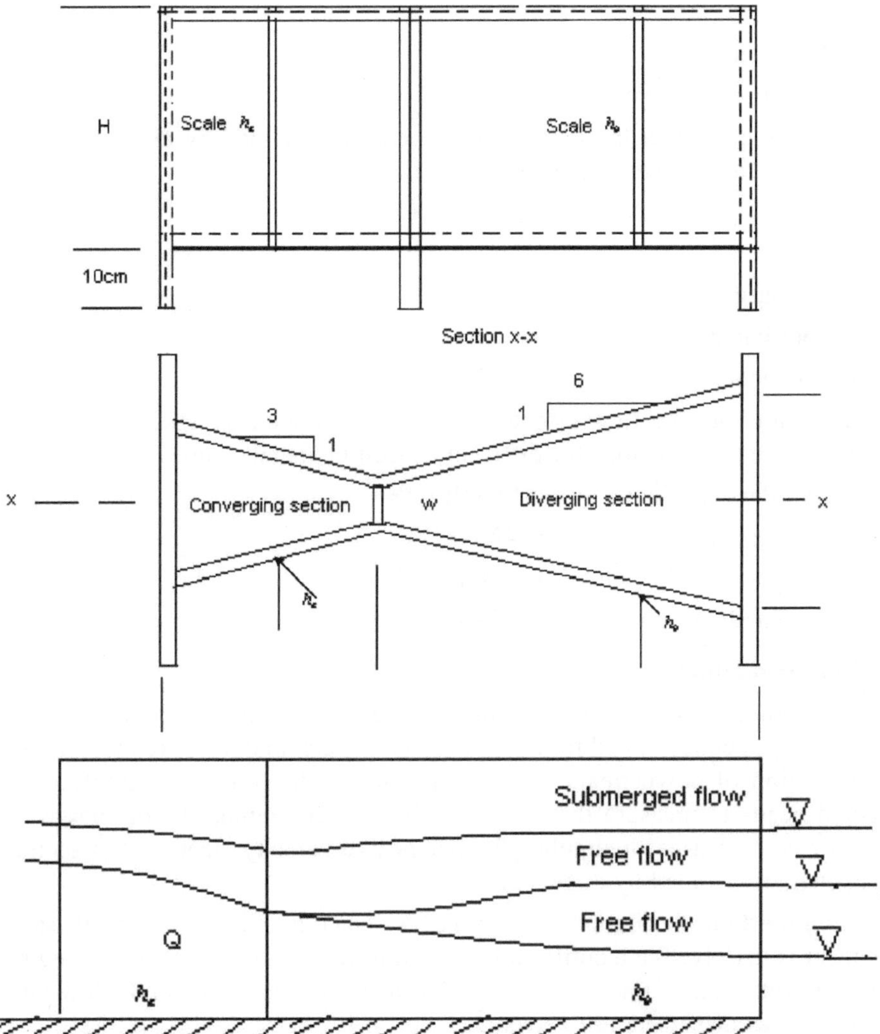

Fig. 2.19 profiles of water surface in a flume showing free flow or submerged flow

The walls are vertical and the floor is flat in free submerged flow. At free flow condition initial depth occurs near the flume neck, which is the point of meet of the converging and diverging sections. The vertical depth provides the basis for determining flow rate knowing the upstream depth (h_a). The upstream depth (h_a) and the flume length ratio should be less than 0.4. The width of flume neck width may vary from 2.5cm to 1.8m and the flume length 45cm to 3m.

The discharge through a cutthroat flume at free flow condition is expressed as by the formula,

$$Q = C_1 h_a^{n_1} \tag{2.18}$$

Where

Q = Flow rate

C_1 = Free flow coefficient

The discharge through the Parshall flume can be obtained by using the similar form of equation to weirs as below.

$Q = Ch^n$

Where

Q = Discharge, l/s

C = Coefficient of free flow discharge

n = Exponent.

By analyzing the free flow discharge data as prescribed by Michael (1978) the following are the equations for different capacity Parshall flumes.

For 7.5cm Parshall flume, $Q = 0.14\ 11h^{1.5493}$ (2.19)

For 15cm Parshall flume, $Q = 0.2496\ h^{1.6067}$ (2.20)

For 23cm Parshall flume, $Q = 0.4676\ h^{1.5296}$ (2.21)

For 30cm Parshall flume, $Q = 0.5599\ h^{1.5689}$ (2.22)

2.4 Tracer method

Tracer methods involve the releasing of tracer (dye, chemicals or radioactive substances) in concentrated form in the flowing water and measure the dilution at same section of downstream where dispersion of the tracer is complete. This method needs to measure the dilution only for determining the discharge. The measurement of usual hydraulic parameters like velocity, depth, cross-sectional area, etc., are avoided by this method.

Dilution method: In dilution method the substance like dye or chemicals said to be tracer is dissolved in a bottle and released in flowing water at a measured rate. The concentration of tracer is measured at a section downstream to flow where the tracer has dispersed 100 percent.

Let,

q_1 = Rate of application of tracer from the bottle

C_0 = Concentration of tracer percent in upstream water

C_1 = Concentration of tracer in bottle

$$= \frac{Weight\ of\ tracer}{Weight\ of\ water}$$

C_2 = Concentration of tracer at downstream

Q = Discharge at the downstream

So, $QC_0 + q_1C_1 = (Q + q_1)\ C_2$

$$\therefore Q = q_1 \frac{(C_2 - C_1)}{(C_0 - C_2)} \tag{2.23}$$

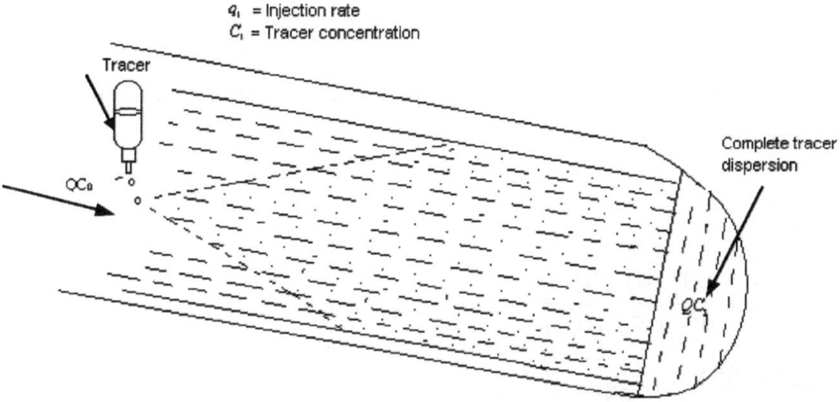

Fig. 2.20 Sketch illustration of the dilution method of measuring water through open channel

Salts are used in dilution method for determining the discharge of flow. The direct measurement of salt concentration is somewhat difficult but the electrical conductivity of salt concentration is easily measured which is directly related to salt concentration is somewhat different but the electric conductivity of salt concentration is easily measured which is directly related to salt concentration. The common salt (NaCl) provides good conduction to electricity. The concentration of sodium dichromate of water sample is determined by chemical analysis or flame photometer determination. The dyes when used as the tracer substances color the water. These colored samples are compared to standard concentrated or dilute solution.

Example 2.17 A 7.5cm flume was used to measure the discharge of an irrigation stream. The depth of water recorded at the convergence section and at the point of divergence was 25cm and 10cm, respectively. Find out the size of the stream. Use C=0.1411 and n = 1.5493.

Solution:

We have, $h_b/h_a = 0.4 < 0.5$

So, the flow is of free flow.

The discharge, $Q = Ch^n$

The discharge, $Q = Ch^n$

$Q = 0.1411 h^{1.5493}$

= 20.67 *l/s*

Example 2.18 In measuring the discharge of an irrigation channel the common salt of dilution 25 percent is applied at the rate of 1.5l/h. the salt concentrations are found $1.2x10^{-3}$ and $1.5x10^{-3}$ percent at upstream and downstream respectively. What is the discharge through the channel?

Solution: Salt concentration at the upstream, C_0=$1.2x10^{-3}$ percent =$1.2x10^{-5}$ in fraction

Similarly, salt concentration at the downstream, C_2 =$1.5 x 10^{-5}$ in fraction

Similarly, salt concentration of tracer in bottle, C_1=25 % = 0.25 in fraction

The discharge of the irrigation channel,

$$Q = q_1 \frac{(C_2 - C_1)}{(C_0 - C_2)}$$

$$= 1.5l/h \frac{(1.5 \times 10^{-5} - 0.25)}{(1.2 \times 10^{-5} - 1.5 \times 10^{-5})}$$

$$= \frac{1.5(-0.249985)}{0.3 \times 10^{-5}} = 124992.5l/h$$

= 34.72 *l/s*

Duty of water

Duty represents the irrigation capacity of a unit of water. It may be stated as the area in hectare or acre to be irrigated by 1.0cumec or 1.0cusec of discharge of a pump or canal sources throughout the base period.

Base period: It is the time period in days between the first irrigation for preparation of land or other purpose for growing crops to the last irrigation.

Delta: It is the depth of water required by a crop during the crop period. It may be represented by the symbol Δ.

Crop period: This is the period in between the sowing to the harvesting of crop.

Kor period & Kor depth: The water requirement in crop is maximum during the growing stage when irrigation water may be required and applied at higher quantity, the other irrigations are done as usual interval, is called as the Kor watering. The depth of water applied this time is called Kor depth and this period within the base period is called Kor period.

Relation between duty and delta: Let the duty in hectare and delta in meter is represented by D & Δ respectively. The volume of water applied in D hectare area for the depth Δ.

= D Δ ha-m

= D Δ $10000m^3$

Let the base period is represented by B. The volume of water supplied during this period = B x 24 x 60 x $60m^3$

$$\therefore D\Delta 10000 = B \times 24 \times 60 \times 60$$

$$Or,\ \Delta = \frac{B \times 24 \times 60 \times 60}{D \times 1000} = 8.64\frac{B}{D}m$$

Paleo: It is the application of water for bringing the soil moisture suitable for sowing the seeds or transplanting.

Example 2.19 Determine the delta of a crop for a base period of 90 days and duty of 1200ha.

Solution: $\Delta = 8.64\frac{B}{D}m = \frac{8.64 \times 90}{1200} = 0.648\text{m} = 64.8\text{cm}$

Example 2.20 A water course has culturable command area of 1000 hectares. There are two crops A & B growing in a season. Crop A & B has intensity of irrigation 50 & 40% and Kor period 15 & 10 days respectively. Determine the discharge of water course if Kor depth for A & B is 12 & 16cm respectively.

Solution:

For crop A:

Area under cultivation: 1000 x 0.5 = 500ha

Kor period = B = 15 days

Δ = 12cm

We have,

$$\Delta = 8.64\frac{B}{D}m$$

We have,

$$Or,\ D = \frac{8.64 \times B}{\Delta} = \frac{15 \times 8.64}{0.12} = 1080\text{ha/cumec}$$

Therefore, required discharge rate for area of 500ha

$$= \frac{500}{15 \times 8.64} \times 0.12 = 0.463\text{cumecs}$$

For area B:

Area = 1000 x 0.4 = 400ha

Kor period B = 10days, Δ = 16cm

$$D = \frac{8.64 \times B}{\Delta} = \frac{10 \times 8.64}{0.16} = 540\text{ha/cumec}$$

Discharge rate = $\frac{400}{540}$ = 0.74 m^3/s

Questions & Problems

2.1 Classify the methods of flow measurements.

2.2 Describe the volumetric method of flow measurement.

2.3 Describe the different structures used in measuring the flow.

2.4 Describe the flow measurement by using current meter. What are the advantages and disadvantages of this method?

2.5 Write the expressions for measurement of water flow by orifice, 90^0 V-notch and trapezoidal weir.

2.6 What is Cipolletti weir? What are the advantages of it over other weirs?

2.7 Derive the expression for discharge through a V-notch weir.

2.8 A barrel of 50cm was used to measure the discharge from a shallow tube well. The discharge recorded 65cm height of water in 15 seconds. What was the rate of discharge? Ans: 8.49l/s

2.9. Venturimeter used in a pipe of 20mm diameter, diameter of constriction of venturi 10mm and discharges at a rate of 7.5l/s. Calculate the deflection of mercury in the venturimeter. Assume coefficient of discharge 0.98. Ans: 7.2 cm

2.10. Calculate the discharge of a 60cm suppressed rectangular weir with 20cm head of water. What is the coefficient of discharge of weir if Francis formula is taken as the reference? Ans: 0.625

2.11. A 60^0 V-notch weir measures the discharge 41l/s at 30cm head of water. What is the coefficient of discharge? Ans: 0.604

2.12 A rectangular weir of width 40cm and head 15cm with no end contraction equals the discharge of Cipoletti weir of length 30cm. What is the head of water in the Cipoletti weir? Ans: 18.13 cm

2.13 A Cipolletti weir has the crest length of 1.0m and crest water level 0.5m. What is the discharge and average velocity of the weir? Ans. 305.63l/s, 94.75cm/s

2.14 A Cipolletti weir has the crest length of 1.2m and the crest height of water over the crest is 0.4m. What is the average approach velocity over the crest? Ans. 108.59 cm/s

2.15 Determine the flow of a channel from the following observed readings taken at 0.6-stream depth by a current meter. Take a = 0.03 & b = 0.50.

Distance at which readings are taken, m	No.of revolution	Channel depth, m	Duration of observation, s
0	0	0.0	0
1	45	0.5	60
3	66	1.1	60
5	95	1.5	60
7	90	1.6	60
9	37	0.7	60
10	0	0.0	0

Ans: 7.113m^3/s

2.16 A Parshall flume measured the discharge 15l/s in a free flow condition. What is the head of water if the coefficient (C) and exponent (n) are 0.15 and 1.5 respectively? Ans: 21.54cm

2.17 The common salt of 30 percent dilution applied to upstream of channel at a rate of 1 litre per hour. The water collected at the downstream recorded $2.0x10^{-3}$ percent salt concentrations. What is the discharge through the channel if the initial salt concentration in channel is $1.5x10^{-3}$ percent? Ans: 16.66l/s

2.18 Select the appropriate answer from the following multiple-choice questions

1. The floor of the throat in Parshall flume is inclined downward at a slope of

   a) 7 vertical to 12 horizontal b) 8 vertical to 21 horizontal

   c) 9 vertical to 24 horizontal d) 9 vertical to 20 horizontal.

2. The velocity of flow in the upstream of the weir notch should not exceed

   a) 10 cm/s b) 15 cm/s

   c) 20 cm/s d) 25 cm/s

3. The depth of water over the rectangular weir should not be less than

   a) 5 cm b) 7.5 cm

   c) 10 cm d) 15 cm

4. A V-notch weir may be

   a) 90^0 b) 60^0

   c) 30^0 d) any degree

5. In a free fall condition of Parshall flume (2.5-7.5cm) the ratio of head in divergence to convergence limits by

   a) 0.2 b) 0.3

   c) 0.4 d) 0.5

6. The head of water over a free fall weir is 30cm. If the coefficient of velocity of flow 0.75, the velocity of flow is

   a) 1.15m/s b) 1.29m/s
   c) 1.82m/s d) 2.43m/s

7. An area of 200m^2 is to be irrigated by 3cm of water. The water enters in to the area through a 90^0 V-notch weir. If the depth of water at weir is 20cm, the time requires to irrigate the area is approximately

   a) 8.3 min b) 11.2 min
   c) 14.5 min d) None of these

8. The depth of water required in kor period of 20 days 15cm. The duty of water is

   a) 11.52 ha/cumec b) 115.2 ha/cumec
   c) 1152 ha/cumec d) 11520 ha/cumec

9. The head of water over a 20cmx30cm orifice is 40cm at 10cm submerged flow. At coefficient of discharge 0.61, the discharge through the orifice is

   a) 62.78l/s b) 88.79l/s
   c) 125.58l/s d) 177.58l/s

10. The velocity of water measured 2.50m/s at a pipe outlet of head of water 0.5m. If the coefficient of contraction is 0.80, the coefficient of discharge is

   a) 0.61 b) 0.62
   c) 0.63 d) 0.64

Ans.

1. c) 2. b) 3. a) 4. d) 5. c) 6. b) 7. a) 8. c)
9. b) 10. d)

2.19 Write **True** or **False** of the following statements

1. Volumetric method of water measurement is suitable for large stream.
2. 1Ha-cm is equal to 10000 liter
3. Horizontal axis current meter is not affected by oblique flows
4. Velocity of water in an open channel is maximum on the surface
5. Pitot tube can be used to measure flow in open channel
6. Dethridge method is widely used in India to measure the irrigation water
7. When flow occurs through the converging section of a venture meter the velocity decreases
8. Coefficient of discharge is more than the coefficient of contraction
9. V-notch weir should have only 90^0 notch

10. Weir crest is the horizontal distance between the ends of the crest to the sides of the weir ponds
11. Parshall flume is a self-cleaning device.
12. Parshall flume gives better result than weirs.
13. Cut throat flume has got very small throat.
14. The throat of a Parshall flume is a level section.
15. Tracer method requires the measurement of upstream velocity.
16. Cipoletti weir discards the consideration of ends contraction..
17. Weir may be used in any small supply of water
18. Parshall flume is an open channel type device.
19. Weir should have upstream uniform flow and maximum velocity 15cm/s.
20. Crop period is the time period between the first and last irrigation.

Ans.

1. False	2. False	3. True	4. False	5. True	6. False	7. False
8. False	9. False	10. False	11. True	12. True	13. False	14. False
15. False	16. True	17. True	18. True	19. True	20. False	

References

Giles, R.V., Evett, J.B, & C.Liu (2004). Theory and Problems of Fluid Mechanics and Hydraulics (Third Edition). Tata McGraw-Hill Publishing Company Limited, New Delhi.

Majumdar, D.K (2000). Irrigation Water Management Principles and Practices. Prentice-Hall of Hall of India Private Limited, New Delhi.

Michael, A.M (1985 Edition). Irrigation Theory and Practice. Vikas Publishing House Ltd., New Delhi

Murty, V.V. (2004). Land and Water Management (Fourth Edition). Kalyani Publishers, Ludhiana.

Subramanya, K (2003). Flow in Open Channels. Tata McGraw-Hill Publishing Company Limited, New Delhi. Second Edition.

3

Irrigation Efficiencies

3.1 Introduction

Irrigation efficiency evaluates how effectively the water has been distributed for the purpose of crop production. Water is conveyed to crop field through canal water courses and channels. During this process considerable amounts of water get lost through run off, deep percolation, evaporation, etc. The losses of water greatly depend on the skill of the irrigator, degree of land preparation, and design and planning of the system. The extent of water losses starting to diversion of water from its source to crop field though conveyance and ultimate application and distribution of water in crop field which can be estimated separately. The estimation of these efficiencies enables one to identify the step(s) where improvement is required. The method of measurement and evaluation for its efficiency is prerequisite for proper use of water. Irrigation efficiency in developing countries in general is poor particularly in gravity irrigation methods. However, water is become more and more scarce. Due to pressing need of it judicious and efficient management is one of the vital issues to be solved on priority basis.

3.2 Various Irrigation Efficiencies

Project irrigation efficiency

It may be defined as the ratio of or percentage of the irrigation water retained in root zone depth and made available for consumptive use of the plants to the amount of water diverted from the source. Thus, it is the consumption of net amount of water by the plants to the water supplied in the project from the source. It is an estimate of water use efficiency of water in the project. The irrigation efficiency said to be farm irrigation efficiency if the water delivered is measured at farm head gate or well and field irrigation efficiency if measured in the field.

Water conveyance efficiency

It is ratio of water delivered to the field to water diverted from the source in percentage. Water is delivered to the field from reservoirs, river, etc, through channel networks. Water conveyance efficiency evaluates the efficacy of this water conveying system. It is expressed as,

$$E_c = \frac{W_f}{W_d} \times 100 \qquad (3.1)$$

Where,

E_c = Water conveyance efficiency, percent

W_f = Water delivered to the field or farm

W_d = Water diverted from the source

Water application efficiency

It is ratio of water stored in the root zone depth to water delivered to the field in percentage. It is a measure of efficiency of water application in the field. Other way, it is an estimate of water which in responsible to increase the soil moisture up to crop root zone out of the amount applied to the field. It is expressed as

$$E_a = \frac{Ws}{Wf} \times 100 \qquad (3.2)$$

Where,

E_a = water application efficiency, percent

W_s = Water stored in the root zone depth

W_f = water applied to the field

Application losses of water due to inefficient application ranges in between 28 to 50 percent (Sharma, 1993). On an overage, other than rice crop, the application loss is reported to be 17 percent (Majumdar, 2000). The common sources of losses are surface run off and deep percolation. The deep protection loss in wet rice varies as from 38 to 80 percent in various soils (Mondal, 1983). Neglecting the evaporation loss, we may have

$$W_f = W_s + D_f + R_f \qquad (3.3)$$

$$or, E_a = \frac{W_f - (D_f + R_f)}{W_f} \times 100$$

Where,

D_f = deep percolation below the root zone depth

R_f = run off water from the field

In general, lower the amount of water applied higher the application efficiency. Application efficiency does not refer a measure how effectively the requirement of water is satisfied but how much of the applied water is retained within the root zone depth.

Water storage efficiency

It is ratio of water storage in the root zone depth to water needed prior to irrigation in percentage. It is expressed as

$$E_s = \frac{W_s}{W_n} \times 100 \qquad (3.4)$$

Where,

E_s = water storage efficiency, percent

W_s = water stored in the root zone depth

W_n = water needed prior to irrigation

The amount of water needed in irrigation is the amount, which has been depleted from the field capacity of the soil. The water application efficiency in irrigation may be as high as 100 percent but the water storage efficiency is a measure of how the requirement of water in the soil has been satisfied.

Water distribution efficiency

It is a measure, which indicates the extent to which water is uniformly distributed. It is expressed as

$$E_d = \left(1 - \frac{\bar{y}}{\bar{d}}\right) \times 100 \tag{3.5}$$

Where,

E_d = water distribution efficiency, percent

$\bar{y}$ = average numerical deviation in depth of water stored from average depth stored during irrigation

$\bar{d}$ = average depth of water stored along the run during the irrigation

Operational efficiency

It is the ratio of actual project efficiency to the theoretical efficiency of an ideally designed and managed system using the same irrigation method and facilities. Low operational efficiency may be due to poor management and system design.

Water use efficiency

It evaluates the benefit of water use through economic crop production. Water use efficiency (WUE) is the ratio of economic crop yield to the amount of water used for this field. The amount of water use includes the evapotranspiration, water requires the metabolic activities of the plants and unavoidable deep percolation loss. Thus,

$$WUE = \frac{Y}{WR} = \frac{Y}{ET + G + D} \tag{3.6}$$

Where,

WUE = water use efficiency (kg/ha/mm)

Y = marketable yield of crops, kg

WR = water requirements of crops, mm

ET = evapotranspiration, mm

G = water used for metabolic activities of the plants, mm

D = unavoidable deep percolation loss, mm

Since, very little amount of water is used for metabolic activities of the plants, it is agreed that this amount may be neglected in calculating water requirement of crops. The WUE of some crops found by conducting experiment at Memari in Bardhaman district of West Bengal are listed in Table 3.1-3.3.

Table 3.1. Water expense and water use efficiency of cauliflower as influenced by the date of planting and irrigation levels

Treatment	Water expense, cm	Yield (kg/ha)	WUE(kg/ha-cm)
A. Date of planting			
D_1 (2nd week of October)	24.64	4437	180.1
D_2 (1st week of November)	25.24	3654	144.7
B. Level of irrigation			
I_1 (IW/CPE=0.6)	23.78	3491	146.8
I_2 (IW/CPE=0.9)	24.49	4687	191.3
I_3 (IW/CPE=1.2)	26.56	3958	149.0

Table 3.2 Water expense and water use efficiency of cabbage as influenced by moisture regime and nitrogen application

Treatment	Water expense (cm)	Curd yield (kg/ha)	WUE (kg/ha-cm)
A. Moisture regime (I)			
I_1 (20cm)	21.11	39503	1871
I_2 (16cm)	19.89	38564	1938
I_3 (12cm)	19.04	38487	2021
I_4 (8cm)	18.72	29697	1586
B. Levels of nitrogen (N)			
N_0 (0 kg/ha)	18.74	30586	1662
N_1 (40 kg/ha)	19.59	38662	1871
N_2 (80 kg/ha)	19.52	38247	1959
N_3 (120 kg/ha)	20.74	40757	1965

Table 3.3 Effect of tillage treatments and irrigation regimes on water use efficiency (kgha^{-1}mm^{-1}) of sunflower (pooled data of 1998 and 1999) *Source:* Gurumurty & Rao (2006)

Tillage	Irrigation regime (I)				
	I_1	I_2	I_3	I_4	Mean
T_1	2.05 (262)	2.16 (320)	2.08 (393)	1.97 (439)	2.07 (354)
T_2	2.45 (278)	2.6 (342)	2.55 (414)	2.45 (440)	2.51 (374)
T_3	2.62 (290)	2.84 (356)	2.73 (438)	2.69 (478)	2.72 (391)
T_4	2.89 (303)	3.11 (373)	2.89 (452)	2.8 (483)	2.92 (403)
T_5	3.14 (313)	3.34 (390)	3.14 (467)	3.11 (495)	3.18 (416)
Mean	2.62 (289)	2.81 (356)	2.68 (433)	2.6 (471)	

CD (P=0.05)

Tillage (T) 0.18

Irrigation regime (I) 0.13

Interaction

T at same I 0.22

I at same T 0.15

T_1 = zero tillage

T_2 = bullock-drawn indigenous plough twice +bullock drawn cultivator twice-farmers' practice

T_3 = bullock-drawn indigenous plough twice +power tiller drawn cultivator

T_4 = tractor-drawn M.B. plough once +tractor-drawn cultivator

T_5 = deep tillage with tractor drawn disc plough +rotavator

I_1 = IW/CPE=0.6, I_2= IW/CPE=0.8, I_3= IW/CPE=1.0, I_4= IW/CPE=1.2

The water use efficiency takes into account the field water requirement of crops. The yield depends on various factors of crop management including insects, pests and disease control, weather hazard and weather condition. The water requirement is influenced by climatic condition and soil and crop management practices.

Economic (irrigation) efficiency

It is the ratio of actual income (net or gross) attained by operating an irrigation system to the income expected in ideal condition.

Example 3.1 A stream of 150 l/s was diverted from a canal and 120 l/s was delivered to a wheat field of 1.75 ha. The irrigation continued for 7.5 hours. The effective root zone depth was 1.8m. The run off loss in the field was 450m^3. The depth of water penetrated linearly from 1.8m at the head end to 1.2m at the tail end. The moisture holding capacity of the soil is 25cm/m depth of soil. Irrigation was given at 50% depletion of available soil moisture. Determine the water conveyance efficiency, water application efficiency, water storage efficiency and water distribution efficiency.

Solution

i) water conveyance efficiency $= \dfrac{\textit{water delivered to the field}}{\textit{water diverted from the source}} \times 100$

$$= \frac{120 l/s}{150 l/s} x 100 = 80\%$$

ii) Water applied to the field = 120 l/s x 7.5 hour = 3240m^3

Run off loss = 450 m^3

Water stored in the root zone depth = 3240-450 = 2790m^3

Water application efficiency $= \dfrac{\textit{water stored in the root depth}}{\textit{water applied to the field}}$

$$=\frac{2790}{3240}=86.11\%$$

iii) Depth of available soil moisture up to root zone depth

= 25cm/m x 1.8m = 45cm

The depth of water available at the time of irrigation

= 45 x 50/100 = 22.5cm

Additional depth of water required for irrigation

= 45-22.5 = 22.5cm

Volume of water required to irrigate = 22.5cm x 1.75 ha

$$=\frac{22.5}{100}x1.75\times10000$$

= 3937.5m^3

$$\text{Water storage efficiency} = \frac{\text{Water stored in the root zone depth}}{\text{Water needed prior to irrigation}}$$

$$=\frac{2990\,m^3}{3937.5\,m^3}x100=70.86\%$$

iv) Average depth of water penetration in soil $\bar{d}=\frac{1.8+1.2}{2}=1.5\text{m}$

Average numerical deviation of depth of penetration from the average depth.

$$\bar{y}=\frac{\left[(1.8-1.5)+(1.5-1.2)\right]}{2}=0.3$$

Water distribution efficiency, $E_d = \left(1-\frac{\bar{y}}{\bar{d}}\right)\times100$

$$= \left(1-\frac{0.3}{1.5}\right)\times100 = 80\%$$

Example 3.2 An area of 20 ha is to be irrigated by a pump working for 15 hours a day. The irrigation is given at 50% depletion of available soil moisture. The available water holding capacity of the soil is 25cm/m. The root zone depth of soil is 75 cm. The maximum daily consumptive use rate of crop is 5 mm/day. Determine the (i) net irrigation requirement, (ii) gross irrigation requirement, (iii) irrigation period and (iv) irrigation system capacity. Assume conveyance, application and pump efficiencies as 70%, 75% and 75% respectively.

Solution

i) Root zone available soil moisture depth = $\frac{25x75}{100}=18.75\text{cm}$

Net irrigation requirement = 18.75 x 50/100 = 9.37cm

(ii) Gross irrigation requirement

$$= \frac{Net\, irrigation\, requirement}{Conveyance\, efficiency \times Application\, efficiency}$$

$$= \frac{9.37}{70/100\, x\, 75/100} = 17.85 \text{cm}$$

(iii) Irrigation period = $\frac{Net\, irrigation\, requirement}{Average\, rate\, of\, consumptive\, use}$

$$= \frac{9.37}{0.5 \text{cm}/\text{day}} = 18.74 \cong 19 \text{days}$$

(iv) Irrigation system capacity

$$= \frac{Gross\ depth\ of\ irrigationxArea}{Irrigation\ period \times Pumping\ hours\ /\ daysxPump\ effi.}$$

$$= \frac{17.85 \text{ cm} \times 20 \text{ha}}{19\, days \times 15\, hours\, /\, day \times 75/100}$$

$$= \frac{17.85 cm \times 20 \times 10{,}10{,}000 \times 100 \times 100 cm^2}{19 \times 15 \times 3600 \sec \times 75/100}$$

$= 46{,}393.76 \text{cm}^3 / \text{s}$

$= 46.39 l / s$

Example 3.3 Determine the irrigation efficiency from the following data:

Water conveyance loss = 30%

Deep percolation and surface run off loss = 25%

Water stored in soil lost by evaporation = 15%

Solution

Let W_d be the water diverted from the source,

Water applied to the field = W_d-0.3W_d = 0.7W_d

Water stored in the soil = 0.7W_d-(0.7W_d) x 25%

= W_d (0.7-0.175)

= 0.525W_d

Water consumed = $\frac{0.525W_d}{100} \times (100-15)$

= 0.444 W_d

So, irrigation efficiency = $\frac{0.446W_d}{W_d} \times 100 = 44.6\%$

Example 3.4 Determine the water use efficiency from the following data:

No. of treatment	Depth of water applied, cm	Effective rainfall, cm	Soil water used, cm	Seed yield, kg/ha
1	4	3.5	2.5	1225
2	6	3.5	2.3	1320
3	8	3.5	2.1	1440
4	10	3.5	1.9	1450

Solution

No. of treatment	Water used by the crop, cm	Seed yield, kg/ha	Water use efficiency, kg/ha/cm
1	4+3.5+2.5 = 10	1225	122.5
2	6+3.5+2.1 = 11.8	1320	111.8
3	8+3.5+2.1 = 13.6	1440	105.88
4	10+3.5+1.9 = 15.4	1450	94.16

Example 3.5 Irrigation is given everyday to 0.75 ha sugarcane and 0.5 ha potato. The readily available moisture for sugarcane and potato after irrigation were 10 cm and 15 cm, respectively. Sugarcane was irrigated by 30 furrows each discharging 20l/min & potato was irrigated by 80 furrows each discharging 7l/min. Assuming uniform application of water over the field and no deep percolation loss, compute the application efficiency for each crop and the farm.

Solution

Irrigation water stored in sugarcane field = 0.75 ha x 10 cm

= 0.75 x 10000 x 10/100 m^3 = 750m^3

Irrigation water stored in potato field = 0.5 ha x 15 cm

= 0.5x 10000m^2 x 15 / 100m = 750m^3

Water applied to sugarcane field by 30 furrows in a day

= 30 x 20 l/minx day

= 30 x 20 l/min x 24 x 60 min

= 648000 l = 864m^3

Water applied to potato field by 80 furrows in a day

= 80 x 7 l/min x 24 x 60 min

=806400l

= 806.4m^3

Application efficiency for sugarcane = $\frac{750}{864}\times 100 = 86.8\%$

Application efficiency for potato = $\frac{750}{806.4}\times 100 = 93\%$

Farm application efficiency = $\frac{(750+750)}{(864+806.4)}\times 100$

= 89.7%

Example 3.6 A soil with available moisture holding capacity of 25cm per meter depth of soil was irrigated after depletion of 60 percent of the available moisture. A stream of 0.04 m^3/s was diverted from a tube well and 0.032 m^3/s was delivered to the field. An area of 1ha was irrigated in 14 hours. The root zone depth was 1m. Moisture penetration varied linearly from 1.2m at the head to 0.8m at the tail of the field. Determine the water conveyance efficiency, application efficiency, storage efficiency and distribution efficiency. (GATE, 1994)

Solution

Available moisture holding capacity = 25 cm/m

Irrigation applied at 60% depletion of available soil moisture.

Root zone depth = 1m

Moisture penetration = 1.2m to 0.8m linearly

Stream = 0.04m^3/s diverted from the source and 0.032m^3/s delivered to the field

$$\text{Water conveyance efficiency} = \frac{0.32\text{m}^3/\text{s}\times 100}{0.04m^3/\text{s}}\times 100 = 80\%$$

Water needed in the root zone depth = 25 x 0.6 = 15 cm

$$\text{Water applied to the field} = \frac{0.032\text{m}^3/\text{s}x14hours}{1ha}$$

$$= \frac{0.032x14x3600x100}{10000} = 16.128cm$$

The root zone depth is 1m. The water penetrated up to 1.2m at the head end & 0.8m at the tail end. The penetration varied linearly with the length. So, it can be said that average depth of 1.2m & 0.8m i.e. 1m penetration of water has taken place up to middle point of run. Next to middle half the water has penetrated linearly 1m to 0.8m. Therefore, average depth of water penetration in the root zone depth.

$$\frac{1+0.9}{2} = 0.95\text{m}$$

So, water stored in the root zone depth from irrigation = 0.95 x 15 =14.25 cm

$$\text{Application efficiency} = \frac{water\ stored\ in\ root\ zone\ depth}{water\ delivered\ to\ the\ field}x100$$

$$\frac{14.25\,\text{cm}}{16.128}\times 100 = 88.35\%$$

$$\text{Storage efficiency} = \frac{water\ stored\ in\ root\ zone\ depth}{water\ needed\ prior\ to\ irrigation}x100 = \frac{14.25}{15} = 95\%$$

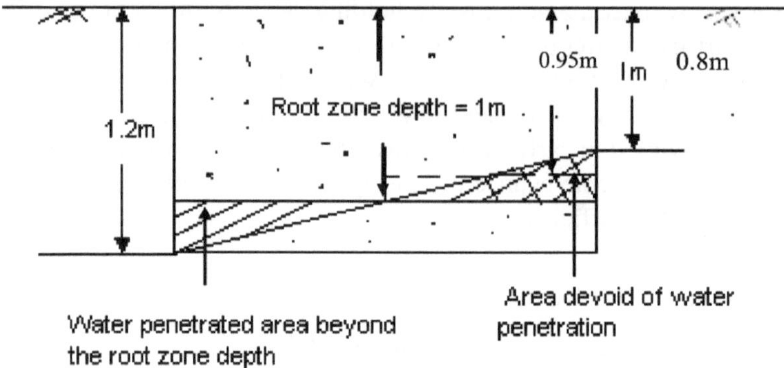

Water distribution efficiency, $= \frac{(1-\bar{y})}{\bar{d}} \times 100$

Where, $\bar{d} = \frac{1.2+0.8}{2} = 1.0m$

$\& \; \bar{y} = \frac{1.2\text{-}1.0+1.0\text{-}0.8}{2} = 0.2m$

$\therefore$ Water distribution efficiency = (1-0.2/1) x 100 = 80%

Example 3.7 Irrigation water was applied at a rate of 10 l/s for 1hour 10 minutes to a cropped border strip of 5m wide and 100m long with 0.05% field slope. It took 1 hour to reach the water at tail end of the border. What was the (a) depth of flow in the border, (b) the depth of standing water at the head end and tail end just after stopping the application of water, (c) time required to penetrate the standing water in to soil profile, (d) distribution, application and storage efficiency of irrigation. Assume average infiltration rate of soil 3cm/h, root zone depth 60cm and porosity of soil 30% of which 50% was depleted at the time of irrigation.

Solution:

a) Time required to irrigate a border strip,

$$t = \frac{d}{f} ln\left(\frac{Q}{Q - fA}\right)$$

$$\therefore Depth\,of\ flow, d = \frac{ft}{ln\left(\frac{Q}{Q-fA}\right)} = \frac{3cm/h x 1.0h}{ln\left(\frac{10\times3600 l/h}{10\times3600\text{-}3cm/hx\,5m\times100m}\right)}$$

$$= \frac{3\times10^{-3}m}{ln\left(\frac{36m^3/h}{36m^3/h - \frac{3}{100}\times500m^3/h}\right)}$$

$$= \frac{3\times10^{-3}m}{ln\left(\frac{36}{36\text{-}15}\right)} = \frac{3\times10^{-3}m}{ln\left(\frac{36}{12}\right)} = \frac{3\times10^{-3}m}{(0.539)} = 5.56\times10^{-3}m$$

b) After stopping the application of water, the water will settle to have level surface. At the time of stopping water application, there were water for infiltration for 1 hour 10 minutes and 10 minutes at head end and tail end respectively. Therefore, the depth of infiltration for that period,

at head end = 3cm / h x 1h 10 min = 3 x 1.165xm = 3.5 cm

at tail end = 3cm / h x 10min = 0.5 cm

Average depth of water penetrated = $\frac{3.5+0.5}{2} = 2cm$

Depth of water applied = $\frac{\text{Volume of water applied}}{\text{Area}}$

$$= \frac{10l/s\times1h10\,min}{5m\times100m}$$

$$= \frac{10l/s\times4200s}{500\text{m}^2} = 0.084m = 8.4cm$$

Average depth of standing water = 8.4 - 2 = 6.4cm

Field slope = 0.05%

Therefore, when water at head end just perish, the water at tail end is 5cm. Say, x was the depth of water at head end at the time of stopping water application,

$$\therefore \frac{x+5+x}{2} = 6.4$$

$\therefore$ 2x = 2x 6.4 - 5

or, x = 3.9cm

Depth of water at tail end = 5 + 3.9 = 8.9cm

c) Time required penetrating this standing water,

at head end = $\frac{3.9cm}{3cm/h} = 1.3h$

at tail end = $\frac{8.9cm}{3cm/h} = 2.97h$

d) Infiltration opportunity time,

at head end = 1.3h + 1h10 min = 2.47h

at tail end = 2.97 h + 10 min = 3.14h

Depth of water penetrated,

at head end = 3cm /hx 2.47h = 7.41cm

at tail end = 3cm / h x 3.14 = 9.42cm

Porosity of soil=30%, moisture depletion=50%

So, depth of water penetration in soil profile,

at head end $= \frac{7.41cm}{0.3\times0.5} = 49.4cm$

at tail end $= \frac{9.42cm}{0.3\times0.5} = 62.80cm$

Average depth of penetration $= \frac{49.4+62.80}{2} = 56.10cm$

Distribution efficiency $= \left\{1 - \frac{\frac{(56.10-49.4)+(62.80-56.10)}{2}}{56.10}\right\}\times100$

$= \left\{1 - \frac{6.7}{56.10}\right\}\times100 = 88.06\%$

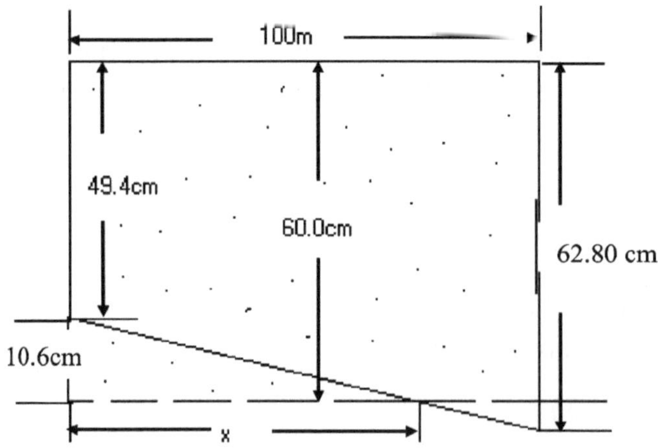

Difference of depth of water penetration from head end to tail end = 62.80 - 49.40 = 13.40cm

From the above figure, $x = \frac{100\times10.6}{13.40} = 79.10\text{cm}$

Average depth of penetration of water within the root zone depth

$$=\frac{\frac{49.4+60}{2}\times 79.10+(100\text{-}79.10)\times 60}{100}=\frac{54.7\times 79.10+20.90\times 60}{100}$$

$$=\frac{4326.77+1254}{100}=55.71\text{cm}$$

Depth of water retained in the root zone zone depth

= 55.81 x 0.3 x 0.5 = 8.37cm

Application efficiency =

$$\frac{water\ stored\ in\ the\ root\ zone\ depth}{water\ applied\ to\ the\ field}=\frac{8.37}{8.4}=99.64\%$$

Storage efficiency

$$=\frac{water\ stored\ in\ the\ root\ zone\ depth}{water\ needed\ prior\ to\ irrigation}=\frac{8.37}{60\times 0.3\times 0.5}=\frac{8.37}{9.0}=93\%$$

Example 3.8: In a cropped field, the following data are observed.

Moisture content at field capacity (weight basis)	=	36%
Current moisture content (weight basis)	=	24%
Bulk density of soil	=	1.5Mg/m^3
Effective root zone depth	=	0.8m
Conveyance efficiency	=	80%
Application efficiency	=	90%

To bring soil moisture content to field capacity, the depth of irrigation in mm will be ______

(GATE, 2016)

Solution:

Soil moisture loss = $\frac{(FC\text{-}M_S)A_sD}{100}$

$$=\frac{(36\text{-}24)1.5\times 80}{100}=14.4cm$$

Depth of irrigation water required to bring the soil in field capacity

$$=\frac{14.4cm}{\text{Conveyance efficiency}\times\text{Application efficiency}}$$

$$=\frac{14.4cm}{0.8\times 0.9}=20cm=200\text{ mm}$$

Questions and Problems

3.1 What do you mean by irrigation efficiency? What is the importance of it?

3.2 Define water conveyance, application, storage and distribution efficiency?

3.3 What is the difference between project and operational efficiency?

3.4 What is water use efficiency? How is it important to cultivation?

3.5 The depth of water applied to a field was 20 cm. The potential evapotranspiration was 10 mm during 2 days after the irrigation. Determine the irrigation efficiency from the following soil moisture data.

Soil depth, cm	Soil moisture prior to irrigation, %	Soil moisture after 2 days of irrigation, %	Apparent specific gravity of soil,
0-30	15.0	26.5	1.45
30-60	17.5	27.0	1.50
60-90	18.2	27.2	1.50
90-120	22.0	26.7	1.50

3.6 An area of 500 ha is to be irrigated by a channel of an outlet. The cultivated command area is 80% of the total area. The intensity of irrigation is 40% for Rabi crop & 50% for Kharif crop. Assuming conveyance loss of channel as 10% of outlet discharge, determine the discharge at the head end of the channel. The discharge factor for Rabi and Kharif season is 1800 ha/m^3/s and 2500 ha/m^3/s, respectively. **Ans. 0.088 cumecs**

3.7 An area of 1 hectare was irrigated in 15 hours with a stream of 20 l/s. Depth of root zone was 80 cm and available moisture holding capacity 20cm/m. Irrigation was applied when 50% of available moisture was depleted. Water application efficiency was 60%. Determine the water storage efficiency. Ans. 81%. (GATE, 1992)

3.8 Select the appropriate answer from the following:

1. Irrigation efficiency considers
   a) Effective water distribution for land preparation
   b) Effective soil moisture distribution in root zone depth
   c) Crop response to irrigation
   d) Water requirement in crop

2. The efficiency of irrigation does not depend on
   a) Skill of irrigator b) Degree of land preparation
   c) Planning of irrigation system d) Season of crop

3. Information of irrigation efficiencies enables one to
   a) Select the appropriate crop rotation
   b) Identify the points where improvements are required

c) Modify the land use pattern

d) Select the suitable source of water

4. The lowest value of the following is

a) Project efficiency b) Conveyance efficiency

c) Storage efficiency d) Distribution efficiency

5. Water use is not related to

a) Crop yield b) Water requirement of crop

c) Irrigation water applied d) Numbers of irrigation.

6. In a small plot of horticultural crop 250 liter of water was applied, of which 30 liter and 20 liter was lost in deep percolation and runoff respectively. The application efficiency of irrigation was

a) 80% b) 88%

c) 92% d) 100%

7. Irrigation water diverted from a STW source @5l/s was conveyed to the field through a lined channel for 15 minutes. After that due to sudden fault in the channel the water was conveyed to the field @ 3.5l/s to satisfy the irrigation water requirement. The conveyance efficiency of water application was

a) 66.67% b) 70%

c) 90% d) 100%

8. The depth of water applied 7.0cm in a field of 200m length and slope 0.05%. If no water is lost in any form during the water application, the depth of water at the tail end just after irrigation was

a) 5.0cm b) 7.5cm

c) 10.0cm d) 15.0cm

9. The water penetrates linearly by 0.80m to 1.1m in a crop field of root zone depth 1.0m and length 100m. Water penetrates just root zone depth at length

a) 33.33m b) 50.0m

c) 66.67m d) 85.0m

10. The infiltration characteristics of a field is expressed by $I = 0.7t^{0.3}$. The irrigation water took 1 hour 21 minutes to penetrate in to the soil. The depth of water applied was

a) 0.76 mm b) 7.6 mm

c) 1.62 cm d) 2.62 cm

Ans.

1 b) 2. d) 3. b) 4. a) 5. d) 6. a) 7. c) 8. d) 9. a) 10. d)

3.9 Write True or False of the followings.

1. Project irrigation efficiency is always more than conveyance or application efficiency.
2. In general application efficiency is more for lower volume of application of water.
3. Higher application and distribution efficiencies do not ensure high storage efficiency.
4. Loss of water through deep percolation in wet rice cultivation is taken into account of water requirement.
5. Uniformity coefficient of distribution and distribution efficiency provides different understanding of the state of water distribution.

Ans.

1. False 2. True 3. True 4. True 5. False

References

Gurumurty, P. & M.S. Rao (2006). Effect of tillage and irrigation regimes on soil physical properties and performance of sunflower in rice fallows II. Water use and crop yield. JISSS, 54 (1). p.18-23.

Sharma, R.K. & T.K. Sharma (1993). Irrigation Engineering (Vol. I). Oxford & IBH Publishing Co. Pvt. Ltd.p.136.

Majumdar,D.K.(2000).Irrigation Water Management-Principle & Practices. Prentice-Hall of India Pvt. Ltd., New Delhi.p.249

Mandal, A.K. & Majumdar, D.K. (1983). Effect of phasic soil submergence and saturation on growth and water use in rice in semi-arid lateritic tract of West Bengal. Proc.70th Indian Sci. Cong. Part III:Sec.X:Abstract. p.7-8.

4

Scheduling Irrigation

Scheduling irrigation is the process of determining when to irrigate and how much to irrigate. Selecting the appropriate time span in between two irrigation and applying the required volume of water in each irrigation, ensures saving of water, minimize energy use, higher crop yield and lower production costs. Irrigation applied earlier than due date causes loss of water through evaporation or deep percolation and later causes water stress to plants and thereby affects the growth and yield of the plants.

Scheduling of irrigation may be required in two field conditions: (i) when available of water is adequate, (ii) when availability of water is limited.

In the places where water is adequate scheduling of irrigation follows full irrigation with time sequence of minimum number of irrigation to ensure maximum yield and high water use efficiency. When water is limited its use is rationalized by using at the sensitive stage of the plants and withholding irrigation at other stages .The sensitive stages so selected that by using the available water provides possible maximum yield.

4.1 When to Irrigate

Most of the plants are efficient in taking water when soil moisture is near to field capacity. As the soil moisture content decreases from field capacity, soil moisture tension increases and plants start to feel stress in up taking water. More and more decrease in soil moisture leads to more and more stress to plants up taking. Plants start to wilt and growth is retarded in such condition. When soil moisture reaches to such a lower limit that plants cannot extract water then permanent wilting of the plants occur. The plants may restore the damage by addition of water before the permanent wilting. However, quantity and quality of yield decreases much depending on the degree of stress experienced by the plants. In most of the crops the irrigation is applied at 50% depletion of available soil moisture. There are some crops which give maximum yield if irrigation is given at 25% depletion of available soil moisture. The Table 4.1 gives the optimum soil water regime for different crops grown in India.

Table 4.1. Optimum soil water regime for different crops in India

Crop	Optimum soil Water regime	Soil depth for Soil water measurement, cm	Soil type	Place
Rice	Submergence to 4cm water height	-	All	Bikramganj, Mandya, Kharagpur, New Delhi, Siruguppa, Bhawanisagar
Sorghum	100-50% A inhot weather, 100-25% A in *kharif*	30	Clayey	Siruguppa
Pearl millet	100-75 A0.2 atm. tension	30	Clayey Sandy loam and loam	Anand
Crowfoot millet	100-50% A	30	Clayey	Siruguppa
Maize	0.65atm. tension 100-50% A 100-50 A	15 60 30	Sandy loam Sandy loam Clayey	New Delhi Sabour Sriguppa
Wheat	100- 50% A 100-20% A	60 60	Sandy loam Sandy loam	New Delhi Pantnagar
Barley	100-50% A	60	Sandy loam	Jobner
Green gram	100-50% A	60	Sandy loam	New Delhi
Groundnut	100-60%	60	Sandy loam To loam	Bhawanisagar
Cotton	100-20%	30	Clayey loam to clayey	Bhawanisagar
Sugarcane	100-75% A at formative stage and 100-25% Aat other stages	30 at formative stage and 90 other stage	Loam	Anakapalle
Tobacco	0.3 atm. tension	18	Sandy loam	New Delhi
Jute	100-50% A	30	Sandy loam	Barrackpore
Berseem	0.25 atm. tension 0.20 atm. tension 100-75% A	22 30 30	Sandy loam Sandy loam Sandy loam	New Delhi Hissar Jhansi
Green fodder of Sorghum, Lucerne and Cluster bean	100-75% A	22	Sandy loam	Jhansi
Potato	0.3 atm. tension	15	Sandy loam	New Delhi Kharagpur
Onion	0.65 atm. tension	8	Sandy loam	New Delhi
Tomato	100-60% A	120	Sandy loam to loam	New Delhi

Root vegetables, Raddish, turnip, beet root	0.2 atm. tension	18	Sandy loam	New Delhi
Sugarbeet	0.24 atm. tension	60	Loam	Padegaon
	0.25 atm. tension	25	Sandy loam	

A = available soil moisture Source: Majumdar, 2000

The range of available soil moisture in which plants do not suffer much due to water stress is called the optimum soil water regime. The upper limit of this regime is field capacity other than rice. Lower limit may be called as critical soil moisture level. Water is suggested to apply at the lower limit soil moisture level.

A thorough knowledge of soil-water-plant-atmosphere is required for proper scheduling of irrigation water since water loss is greatly influenced by evaporative demand, soil moisture status and plant characteristics .The basis of scheduling irrigation attempted from time to time are based on soil, plant and atmospheric parameter.

Soil Parameter

Scheduling irrigation of this method based on the status of the moisture content in the soil. Knowing the deficit of soil moisture content irrigation is applied at predetermined minimum moisture level to bring the soil moisture at field capacity. The soil moisture is determined either by directly or indirectly through inference from other soil characteristics such as soil water potential or electric conductivity. Irrigation based on soil moisture content is the most accurate and dependable method.

Gravimetric method

In this direct method of determining soil moisture the soil samples are collected from different depth of root zone area .The samples are weighed, dried in oven at 105^0 to 110^0 for about 24 hours and reweighed .The difference of weight before and after the drying gives the measure of moisture content in a definite soil mass and can be expressed in percentage of soil. This method is most accurate and reliable but requires laboratories and time consuming.

Feel and appearance

The farmers through experiences use to irrigate their fields on estimation the soil moisture content by the feel and appearance of the soil. The soil samples are collected from different root zone depth by the soil augur. The soils are then formed into balls tossed in air and caught in one hand. The characteristics of the ball on squeezing, pressing or appearance gives the idea of soil moisture content. The Table 4.2 gives the description of soil moisture for different soils.

Table 4.2. Guide for judging the amount of available soil moisture range

Available soil Moisture range	**Coarse textured (loamy sand)**	**Moderately coarse (loamyAnd silt loamy**	**Medium texture (loamy and siltloamy)**	**Fine texture (clay, loamy and silty clay loamy)**
Field capacity (100%)	On squeezing, no free water appears on soil, but wet outline is left on hand	Similar symptoms	Symptoms for all type of soils	Symptoms for all type of soils
75-100%	Tends to stick together slightly, sometimes forms a very weak ball under pressure	Forms weak ball, breaks easily, do not slick	Forms a ball, is very pliable, slicks readily	Easily ribbons out between fingers, has slick feeling
50-75%	Appears to be dry, do not form a ball with pressure	Tends to form a ball under pressure but seldom holds together	Forms a ball somewhat plastic, sometimes slick slightly with pressure	Forms a ball, ribbons out between thumb and fore-finger
25-50%	As above, but ball is formed by squeezing very firmly	Appears to be dry, do not form a ball unless squeezed very firmly	Somewhat crumbly holds together with pressure	Somewhat pliable, forms a ball under pressure
0-25%	Dry, single grained flows through fingers	Dry, loose, flows through fingers	Powdery dry, sometimes slightly crusted but easily broken down in to powdery conditions	Hard, cracked, sometimes has loose crumbs on surface

Source: Reddy & Reddy, 1995

Tensiometer

Tensiometers are used to determine the soil moisture tension relating to soil moisture content (Fig.4.1).The available soil moisture versus tension curve i.e. characteristic curves made earlier for the required soils are used to estimate the moisture content. Irrigation is done for the particular tension indicated by the tensiometer or otherwise.

The farmers do not favour much the use of tensiometer in scheduling irrigation since the device works well within the higher suction range of 0.85 bars only. The use of tensiometer is limited to light soils where most of the available water is held at low tension. This device also takes much time to reach at tension equilibrium between the porous block and the surrounding soil.

(a) Conventional tensiometer

(b) Tensiometers used in research field

Fig. 4.1. Tensiometers

Electrical resistance

The electrical resistance varies inversely with the moisture content in soil. This idea is utilized in scheduling of irrigation. However, there are few draw backs in using this method. The resistance block can not be used at low tension at which most of the available water is held; resistance do not directly indicates the prevailing time-lag for tension equilibrium between the porous block and the surrounding soil. Gypsum and plaster of paris are the common materials used as porous block.

Water budget equation

This method computes the everyday ET, effective rainfall, soil moisture content, etc. to determine the prevailing moisture status in soil. This is very cumbersome and practically not in use.

Plant Parameter

The irrigation is done to supply water to plants needs. The state of the plants may be a direct method for determining the time and amount of irrigation.

Appearance and growth

Under water plants use to show some characteristics change in appearance. These may be change of normal colour of the plants leaves, leaf and shoot wilting, curling or rolling of leaves and dropping of leaves. How quickly the change of appearance of plants will occur depend on the type of plants. Some plants are very sensitive to water and show the stress of water very early. Some plants like wheat changes the normal deep green to light green and then yellow at water stress. Wheat and rice also shows the water stress by rolling the leaves. Sunflowers and sugar beat indicates the water stress by temporary wilting of plants during the hottest period of the day. Fruits and orchards do not easily show the symptom of water stress until serious retardation in growth. The growth rate of plant in the forms of diameter and height can also be the indicator of water stress of plants. However, growth and appearance often cannot be considered as the effective parameter for deciding the time of irrigation owing to unavailability and high cost of equipments, inadequate standardization and difficulties in precise growth measurement. The plants show visible symptom of water deficiency gradually long after they are under water stress. The change of colour is some time misleading due to nutritional disorder, insects, pests or diseases attack and varietal character.

Leaf temperature

Due to partial or full stomata closure of the plant leaves during the reduced rate of availability of water the leaf temperature rises .The difference of temperature between plant canopy and ambient air can be measured by hand –held infrared thermometer. Depending on the difference of their temperature the moisture stress is determined. Some of the workers suggested difference between the stressed and unstressed canopy temperature as a better index for water deficit than the difference between plant canopy and the ambient temperature.

Leaf water potential

Some plants show good correlation between leaf water content or potential to available soil water .The relation is suitably used in determining the time of irrigation. Leaf water potential can be measured by placing the leaf in a pressure chamber .The entire water of leaf is forced out by slowly increasing the pressure in the chamber. This is the pressure, which is taken as the leaf water potential. Higher is the leaf potential higher is water stress in plants and lower soil moisture availability. The leaf water potential depends on the leaf age; leaf exposure to solar radiation and the time of the day the leaf is collected. The crops like sugarcane, cotton, wheat etc. can be irrigated by fixing the time of irrigation guided by leaf water potential. However, sophisticated and costly equipments and lack of standardization has restricted large-scale use of this technique.

Stomatal resistance

The vapour diffusion to the atmosphere through the plant leaves are governed by the degree of stomatal closure which under sufficient day light is mainly regulated by the leaf water deficit. The closure of the stomata can be said the resistance of stomata in diffusion of water through it. Stomatal resistance is therefore an index to the need of water. The degree of closure of the stomata gives the measure of the level of water stress. Stomata remains fully open when water supply is adequate and start to close it with scarcity of water.

Indicator plant

There are some plants, which are very sensitive to water stress. These plants may be used as the indicator plants in the crops, which do not show easily the symptom of water stress. The sunflower is a good indicator plant. It is often used as an indicator plant in onion. The indicator plants must have the character of showing the water stress before the crop suffered from it.

Critical stage of growth

The crops of which physiological stages are very distinct can be used for selecting the time of irrigation. The crown root initiation, tillering, flowering, milk and dough stages in wheat are very critical. Similar are the pegging, pod setting and development in mustard and tasseling, silking and grain filling stages in maize. The plants suffer much if water is scarce at the critical stages. The common farmers can easily identify the critical stages of crops for irrigation. However, there are the crops for which critical stages are not well defined.

Meteorological Parameter

Empirical formulae

Penman (1948) & Thornthwaite (1948), Blaney – Criddle (1950) and Christiansen (1968) developed empirical formulae for estimating the potential evapotranspiration by using the meteriological parameters. The knowledge of potential evapotranspiration may be used in estimating the evapotranspiration of the crop and thereby the soil moistuire depletion. Where the soil moisture depleted to a predetermined level the irrigation is applied to replenish it. This method considered being complicated for ordinary farmers' use.

Evaporimeter

The evaporation data of evaporimeter like USDA class A pan are used to determine the time of irrigation. The evaporation data for the specified period when multiplied by the pan factor and crop coefficient of that period represents the evapotranspiration of the crop at that period. Irrigation is applied when depletion of soil moisture through evapotranspiration reaches to a predetermined level.

Irrigation water/Cumulative pan evaporation ratio (IW/CPE ratio)

The basis of irrigation in this method is the ratio of fixed depth of irrigation water to cumulative pan evaporation (E_{pan}). The ratio usually varies in between 0.5 to 1.2. The pan evaporation values are added every day till it reaches to predetermined IW/CPE ratio. Irrigation water depth is prefixed for a crop. Generally, the vegetable crops are given 3cm of water and 5cm in the field crops in each irrigation. Experiments are conducted to select the appropriate IW/CPE ratio for a crop for a particular agro-climatic zone, which gives the maximum yield. The performances of irrigation experiments following IW/CPE ratio conducted by All India Coordinated Research Project (AICRP) on Water Management are presented in Table 4.3-4.6.

Table 4.3. Effect of irrigation regimes and different agro-techniques for tillage/sowing on grain yield of wheat during 1998-99 and 1999-2000 at Faizabad

Treatments	**Wheat grain yield (t/ha)**		**Mean**	**No. of irrigation**	
98-99	**99-00**		**98-99**	**99-00**	
Irrigation schedules					
• IW/CPE=0.8	2.96	3.12	3.04	4	4
• IW/CPE=0.8 up to LJ + IW/CPE=1.0 up to dough	3.20	3.38	3.29	5	5
• IW/CPE=1.0	3.57	3.65	3.601	5	5
• IW/CPE=1.0 LJ +IW/CPE =1.2 up to dough	3.88	3.90	3.89	6	6
$CD_{(0.05)}$	0.108	0.21	3.35		
Agro-techniques					
• Tractor drawn implements + Broadcasting	3.27	3.44	3.35		
• Tractor drawn implements + Deshi plough sowing	3.58 3.63	3.66 3.83	3.62 3.73		
• One pass Chinese planter	3.13	3.13	3.13		
• One pass Pantnagar planter $CD_{(0.05)}$	0.135	0.24			

Table 4.4 Effect of irrigation schedules and fertility levels on grain yield and water-use of Rajmash of Faizabad during *Rabi* 1999-2000

Treatments	**Seed yield (t/ha)**	**Irrigation No. (cm)**	**Irrigation Depths (cm)**	**Effective rainfall**	**Total water supplied**	**WUE (kg/ha-cm)**
Irrigation levels						
• IW/CPE=0.5	1.40	2	10	6.30	16.3	86.13
• IW/CPE=0.7	1.68	3	15	6.30	21.3	78.84
• IW/CPE=0.9	2.26	4	20	6.30	26.3	86.08
$CD_{(0.05)}$	0.08					
Nitrogen						
• 50kg/N/ha + 5t FYM/ha	1.42					
• 100kg/N/ha + 5t FYM/ha	1.97					
• 150kg/N/ha + 5t FYM/ha	2.07					
• 50kg/N/ha	1.39					
• 100kg/N/ha	1.83					
• 150kg/N/ha	2.03					
$CD_{(0.05)}$	0.13					

Effective rainfall=6.3 cm

Table 4.5 Effect of irrigation and sulphur levels on grain yield and oil content of mustard during Rabi (1999-2000) at Patna

Treatments	Grain yield (t/ha)	Oil content (%)	Irrigation		WUE(kg/ ha-cm)
			No.	Depths (cm)	
Irrigation levels					
Rainfed	1.07	41.2			
IW/CPE=0.4	1.45	41.6	1	5	290.0
IW/CPE=0.8	1.57	41.8	2	10	1.57.9
$CD_{(0.05)}$	0.13	0.20			
Sulphur level	1.17	41.2			
Control	1.36	41.6			
15kg S/ha	1.43	41.7			
30kg S/ha	1.50	41.7			
45kg S/ha	0.14	0.25			
$CD_{(0.05)}$					

Interaction: Irrigation x Sulphur = NS

Table 4.6 Effect of irrigation schedules and nitrogen levels on the yield and WUE of red chillies during 1999-2000 at Kota

Treatments	Yield of dried chillies	Irrigation requirements		WUE
		No.	Depths (cm)	(kg/ha-cm)
Irrigation				
IW/CPE=0.6	1.46	1*+4	34	57.9
IW/CPE=0.8	2.01	1*+5	40	56.5
IW/CPE=1.0	2.08	1*+7	52	56.0
$CD_{(0.05)}$	0.08			
Nitrogen levels (kgN/ha)				
40	1.37			
80	1.67			
120	1.92			
160	2.02			
$CD_{(0.05)}$	0.10			

Rainfall = 0.4 cm

* Includes one common irrigation of 10cm depth. Depth of post planting irrigation = 6.0 cm

This method of scheduling of irrigation is mainly practiced in research stations. The farmers also can be trained to adopt the method since the only equipment required in this method is a simple evaporimeter. The limitation of this method is that it does not give due importance to the critical stages of plants .It might happened that to reach at the desired CPE value for an irrigation the critical stage of the crops are bypassed. In most of the crops the flowering and pod formation stages are very critical. If water is not available in plant at these stages yield suffer much.

Rough method of farmers

Most of the farmers are unaware about the scheduling of irrigation as discussed earlier except the feel and appearance method. They have developed some techniques, which give rough guidance for time of irrigation.

Soil-cum-sand mini plot technique

In this method the water holding capacity of a mini-plot is reduced artificially by mixing about 5% sand in it. In excavating the soil from mini plot soils of each 15 cm layer is kept separated and after mixing the sand the pit is refilled by soil layer wise and compacted to its original bulk density .The crops are cultivated as usual to the mini plot and the surrounding area. Since the soils in the mini plot are of low water holding capacity the crops in it show water deficiency symptom earlier than the surrounding area. The irrigation is given to the entire area. Since the soils in the mini plot are of low water holding capacity the crops show water deficiency earlier than the surrounding area. The irrigation is given in this time. This way the crops are irrigated in time before it is much late when it shows the water stress symptom at relatively advance stage of water deficiency.

Increased population

This is a simple method in which population of the plants of an area of 1mx1m in the crop field is increased by 4 times than the surrounding area. Due to more use of water by the increased number of plants in the demarcated area, water deficiency symptom are shown earlier than the rest of the area indicates the time of irrigation.

Can evaporimeter

There is good correlation between the evaporation from one litre can with evapotranspiration from a crop. The dimension of the can is taken as 14.3 height and 10 cm diameter. It is painted while to prevent rusting and has a screen cover of 6/20 size mesh. Paint is marked below the rim of the can to facilitate measurement of water. The can is filled up to that point when the soil is at field capacity. The can is fixed at crop height and is raised as the crop grows. When the water level in the can drops to a predetermined depth indicates the time of irrigation. The depth of water applied in the irrigation is same to the depth dropped in the can. This depth of water is equal to the amount of water, which has been depleted from the field capacity of the soil. The can is again filled up to the initial level after the irrigation.

4.2 How Much to Irrigate

The amount of water to be applied in an irrigation depend on the level of moisture depletion from the field capacity of the root zone depth of the soil at which irrigation is proposed to be given .The amount of water available from rainfall during the period of up to next irrigation and amount requires for leaching the salts have to be adjusted to determine the net amount of water to be applied for the irrigation .In each irrigation the aim is to bring the soil moisture content at field capacity.

Net Irrigation Requirement

This is the net depth of water to be irrigated after adjusting the amount available from rainfall, groundwater contribution, antecedent soil moisture or any other gain or loss of moisture. The net depth of water equals to the amount, which raises the soil moisture level to field capacity. Assuming no other contribution to soil moisture the net irrigation requirement is the difference between the field capacity and the soil moisture content in the root zone depth at the time of irrigation.

$$NIR = \sum_{i=1}^{n} \frac{(M_{fri} - M_{bi})\, A_s D_i}{100} \tag{4.1}$$

NIR = Net irrigation requirement, cm

M_{fci} = Field capacity of the soil in i th layer, percent

Mbi = Moisture content in the i th layer at the time of irrigation, percent

As = Apparent specific gravity of the soil in i th layer

D_i = Depth of the i th soil layer with in root zone depth, cm

N = Number of soil layers within the root zone depth.

Gross Irrigation Requirement

Gross irrigation requirement is the total of net irrigation requirements plus the other losses of water incurred during the irrigation. Gross irrigation requirement may be estimated for a small field to any farm area or command to any channel to entire project.

$$GIR = \frac{Net\ irrigation\ requirement}{Field\ efficiency\ of\ the\ system}$$

Irrigation Frequency or Irrigation Interval

Irrigation frequency or irrigation interval refers to the days between two irrigations during the period of highest consumptive use of the crop. The consumptive use rate depends on plants, soil and climate. Usually the rate is high when the days are longer and hotter, soils are sandy and plants are grown to maximum canopy.

Irrigation frequency is used to design the irrigation system, the time between two irrigations during the period of highest consumptive use. The irrigation system must have the capacity to supply the water during this period. Irrigation is usually given at 50% depletion of available soil moisture.

$$Irrigation\ frequency = \frac{M_{fc} - M_b}{P_{cu}} \tag{4.2}$$

M_{fc} = Field capacity of the soil in crop root zone depth, cm

M_b = Moisture content of the soil at the time of irrigation, cm

P_{cu} = Peak period consumptive use rate, cm

Irrigation Period

Irrigation period is the number of days that may be allowed to apply water in a design given area during the peak consumptive use of the crop. Irrigation period is the basis of irrigation system design. Irrigation period is never greater than the irrigation frequency. Say, irrigation frequency for an area is 15 days. The capacity of the irrigation system should be such that the entire area needs to be irrigated within 15 days. However, anybody may suggest the irrigation system, which irrigates the entire area in13, or 14 days and keep 1 or 2 days for maintenance of the system and leisure time of the operator. In this case the irrigation period 13 or 14 days. Thus, irrigation period is the actual days of irrigation within the period of irrigation interval.

Example 4.1 Calculate the cumulative evaporation required for scheduling irrigation at IW/CPE ratio of 0.5, 0.6 &0.8 with 5 cm of irrigation water.

Solution

$$IW / CPE = 0.5 \therefore CPE = \frac{\text{IW}}{0.5} = \frac{5cm}{0.5} = 10cm$$

$$IW / CPE = 0.6 \therefore CPE = \frac{IW}{0.6} = \frac{5cm}{0.6} = 8.33cm$$

$$IW / CPE = 0.8 \therefore CPE = \frac{IW}{0.8} = \frac{5cm}{0.8} = 6.25cm$$

Example 4.2 The following are the weekly average daily pan evaporation data in millimeters for the irrigation days of a crop which is scheduled for irrigation at IW/CPE = 0.8 and 5cm of water in each irrigation.

2.65, 2.78, 3.0, 3.2, 3.8, 4.2, 4.5, 5.0, 5.5, 5.2, 5.1, 5.3

The first irrigation was given on the first day of the first week. Determine the dates of other irrigation and the irrigation water requirement of the crop.

Solution:

IW/CPE = 0.8, CPE = IW/0.8 = 5/0.8 = 62.5mm

Therefore, the date of any irrigation comes after the cumulative pan evaporation of 62.5mm from the previous irrigation.

2nd irrigation

CPE of first 3 weeks = 2.65x7+2.78x7+3.0x7= 59.01mm

Remaining pan evaporation to satisfy 62.5mm CPE = 62.5-59.01 = 3.49mm

Weekly average daily pan evaporation of 4th week = 3.2mm. This is very close to 3.49 mm.

Therefore, the second day of the 4th week is the day of second irrigation.

3rd irrigation

CPE of 4th (part) & 5th week = 3.2x6 + 3.8x7 = 45.8mm

Remaining pan evaporation to satisfy 62.5mm CPE = 62.5-45.8 = 16.7mm

Weekly average daily pan evaporation of 6th week = 4.2mm

So, 16.7mm evaporates in 16.7/4.2 = 3.97 @ 4 days

Thus, the 5th day of 6th week is the day of 3rd irrigation.

4th irrigation

CPE of 6th (part) & 7th week = 4.2x3 + 4.5x7 = 44.1mm

Remaining pan evaporation to satisfy 62.5mm CPE = 62.5-44.1 = 18.4mm

Weekly average daily pan evaporation of 8th week = 5.0mm

So, days required to evaporates 18.4mm = 18.4/5 = 3.68 ≅ 4

Thus, 5th day of 8th week is the day of 4th irrigation.

5th irrigation

CPE of 8th (part) & 9th week = 5.0x3 +5.5x7 = 54.9mm

Remaining pan evaporation to satisfy 62.5mm CPE = 62.5-53.5 = 9mm

Weekly average daily pan evaporation of 10th week = 5.2mm

So, days required to evaporates 9 mm = 9/5.2 = 1.73 ≅ 2

Thus, 3rd day of 10th week is the day of 5th irrigation.

6th irrigation

CPE of 10th (part) & 11th week = 5.2 x4 +5.1x7 = 56.5mm

Pan evaporation required = 62.5-56.5 = 6.0mm

Weekly average daily pan evaporation of 12th week = 5.3mm

So, days required to evaporates 6.0 mm = 6.0/5.3 = 1.13 ≅ 1

Thus, 2nd day of 12th week is the day of 6th irrigation.

After the 6th irrigation days remain only 6. There is no scope to satisfy the required CPE of 62.5mm within this period. Therefore, 6th irrigation is the last irrigation.

Total irrigation requirement of the crop = 5cm x 6 = 30 cm.

Example 4.3 Weekly average daily pan evaporations were 3.11, 3.3, 4.0, 3.7, 2.9, 2.8, and 2.6mm for the irrigation days of a vegetable crop which is scheduled for irrigation at IW/CPE = 0.8 and 3cm of water in each irrigation. A rainfall of 12mm occurs on 5th day of 4th week. The first irrigation was given on the first day of the first week. Determine the dates of other irrigations. What will be the dates of irrigations if rainfall of 30mm occurs instead of 12mm?

Solution:

IW/CPE = 0.8, CPE = IW/0.8 = 3/0.8 = 3.75cm = 37.5mm

Therefore, the date of any irrigation comes after the cumulative pan evaporation of 37.5mm from the previous irrigation.

2nd irrigation

CPE of first week = 3.11x7= 21.77mm

Remaining pan evaporation to satisfy 37.55mm CPE =37.5-21.77 = 15.73mm

Weekly average daily pan evaporation of 2nd week = 3.3mm

So, days required to evaporates 15.73 mm = 15.73/3.3 = 3.75 ≅ 4

Thus, 5th day of 2nd week is the day of 2nd irrigation.

3rd irrigation

CPE of 2nd and 3rd week = 3.3x3 + 4.0x7 = 37.9mm@ 37.5mm

Thus, 1st day of 4th week is the day of 3rd irrigation.

There was the rainfall of 12mm on =5th day of 4th week.

At the time of rain the supply need of water in soil profile

=3.7x4x0.8=11.84mm

Therefore, it may be assumed that the crop has effectively used the entire rainwater. The deferred irrigation date may be fixed assuming 49.5mm (37.5 + 12 =49.5mm) CPE.

4th irrigation

CPE of 4th and 5th week = 3.7x7 + 2.9x7 = 46.2mm

Remaining pan evaporation to satisfy 49.34mm CPE =49.5-46.20 = 3.3mm

Weekly average daily pan evaporation of 6th week = 2.8mm

So, days required to evaporates 3.14 mm = 3.14/2.8 = 1.17 ≅ 1

Therefore, 4th irrigation is on 2nd day of 6th week.

For the remaining 6 days of 6th week and 7 days of 7th week provide the CPE

= 6x2.8 + 2.6x7 = 35mm

This CPE of 35mm is very close to 37.5mm. Therefore, the last irrigation may be applied on last day of last week.

If the rainfall of 30mm occurs on 5th day of 4th week when there was a supply need of 11.84mm only, the excess 18.16mm may have lost through direct or indirect runoff (percolation). If direct runoff does not take place , the indication of rate of percolation of water from the root zone depth is required. However, without this information, the date from which soil sampling could be possible may be taken as effectively the day of the previous irrigation for the next irrigation. The possible sampling date usually comes after 2 to 4-5 days depending on the soil characteristics. In the present case, the further calculation of irrigation dates may be fixed as if the 3rd irrigation is given on 6th or 7th day of 4th week.

Example 4.4 The pump discharges at a rate of 10 l/s for 10 hours in a day. What is the command of this pump if irrigation water applied is 7.5 cm and irrigation interval is 15 days?

Solution:

Total discharge in 15 days = 15 days x 10 hours / day x 10l/s

= 15 x 10 x 3600 x 10

= 5400 m^3

$$\text{Command area of the pump} = \frac{Volume\ of\ water\ applied}{Depth\ of\ water\ applied}$$

$$\frac{5400m^3}{7.5cm} = \frac{5400m^3}{0.075m} = 72000m^2 = 7.2ha$$

Example 4.5 A field crop is irrigated when the available soil moisture reduces to 60%. The moisture content at field capacity and wilting point are 32% and 12% respectively. The bulk density of the soil is 1.5g/cm³. The field water application efficiency is 75% and the crop root zone depth is 50cm. The gross depth of irrigation required to bring soil moisture content to field capacity in cm is 6,8,9,12 (GATE, 2020)

Solution:

Gross depth of irrigation requirement

$$GIR = \sum_{i=1}^{n} \frac{(M_{fci}\ 'M_{bi})\ A_s D_i}{100xIe} x\%\,reduction$$

$$= \frac{(32-12)\times 1.5\times 50}{100\times 0.75} \times (100-60)/100$$

$$= \frac{20\times 75\times 40}{100\times 100\times 0.75} = 8\text{cm}$$

Example 4.6 Unsaturated soil sample is collected from a field when the soil moisture is at field capacity. The inside diameter of the core sampler is 7.5 cm with a height of 15cm. Weight of the core sampling cylinder with moist soil is 2.81 kg and that of the oven dry soil is 2.61 kg. The weight of the core sampling cylinder is 1.56 kg. Assuming $\pi = 3.14$, the water depth in centimeter per meter depth of the soil *(rounded off to two decimal places)* is…….. (GATE, 2019)

Solution

$$\text{Cross-sectional are of core sampler} = \pi\frac{D^2}{4} = 3.14x\frac{(7.5)^2}{4} = 44.156cm^2$$

Weight of water in soil = 2.81-2.61 = 0.2kg = 200cm^3

$$\text{Depth of water in sampler of 15cm height} = \frac{200\text{cm}^3}{44.156cm^2} = 4.529cm$$

$$\text{Depth of water per meter depth of soil} = \frac{4.529}{15} x100 = 30.20cm$$

Example 4.7 An irrigation stream of 27l/s is diverted to a check basin of size 12m x 12m. The water holding capacity of the soil is 15% and the average soil moisture content in the crop root zone prior to applying water is 7.5%. The depth of root zone is 1.2m and the apparent specific gravity of the soil is 1.5. Assuming no loss of due to deep percolation, irrigation time (in minute) to which water (density =1000 kg/m^3) can be filled in this bin *(rounded off to two decimal places)* is........

(GATE, 2019)

Solution

$$\text{Depth of irrigation} = \frac{(M_{fc}\ 'M_b)A_s D}{100} = \frac{(15-7.5)1.5x120}{100} = 13.5cm$$

$$\text{Volume of water required to irrigate} = 12mx12mx13.5\text{cm} = 19.44m^3$$

$$\text{Irrigation time} = \frac{19.44m^3}{27/1000m^3/s} = 718.52s = 11.96\min$$

Example 4.8 The elevations of pressure gauge and porous cup of the tensiometer installed in unsaturated zone are 145.8m and 144.2m, respectively. Pressure measured at the gauge is -19.62 x 10^{3}N/m^2, the specific weight of water is 9810 N/m^3. The estimated pressure at the porous cup is .... N/m^2. (GATE, 2018)

Solution

The elevation difference between pressure gauge and porous cup of tensiometer = 145.8-144.2 =1.6m

Pressure equivalent of 1.6m pressure head = 1.6 x 9810 N/m^2 = 15.696 x 10^3 N/m^2

Therefore, the estimated pressure at porous cup = -19.62 x 10^3 N/m^2 - 15.696 x 10^3 N/m^3 = -3.924 x 10^{3}N/m^3 = -3924 N/m^2

Example 4.9 The soil of a crop field has field capacity of 25% and wilting point of 13% on weight basis. The effective rootzone depth of the crop is 0.70m and the consumptive use of water by the crop is 5 mm/day. Apparent specific gravity of the soil is 1.50. If the allowable soil moisture depletion is 40%, the permissible moisture depletion between irrigations and the frequency of irrigation are:

(a) 5cmm, 10 days (b) 10 cm, 7 days (c) 8 cm, 12 days (d) 4 cm, 8 days

(GATE, 2018)

Solution

Field capacity, FC = 25%

Wilting point, PWP = 13%

Effective root zone depth, D = 0.7m = 70 cm

Apparent specific gravity, A_s = 1.50

Soil moisture depletion = 40%

$$\text{Irrigation Requirement} = \frac{(FC - PWP)A_s D}{100} x\ depletion$$

$$= \frac{(25-13)1.5x70x0.4}{100} = 5.04cm \cong 5cm$$

$$\text{Frequency of irrigation} = \frac{5cm}{5mm/day} = \frac{5cm}{0.5/day} = 10\ days$$

Example 4.10 5ha field under wheat crop is irrigated at 40% depletion of available moisture content. The field capacity, wilting point, and bulk density of the soil in the field are 32% (on weight basis), 20% (on weight basis) and 1400kg/m^3 respectively. To irrigate the field in a day of 10 hours, a pump is used which lifts 270m^3 of water per hour against a total head of 20m.

i) If the root zone depth is 0.8m, the volume of water required to irrigate

a) 2700	b) 3150
c) 3600	d) 4050

ii) The pump is driven by an electric motor. If the pump, drive and motor efficiencies are 80%, 82% and 90% respectively, the required size of electric motor in horse-power will be

a) 20.0	b) 24.5
c) 30.0	d) 33.5

(GATE, 2013)

Solution

i) Area of field, A=5ha

Field capacity, FC = 32 (on weight basis)

Wilting point, PWP = 20% (on weight basis)

Soil moisture depletion = 40%

Bulk density, A_s = 1400 kg/m^3 = 1.4g/cm^3

Root zone depth, D = 0.8m = 80 cm

Depth of irrigation water, d

$$= \frac{(FC - PWP)A_s D}{100} x\ depletion\ (fraction)$$

$$= \frac{(32\text{-}20)1.4x80}{100} x0.4$$

$$= 5.38cm$$

Volume of water applied, V = A x d = 5ha x 5.38 cm = 5 x 10000m^2 x 5.38/100m = 2690m^3

ii) Pump, drive and motor efficiencies = 80%, 82% and 90%

Volume of water pumped = 270m^3 in 1 hour

Lift of water, H = 20m

$$Pump\ discharge, Q = \frac{270m^3}{1hour} = \frac{270x1000\text{kg}}{3600s} = 75\text{kg} / s$$

$$\text{Pump horsepower, HP} = \frac{75kg\ /s\ x20\text{m}}{75kg - m / sx0.8x0.82x0.90} = 33.87$$

Example 4.11 Irrigation was given to bring the soil in field capacity. Determine the field capacity of the soil from following data:

Root zone depth =1.5 m

Moisture content in the soil =7.5%

Dry density of the soil =1.5 g /cc

Water applied to plot =50 m^3

Water lost due to evaporation, etc. =15%

Area of plot = 100 m^2

Solution: Water applied in the plot= 50m^3

$$\text{Water lost} = \frac{50x15}{100} = 7.5m^3$$

Water retained in soil = 50-7.5 = 42.5m^3

Dry weight of total soil up to root zone depth= 100x 1.5 x1.5x10 ^{6} gm

$$\text{Percent of water added} = \frac{42.5x10^6\text{x}100}{100xl.5xl.5x10^6} = 18.89$$

Field capacity = 18.89 +7.5=26.39%

Example 4.12 The following data were obtained in different layers in the root zone before the irrigation.

Depth of sampling layer (cm)	Moisture content (%)	Bulk density (g/cm^3)	FC(%)	PWP (%)
0-30	16.20	1.44	30.50	11.85
30-60	15.56	1.44	30.50	11.10
60-90	17.34	1.47	31.20	12.50
90-120	20.87	1.48	32.00	13.40

Determine (i) the available soil moisture in the root zone depth, (ii) the moisture content in different layers of the root zone depth, and (iii) the depth of water to be applied to bring the soil at field capacity.

Solution: (i) Available soil moisture in different layers,

$$0-30cm=(30.50-11.85)x1.44x\frac{30}{100}=6.99cm$$

$$30-60cm=(30.50-11.10)x1.44x\frac{30}{100}=8.38\text{cm}$$

$$60-90cm=(31.20-12.50)x1.47x\frac{30}{100}=8.25cm$$

$$90-120cm=(32.00-13.40)x1.48x\frac{30}{100}=8.26cm$$

Total of 120cm = 30.63cm

ii) Moisture content of soils in different layers,

$$0-30cm=16.20x1.44x\frac{30}{100}=6.70cm$$

$$30-60cm=15.56x1.44x\frac{30}{100}=6.72cm$$

$$60-90cm=17.34x1.47x\frac{30}{100}=7.65cm$$

$$90-120\text{cm}=20.87x1.48x\frac{30}{100}=9.27cm$$

Total = 30.34cm

iii) Moisture content at field capacity in different layers of soil,

$$0-30cm=30.50x1.44x\frac{30}{100}=13.18cm$$

$$30-60cm=30.50x1.44x\frac{30}{100}=13.18cm$$

$$60-90cm=31.20x1.47x\frac{30}{100}=13.76cm$$

$$90-120cm=32.00x1.48x\frac{30}{100}=14.21cm$$

Total = 54.33cm

Net depth of water to be applied to bring the soil in field capacity

= 54.33-30.63 = 23.97cm

Example 4.13 A 2ha field crop is irrigated by a tube well having a discharge of 25l/ s. the moisture content at successive depth in the root zone prior to irrigation is given.

Depth, cm	0-25	25-50	50-75	75-100
Moisture content, %	5.5	6.0	6.5	7.5

The bulk density of the soil is 1.5g/cc. The moisture content at field capacity is 18cm per meter depth of soil. Determine the (i) moisture content of the soil in the root zone at the time of irrigation (ii) net irrigation requirement (iii) time required to operate the tube well to bring the soil moisture to field capacity

(GATE, 1999).

Solution

Average moisture content of soil (up to 100 cm)

i) Depth of initial moisture per metersoil profile = $WA_s = 6.375x1.5 = 9.56cm$

ii) Irrigation depth = 18 - 9.56 = $8.44cm$

iii) Time required to operate the tube well

$$= \frac{8.44cmx2ha}{25l/s} = \frac{8.44/100mx2x10000m^2}{25/1000m^3/s}$$

$= 18.75h$

Example 4.14 A river lift pump of capacity 10l/s working for 15 hours in a day is used to irrigate a wheat field. The irrigation is done at 50% depletion of soil moisture and the available soil moisture is 15cm/m. The conveyance and application efficiencies are 75% and 80% respectively. The peak consumptive use rate is 5mm/day. The operator of the pump is allowed one-day leave in a week. Determine the irrigation period, irrigation interval and the area of the field to be covered by using this pumping system.

Solution

Root zone depth of wheat crop = 90 cm

$$\text{Net depth of water application} = \frac{90cmx15cmx0.5}{100cm} = 6.75cm$$

$$\text{Irrigation int erval} = \frac{6.75cm}{0.5cm} = 13.5 \cong 14 days$$

Irrigation period = 14 days - one-day/week

= 14 days - 2 days

= 12 days

Volume of water pumped in an irrigation cycle $= 10l/sx15hx12days$

$= 6480m^3$

$$\text{Volume of water required/irrigation/ha} = \frac{10000m^2x6.75cm}{0.75x0.80} = 1125m^3$$

$$\text{Area to be irrigated} = \frac{6480m^3}{1125m^3/ha} = 5.76ha$$

Example 4.15 Wheat crop is grown in an area of 3 ha .The crop is irrigated at IW/CPE ratio of 0.8 and requires an interval of 20 days. The average pan evaporation during the period is observed to be 2.5 mm/day. The irrigation source located at a distance of 300m is a confined well (diameter 10 cm). The thickness of the aquifer is 10 m and hydraulic conductivity is 4 cm/min. Initial piezometric level is 4 m below the pump level and during pumping it should be restricted to 7m. The pump has to work against a total head of 10 m at an overall efficiency of 50%. Compute the cost of electricity consumed for one irrigation assuming the unit electricity charge of Rs. 2.00 .The radius of influence is 300m and conveyance efficiency of the channel is 75%.

Solution: IW/CPE = 0.8

Therefore, IW= 0.8x20x 2.5= 40 mm

$$\text{Discharge from the pump}, Q = \frac{2\pi Kbs}{\ln (r_0/r_w)}$$

$$= \frac{2\pi x4cm/\min x10mx(7-4)m}{\ln(300m/0.05m)}$$

$= 0.0144m^3/s$

Water requirement to crop during the irrigation period $3hax40mm = 1200m^3$

$$\text{Water need to be diverted from the pump} = \frac{1200}{0.75} = 1600m^3$$

$$\text{The time required pumping } 1600m^3 \text{ of water} = \frac{1600m^3}{0.0144m^3/s} = 30.86h$$

Since the pump efficiency is 50%, the pump horsepower, $HP = \frac{QH}{75}x2$

$$= \frac{0.0144x10x2x1000}{75}$$

= 3.84

Pump capacity in kW = 3.84x0.736=2.83

The electricity units consumed by the pump during the irrigation period

= 30.86x2.83kW - h = 87.33

Cost of the electricity = 87.33x2 = Rs.174.67

Example 4.16 In a research field of rice irrigations were given in a plot (A) keeping in view always some standing water in the field and in an another plot (B) three days after disappearance of water. Each irrigation uses 5cm of water. The results are as below.

Plots	Nos.of irrigation	Cost of weeding/ha,(Rs.)	Yield,Kg/ha
A	15	2100	5600
B	12	2450	5545

Determine the water use efficiencies. Which scheduling of irrigation is preferable? Assume cost of water and produce as Rs.1.25/m^3 and Rs.375/quintal respectively.

Solution

Let the area cultivated under each plot = 1ha

Water use efficiency of plot A $= \dfrac{5600 kg/ha}{15x5cm} = 74.87 kg/ha/cm$

Water use efficiency of plot B$= \dfrac{5545 kg/ha}{12x5cm} = 92.42 kg/ha/cm$

Volume of water applied in plot A = 15 x 5 cm x 1ha = 7500m^3

Volume of water applied in plot B = 12 x 5 cm x 1ha = 6000m^3

Cost of water for plot A = 7500m^3 x Rs.1.25/m^3 = Rs.9375/-

Cost of water for plot B = 6000m^3 x Rs.1.25/m^3 = Rs.7500/-

Cost of produce (rice) of plot A = 5600 kg x Rs. 375/quintal = 5600 x 375/100 = Rs. 21000/-

Cost of produce (rice) of plot B = 5545 kg x Rs. 375/quintal = 5545 x 375/100 = Rs. 20794/-

Cost of irrigation and weeding in plot A = Rs. 9375 + Rs. 2100 = Rs.11475/-

Cost of irrigation and weeding in plot B = Rs. 7500 + Rs. 2450 = Rs. 9950

Net benefit in plot A (excepting other costs) = Rs.21000 - Rs.11475 = Rs.9525/-

Net benefit in plot B (excepting other costs) = Rs.20794 - Rs.9950 = Rs.10844/-

Therefore, in respect of irrigation practices the second method of applying water i.e., three days after disappearance of water is preferable.

Questions & Problems

4.1 What do you mean by scheduling irrigation? In what principle scheduling is done in water scarce area?

4.2 Classify the parameters on which scheduling of irrigation may be done. Discuss the plant parameters in scheduling irrigation.

4.3 Explain the soil parameter in scheduling irrigation.

4.4 Discuss a few farmers' practice in scheduling irrigation.

4.5 Explain the different meteorological approaches for scheduling irrigation.

4.6 Discuss the IW/CPE ratio as the method of scheduling irrigation.

4.7 Define net and gross irrigation requirement. How do you calculate them?

4.8 Define irrigation interval and irrigation period. What is the difference between these two?

4.9 What are the advantages and disadvantages of soil-water potential, electrical resistance and evaporimeter as the parameter for scheduling the irrigation?

4.10 Determine the irrigation interval from the following data:

Field capacity of the soil = 30%

Permanent wilting point = 14%

Apparent specific gravity of soil = 1.45

Root zone depth = 60cm

Daily consumptive use rate = 5mm

Time of irrigation = at 50% depletion of available soil moisture

Ans: 14 days

4.11 A soil has 30% field capacity and 15% permanent wilting point. The apparent specific gravity of the soil is 1.45. The root zone depth is 90cm. Irrigation water is applied when moisture content is reduced to 20%. Determine the storage capacity of the soil, depth of water to be applied if application efficiency is 80%. How much area to be irrigated in 1 hour at a discharge rate of 25l/s?

Ans: 19.58 cm, 16.31 cm & 551.8m^2.

4.12 The following are the weekly average daily pan evaporation data for the irrigation days of a crop which is scheduled for irrigation at IW/CPE = 1.2 and 5cm of water in each irrigation.

4.60, 4.73, 51, 5.2, 5.8, 6.2, 4.5, 6.0, 6.5, 6.2, 6.1, 6.5

The first irrigation was given on the first day of the first week. Determine the dates of other irrigation and the irrigation water requirement of the crop. What will be the changed dates of irrigations if a rain of 30mm occurs on second day of sixth week?

4.13 Wheat crop is irrigated by using 6 cm of water in each irrigation following IW/CPE = 1.2 from a water source of 7.5l/s available for 18h/day. The average pan evaporation during the period is 2.75mm/day. Assuming the irrigation days of 80 during the cropping period, determine the irrigation period, area to be irrigated and the irrigation cost per ha. The cost of water is Rs.1.25/m^3.

Ans: 18 days,17.64 ha, Rs.2747

4.14 In an irrigation research water was applied following the treatments IW/CPE = 0.8,0.9 & 1.0. The irrigation days in the cropping period was 60.

The average pan evaporation during the period was 2.65mm/day. The produces of the treatments were 7500, 8000, 8100kg/ha respectively. Determine the water use efficiencies. Ans;58.96kg/ha/mm, 55.90kg/ha/mm, 50.94 kg/ha/mm

4.15 Select the appropriate answer from the following:

1. Scheduling irrigation is the process of determining
   a) Where to irrigate and when to irrigate
   b) Where to irrigate and how much to irrigate
   c) When to irrigate and how much to irrigate
   d) Where to irrigate and how much to irrigate
2. Scheduling of irrigation is done
   a) For adequate water availability
   b) For inadequate water availability
   c) For adequate and inadequate water availability
   d) For none of the above
3. Which one of the following is not a farmers' rough method of scheduling irrigation
   a) Pan evaporation b) Increased population
   c) Can evaporimeter d) Soil-cum-sand mini plot technique
4. Which of the following IW/CPE ratio may requires more water
   a) 0.5 b) 0.7
   c) 0.9 d) 1.2
5. Which of the following depth of water per application close to vegetable irrigation
   a) 3 cm b) 4 cm
   c) 5 cm d) 6 cm
6. Water applied 3cm in each irrigation following the treatments IW/CPE=0.5 & IW/CPE=1.2. The CPE in the irrigation days was 25cm. The difference of water application between the two irrigation treatment was
   a) 12.5 cm b) 14.5 cm
   c) 15 cm d) 17.5 cm
7. An amount of $30m^3$ water was applied in a field of soil moisture content 10%, root zone depth 1.0m, dry density of the soil 1.5g/cc, and area of the field $150m^3$ to bring it to field capacity. The field capacity of the soil was
   a) 13.33% b) 23.33%
   c) 30.0% d) 33.33%.

8. A river lift pump of capacity 8.5l/s operates for 16 hours per day. The peak consumptive use rate per day is 4mm, irrigation interval is 10 days, conveyance & application efficiencies are 70% & 75% respectively. The area it irrigates in a day is

   a) 0.33 ha b) 0.48 ha

   c) 0.64 ha d) 0.81 ha

9. A tube well pumps at the rate of 15l/s at a total head of 10m and pump efficiency is 60%. If the power charge is Rs.4.0/unit, the cost of pumping for irrigation period of 30 days is approximately

   a) Rs.7,000 b) Rs.8,000

   c) Rs.9,000 d) Rs.9,500

10. Daily $200m^3$ water is applied in a farm at 10 days irrigation interval. If 2 days leisure period is allowed to pump operator in an irrigation cycle, the daily water application required is

   a) $160m^3$ b) $180m^3$

   c) $250m^3$ d) $300m^3$

Ans.

1. c) 2. c) 3. a) 4. d) 5. a) 6. d) 7. b) 8. c)
9. a) 10. c)

4.16. Write True or False of the following:

1. IW/CPE ratio of scheduling irrigation gives weightage to critical stage of plants.
2. Irrigation is always done at 50% depletion of soil moisture.
3. Sunflower is a good indicator plant.
4. Usually the irrigation interval is more in light soil.
5. Tensiometer is the device for measuring the soil moisture potential up to 5 bar.
6. Electrical resistance method is a direct method to determine the soil moisture content.
7. Gravimetric method is the most accurate method for determining the soil moisture.
8. Irrigation period is less or equal to irrigation interval.
9. Irrigation interval is usually less in shallow rooted crops.
10. In water budget technique of scheduling irrigation the contribution of ground water is taken in to account.

Ans

1. True 2. False 3. True 4. False 5. False 6. False 7. True
8. True 9. True 10. True.

References

Annual Report (2000-2001). Directorate of Water Management Research, Patna-801505 (Bihar).

Majumdar, D.K. (2000). Irrigation Water Management. Principle & Practices. Prentice-Hall of India Pvt. Ltd., New Delhi. p.263

Reddy, G.H.S & T.Y Reddy. Efficient Use of Irrigation Water. Kalyani Publishers. p.137

5

Design of Irrigation Channels

5.1 Irrigation Channels

Canal or channel is an artificially made conduit through which water is supplied to the field from river, tank, reservoir or wells. The open channel is the most common type of water conveyance system. If the pipes and drains are partially full behave like open channel. The irrigation channel may be lined or unlined and the flow is either uniform or varied.

The canal may be classified in following ways.

a) Classification based on the source of supply
   i) Permanent or perennial canal
   ii) Inundation canal

A canal is said to be permanent when it is fed by a permanent source of supply and has permanent networks of distribution of supplies. Inundation canals draw their supplies from river usually when water stage exceeds a predetermined level. Headworks are not provided for diversion of river water to canal. However, canal head regulators are provided to regulate the canal head as required time to time to accommodate the changing pattern of river course.

b) Classification based on the function of the canal
   i) Irrigation canal
   ii) Carrier canal
   iii) Feeder canal
   iv) Navigation canal
   v) Power canal

Irrigation canal carries water for crop field. Carrier canal carries water for other canal besides doing irrigation. Feeder canal feed water to two or more canal. Navigation canal permits to water transport. Power canal provides the water for generating power.

c) Classification based on the relative importance of the canal in a given network of canals
   i) Main canal
   ii) Branch canal

iii) Major distributor

iv) Minor distributor

v) Water courses

Main canal diverts the water from the source. Usually it does not supply the irrigation. It delivers the water to branch canal or major distributaries. Branch canal originated from main carries supply for major or minor distributaries. Usually branch canal water is not directly used for irrigation. However, some direct outlets are sometime provided. Major distributaries are usually less than the branch canal. Major distributaries take off from branch or main canals. These are the canals, which supply the irrigation water directly through outlet or convey to minor distributaries. Minor distributaries receive the water from major distributaries or branch canals provide water to the watercourses through outlet provided along them. Watercourses are the tail channels of an irrigation canal network feed water to the irrigation field. Watercourses usually receive water from minor distributaries. However, it may take off from major distributaries or branch canals considering the convenience in irrigating the field.

Open channel

It is the conduit in which the liquid flow occurs under the influence of gravity. The liquid should have the free surface and is not under pressure other than its own weight. Open channel flow occurs in rivers, canals, flumes and the partially flowing pipes (sewer, drain, etc.)

Steady flow: It is the flow condition in which the flow at any point does not change with time $\left(\frac{\partial v}{\partial t}=0, \frac{\partial y}{\partial t}=0, etc.\right)$. Lined channel flow is usually steady.

Unsteady flow: It is the flow condition in which the flow at any point change with time $\left(\frac{\partial v}{\partial t}\neq 0, \frac{\partial y}{\partial t}=0, etc.\right)$. The intermittent runoff into terrace outlets, diversion channels, etc., is unsteady.

Uniform flow: It is the flow condition in which the depth, slope, velocity, and cross-section remain constant throughout the length of channel $\left(\frac{\partial y}{\partial L}=0, \frac{\partial v}{\partial L}=0, etc.\right)$. Field channels are designed assuming uniform flow at maximum discharge.

Non-uniform flow: It is the flow condition in which the flow changes along the length of open channel $\left(\frac{\partial y}{\partial L}\neq 0, \frac{\partial v}{\partial L}\neq 0, etc.\right)$. Non-uniform may be steady or unsteady. The flow in a channel of variable cross-section is non-uniform.

Shear stress at a cross-section in a horizontal circular pipe under steady flow

Referring to Fig. 7.1 and since the flow is steady it satisfies the following condition,

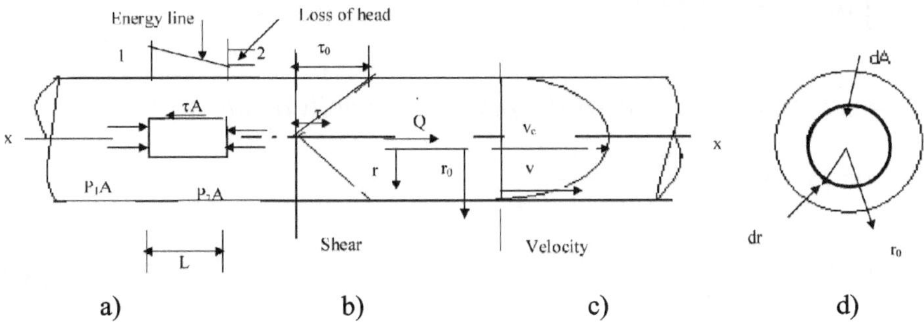

Fig. 5.1. Shear stress in circular pipe

$$p_1(\pi r^2) - p_2(\pi r^2) - \tau(2\pi r L) = 0$$

$$or, \tau = \frac{(p_1 - p_2)r}{2L} \tag{5.1}$$

Where, p_1 and p_2 = water pressure in section 1 & 2.

r = radius of pipe

τ = shear stress of fluid flow at any radius, r. τ is zero when r = 0 and

τ is maximum (τ_{max}) where r is maximum (r_{max}).

Multiplying Eq.5.1 by $\frac{w}{w}$ (w = unit weight of water)

$$\tau = \frac{w}{2L}\left(\frac{p_1}{w} - \frac{p_2}{w}\right)r,$$

$$= \frac{w}{2L}(h_1 - h_2)r$$

$$= \frac{wh_L r}{2L} \tag{5.2}$$

Variation of velocity of flow in a pipe

Shear stress in a pipe flow

$$\tau = -\mu \frac{\partial v}{\partial r} \tag{5.3}$$

Using the value of τ in Eq 5.1

$$-\mu\frac{\partial v}{\partial r}=\frac{(p_1-p_2)r}{2L}$$

$$\therefore -\int_{v_c}^{v} dv=\frac{p_1-p_2}{2L\mu}\int rdr$$

(v_c = *velocty at the centre of the pipe*, v = *velocity at any po*int)

$$or,(v-v_c)=\frac{(p_1-p_2)}{2L\mu}\frac{r^2}{2}$$

$$or, v=v_c=\frac{wh_L}{4\mu L} \tag{5.4}$$

Since velocity at boundary=zero,

$$v_c=\frac{wh_L r_0^2}{4\mu L} \tag{5.5}$$

General equation in steady uniform flow in open channel

Referring to Fig.5.2 and taking summation of forces acting on section A & B,

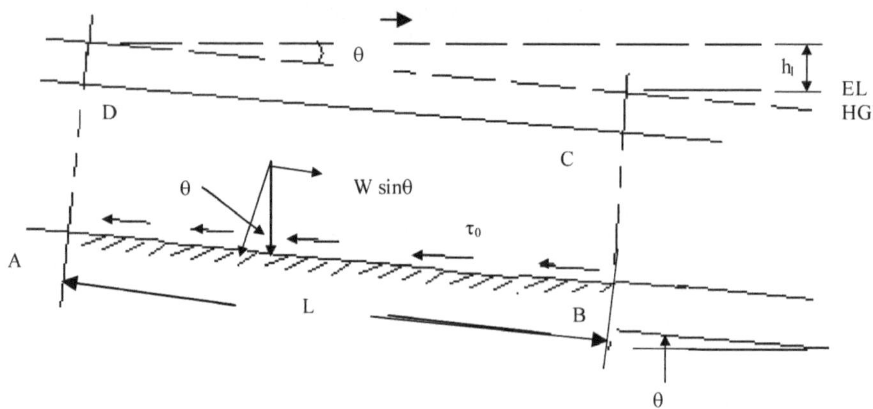

Fig. 5.2. Steady uniform flow in open channel

p_1A - p_2A + wALSin θ - τ_0 PL = 0. (5.6)

Where,

p_1, p_2 = Average pressure of water at section A & B

τ_0 = Boundary shear stress

From Eq. 5.1, *w*ALSin θ = τ_0 PL

or, wAL tan θ = τ_0 PL

For small value of θ, Sin θ = tan θ = S (slope of the channel)

$$\therefore \tau_0 = wRS, \left(\tau_0 = \frac{wAL\,tan\,\theta}{PL} = \frac{wA\,tan\,\theta}{P} = wRS \right)$$

Using the equation derived earlier,

$$h_L = \frac{2L\tau_0}{wr}$$

In Darcey-Weisbach equation, $h_L = f\frac{L}{D}\frac{v^2}{2g}$

$$\therefore h_L = \frac{2L\tau_0}{wr} = h_L = f\frac{L}{D}\frac{v^2}{2g}$$

$$\therefore \tau_0 = f\rho\frac{v^2}{8}, (w = \rho g)$$

So, wAL tan θ = $f\rho\frac{v^2}{8}\frac{A}{R}L$

or , $v^2 = \frac{8wLR\,tan\,\theta}{f\rho} = \frac{8g\,R\,tan\,\theta}{f}$

$$\therefore v = \sqrt{\frac{8g}{f}RS}$$

$$= C\sqrt{RS} \qquad (5.7)$$

For laminar flow, $f = \frac{64}{R_e}, R_e = \text{Reynold's number}$

$$\therefore C = \sqrt{\left(\frac{8g}{64}\right)R_e} \qquad (5.8)$$

Expression for critical depth, critical energy, and critical velocity for (a) rectangular channel and (b) any channel.

a) For rectangular channel

Specific energy as per definition,

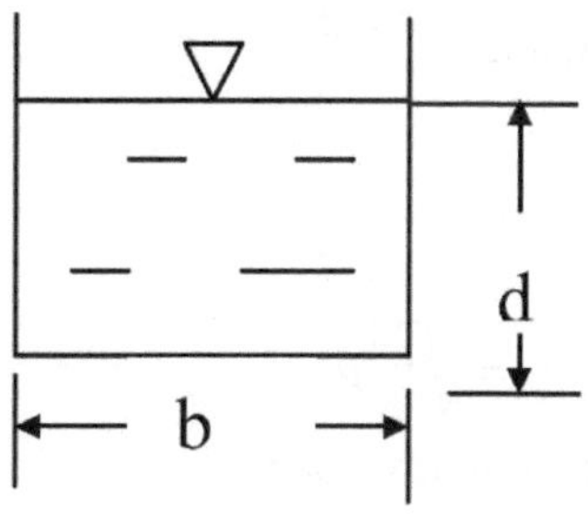

$$E = y + \frac{v^2}{2g} = y + \frac{1}{2g}\left(\frac{Q}{A}\right)$$

$$= y + \frac{1}{2g}\left(\frac{Q}{by}\right)^2$$

$$= y + \frac{1}{2g}\left(\frac{q}{y}\right)^2$$

$$q = \frac{Q}{b}, \textit{flow per unit bottom width} \qquad (5.9)$$

The critical depth occurs when the specific energy is minimum for a given discharge,

$$So, \frac{dE}{dy} = 0$$

$$Or, \frac{dE}{dy} = \frac{d}{dy}\left[y + \frac{1}{2g}\left(\frac{q}{y}\right)^2\right] = 1 + \frac{q^2}{2g}\left(-2y^{-3}\right)$$

$$= 1 - \frac{q^2}{gy^3} = 0 \text{ or, } gy^3_c = q^2$$

$$\therefore y_c = \sqrt[3]{\frac{q^2}{g}}$$

$$Or, q = g^{1/2} y_c^{3/2} \tag{5.10}$$

So, putting the value of $q = g^{1/2} y_c^{3/2}$ of Eq.5.10 in Eq.5.9, the critical specific energy,

$$E_c = y_c + \frac{1}{2g}\left(\frac{g^{1/2} y^{3/2}}{\sqrt[3]{\frac{gy_c^3}{g}}}\right)^2$$

$$= y_c + \frac{1}{2g} gy_c = y_c + \frac{y_c}{2}$$

$$= \frac{3}{2} y_c \tag{5.11}$$

$$\text{Again, } y_c^3 = \frac{q^2}{g}$$

$$Or, \ y_c^3 = \frac{v_c^2 . y_c}{g}$$

$$\therefore v_c = \sqrt{y_c^2 g} \tag{5.12}$$

b) For any channel

$$E = y + \frac{v^2}{2g} = y + \frac{1}{2g}\left(\frac{Q}{A}\right)^2$$

For a constant Q the area A varies with the depth y, differentiating with respect to y,

$$\frac{dE}{dy}=1+\frac{Q^2}{2g}\left(-\frac{2}{A^3}\right)\frac{dA}{dy}=1-\frac{Q^2}{A^3 g}\frac{dA}{dy}=0$$

The area dA may be substituted by the multiplication of differential depth dy and surface width b'.

$$\therefore 1=\frac{Q^2}{A_c^3 g}\frac{dA}{dy}=\frac{Q^2}{A_c^3 g}\frac{b'dy}{dy}$$

$$Or, \frac{Q^2}{g}=\frac{A_c^3}{b'} \tag{5.13}$$

The right hand side of Eq. 5.13 contained A_c, which is a function of $y_{c;}$ the value of y_c may be determined by trial and error which gives LHS=RHS.

From Eq. 5.13

$$\frac{Q^2}{A_c^2 g}=\frac{A_c}{b'}$$

$$Or, \frac{v_c^2}{g}=\frac{Ac}{b'}, \quad \therefore\ v_c=\sqrt{\frac{gA_c}{b'}} \tag{5.14}$$

Considering the mean depth y_m equal to the area A_c divided by the surface dimension b'

$$v_c=\sqrt{gy_m} \tag{5.15}$$

The critical specific energy,

$$E_c=y_c+\frac{v_c^2}{2g}$$

$$=y_c+\frac{gy_m}{2g}=y_c+\frac{y_m}{2} \tag{5.16}$$

5.2 Design Considerations

Open channel design consists of designing the cross-section and velocity of flow for a given discharge. The cross-section should be adequate and velocity should be within the permissible limit of the material on which the channel is made. The elements of an open channel are described as shown in Fig.5.3.

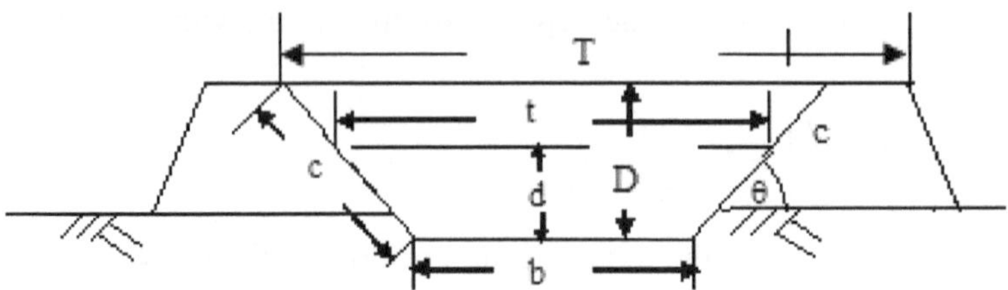

Fig. 5.3 Cross-section of an open channel

T = Top width

t = Top width of the water surface when water is at depth d

b = Bottom width

D = Depth of water in channel

c,c= Wetted sides

θ = Angle between the sloping sides and the horizontal

A = Cross-sectional area

P = Wetted perimeter=b+c+c

R = Hydraulic radius.

Definitions

Wetted perimeter (P): It is the sum of the length of that part of the channel sides and bottom that are in contact with water. The surface area of a channel provides frictional resistance to flow which in turns proportional to wetted perimeter.

Wetted cross-section (A): It is the cross-sectional area of channel through which water is flowing. Referring to Fig 5.1,

$$a = \frac{(b+t)d}{2}$$

Hydraulic radius (R): It is the ratio of wetted cross-sectional area to wetted perimeter.

$$R = \frac{A}{P}$$

The hydraulic radius or hydraulic depth is important parameter in channel flow. Velocity of flow is proportional to square root of hydraulic radius $(V \alpha \sqrt{R})$.

Hydraulic slope (S): It is the slope of the water surface. It is measured by the vertical drop h for a certain channel length l.

$$S = \frac{h}{l}$$

It is expressed either in fraction or percentage. The land slope of the channel often used as hydraulic slope as because the flow is assumed steady uniform.

Free board: It is the vertical distance between the top of the channel bank to the anticipated depth of water in the channel (D-d). This is the additional capacity provided to the channel to check the overtopping due to unforeseen situation. Twenty percent increase to anticipated depth of water is usually sufficient for free board. The approximate selection of free board depends on the specific situation on which the channel is made. However, a minimum depth of 15cm is to be provided to small irrigation channels.

Discharge (Q): It is the volume of water passes in channel section in unit time.

$Q = AV$

The discharge may be expressed in m^3/s; l/s etc. for small irrigation channel l/s is preferred. The velocity of flow in the earthen channel should be within the permissible limits of various soil characteristics (Table 5.2). Higher velocities are allowed in case of lined channels. The banks of the earthen channel to be stable enough to carry the desired flow safely. The side slopes should be designed in conformity to the soil conditions (Table 5.1). The side slopes should not be steeper than the angle of repose of the type of materials on which the channel is made.

Table 5.1 Suitable slopes for different type of materials

Materials	Side slope (H: V)
Clay	1:1
Silt loam	1.5:1
Sandy loam	2:1
Coarse sandy loam	3:1
Rock	Nearly vertical
Muck & pear soil	0.25:1
Stiff clay or earth with concrete lining	0.5:1 to 1:1
Earth with stone lining	1:1

Table 5. 2 Permissible velocities for various soil textures

Type of soil	Maximum permissible velocity, cm/s
Bare channels	
Sand	45
Loam, sandy loam and silt loam	60
Clay loam	65
Clay	70
Gravelly clay	100
Vegetated channels	
Poor vegetation	90
Fair vegetation	120
Good vegetation	150

Design formulae

1. Chezy's formula (1775)

   $V \alpha \sqrt{R} \sqrt{S}$

$$\text{Or,}\ \ V = C\sqrt{RS} \tag{5.17}$$

Where

V = Mean velocity, m/s

R = Mean hydraulic radius or hydraulic mean depth, m

$$= \frac{\text{Wetted cross-sectional area}}{\text{Wetted perimeter}}$$

S = Water surface slope

C = Chezy's constant, it is a factor of channel resistance.

This is a fundamental formula for uniform flow in open channel and is the basis for modern velocity formulae. Thus the formula expresses the mean velocity of flow as a function of the hydraulic roughness of the channel i.e., the hydraulic radius and the hydraulic slope.

It is based on the assumption that (i) force resisting per unit area of channel bed is proportional to the square of mean velocity and (ii) the effective component of gravity force causing the flow must be balanced by total force of resistance of flow (Sharma, 1993). Chezy's formula is recognized as a classical equation of flow and applicable to all cases where uniform flow is either assumed or in existence. The limitation of Chezy's formula are: (i) it is not dimensionally correct, (ii) it does not take into account physical character of the channel, (iii) it aims to application to all type of rugosity, (iv) it assumes that hydraulic gradient runs parallel to energy gradient. This is a condition, which may occur, in fairly straight channels in steady or nearly steady discharge rates.

2. Manning's formula (1889)

   Manning, an Irish engineer evolved the formula

$$V = \frac{1}{n} R^{2/3}\ S^{1/3} \tag{5.18}$$

Manning's formula gives the better result than Chezy's formula. The Chezy's C is equal to $\frac{1}{n}R^{1/6}$ in Manning's formula. This n is the coefficient of rugosity and its value is close to Kutter's N for practical purpose in normal ranges of slopes and hydraulic radius. Manning's formula is frequently used in channel design. The experimentally verified values of n for different type of channels are given in Table 5.3.

Table 5.3 Values of Manning's 'n'

Types and description of channel	**Manning's 'n'**
Earth channels	
Straight and uniform	0.023
Winding, sluggish	0.025
Stony bed, weeds on bank	0.035
Small drainage ditches	0.035-0.04*
Lined channels	
Concrete	0.015
Masonry, rubble	0.017 to 0.030
Metal, smooth	0.011 to 0.015
Wooden	0.011 to 0.014
Vegetated waterways	0.020 to 0.040
Pipes	
Cast iron	0.012 to 0.013
Clay or concrete drain tile	0.011
Steel	0.015 to 0.017
Vitrified sewer pipe	0.013 to 0.015

*b<1.2m, n=0.04;b>1.2m,n=0.035

Courtesy: Michael, 1985

3. Darcy-Weisbach formula

$$V=\sqrt{\frac{8gRS}{f}} \tag{5.19}$$

Where

g = Acceleration due to gravity, m/s^2

f = Darcy-Weisbach roughness coefficient or friction factor.

The Chezy's constant $C=\sqrt{\frac{8g}{f}}$ in Darcy-Weisbach formula. Thus, the resistance coefficients C, n and f in Eq.5.17-5.19 are related as below.

$$C=\frac{1}{n}\mathrm{R}^{1/6}=\sqrt{\frac{8g}{f}} \tag{5.20}$$

4. Bazin's formula (1897)

Bazin's considered Chezy's C is a function of R but not of S. He gave the value as below:

$$C=\frac{87}{1+\frac{M}{\sqrt{R}}} \tag{5.21}$$

Where, M = coefficient of roughness which depends on the roughness of channel surface. The values of M are nearly constant than Kutter's N for fairly smooth channels. The formula is derived from small experimental channels and less dependable than Kutter's formula. The values of M for some channel section given as below.

Type of channel section	M
Earthen channel (i) in bad condition	1.75
(ii) in usual condition	1.30
Plaster, very smooth earth	0.85
Smooth rubble masonry	0.46
Timber, neat brickwork	0.16
Smooth plaster, planed timber	0.06

Courtesy: Sharma & Sharma, 1993

5. Kutter,s formula (1869)

Kutter gave the value of Chezy's C as below:

$$C = \frac{\frac{1}{N} + \left(23 + \frac{0.00155}{S}\right)}{1 + \left(23 + \frac{0.00155}{S}\right)\frac{N}{\sqrt{R}}} \quad (5.22)$$

Where, N= Kutter's roughness coefficient. The values of N for different soils are given in Table 5.4.

Kutter's formula is widely used but it has got some limitation. It is unnecessary complicated, coefficient C increases with slope for hydraulic radius of less than 1m but drops when the slope rises for R greater than 1m, the value of N assumed to be depend only on the nature of the material and overall it is not more accurate than Manning's formula.

Table 5.4 Values of Kutter's roughness factor for N

Nature of the channel surface	Perfect	Good	Fair	Bad
Alluvium regime	0.017	0.020	0.0225	0.025
Loose gravel	0.0275	0.030	0.033	0.035
Rock cut, smooth and irregular	0.025	0.030	0.035	0.035
Rock cut and irregular	0.035	0.040	0.045	0.050
With cobble bed and earthen banks	0.025	0.030	0.035	0.040
With earthen bed and rubble sides	0.025	0.030	0.033	0.035

Courtesy: Sharma & Sharma, 1993

6. Pavlovskey's formula (1925)

Pavlovskey proposed the following equation as an improvement to Manning's equation

$$C = \frac{1}{n} R^{y} \quad (5.23)$$

Where, $y = 2.5\sqrt{n} - 0.13 - 0.75\sqrt{R}\left(\sqrt{n} - 0.10\right)$

The formula is widely used in former USSR

The limitation of the formula is that it is valid for values of R between 0.1m and 3.0m and n between 0.011 and 0.040. The formula does not use the reference of silt in water in designing the earthen irrigation channels. As a result the silt clearance from channels incurs large expenditure.

Most economical channel section

The most economical cross-section of a channel is that which offers least resistance to the flow i.e., permits maximum flow for a given cross-section. For a given section the discharge is maximum only when velocity is maximum. According to Chezy's formula, velocity is maximum when hydraulic radius R is maximum, and R is maximum if the wetted perimeter P is minimum. Theoretically, the semicircular section complies best with this condition. The most economical section for different geometrical shape can be derived. For an example, the trapezoidal section is derived as follows (Fig.5.3).

$$\text{Hydraulic radius (R)} = \frac{\text{wetted area (A)}}{\text{wetted perimeter (P)}}$$

When R is maximum, P is minimum. For a minimum value of P, $\frac{dP}{dd} = 0$ where d = depth of water.

In a trapezoidal section, $P = b + 2d\sqrt{z^2+1}$

and A = (b + zd) d

$$Or,\ b = \frac{A}{d} - zd$$

$$Therefore,\ P = \frac{A}{d} - zd + 2d\sqrt{z^2+1}$$

$$\therefore \frac{dp}{dd} = -\frac{A}{d^2} - z + 2\sqrt{z^2+1} = 0$$

$$Or,\ \frac{A}{d^2} + z - 2\sqrt{z^2+1} = 0$$

$$Or,\ \frac{(b+zd)d}{d^2} + z - 2\sqrt{z^2+1} = 0$$

$$Or,\ b + zd + zd - 2d\sqrt{z^2+1} = 0$$

$$Or,\ \frac{b}{2} + zd = d\sqrt{z^2+1} \qquad (5.24)$$

So, half of the top width=length of the side

Again, $R = \frac{A}{P}$

$$=\frac{(b+zd)d}{b+b+2zd}=\frac{(b+zd)d}{2(b+zd)}=\frac{d}{2} \tag{5.25}$$

So, hydraulic radius=half of the depth of water.

Again, putting $z=cot\theta=\frac{cos\theta}{sin\theta}$ in Eq. 5.24

$$b+2\frac{\cos\theta}{\sin\theta}d=2d\sqrt{1+\frac{\cos^2\theta}{\sin^2\theta}}$$

$$or, b\,sin\theta+2d\cos\theta=2d\sqrt{\sin^2\theta+\cos^2\theta}=2d$$

$$\text{or}, b\sin\theta=2d(1-\cos\theta),$$

$$\left[\cos(\theta)=\cos\left(\frac{\theta}{2}+\frac{\theta}{2}\right)=cos^2\frac{\theta}{2}-sin^2\frac{\theta}{2}=1-sin^2\frac{\theta}{2}-sin^2\frac{\theta}{2}=1-2\,sin^2\frac{\theta}{2}\right]$$

$$or,\ b=\frac{2d.2\,sin^2\frac{\theta}{2}}{2\,sin\frac{\theta}{2}cos\frac{\theta}{2}}$$

$$\left[sin(\theta)=sin\left(\frac{\theta}{2}+\frac{\theta}{2}\right)=sin\frac{\theta}{2}cos\frac{\theta}{2}+cos\frac{\theta}{2}sin\frac{\theta}{2}=2\,sin\frac{\theta}{2}cos\frac{\theta}{2}\right]$$

$$=2d\frac{sin\frac{\theta}{2}}{cos\frac{\theta}{2}}=2d\,tan\frac{\theta}{2} \tag{5.26}$$

Similarly for rectangular section most efficient section is obtained when b=2d.

For designing the channel the most economic section is tried first. If the economic section can not be obtained in prevailing condition, then section is designed on trial and error assuming the value of b and d.

Example 5.1 Compute the average shear stress of a rectangular channel of 2m wide and flowing 1.5m deep and bed slope 0.15%.

Solution:

Bed width, b = 2m

Depth, d = 1.5m

$$\text{Channel slope, S}=\frac{0.15}{100}$$

$$\tau_0 = wRS$$

$$= w.\frac{A}{P}.S$$

$$= w.\frac{bd}{b+2d}.S$$

$$= 9.8\times\frac{2\times1.5}{2+2\times1.5}\times\frac{0.15}{100}$$

$$= 9.8\times\frac{3}{5}\times\frac{0.15}{100}$$

$$= 0.0088\text{kPa}$$

Example 5.2 What is the expected flow in rectangular lined channel of 2.5m wide and flowing 1.5m deep with bed slope 1 in 5000 calculated by using Manning's equation? What is the expected value of Chezy's C and Darcy –Weisbach's f for this flow? Assume Manning's n=0.01.

Solution:

b = 2.5m

d = 1.5m

S = 1in 5000 = 0.0002

Meaning's n = 0.01

∴ Area, A = 2.5 x 1.5 = 3.75m²

Perimeter, P = 2.5 + 2 x 1.5 = 5.5m

Hydraulic radius, $R = \frac{A}{P} = \frac{3.75}{5.5} = 0.68$

Discharge, Q = AV

$$= A\times\frac{1}{n}R^{2/3}S^{1/2}, \text{(Using Manning's equation)}$$

$$3.75\times\frac{1}{0.01}\times(0.68)^{2/3}\times(0.0002)^{1/2}$$

= 3.75 x 100 x 0.010954 = 4.11m³/*s*

Discharge, $Q = AV$

$$= AC\sqrt{RS}, \text{(Using Chezy's equation)}$$

$$Or,\ 4.11 = 3.75C\sqrt{0.68\times0.0002}$$

$$\therefore C = \frac{4.11}{0.04373} = 93.98$$

In Darcy-Weisbach, $V = \sqrt{\frac{8gRS}{f}} = \sqrt{\frac{8x9.81x0.68x1/5000}{f}}$

$\therefore f = 8.98 \text{ x } 10^{-3}$

Example 5.3 A 200m long horizontal pipe carries a discharge of 50l/s. The center line of the pipe is 5m above the datum. The diameter of the pipe tapers from 200mm to 100mm. Using g=9.81m/s^2 and neglecting losses in the pipe, if the pressure at the larger end of the pipe is 100kPa, the pressure at the other end of the pipe in kPa is – (GATE , 2014)

Solution:

Discharge, Q = 50l/s = 0.05m^3/s

Diameter of pipe, D_1= 200mm = 0.2m & D_2=100mm=0.1m

1m head = 1000x9.81 = 9810N/m^2, 100kPa = 100000N/m^2

Pressure head at larger end = 100000/9810 = 10.19m

$Q = A_1 V_1 = A_2 V_2$

$$V_1 = \frac{Q}{A_1} = \frac{0.05}{\pi\frac{0.2^2}{4}} = \frac{0.05}{0.03} = 1.59 m/s$$

$$V_2 = \frac{Q}{A_2} = \frac{0.05}{\pi\frac{0.1^2}{4}} = \frac{0.05}{6.57x10^{-3}} = 6.37 m/s$$

$$\text{We have, } P_1 + \frac{v_1^2}{2g} = P_2 + \frac{v_2^2}{2g}$$

$$\text{Or, } P_1 - P_2 = \frac{v_2^2}{2g} - \frac{v_1^2}{2g} = \frac{(6.37)^2}{2g} - \frac{(1.59)^2}{2g} = 2.07 - 0.13 = 1.94$$

$$P_2 = P_1 - 1.94 = 10.19 - 1.94 = 8.25m$$

$$\text{Pressure head of 8.25m} = \frac{8.25x100}{10.19} = 80.96kP_a$$

Example 5.4 A trapezoidal canal, having a bottom width of 5.0m and a slope of 1H:1V, is carrying a discharge of 20m^3/s. The critical depth, in m, is

a) 1.09 b) 1.18

c) 2.06 d) 2.62

Solution:

Bottom width, b = 5.0m

Discharge, Q = $20m^2/s$

Side slope = 1H: V, or, z = 1

In critical depth of trapezoidal channel, $\frac{Q^2}{g} = \frac{A^3}{T}$

Where, A=Cross-sectional area of channel=bd+zd^2=5d+d^2

d= depth of water in the channel

T=top width of the channel= b+2zd=5+2d

$$\therefore \frac{Q^2}{g} = \frac{(5d+d^2)^3}{5+2d}$$

$$LHS\ of\ the\ equation = \frac{20^2}{9.18} = 40.77$$

Let, d=1.09m

RHS of the

$$\text{equation} = \frac{(5d+d^2)3}{5+2d} = \frac{(5x1.09+1.09^2)^3}{5+2x1.09} = \frac{(5.45+1.19)^3}{5+2.18}$$

$$= \frac{(6.64)^3}{7.18} = \frac{292.75}{7.18} = 40.77$$

Thus LHS=RHS

$$\therefore \text{d} = 1.09\text{m}$$

Example 5.5 What is the size of the concrete pipe if it flows 0.8 full to carry a discharge of $5m^3/s$. The pipe is laid on a slope of 0.0002. Assume Manning's n as 0.013.

Solution:

Q = $5m^3/s$

S = 0.0002

n = 0.013

$$\cos\theta = \frac{0.3d}{0.5d} \therefore \theta = 53.13^\circ$$

$$Area\ AOCE = \pi\frac{d^2}{4} \times \frac{2\times 53.13^\circ}{360^\circ}$$

$$= 0.23d^2$$

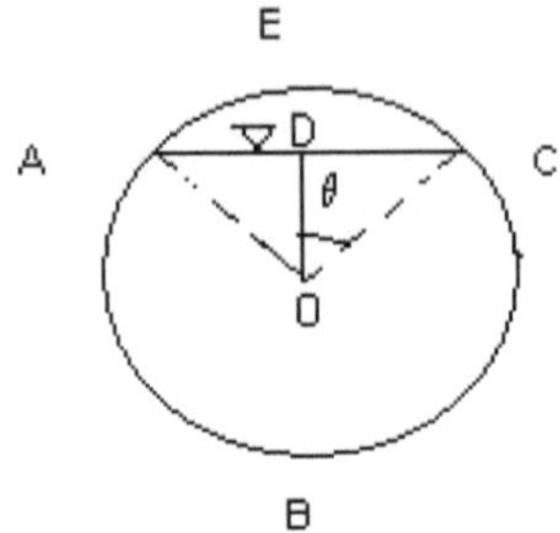

Referring to diagram., and assuming d the diameter of pipe

$$\text{Length of } ABC = \pi d - \frac{\pi d \times 2 \times 53.13^\circ}{360^\circ}$$
$$= 0.705\pi d$$
$$= 2.21d$$

$$\text{Area of triangle } AOC = \frac{1}{2} OD \times AC$$

$$= \frac{1}{2} \times 0.3d \times 0.3d \times \tan 53.13^\circ \times 2$$
$$= 0.12d^2$$

$$\therefore R = \frac{A}{P} = \frac{\pi d^2/4 - area\ ACE}{2.21d}$$

$$= \frac{\pi d^2/4 - (area\ AOCEA \text{ - } area\ AOCA)}{2.21d}$$

$$\frac{\pi d^2/4 - (0.23\text{d}^2 - 0.12\text{d}^2)}{2.21d}$$

$$= \frac{\pi d^2/4 - 0.12d^2}{2.21d}$$

Using Manning's equation,

$$Q = A \times \frac{1}{n} R^{2/3} \text{S}^{1/2}$$

$$\text{or, } 5 = 0.675d^2 \times \frac{1}{0.013} (0.3d)^{2/3} (0.0002)^{1/2}$$

$$\therefore d^{8/3} = \frac{5 \times 0.013}{0.675(0.3)^{2/3} (0.0002)^{1/2}}$$

$$= \frac{0.065}{0.00427} = 15.22$$

$$\therefore \text{d} = 2.78m$$

Example 5.6 Design the most efficient trapezoidal section for a fairly vegetated irrigation channel in medium soil to carry a discharge of $10m^3/s$ on a bed slope of 0.15%. Assume Manning's n as 0.04.

Solution:

$$Q = 10m^3/s, z=1.5\,(\text{Table}\,5.1)\,\text{or}\,\theta=33.69^0, n = 0.04, V_{max} = 1.2m/s\,(\text{Table}\,5.2)$$

$$Let\, d=2.2m \therefore b=2d\tan\theta/2=2x2.2x0.3=1.3m$$

$$A = bd + zd^2 = 1.33x2.2 + 1.5x(2.2)^2 = 10.19m^2$$

Manning' s equation,

$$V = \frac{1}{n}R^{2/3}S^{1/2} = \frac{1}{0.04}\left(\frac{10.19}{9.26}\right)^{\frac{2}{3}}(0.0015)^{1/2} = 1.03m/s$$

$$Q = AV = 10.19x1.03 = 10.49m^3/s$$

The discharge $10.49m^3/s$ is slightly higher than the required discharge ($10m^3/s$). Thus, the assumption for d = 2.2m is OK.

Example 5.7 Determine the most efficient cross-section in an open channel to carry a discharge of $1.5m^3/s$ for (a) a semicircular section (half full), (b) a rectangular section, (c) a triangular section (90^0) and (d) a trapezoidal section. Assume the channel bed slope of 0.005 and n=0.02.

Solution:

$$V = \frac{1}{n}R^{2/3}S^{1/2}$$

$$or, \frac{Q}{A} = \frac{1}{n}\left(A/P\right)^{2/3}S^{1/2}$$

$$or, \frac{A^{5/3}}{P^{2/3}} = QnS^{-1/2}$$

$$= (1.5)(0.02)(0.005)^{-1/2}$$

$$= 0.424$$

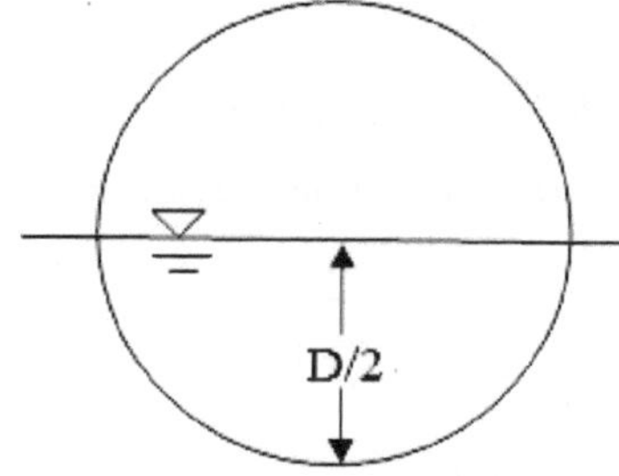

a) For a semicircular section (half full)

$$A = \frac{\pi D^2}{4\times 2}, and\, P = \frac{\pi D}{2}$$

$$\therefore \frac{\left(\pi D^2/8\right)^{5/3}}{\left(\pi D/2\right)^{2/3}} = 0.424$$

$$or, \frac{0.21D^{10/3}}{1.35D^{2/3}} = 0.424$$

$$or, 0.156D^{5/3} = 0.424$$

$$\therefore D = (2.725)^{3/8} = 1.45m$$

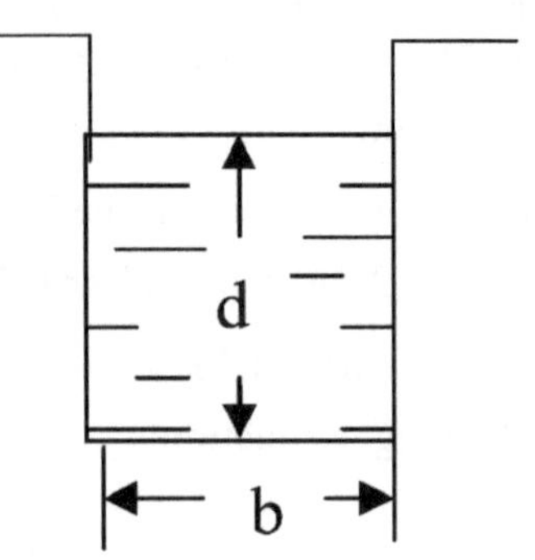

b) For a rectangular section

$b = 2d$

$A = 2d^2$

$p = d + 2d + d = 4d$

$$\frac{A^{5/3}}{P^{2/3}} = 0.424$$

$$\text{or,} \frac{(2d^2)^{5/3}}{(4d)^{2/3}} = 0.424$$

$$\therefore \frac{3.18}{2.53} d^{5/3} = 0.424$$

$$\therefore d^{8/3} = 0.338,$$

$$or, d = 0.665m$$

$$\therefore b = 0.665 \times 2 = 1.33m$$

c) For a triangular section (90^0)

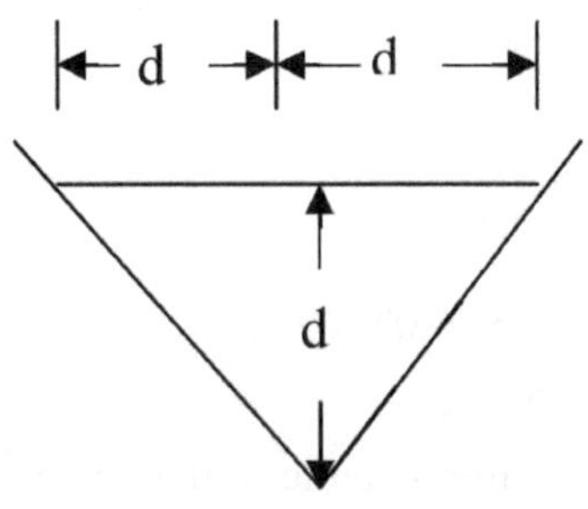

$$P = \sqrt{d^2 + d^2}.2$$

$$= d2\sqrt{2}$$

$$= 2.83d$$

$$A = \frac{1}{2}.d.d.2 = d^2$$

$$\frac{A^{5/3}}{P^{2/3}} = 0.424$$

$$\therefore \frac{(d^2)^{5/3}}{(2.83d)^{2/3}} = 0.424$$

$$or, 0.498d^{8/3} = 0.424$$

$$\therefore d^{8/3} = 0.85$$

$$or, d = 0.94m$$

d) For a trapezoidal channel

Assuming the angle θ = 60° as shown in Fig. 5.1

$$b = 2d \tan \theta/2 = 1.155.d$$

$$P = b + 2d\sqrt{z^2 + 1}$$

$$= 1.155d + 2.309d$$

$$= 3.464d$$

$$A = bd + zd^2$$

$$= 1.155d^2 + 0.577d^2$$

$$= 1.732d^2$$

$$\frac{A^{5/3}}{P^{2/3}} = 0.424$$

$$\text{or,} \frac{(1.732d^2)^{5/3}}{(3.464d)^{2/3}} = 0.424$$

$$\text{or,} 1.09d^{8/3} = 0.389$$

$$\therefore d = 0.70m$$

$$\therefore b = 1.155d = 0.81m$$

Example 5.8 Determine the discharge of a channel of bed width=2.0m, depth of water =1.5m, side slope=1.5:1 and longitudinal slope=1 in 1000. Assume Manning's n = 0.04.

Solution:

Bed width, b=2.0m

Depth of water, d=1.5m

Side slope, s=z: 1=1.5:1

Bed slope, S=1/1000

Area of cross-section, $A = bd + zd^2$

$= 2.0 \times 1.5 + 1.5 \times 1.5^2$

$= 3.0 + 3.375$

$= 6.375 m^2$

Wetted perimeter, $P = b + 2d\sqrt{z^2} + 1$

$= 2.0 + 2 \times 1.5\sqrt{z^2 + 1}$

$= 2.0 + 5.408$

$= 7.408 m$

Hydraulic radius, $R = \frac{A}{P} = \frac{6.375}{7.408} 0.86 m$

Hydraulic slope, $S = \frac{1}{1000} = 0.001$

Using Manning's equation,

Mean velocity, $V = \frac{1}{n} R^{2/3} S^{1/2}$

$= \frac{1}{0.04} (0.86)^{2/3} (0.001)^{1/2}$

$= 22.61 \times 0.032$

$= 0.715 m/s$

Discharge, $Q = AV$

$= 6.375 \times 0.715$

$= 4.56 m^3/s$

Example 5.9 Compute the most efficient bottom width of an open channel to carry a flow 2.2m deep in silty loam soil. What is the discharge of channel if hydraulic gradient (slope) is 0.05%? Assume Manning's n as 0.04.

Solution:

d = 2.2m

S = 0.05/100

Soil = silty loam

Referring to Table 5.1, for silty loam soil, side slope of the channel = 1.5:1

$\therefore z = 1.5$

$\therefore \cot\theta = 1.5, \tan\theta = \frac{1}{1.5} 0.667, and\, \theta = 33.69^0$

For most economic section,

$$b = 2d\tan\frac{\theta}{2}$$

$$= 2\times2.2\times\tan\left(\frac{33.69}{2}\right)^0$$

$$= 1.33m$$

Average cross-sectional area, $A = bd + zd^2$

$$= 1.33\times2.2 + 1.5\times2.2^2$$

$$= 10.186m^2$$

Wetted perimeter, $p = b + 2d\sqrt{z+1}$

$$= 1.33 + 2\times2.2\sqrt{1.5^2+1}$$

$$= 1.33 + 4.4\times1.8$$

$$= 9.26m$$

Hydraulic radius, $R = \frac{A}{P} = \frac{10.186}{9.26} = 1.1m$

Velocity, $V = \frac{1}{n}R^{2/3}S^{1/2}$

$$= \frac{1}{0.04}(1.1)^{2/3}(0.0005)^{2/3}, (\text{n} = 0.04\,\text{from Table}\,5.3)$$

The velocity is within permissible velocity of 60cm/s.

Discharge, Q = AV

$$= 10.186x0.596$$

$$= 6.07m^3/s$$

Example 5.10 Calculate the dimension of most economic rectangular and trapezoidal section of a lined channel at bed slope 1 in 1000 carrying 130l/s. The maximum permissible velocity in channel is 1.5m/s. Assume side slope of trapezoidal section as 1:1 and Manning's n as 0.012.

Solution:

Discharge, Q = 130l/s = 0.13m^3/s

Side slope, s = 1:1, i.e., z = 1

Hydraulic slope, S=1 in 1000 = 0.001

The dimension of the most economic section for rectangular or trapezoidal channel can be determined by assuming a depth of channel and calculating the width which satisfies the requirement of discharge and same time the velocity remains within the permissible limit.

For rectangular channel

Let d = 0.2m

For most economic section in rectangular channel

$$b = 2d$$

$$\therefore\ b = 0.3x2 = 0.6$$

Wetted perimeter, $p = b + 2d = 0.6 + 2x0.3 = 1.2m$

$A = bxd$

Wetted area, $A = 0.3x0.6 = 0.18m^2$

$$Mean\ velocity, V = \frac{1}{n} R^{2/3} S^{1/2}$$

$$= \frac{1}{0.012}\left(\frac{0.18}{1.2}\right)^{2/3}(0.001)^{1/2}$$

$$= \frac{1}{0.012} x0.282x0.0316$$

$$= 0.74m/\sec$$

$$Discharge, Q = AxV$$

$$= 0.18x0.74$$

$$= 0.134m^3/s$$

Therefore, the section of d = 0.3m and b = 0.6m is suitable for carrying the discharge of 0.13m^3/s.

For trapezoidal section

Trial 1 Let d=35cm

$$b = 2d\tan\frac{\theta}{2}, \text{since}\ \theta = 45^0\ \text{for side slope}\ 1:1$$

$$b = 2x35x0.4142 = 28.99cm = 0.29m$$

$$A = bd + zd^2$$

$$= 0.29x0.4 + 0.2^2 = 0.156m^2$$

$$P = b + 2d\sqrt{z^2 + 1}$$

$$=0.29+2x0.35\sqrt{1^2+1}$$

$$=0.29+0.7\sqrt{2}=1.28$$

$$R=\frac{A}{P}=\frac{0.156}{1.28}=0.12$$

$$V=\frac{1}{n}R^{2/3}S^{1/2}$$

$$=\frac{1}{0.012}(0.12)^{2/3}(0.001)^{1/2}=0.64\,m/s$$

$$Discharge, Q=AxV=0.156x0.64=0.998m^3/s=99.84l/s$$

The discharge 99.84l/s is less to desired discharge 130l/s.

Trial 2 Let d=40 cm

$$b=2d\tan\frac{\theta}{2}$$

$$=2x0.4x0.4142=0.33m$$

$$A=bd+zd^2$$

$$=0.33x0.40+0.2^2=0.17\text{m}^2$$

$$P=b+2d\sqrt{z^2+1}$$

$$=0.33+2x0.4\sqrt{1^2+1}$$

$$=0.33+0.8\sqrt{2}=1.46\text{m}$$

$$R=\frac{A}{P}=\frac{0.17}{1.42}=0.12$$

$$V=\frac{1}{n}R^{2/3}S^{1/2}$$

$$=\frac{1}{0.012}(0.12)^{2/3}(0.001)^{1/2}=0.64\,m/s$$

$$Discharge, Q=AxV=0.17x0.64=0.1088m^3/s=108.80l/s$$

The discharge 108.80l/s is less to desired discharge 130l/s.

Trial 3 Let d = 45 cm

$$b=2d\tan\frac{\theta}{2}$$

$b = 2x45x0.4142 = 37.28cm = 0.37m$

$A = bd + zd^2$

$= 0.37x0.45 + 0.2^2 = 0.21m^2$

$P = b + 2d\sqrt{z^2 + 1}$

$= 0.37 + 2x0.45\sqrt{1^2 + 1}$

$= 0.37 + 0.9\sqrt{2} = 1.64m$

$$R = \frac{A}{P} = \frac{0.21}{1.64} = 0.128$$

$$V = \frac{1}{n} R^{2/3} S^{1/2}$$

$$= \frac{1}{0.012}(0.128)^{2/3}(0.00.1)^{1/2} = 0.67m/s$$

$Discharge, Q = AxV = 0.21x0.67 = 0.1407m^3/s = 140.7l/s$

The discharge 140.07l/s is close to desired discharge 130l/s and acceptable.

Example 5.11 Design an earthen channel with the following data:

Discharge = 1.3m^3/s

Soil = silty loam

Land slope = 0.15%

Manning's n = 0.04

Solution: The soil is silty loam. Therefore, the side slope of the channel may be taken as 1.5:1 and the limiting velocity of flow 0.6m/s. In designing the earthen channel economic section to be tried first.

Trial 1. Let d = 1.2m,

$\therefore b = 2d \tan \theta/2$

$= 2x1.2 \tan(33.69/2)^0 = 2.2x0.303$

$= 0.73m/s$

$A = bd + zd^2 = 0.73x1.2 + 1.5x(1.2)^2$

$= 3.036m^2$

$P = b + 2d\sqrt{z^2 + 1} = 0.73 + 2x1.2\sqrt{1.5^2 + 1}$

$= 5.05m$

$R = A/P = 3.036/5.05 = 0.6m$

Manning's equation, $V = \frac{1}{n} R^{\frac{2}{3}} S^{\frac{1}{2}}$

$= \frac{1}{0.04} (0.6)^{\frac{2}{3}} (0.0015)^{\frac{1}{2}}$

$= 0.68 m/s$

$Q = AV = 3.036x0.68 = 2.09 m^3/s$

The discharge and the velocity both exceed the limit.

Trial 2 Let d = 1.0m,

$\therefore$ b $= 2d \tan \theta/2$

$= 2x1.0 \tan(33.69/2)^0 = 2x0.303$

$= 0.6 \text{m/s}$

$A = bd + zd^2 = 0.6x1.0 + 1.5x(1.0)^2$

$= 2.1 m^2$

$P = b + 2d\sqrt{z^2+1} = 0.6 + 2 \times 1.0 \sqrt{1.5^2+1}$

$= 4.2 m$

$R = A/P = 2.1/4.2 = 0.5 m$

Manning's equation, $V = \frac{1}{n} R^{\frac{2}{3}} S^{\frac{1}{2}}$

$= \frac{1}{0.04} (0.5)^{\frac{2}{3}} (0.0015)^{\frac{1}{2}}$

$= 0.61 m/s$

$\text{Q} = \text{AV} = 2.1 \times 0.61 = 1.28 m^3/s$

The velocity exceeds the limit (0.6m/s) and discharge is less than 1.3m³/s.

Trial 3 Let d = 0.9m,

$\therefore$ b $= 2d \tan \theta/2$

$= 2 \times 0.9 \tan(33.69/2)^0 = 1.8 \times 0.303$

$= 0.54 \text{m/s}$

$A = bd + zd^2 = 0.54 \times 0.9 + 1.5 \times (0.9)^2$

$= 1.701 m^2$

$P = b + 2d\sqrt{z^2+1} = 0.54 + 2\times 0.90\sqrt{1.5^2+1}$

$= 3.78m$

$R = A/P = 1.701/3.78 = 0.45m$

Manning's equation, $V = \frac{1}{n}R^{\frac{2}{3}}S^{\frac{1}{2}}$

$= \frac{1}{0.04}(0.45)^{\frac{2}{3}}(0.0015)^{\frac{1}{2}}$

$= 0.57m/s$

$Q = AV = 1.701x0.57 = 0.97m^{3/}s$

It appears from the above trials that if d is taken less than 1.0m, the discharge through the channel is less than 1.25m^3/s and if d is taken more than 1.0m; the velocity of flow exceeds the limit (0.6m/s). Therefore, the economical section of the channel is not possible. It may be done through trial and error.

Trial 1 Let b = 2m, d = 0.75m,

$A = bd + zd^2 = 2.0\times 0.75 + 1.5\times(0.75)^2$

$= 1.5{+}0.84 = 2.34m^2$

$P = b + 2d\sqrt{z^2+1} = 2 + 2\times 0..75\sqrt{1.5^2+1} = 4.7m$

$R = A/P = 2.34/4.7 = 0.49m$

$V = \frac{1}{n}R^{\frac{2}{3}}S^{\frac{1}{2}}$

$= \frac{1}{0.04}(0.49)^{\frac{2}{3}}(0.0015)^{\frac{1}{2}}$

$= 0.6m/s$

$Q = AV = 2.34\times 0.6 = 1.40m^3/s$

Trial 2 Let b = 3m, d = 0.6m,

$A = bd + zd^2 = 3.0\times 0.6 + 1.5\times(0.6)^2$

$= 1.8 + 0.54 = 2.34m^2$

$= P = b + 2d\sqrt{z^2+1} = 3 + 2\times 0.6\sqrt{1.5^2+1} = 5.16m$

$R = A/P = 2.34/5.14 = 0.45m$

$V = \frac{1}{n}R^{\frac{2}{3}}S^{\frac{1}{2}}$

$$=\frac{1}{0.04}(0.45)^{\frac{2}{3}}(0.0015)^{\frac{1}{2}}$$

$$=0.57 m/s$$

$$Q = AV = 2.34 \times 0.57 = 1.33 m^3/s$$

The velocity is within the permissible limit and the discharge is satisfactory. Hence, the assumptions for b = 3m and d = 0.6m are acceptable.

5.3 Regime Approach

1. Kennedy's formula (1895)

Kennedy(1895) a Punjab engineer did pioneering research for a stable non-silting, non-scouring irrigation canal system. He worked on some selected reaches of Upper Bari Doab canal system in Punjab (Pakistan) and concluded as under.

1. The flowing water has to counteract friction against the bed of the channel resulting in generation of vertical eddies rising up gently to the surface. These eddies are responsible to keep the most of the silt in suspension. Some eddies also develop from the sides of the channel which are far most of its horizontal and do not have any silt supporting power. Therefore, the silt supporting power is proportional to the bed width of the channel and not to its wetted perimeter.
2. The velocity that is sufficient to generate these eddies to keep the sediments in suspension and thereby avoiding the silting up is designated as critical velocity. Thus, the critical velocity is non-silting-non-scouring. The silt power of a channel is proportional to $V_0^{5/2}$. Safe velocity against erosion of channel in Punjab is 1m/sec corresponding to a depth of not more than 3m/sec.
3. The coefficient of roughness n for all channels is 0.0225. However, later he found that this value is not constant and suggested value of 0.02 for large canals and 0.025 for small channels.

Kennedy on his observation on Upper Bari Doab Canal system gave the classic empirical equation as below.

$$V_0 = 0.55 D^{0.64} \tag{5.28}$$

In general $V_0 = CD^n$ (5.29)

Where, V_0 = critical (non-silting, non-scouring) velocity (m/s). Fine silt has lower critical velocity than coarse silt.

D = full supply depth of water over bed portion of a channel, m.

n = any index number.

Kennedy studied only the Upper Bari Doab Canal system, which were flowing in sandy silts. Therefore, the equation was derived by him on the basis of observation on those canals and is applicable only to similar condition. Kennedy later realized the importance of silt grade on critical velocity ratio (CVR) and modified the equation as below.

$$V = 0.55mD^{0.64} \quad (5.30)$$

Limitation of Kennedy's theory

1. Kennedy's theory does not provide the clue for the channels that best suits for a particular discharge. The importance of relation between bed width and depth is ignored. Various state departments developed the B/D ratio on the basis of their experience.
2. Kennedy used the Kutter's equation for finding the mean velocity of flow in the channel. Therefore, the limitations of Kutter's equation are incorporated in design process of Kennedy. The arbitrary assumed value of 0.0225 of n is not correct.
3. Silt concentration and bed width are not considered.
4. Silt, grade and silt charge are not defined.
5. Design of channels involves trials and error to find out the velocity and depth for the required discharge.
6. No method has been suggested to measure the value of m or CVR for the channels of different silt grade.

Design procedure following Kennedy's method

Case 1. Given Q, N, m and S.

1. Assume a trial value of D in meters. Calculate the velocity V by using the equation

$$V = 0.55D^{0.64}$$

2. Calculate the cross section A from $A = \frac{Q}{V}$.
3. Calculate the bed width B by using known value of D and A. The side slope of the channel is assumed to be ½:1 when the channel has run for some time. Therefore,

$$A = BD + ZD^2$$

$$= BD + \frac{D^2}{2}$$

4. Calculate the perimeter P and hydraulic radius R by the following equations

$$P = B + 2\sqrt{\left(D/2\right)^2 + D^2}$$

$$= B + 2D\sqrt{5/4}$$

$$= B + D\sqrt{5}$$

$$R = A/P$$

$$= \frac{BD + \frac{D^2}{2}}{B + D\sqrt{5}}$$

5. Calculate the actual mean velocity following Kutter's equation. If the velocity is same as found out in step 2, the assumed depth is correct. Otherwise, change the value of D till the two velocities are same.

Example 5.12 Design the channel with the following data by using Kennedy's formula.

Discharge = 10 cumecs
Slope = 0.0002
N = 0.0225
m = 1

Solution:

Trial 1 Assuming D = 1.2m

$$V = 0.55mD^{0.64}$$

$$= 0.55 \times 1 \times 1.2^{0.64}$$

$$= 0.62 m/s$$

$$A = \frac{Q}{V} = \frac{10}{0.62} = 16.13m^2$$

The side slope of a canal in alluvial soil is assumed as $\frac{1}{2}:1$ if it runs for some time.

$$Again, A = BD + \frac{D^2}{2}$$

$$or, 16.13 = B \times 1.2 + \frac{1.2^2}{2}$$

$$= 1.2B + 0.72$$

$$\therefore B = \frac{16.13 - 0.72}{1.2} = 12.84m$$

$$P = B + D\sqrt{5}$$

$=12.84+1.2\sqrt{5}$

$=12.84+2.68$

$=15.52m$

$R=\dfrac{A}{P}$

$=\dfrac{16.13}{15.52}=1.04m$

$$\text{Kutter's } C=\frac{\frac{1}{N}+\left(23+\frac{0.00155}{S}\right)}{1+\left(23+\frac{0.00155}{S}\right)\frac{N}{\sqrt{R}}}$$

$$=\frac{\frac{1}{N}+\left(23+\frac{0.00155}{0.0002}\right)}{1+\left(23+\frac{0.00155}{0.0002}\right)\frac{0.0225}{\sqrt{1.04}}}$$

$$=\frac{\frac{1}{0.0225}}{1+(30.75)\frac{0.0225}{1.02}}+30.75=44.8$$

$V=C\sqrt{RS}$

$=44.8\sqrt{1.04\times 0.0002}$

$=44.8\times 0,0144$

$=0.65m/s$

$CVR=\dfrac{V_0}{V}=\dfrac{0.62}{0.65}=0.95$

Trial 2

$D=1.25m$

$V_0=0.55mD^{0.64}$

$=0.634m/s$

$A=\dfrac{Q}{V}=\dfrac{10}{0.634}=15.77m^2$

$$A = BD + \frac{D^2}{2}$$

or, $15.77 = B \times 1.25 + \frac{1.25^2}{2}$

$$\therefore\ B = \frac{15.77 - 0.78}{1.25} = 11.99$$

$$P = B + D\sqrt{5}$$

$$= 11.99 + 1.25\sqrt{5}$$

$$= 14.79m$$

$$\text{Kutter,s}\ C = \frac{\frac{1}{N} + \left(23 + \frac{0.00155}{S}\right)}{1 + \left(23 + \frac{0.00155}{S}\right)\frac{N}{\sqrt{R}}}$$

$$= \frac{\frac{1}{N} + \left(23 + \frac{0.00155}{0.0002}\right)}{1 + \left(23 + \frac{0.00155}{0.0002}\right)\frac{0.0225}{\sqrt{1.066}}}$$

$$= \frac{\frac{1}{0.0225} + 30.75}{1 + (30.75)\frac{0.0225}{\sqrt{1.066}}} = \frac{75.19}{1.67} = 45.02$$

$$\therefore V = C\sqrt{RS}$$

$$= 45.02\sqrt{1.066 \times 0.0002}$$

$$= 0.657m/\text{s}$$

$$CVR = \frac{V_0}{V} = \frac{0.634}{0.657} = 0.965$$

Trial 3

$$D = 1.35m$$

$$V_0 = 0.55mD^{0.64}$$

$$= 0.67m/sec$$

$$A = \frac{Q}{V_0} = \frac{10}{0.67} = 14.93m^2$$

$$A = BD + \frac{D^2}{2}$$

$$or,\ 14.93 = B \times 1.35 + \frac{1.35^2}{2}$$

$$\therefore\ B = \frac{14.93 = 0.911}{1.35} = 10.38m$$

$$P = B + D\sqrt{5}$$

$$= 13.40m$$

$$R = \frac{A}{P} = \frac{14.93}{13.40} = 1.114m$$

$$\text{Kutter}'C = \frac{75.19}{1 + (30.75)\frac{0.0225}{\sqrt{1.14}}} = \frac{75.19}{1.656} = 45.42$$

$$\therefore\ V = C\sqrt{RS}$$

$$= 45.42\sqrt{1.114 \times 0.0002}$$

$$= 0.68m/\text{s}$$

$$CVR = \frac{V_0}{\text{V}} = \frac{0.67}{0.68} = 0.985$$

Trial 4

$$D = 1.40m$$

$$V_0 = 0.55mD^{0.64}$$

$$= 0.68m/\text{s}$$

$$A = \frac{Q}{V_0} = \frac{10}{0.68} = 14.71m^2$$

$$A = BD + \frac{D^2}{2}$$

$$or,\ 14.71 = B \times 1.40 + \frac{1.40^2}{2}$$

$$\therefore\ B = \frac{14.71 - 0.98}{1.40} = 9.81m$$

$$P = B + D\sqrt{5}$$

$=9.81+1.40\sqrt{5}$

$=12.94m$

$R=\frac{A}{P}=\frac{14.71}{12.94}=1.137m$

$\text{Kutter'}C=\frac{75.19}{1+(30.75)\sqrt{\frac{0.0225}{1.137}}}=\frac{75.19}{1.65}=45.6$

$\therefore\ V=C\sqrt{RS}$

$=45.6\sqrt{1.137\times0.0002}$

$=0.688m/sec$

$CVR=\frac{V_0}{V}=\frac{0.68}{0.688}=0.99$

Studying the above trials it appears that the CVR values approaches to 1.0 with the increase of the value of D from 1.2 to 1.4m though all the CVR values of the trials are within the allowable limit of 0.9-1.1.Therefore, the value of 1.4m for D in Trial 4 will be better selection.

Example 5.13 Design an irrigation channel in clayey alluvial soil to carry a discharge of 50 cumecs. Assume N= 0.0225, m =1, channel side slope =1:1 and a bed slope of 0.15m per kilometer.

Solution:

Trial 1

$\text{Assuming D} = 2,0\text{m}$

$V=0.55mD^{0.64}$

$=0.55\times1\times1.2^{0.64}$

$=0.86m/s$

$A=\frac{Q}{V}=\frac{50}{0.86}=58.14m^2$

$Again, A=BD+D^2$

$\text{or,}\ \ 58.14=B\times2+2^2$

$=1.2B+0.72$

$\therefore\ B=\frac{58.14-4}{2}=\frac{54.14}{2}=27.07m$

$P=B+2.828D$

$= 27.07 + 5.656$

$= 32.635m$

$$R = \frac{A}{P}$$

$$= \frac{58.14}{32.635} = 1.78m$$

$$\text{Kutter's } C = \frac{\frac{1}{N} + \left(23 + \frac{0.00155}{S}\right)}{1 + \left(23 + \frac{0.00155}{S}\right)\frac{N}{\sqrt{R}}} = \frac{\frac{1}{N} + \left(23 + \frac{0.00155}{0.00015}\right)}{1 + \left(23 + \frac{0.00155}{0.00015}\right)\frac{0.0225}{\sqrt{1.78}}}$$

$$= \frac{\frac{1}{N} + \left(23 + \frac{0.00155}{0.00015}\right)}{1 + (33.33)\frac{0.0225}{1.334}} = \frac{77.77}{1.562} = 49.78$$

$V = C\sqrt{RS}$

$= 49.78\sqrt{1.78 \times 0.00015}$

$= 49.78 \times 0.0144$

$= 0.717 m/s$

Trial 2

Assuming $\text{D} = 1.8m$

$V = 0.55mD^{0.64}$

$= 0.55 \times 1 \times 1.8^{0.64}$

$= 0.80 m/s$

$$A = \frac{Q}{V} = \frac{50}{0.80} = 62.5m^2$$

Again, $A = BD + D^2$

or, $62.5 = B \times 1.8 + 1.8^2$

$$\therefore\ B = \frac{62.5 - 1.8^2}{1.8} = 32.92m$$

$P = B + 2.828D$

$= 32.92 + 5.09$

$= 38.01m$

$$R = \frac{A}{P}$$

$$= \frac{62.5}{38.01} = 1.64m$$

$$\text{Kutter's}\, C = \frac{\frac{1}{N} + \left(23 + \frac{0.00155}{S}\right)}{1 + \left(23 + \frac{0.00155}{S}\right)\frac{N}{\sqrt{R}}}$$

$$= \frac{\frac{1}{N} + \left(23 + \frac{0.00155}{0.00015}\right)}{1 + \left(23 + \frac{0.00155}{0.00015}\right)\frac{0.0225}{\sqrt{1.64}}}$$

$$= \frac{\frac{1}{0.0225} + 33.33}{1 + (33.33)\frac{0.0225}{1.334}} = \frac{77.77}{1.56} = 49.85$$

$$V = C\sqrt{RS}$$

$$= 49.85\sqrt{1.64 \times 0.00015}$$

$$= 49.85 \times 0.0157$$

$$= 0.78m/s$$

$$CVR = \frac{V_0}{V} = \frac{0.80}{0.78} = 1.02$$

The CVR is very close to 1.0, hence OK.

Case II Given Q, N, m and B/D ratio from Wood's table.

1. Calculate A in terms of D

$$B/D = x$$

$$\therefore\ B = Dx$$

$$\text{Let} \therefore\ A = BD + \frac{D^2}{2} = xD^2 + \frac{D^2}{2}$$

$$= D^2\left(x + \frac{1}{2}\right)$$

2. Calculate the velocity V in terms of D by using Kennedy's equation

$$V = 0.55mD^{0.64}$$

Calculate D using the continuity equation of flow. Thus,

$$Q = AV = D^2(x+\frac{1}{2})0.55mD^{0.64}$$

$$\therefore \text{ D}\left[\frac{Q}{0.55m\left(x+\frac{1}{2}\right)}\right]^{\frac{1}{2.64}}$$

3. Calculate B and R using the known value of D

$$\text{B} = xD$$

$$and\ R = \frac{BD+\frac{D^2}{2}}{B+D\sqrt{5}}$$

4. Calculate V by using Kennedy's equation

$$V = 0.55mD^{0.64}$$

5. Calculate the slope S from known values of V and R and using Kutter's equation.

Table 5.5 Wood's normal design table (N=0.0225)

Discharge (cumecs)	B/D ratio	B (m)	D (m)	Slope 1 in...	V/V_0	Mean V
0.283	2.9	1.45	0.49	3333	0.92	0.344
0.708	3.4	2.21	0.66	3636	1.01	0.424
1.416	3.7	3.13	0.84	4000	1	0.476
2.832	4.2	4.42	1.04	4444	1	0.555
7.079	4.5	6.7	1.43	4444	1.01	0.7
14.158	5.7	9.75	1.72	5000	1	0.775
28.315	7.6	15.28	1.98	5000	1.03	0.88
56.63	11.3	25.41	2.26	5714	1.03	0.945
141.575	22.5	56.5	2.5	6666	0.98	0.975
283.15	41	105	2.59	6666	1.02	1.03
566.3	78	212	2.72	8000	0.98	0.975

Source: Punmia, 1992

Table 5.6 Bed width to depth ratio for different discharge in unlined channel

Q	B/D
0.28	2.9
0.56	3.1
0.85	3.25
1.41	3.45
1.98	3.82
28.32	7.00
283.17	16.5

Source: CWPC Manual on Irrigation, 1960

The B/D ratio for different Q and B for B/D ratio can also be obtained by using the equations as stated below. These equations have been developed by using the data in Table 11.5.

$$B/D = 0.1317Q + 3.6046 \tag{5.30}$$

$$B = 2.7895B/D - 6.6961 \tag{5.31}$$

Example 5.14 Design an irrigation channel to carry a discharge of 2.8 cumecs. Assume N=0.0225, m=1 and B/D=4.2.

Solution:

$$B/D = 4.2$$

$$B = 4.2D$$

$$\therefore\ A = BD + \frac{D^2}{2}$$

$$= 4.2D^2 + \frac{D^2}{2}$$

$$= D^2(4.2 + 0.5)$$

$$V = 0.55mD^{0.64}$$

$$= 0.55D^{0.64},\ m = 1$$

$$Q = AV$$

$$= 4.7\text{D}^2x0.55D^{0.64}$$

$$or,\ D = \left[\frac{Q}{2.585}\right]^{\frac{1}{2.64}}$$

$$= \left(\frac{2.8}{2.585}\right)^{\frac{1}{2.64}}$$

$$= (1.08)^{\frac{1}{2.64}}$$

$$= 1.03m$$

$$\therefore\ B = 4.2D$$

$$= 4.2x1.03 = 4.32m$$

$$R = \frac{BD + \frac{D^2}{2}}{B + D\sqrt{5}}$$

$$= \frac{4.33x1.03 + (1.03)^2}{4.32 + 1.03\sqrt{5}}$$

$$= \frac{4.46 + 0.53}{4.33 + 2.303}$$

$$= \frac{4.99}{6.63}$$

$$= 0.75m$$

$$V = 0.55mD^{0.64}$$

$$= 0.55 \times (1.03)^{0.64}$$

$$= 0.55 \times 1.02$$

$$= 0.56m/s$$

$$V = \frac{1}{N} + \left(23 + \frac{0.00155}{S}\right)\frac{N}{\sqrt{R}}\sqrt{RS}$$

$$= \frac{\frac{1}{N} + \left(23 + \frac{0.00155}{S}\right)}{1 + \left(23 + \frac{0.00155}{S}\right)\frac{0.0226}{\sqrt{0.75}}}\sqrt{0.75S}$$

$$= \frac{23 + 44.44 + \frac{0.00155}{S}}{1 + \left(23 + \frac{0.00155}{S}\right)0.026} \times 0.866S^{1/2}$$

$$0.56 = \frac{67.44 + \frac{0.00155}{S}}{1 + 0.598 + \frac{2.99 \times 10^{-5}}{S}} \times 0.866S^{1/2}$$

$$or, \frac{0.56}{0.866} = \left(1.598 + \frac{2.99 \times 10^{-5}}{S}\right) = 67.44S^{1/2} + 0.00155S^{-1/2}$$

$$or, 1.033 + \frac{1.93 \times 10^{-5}}{S} = 67.44S^{1/2} + 0.00155S^{-1/2}$$

Let $S = 1$ *in* 5000

LHS = 1.13, *RHS* = 0.95+0.081 = 1.03

$LHS \neq RHS$

Let S = 1*in* 4500

LHS = 1.12, *RHS* 0.005 + 0.077 = 1.08

LHS $\neq$ *RHS*

Let S = *1in* 4200

LHS = 1.114, *RHS* = 1.04 + 0.0745 = 1.114

LHS = *RHS*

$\therefore$ *S* = 1*in* 4200 *is the designed slope*

Silt carrying capacity of channel following Kennedy's theory

As per Kennedy's theory, the silts held in suspension depends on the vertical component of the eddy, which varies as the bed width B and some power of velocity of flow.

Let V_0 = critical velocity in the channel

Q_1 = quantity of silt transported by the channel

Q = discharge in the channel.

$\therefore Q_1 \propto BV_0^n$

$$or,\ Q_1 = c^1\, BV_0^n \qquad (5.32)$$

Where c^1 = some constant.

Let p = proportion of silt in water

$= \frac{Q_1}{Q}$

$\therefore Q_1 = pQ$

Considering a wide channel with vertical slopes

$Q = BDV_0 (approx)$

$$\therefore Q_1 = pBDV_0 \qquad (5.33)$$

From Eqs.5.32 and 5.33

$cBV_0^n = pBDV_0$

$V_0^{n-1} = \frac{p}{c^1} D$

$$or,\ V_0 = cD^{\frac{1}{n-1}} \qquad (5.34)$$

The equation is similar to Kennedy's equation

$$V_0 = c\, D^y = c\, D^{0.64} \qquad (5.35)$$

From Eqs.5.34 and 5.35

$$\frac{1}{n-1}=0.64$$

$$or, \frac{1}{n-1}=0.64$$

$$or, n=\frac{1+0.64}{0.64}=2.56=\frac{5}{2}(approx.)$$

Therefore, the silt supply in the channel

$$Q_1=cBV_0^{5/2} \tag{5.36}$$

Garret's Diagram (1913)

Garret (1913) in India presented hydraulic diagrams by using Kennedy's and Kutter's equation for non-silting channels of any discharge to over 335 cumecs, bed slopes 1/100 to 1/1000 and for the value of Kutter's n from 0.018 to 0.03.

Lindley's formula (1919)

Lindley presented the formula relating velocity and depth, velocity and bed width and bed width and depth as below:

$$V=0.567\ D^{0.57}$$

$$V=0.274\ B^{0.355}$$

$$B=7.86\ D^{1.67}$$

Lindley was the first to postulate the regime concept. He defined the regime channel as follows (Sharma & Sharma, 1993).

i) When an artificial canal is constructed in alluvium to carry silt water, it beds and banks would silt to scour until the depth, slope and width attain a state of balance, to which he designated as regime.
ii) Regime dimension depend on discharge, nature of bed and quantity of silt.
iii) For a channel to carry a certain discharge and sand charge will ultimately develop a cross section and a slope so that equilibrium is established between the force acting; it is presumed that the channel can freely develop its regime.

Limitations

i) The Lindley observation did not relate to channels in regime. He simply selected straight and regular reaches.
ii) No attempt was made to correlate rugosity and silt grade.
iii) *A formal cross section consisting of a horizontal bed and* $\frac{1}{2}:1$ *side slope was assumed and it is not practicable to apply his equations to natural channels.*

Lacey's formula (1929)

Lacey improved the Kennedy and Lindley's formula by considering the slope, size and bed material as designed parameters. He stated that the dimensions, width, depth and slope of a regime channel to carry a given water discharge loaded with a given sediment discharge are all fixed by nature. The regime channel concept i.e., which neither silt nor scours given by Kennedy was found inadequate to Lacey. According to Lacey a channel will be in regime if it flows incoherent unlimited alluvium of the same character as that transported and the silt grade and silt charge are all constants. The loam granular material, which can be scoured with the same ease with which it is deposited, is called incoherent alluvium. The following condition has to be attained in a perfect regime channel.

i) The channel flows in incoherent alluvium at a certain grade and quantity of suspended discharge under non-silting, non-scouring equilibrium condition.

ii) The channel running indefinitely and discharge at constant rate and attained a definite stable section and slope after silting and scouring.

If the channel is designed for smaller section and steeper slope than required to accommodate certain discharge it will scour till regime is developed. Similarly, if the large section and flatter slope are designed than required for a discharge it will silt till free regime is attained.

The perfect regime condition is rare occurrence. A channel reaches to final regime after the completion of initial regime. The channel constructed for width and longitudinal slope smaller than those required will tend to widen its width and steeper its slope if the bed and bank soils are incoherent alluvium and non-rigid. If the bank of the channel is rigid i.e., not free to erode then width is not widened but the longitudinal slope becomes steeper. These are the situation of initial regime. When a channel is provided with same type material on the bed and sides of the channel to protect it from scouring the channel is said to be of permanent regime. Regime theory is not applicable for this type of channel.

Lacey assumed that the silt in the cahnnel is kept in suspension due to vertical component of eddies generated from all the points along the wetted parameter. Lacey, therefore, using the available data of regime channels plotted the (i) velocity (V) and hydraulic radius (R) and (ii) cross-sectional area (A) and velocity as variables unlike the Kennedy who assumed the depth (D) as variable to reach at relationship in regime channel flow. He, thus, found

$$V \propto \sqrt{R}$$

$$or, V = C\sqrt{R} \qquad (5.37)$$

Where C is a constant.

After including the silt factor, Lacey obtained

$$V = \sqrt{\frac{2}{5} fR} \tag{5.38}$$

For a normal sediment of relative density equal to 2.65, the value of f in Lacey's relation

$$f = 1.76\sqrt{d}$$

Where d is the mean diameter of sediment in millimetre. He also obtained the relation

$$Af^2 = 140V^5 \tag{5.39}$$

After analysing the Lindley's data and data from other sources of regime channels, Lacey found,

$$V = 10.8R^{2/3}S^{1/3} \tag{5.40}$$

Where, S = slope of the water surface.

The Eq.5.40 has considerable use in determining flood discharge.

The Eq.5.38 can be rewritten as $V^4 = \frac{4}{25} f^2 R^2$

Using this equation in Eq.5.39, f^2 is eliminated

$$\frac{25V^4}{4R^2} = 140\frac{V^5}{A}$$

$$or,\ A\left(\frac{25}{4R^2}\right) = 140V$$

Multiplying both sides by A

$$A^2\left(\frac{25}{4R^2}\right) = 140Q$$

$$or, \frac{25}{4}\mathrm{P}^2 = 140Q, \text{since } hydraulic\ perimeter, \mathrm{P} = A/\mathrm{R}$$

$$or, \mathrm{P}^2 = \frac{140 \times 4 \times Q}{25}$$

$$\therefore \mathrm{P} = 4.75\sqrt{Q} \tag{5.41}$$

Multiplying by V in both sides of Eq.5.39

$$AVf^2 = 140V^6$$

$$\therefore Qf^2 = 140V^6$$

$$\therefore V = \left(\frac{Qf^2}{140} \right)^{1/6} \tag{5.42}$$

Eq.5.42 is used in channel design following Lacey's theory.

Raising both sides of Eq. 5.40 to the power 3,

$$V^3 = 1260 R^2 S \tag{5.43}$$

Similarly cubing both sides of Eq.5.38

$$V^3 = \left(2/5 \right)^{3/2} f^{3/2} R^{3/2}$$

$$Hence, 1260\, R^2 S = \left(2/5 \right)^{3/2} f^{3/2} R^{3/2}$$

$$or,\ S = \frac{f^{3/2}}{4980 R^{3/2}} \tag{5.44}$$

Again using Eq.5.43

$V^3 = 1260\ R^2 S$

$$S = \frac{V^3}{R^2} \times \frac{1}{1260}$$

$$= \frac{1}{1260} \left(\frac{V^2}{R} \right)^{\frac{5}{3}} \cdot \frac{1}{V^{\frac{1}{3}} R^{\frac{1}{3}}}$$

$$= \frac{1}{1260} \left(2/5\, f \right)^{5/3} \frac{1}{(RV)^{1/3}} \tag{5.45}$$

$$= \frac{1.723 \times 10^{-4} f^{5/3}}{(RV)^{1/3}},\ \text{where,} \frac{V^2}{R} = 2/5\, f$$

$$= 0.000178 \frac{f^{5/3}}{(RV)^{1/3}} \tag{5.46}$$

Again using the Eq.5.45 and putting R=A/P

$$S = \left(V^2/R \right)^{5/3} \frac{1}{1260 \left(AV/p \right)^{1/3}}$$

$$=\left(V^2/R\right)^{5/3} f^{5/3} \frac{\left(4.75\sqrt{Q}\right)^{1/3}}{1260Q^{1/3}},$$

Putting $P = 4.75\sqrt{Q}$

$$S = \left(2/5\right)^{5/3} \frac{f^{5/3}(4/75)^{1/3} Q^{1/6}}{1260Q^{1/3}}$$

$$= \frac{f^{5/3}}{3340Q^{1/6}} \tag{5.47}$$

Lacey's non-regime flow equation

The Eq.5.40 and the other equations derived by using this equation applicable only for regime flow. The following equations are used for both regime and non-regime channels. Using the Eq. 5.43

$$V^3 = 1260R^{2/3}S, \left(V^2 = \frac{1260R^2S}{V} \therefore V = (1260)^{1/2}\left(R/V\right)^{1/2}\sqrt{RS}\right)$$

$$or, V = 35.5\left(R/V\right)^{1/2}\sqrt{RS} \tag{5.48}$$

Eq. 5.48 is similar to Chezy's equation.

$$\text{Therefore}, C = 35.5\left(R/V\right)^{1/2} \tag{5.49}$$

Putting $V = \mathrm{k}\sqrt{fR}$

$$C = 35.5\left(\frac{R}{k\sqrt{fR}}\right)^{1/2}$$

$$= \frac{35.5R^{1/4}}{\mathrm{k}^{1/2}f^{1/4}} \tag{5.50}$$

$$\text{Taking } \frac{35.5}{k^{1/2}f^{1/4}} = K'/N_{\alpha}. \tag{5.51}$$

Where, Na= Lacey's absolute coefficient of rugosity which depend only upon grade and density of the channel. However, this is not proved by sufficient observations. Using the value of C from Eq.5.50 in Chezy's equation

$$V = \frac{35.5R^{1/4}\sqrt{RS}}{K^{1/2}f^{1/4}}$$

$$= \frac{K'R^{3/4}S^{1/2}}{N_\alpha} \tag{5.52}$$

When this equation is compared to Manning's equation

$$V = \frac{1}{N}R^{2/3}S^{1/2}$$

We get, $K'=1, R=1\, if\;\; N=N_\alpha$.

Thus, Lacey's flow equation is

$$V = \frac{1}{N_\alpha}R^{3/4}S^{1/2} \tag{5.53}$$

The Eq.5.51

$$\frac{35.5}{k^{1/2}f^{1/4}} = K'N_\alpha$$

So, $N_\alpha \propto f^{1/4}$

$f = 1$ and $N_a = 0.0255$ for standard grade silt.

$$\therefore\; N_\alpha = 0.0225 f^{1/2} \tag{5.54}$$

Lacey's shock theory

Lacey's idea of absolute coefficient of rugosity N_a depends only on the sediment size was found incorrect. Lacey introduced a new concept of 'shock' to explain the variation in N_a. The shock resistance is due to the channel bed irregularities. His argument was that the channel requires large slope i.e., the large N_a to overcome the shock resistance. If this argument is guaranteed then one has to conclude that a regime channel is free from undulation of channel bed or shock, which cannot be accepted. However, followings is the modified non-regime flow formula of Lacey.

$$V = \frac{1}{N_\alpha}R^{3/4}(S - s)^{1/4} \tag{5.55}$$

Where, s= slope required to overcome the friction

S= slope required to withstand the shock losses.

The value of N_a of above equation remains independent of channel condition. Buckley suggested the following value of N for earthen channel

Channel condition	N
Very good	0.0225
Good	0.025
Indifferent	0.0275
Poor	0.03

By using the value of N_a the slope (S and s) can be computed. For a good condition channel N_a= 0.025 and for that in very good condition.

$$\frac{1}{0.025} R^{3/4} S^{1/2}$$

$$= \frac{1}{0.0225} R^{3/4} (S - s)^{1/4}$$

$$or.(S - s)^{1/4} = \frac{0.0225}{0.025} s^{1/2} = 0.9S^{1/4}$$

$$or, S - s = 0.81S$$

$$or, 0.19S = s$$

This states that in a good condition channel 19% of the gross slope is lost to overcome the shock resistance and remaining 81% is frictional resistance. Though Lacey's idea of shock losses is said to be unique but cannot be used in practice. Using of different value N_a in different channel condition is more practical.

Limitation of Lacey's theory

1. True regime channel is rare.
2. The various formulae are based on single silt factor f. Due to different phase of flow on bed and sides choice of appropriate value of f is difficult.
3. Lacey's equation does not include the concentration of silt load as a variable for non-silting, non-scouring velocity.
4. Lacey's assumption that coefficient of rugosity N_a is constant for a given size material is not accepted. The value of N_a will change due to continued change in bed ripple formation in a sediment-carrying channel.
5. Lacey did not properly define the silt grade and silt charge.
6 Silt left in channel due to loss of 12-15% water of total discharge by absorption is not considered by Lacey.
7. Artificial channel is not a regime channel. Regime channel theory is not applicable to it.
8. Lacey's assumption of ellipse as the ideal shape of a regime channel is not accepted.

9. The wetted perimeter, $P = 4.75\sqrt{Q}$ is taken as the minimum in incoherent alluvium. The value is not fixed and it varies 3.7 to 5.7 in Punjab soils.
10. Lacey's formulae are not applicable to all discharges.

Design of channel by Lacey's theory

Trial and error practice is avoided in designing channel by Lacey's theory. The discharge Q and mean diameter of particles d and thereby the silt factor f are to be known.

Design procedure

1. Calculate the factor $f = 1.76\sqrt{d}$
2. Calculate velocity, $V = \left(\frac{Qf^2}{140}\right)^{1/6}$
3. Calculate hydraulic mean radius, $R = 4.75\sqrt{Q}$
4. Calculate, $A = Q/V$
5. Calculate wetted perimeter, $P = 4.75\sqrt{Q}$
6. Calculate bed width B and depth D of the channel section since A and P are known. Usually the side slope of the channel is assumed ½:1.

$$A = BD + 0.5D^2$$

$$P = B + D\sqrt{5} \quad or, B = P - D\sqrt{5}$$

$$A = R \times P$$

Hence, $A = D\left(P - D\sqrt{5}\right) + 0.5D^2$

Or, $0.5D^2 - \sqrt{5}D^2 + DP - A = 0$

Or, $-1.736D^2 + DP - A = 0$

$$\therefore D = \frac{P \pm \sqrt{P^2 - 6.944A}}{3.472}$$

and $B = P - 2.236D$

7. Calculate, $R = \frac{BD + 0.5D^2}{B + D\sqrt{5}}$

The calculated value of R should be same to the value obtained in step 3. if R value differs then the value of R in step 3 to be used to find B and D.

8. Calculate, slope $S = \dfrac{f^{5/3}}{3340Q^{1/6}}$

If in the discharge (Q) and slope (S) are known, the Lacey's f (within the range of Lacey's f= 1.0 to 0.8) is determined by using the relation,

$$S = \frac{f^{5/3}}{3340Q^{1/6}}$$

Example 5.15 Design a canal for discharge 50 cumecs and f=0.9 by using Lacey's theory.

Solution:

i) $Q = 50m^3 / \sec$

ii) $f = 0.9$

iii) $V = \left(\dfrac{Qf^2}{140}\right)^{1/6} = 0.813m/\sec$

iv) $R = \dfrac{2.5V^2}{f} = \dfrac{2.5(0.813)^2}{0.9}$ =1.836m

v) $A = \dfrac{Q}{V} = 61.5m^2$

vi) $P = 4.75\sqrt{Q} = 4.75 \times (50)^{1/2} = 33.587m$

vii) $D = \dfrac{P \pm \sqrt{P^2 - 6.944A}}{3.472} = \dfrac{33.587 \pm \sqrt{(33.587)^2 - 6.944 \times 61.50}}{3.472}$

$= \dfrac{33.587 \pm \sqrt{26.477}}{3.472}$

$= 17.26m \text{ or } 2.049 \text{ } i.e., 2.049m$

viii) $B = P - 2.236D = 33.587 - 2.236 \times 2.049 = 28.99$

ix) $R = \dfrac{BD + 0.5D^2}{B + D\sqrt{5}} = \dfrac{28.99 \times 2.049 + 0.5 \times (2.049)^2}{28.99 + 2.049\sqrt{5}} = \dfrac{61.58}{33.578} = 1.833$

The value of R provisionally calculated is same to this R.

x) $S = \dfrac{f^{5/3}}{3340Q^{1/6}} = 1.308 \times 10^{-4}$

Example 5.16 Design an irrigation channel by Lacey's theory for the discharge 15 cumecs and f = 1.0.

Solution:

i) $Q = 15m^3/s$

ii) $f = 1.0$

iii) $V = \left(\dfrac{Qf^2}{140}\right)^{1/6} = 0.689m/s$

iv) $A = \dfrac{Q}{V} = \dfrac{15}{0.689} = 21.77m^2$

v) $R = 2.5\dfrac{V^2}{f} = 2.5(0.689)^2 = 1.186m$

vi) $P = 4.75\sqrt{Q} = 4.75(15)^{1/2} = 18.396m$

vii) $D = \dfrac{P \pm \sqrt{P^2 - 6.944A}}{3.472}$

$= \dfrac{18.396 \pm \sqrt{(18.396)^2 - (6.944 \times 21.77)}}{3.472} = 1.357$

$B = P - 2.236D = 18.396 - 2.236 \times 1.357$

$= 15.351m$

viii) $R = \dfrac{BD + 0.5D^2}{B + D\sqrt{5}} = \dfrac{15.361 \times 1.357 + 0.5 \times (1.357)^2}{14.36 + 1.357\sqrt{5}} = \dfrac{61.58}{33.578} = 1.183m$

The value of R provisionally calculated is same to this R.

ix) $S = \dfrac{f^{5/3}}{3340Q^{1/6}} = 1.906 \times^{10-4}$

Questions and Problems

5.1 Define a canal or channel. How the canals are classified?

5.2 Prove that the coefficient in Manningis, Chezy's and Darcy-Weisbach equations can be related as:

$$\frac{C}{\sqrt{g}}=\sqrt{\frac{8}{f}}=\frac{R^{\frac{1}{6}}}{n\sqrt{g}}$$

5.3 Define steady, unsteady and uniform flow. Derive the expression for shear stress in a pipe under steady flow.

5.4 Prove that the critical specific energy, $E_c=\frac{3}{2}y_c$ in a rectangular channel.

5.5 Draw the sketch of an open channel and label the different elements of it.

5.6 Prove that for most economic trapezoidal channel section $b=2d\tan\frac{\theta}{2}$.

5.7 What is the Kennedy's formula with respect to regime channel? What are the limitations of his theory?

5.8 Discuss the procedure of designing regime channel following Kennedy's theory when Q, N, m and S are given.

5.9 Prove that silt supply in the channel is $Q_1=cBV_0^{5/2}$.

5.10 What is Lacey's formula? Describe the procedure of designing channel following Lacey's formula.

5.11 What are the limitations of Lacey's theory? What is Lacey's shock theory?

5.12 Determine the velocity of flow in a channel of depth 1.5m, breadth 1.0m, side slope 1.5:1 and land slope 0.1%. Assume Manning's n = 0.04.

Ans: V= 0.74m/s.

5.13 Design the earthen channel to carry a discharge of 2.0m^3/s in silty loam soil of land slope 0.15%. Assume Manning's n = 0.04. The velocity of flow should not exceed 0.7m/s.

Ans: B = 0.73m, D = 1.2m, z = 1.5

5.14 Design the most economic channel section to carry discharge of 2 cumecs for (a) semi-circular section, (b) a rectangular section, (c) a triangular section (90^0), and (d) a trapezoidal section of side slope 1.5:1.

Ans: (a) d = 1.84m (b) d =1.07m (c) d = 1.54m (d) 1.2m

5.15 Design the earth channel in clay soil of field slope 0.001 for carrying 2.1m^3/s flow. Assume n = 0.025.

Ans: b = 6.5m, d = 0.45m

5.16 Compute the efficient bottom width for an open channel to carry a flow 2.2m deep in sandy loam soil with the vegetation. What are the velocity and carrying capacity of the channel when the hydraulic gradient is 0.2%?

Ans: b = 0.95m

5.17 Compute the most efficient bottom width and depth of a channel to carry a discharge of 3.2m^3/s in land slope of 15 in 10000. The maximum velocity of

1. the channel should not exceed 85cm/s. calculate the dimension of the channel if the excavated soils are utilized to make pathway through the two sides of the channel. Assume side slope of the channel 1:1, side slope of the pathway 2:1, and Manning's n = 0.04.

Ans: Channel: b= 1.25m, d= 1.5m, D= 1.8m Spoil bank: d= 0.77m, b= 2m

5.18 Design the channel with the following data by using Kennedy's formula.

Discharge = 10 cumecs

Slope = 0.00015

N = 0.0225

m = 1

Ans: D = 0.85m, B = 23.10m

5.19 Design the channel with the following data by using Kennedy's formula.

Discharge = 3.5 cumecs

B/D = 4.5

N = 0.0225

m = 1

Ans: S = 1 in 5000.

5.20 Design a canal for discharge 10 cumecs and f = 1.0 by using Lacey's theory.

Ans: B = 12.34m, D = 1.2m

5.21 Select the most appropriate one of the followings:

1. A main canal delivers water to
   a) Branch canal and water courses
   b) Branch or major distributor
   c) Branch or minor distributor
   d) Major or minor distributor

2. The discharge in a rectangular channel at critical depth
   a) $q = g^{1/2} y_c^{3/2}$ b) $q = g^{1/2} y_c^{5/2}$
   c) $q = g^{1/3} y_c^{3/2}$ d) $q = g^{1/3} y_c^{3/2}$

3. Hydraulic radius of an open channel is
   a) Less than 1 b) More than 1
   c) More or less than 1 d) None of these

4. Hydraulic slope and land slope of channel is usually taken same because
   a) Flow is assumed steady uniform
   b) Flow is assumed steady

c) Flow is assumed unsteady

d) Flow rate is assumed constant.

5. The bottom of an irrigation channel may be

a) Above the ground level b) Below the ground level

c) On the ground level d) Any of the above

6. Most economic section of channel should have in principle

a) Minimum depth b) Minimum bottom width

c) Minimum hydraulic radius d) Minimum wetted perimeter

7. The minimum free board to be provided in small irrigation channel is

a) 7.5 cm b) 15 cm

c) 20 cm d) 25 cm

8. The maximum permissible velocity in a good vegetated channel may be

a) 75 cm/min b) 100 cm/min

c) 120 cm/min d) 150 cm/min

9. Manning's formula was evolved in the year

a) 1779 b) 1889

c) 1939 d) 1979

10. According to Bazin, Chezy's C is a function of

a) Hydraulic radius b) Hydraulic slope

c) Hydraulic radius and slope d) None of the above

Ans.

1. b) 2. a) 3. c) 4. a) 5. d) 6. d) 7. b) 8. d)
9. b) 10. a)

5.22 Write True or False of the following:

1. Inundation canals are fed by permanent source of supply
2. Major distributaries deliver the water to branch canal
3. A carrier canal carries water for other canal besides doing irrigation
4. If the flow of water takes place in a rectangular channel with depth of water 1.5m and velocity 1.2m/s, the specific energy will be 1.75m.
5. Free board is an essential component of an irrigation channel
6. For most economic trapezoidal channel half of the top width is equal to length of the side
7. For most economic trapezoidal channel hydraulic radius is equal to depth of water
8. Chezy's constant is a factor of channel resistance

9. Manning's n has a dimension of $L^{-1/3}T$

10. Chezy's C is equal to

Ans.

1. False 2. False 3. True 4. False 5. True 6. True 7. False
8. True 9. True 10. False

References

R.V. Giles, J.B. Evert & C. Liu (2004). Theory and Problems of Fluid Mechanics and Hydraulics (Third edition). Tata McGraw-Hill Edition, New Delhi.

R.K. Sharma & T.K. Sharma (1993). Text Book of Irrigation Engineering (Volume I). Oxford & IBH Publishing Co. Pvt.Ltd., Laxmi Nagar, New Delhi.

6

Soil-Water-Plant Relationship

The plants collect water by its root system available in the soil medium. A soil rich in nutrients cannot ensure success of crops unless the presence of satisfactory state of water and air, mechanical attributes and thermal regime. The physical properties of soil and plants greatly influence the movement, retention and use of water. The soil should provide sufficient pore spaces and size distribution and decomposition of rocks and for the movement and retention of water for plant needs. Rainwater, irrigation, groundwater, seepage, etc., contribute in replenishing the water in dried up soils. In all the cases thorough understanding of behavior of water is required to ensure the need of water to plants and optimize the use of irrigation water.

6.1 Soils

Soil is initially formed by the disintegration and decomposition of rocks and minerals and accumulation of biological residues. Usually the outer layer of the land surface is referred as soil. Soil is a heterogeneous, polyphasic, particulate, disperse and porous system. The soil consisting of the three ordinary phases in nature, viz. solid, liquid and gaseous. The solid phase consisting of the solids of the soil, comprising of minerals and organic matter and various chemical compounds; the liquid phase, consisting of soil water; and the gaseous phase, consisting of soil air.

Classification of soils

Soil may be classified in different ways, such as on the basis of particles, geological process of formation, etc. Depending on the size of the soil particles they are named as gravel, silt and clay. The following are some common methods to classify the soil particles.

i) U.S Bureau of Soil and Public Road Administration (PRA) system of United States.

ii) International soil classification, proposed at the International Soil Congress at Washington D.C. in 1927.

iii) The M.I.T. classification proposed by Prof. Gilboy of Massachusettes Institute of Technology as a simplification of Bureau of Soil Classification.

iv) Indian Standard Classification based on the M.I.T. system (Table 6.1).

Table 6.1. Particle size classification of soils

Fraction	Diameter, mm
Clay	<0.006
Silt	0.006-0.06
Very fine sand	0.06-0.10
Fine sand	0.10-0.25
Medium sand	0.25-0.50
Coarse sand	0.50-1.0
Fine gravel	1.0-2.0
Gravel	>2.0

a) U.S. Bureau of Soil and PRA classification

Fraction	Diameter, mm
Ultra clay colloids	<0.0002
Fine clay	0.0002-0.0006
Coarse clay	0.0006-0.002
Fine silt	0.002-0.006
Coarse silt	0.006-0.02
Fine mo (majla)	0.02-0.0.05
Coarse mo (majla)	0.05-0.10
Fine sand	0.10-0.20
Medium sand	0.20-0.50
Coarse sand	0.50-1.0
Very coarse sand	0.10-2.0
Gravel	>2.0

b) International classification

Fraction	Diameter, mm
Clay	<0.002mm
Fine silt	0.002-0.006
Medium silt	0.006-0.02
Coarse silt	0.02-0.06
Fins sand	0.06-0.20
Medium sand	0.20-0.60
Coarse sand	0.60-1.0
Gravel	>1.0

c) M.I.T. classification

Fraction	Diameter, mm
Clay	<0.002mm
Silt	0.002-0.075
Fine sand	0.075-0.425
Medium sand	0.425-2.0
Coarse sand	2.0-4.75
Fine gravel	4.75-20
Coarse gravel	20-80
Cobble	80-300
Boulder	>300

d) Indian standard classification

Courtesy: Punmia, B.C. (1994)

The most commonly used method for classification of soils based on grain size is those proposed by the International Soil Science Society (ISSS) which are as follows:

Fraction	Particle diameters USDA	ISSS
Gravel	>2 mm	>2 mm
Very coarse sand	1 to 2 mm	-
Coarse sand	0.5 to 1 mm	0.2 to 2 mm
Medium sand	0.25 to 0.5 mm	-
Fine sand	0.1 to 0.25 mm	0.02 to 0.02 mm
Very fine sand	0.05 to 0.1 mm	-
Silt	0.002 to 0.05 mm	0.002 to 0.02 mm
Clay	<0.002 mm	<0.002 mm

Courtesy: Michael (1978)

Based on the process of formation (origin) soils may be classified as below (Aswa, 1993).

i) **Residual soils**: They are formed due to disintegration of natural rocks due to action of air.

ii) **Alluvial soils**: These are formed due to deposition of sediment materials along the bank of the overflowing rivers.

iii) **Aeolian soils**: These are the deposition of soil due to wind action.

iv) **Glacial soils**: These are formed as the products of glacial erosion.

v) **Colluvial soils**: They are formed at the foothills by the deposition due to rain wash.

vi) **Volcanic soils**: These are formed due to volcanic eruption.

Indian soils are classified as below.

i) **Alluvial soils**: Indian alluvial soils are the deposition of the river Ganga, Indus and Brahmaputra system. These rivers carry the products of the weathering of mountain rocks of different sizes and deposit them as they traverse the plains. These are very fertile soils and constitute the largest and important soil group in India.

ii) **Black soils**: The typical soils derived from Deccan trip having a thin layer to thick stratum. These are the heavy soils of clay to loam and so called as heavy soils. Drainage of these soils are poor and suitable for cultivation of rice and sugarcane. Black soils are found in large area of Maharashtra, western part of M.P. and in some parts of A.P., Gujarat and Tamil Nadu.

iii) **Red soils:** These are the soils formed due to meteoric weathering of ancient crystalline and metamorphic rocks. Red soils are available in Tamil Nadu, Karnataka, Goa, south-eastern Maharashtra, A.P., M.P.,Orissa, Bihar, and some districts of West Bengal and Uttar Pradesh.

iv) **Lateritic soils:** Lateritic soils are found by the deposition of silicious matter of lateritic rocks leached out during weathering under the monsoon. It is the peculiar soils of India and some other tropical countries. These are formed in India on the hills of Karnataka,Kerala, M.P., the Eastern Ghats of Orissa, Maharashtra, West Bengal, Tamil Nadu and Assam.

v) **Desert soils**: The geologically recent origin soils of arid regions in western part of Rajasthan, Haryana, Punjab, lying between Indus river and the Aravalli range is called as desert soil. Soils are poorly developed and called as blown sand.

vi) **Forest soils:** These soils are very rich in organic and vegetable substrances and found in forest and foothills.

Soil Texture

The relative proportion of sand , silt and clay determines the soil texture. Soil particles differ widely in their sizes. Some are easily seen by naked eye and there are some which posses the colloidal properties. The term "colloidal" means the size of intermediate between visible under an optical microscope and invisible molecules. The soil texture has got both quantitative and qualitative significance. Qualitatively it indicates the type of soil and quantitatively it gives the relative proportion of various sizes of soil particles. Soil particles are traditionally divided into three ranges, viz., sand, silt and clay. These are also called as textural fraction or separates. Soils with different proportion of sand, silt and clay are given different designations. There are 12 main such designation or textural classes as shown in the triangular diagram (Fig. 6.1).

The triangular diagram may be considered in modified manner for the determination of texture in feel method (Fig. 6.2). The triangle in Fig.6.2 is divided into three major textural classes.

i) **Clay soils** which include sandy clay loam, clay loam and silty clay loam. The soils form good ribbons.

ii) **Clay loam soils** which include sandy clay loam, clay loam and silty clay loam. These soils form poor ribbons.

iii) **Loam soils** which include sandy loam, loam, and silt loam. These soils form poor ribbons.

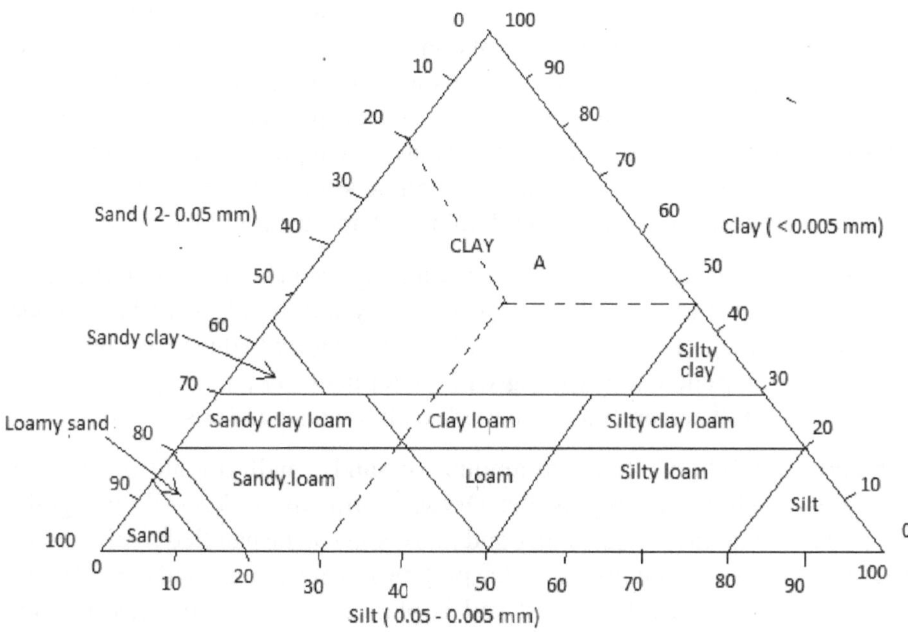

Fig.6.1 Textural triangle showing the percentage of clay, silt, and sand

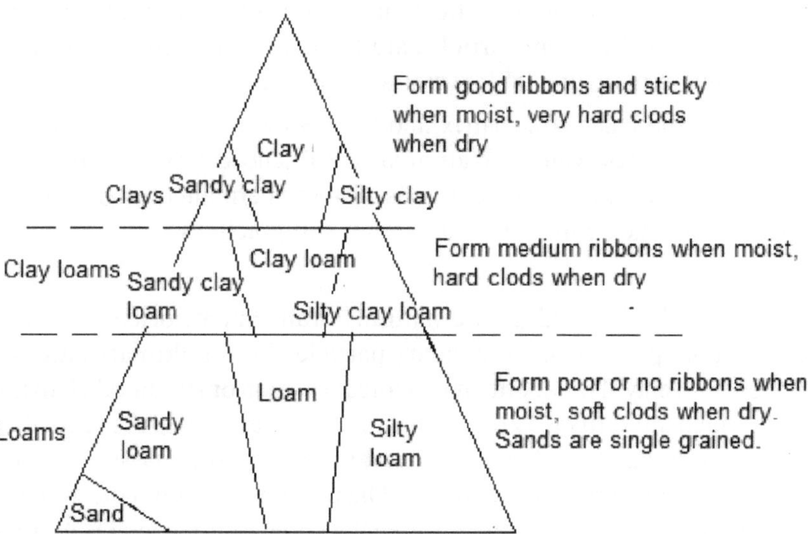

Fig.6.2 Triangular texture for determining soil texture by feel method

Sand is the least complex textural group where the individual grains can be seen and felt easily. If moist sand is squeezed it will form a cast and it can retain its shape even after the pressure is released but crumbles when touched. Dry sands fall apart at the release of pressure. Sands are having relatively simple capillary system with a large volume of non-capillary pore spaces which provide good drainage and aeration. Sandy soils are loose, non-cohesive, low-water holding capacity and relatively inert chemically with much low cation exchange capacity. Sandy loam consists of silt and clay which provide it some amount of stability. Like the sand individual grain can be seen and felt readily. The cast form by squeezing the moist soil will hold its shape at careful handling.

Loam consists of even mixture of approximately equal amount of sand, silt and clay. It is slightly plastic, easily crumble when dry and slightly gritty but provides fairly smooth feel. This soil is favourable for plant growth due to its good water holding capacity and cations along with better aeration than clay. It is neither heavy nor light, therefore, easier to work with it than clay soils.

Silt loam consists of moderate amount of sand, small amount of clay and comparatively large quantity of silts. Dry soil form lumps. Wet soil run together and puddles. The casts made of dry and moist soil can be handled freely without breaking. Clay soil contains more clay particles than silt loam and considered to be fine textured soils. The moist soil is plastic and forms the cast that can withstand considerable handling. Sands are more in **sandy clay loam** and **sandy clay** than the loam. Silts are much less in sandy clay. **Silty clay** forms very hard clods or lumps when dry and usually sticky in wet. It is very complex in nature. Due to plate like shape, the clay particles are having much more surface area and therefore contain more available water and minerals.

Soil texture forms the basic soil matrix and the geometry of voids in the soil matrix. The availability of water and air greatly influenced by the geometry of voids vis-à-vis the soil texture. Soil texture is almost a permanent character of a soil and does not change by the tillage and other soil practices.

Soil structure

Soil structure may be defined as the natural arrangement, orientation and organization of both primary and secondary particles in soil. Primary particles are the sand, silt and clay. The smaller mechanical fraction or structural elements or aggregates which are formed by the primary particles are called the secondary particles. Secondary particles further aggregate to form the masses of soils. Therefore, there are two type of pore spaces. The micropores within the granules and macropores between the granules. Unlike the soil texture, which is more or less constant, the structure is a continuously changing character in response to natural, biological and soil management practices. Soil structure some time means the geometry of the pore spaces which is so complex that it is not possible to measure some attributes which are affected by the soil structure. Soil structure greatly influences the availability of water, air and heat regime in the field. Soil structure affects the seed germination, root growth and establishment of seedlings.

Soil structure may be classified on the basis of shape, size, hardness and stability of aggregates. Most commonly used classification is based on shape of aggregates. These are two groups.

a) **Simple structure:** The structure in which the natural cleavage planes are absent or indistinct.

i) Single grain structure-structureless soil: Usually this occurs in sandy and silty soils of low organic matter. Many sandy soils are referred to as single grain structure or structureless because very less particles are adhere to each other when the soil is dry. However, they adhere to each other when they are moist and behave like aggregated particles of finer texture. Therefore, they can not be considered structureless.

ii) Massive structure: It is similar to simple grain structure except that it is coherent. The example of massive structures is dense soil crusts, plough pans and fargi-pans.

b) **Compound structure:** In compound structure the cleavage planes are distinct. It is described according to the relative length of the vertical and horizontal axes and by the contours of their edges. Accordingly, they are grouped as the following types.

i) Cube like structure: in this structure the vertical and horizontal axes are almost of equal length, the corners are angular and edges are sharp and distinct.

ii) Columnar structure: The vertical axes are longer than horizontal axes. The vertical cleavage planes are predominant, corners are subangular and rounded edges are sharp.

iii) Platey structure: The horizontal axes are longer than vertical axes and horizontal cleavage planes are predominant.

iv) Angular structure: The corners and edges are sharp.

v) Sub-angular structure: Sharp edges but rounded corner

vi) Granular structure: Both corners and edges rounded having less pore space.

Genesis of soil structure

Soil structures are formed by the process of flocculation and cementation of floccules. The primary particles with a high electrokinetic (zeta) potential repel each other when they collide in a suspension. When the potential is lowered sufficiently a collision between the particles results in a mutual attraction and formation of floccules. This is 'salt type' flocculation. Flocculation also occurs due to electrostatic attraction between the positive edges and faces of clay minerals. This edge-to-face type of flocculation provides more floccules than 'salt type'.

The formation of aggregates requires a cementation or binding together the flocculated particles. The flocculation and aggregates are different. Flocculation aids in the formation of aggregates but itself not the aggregate. Aggregate formation depends on coagulation and pressure. These are further dependant on the capacity of cation, clay particles interaction in relation to moisture content and temperature, clay-organic matter interaction and soil flora and fauna. The porosity, aggregation, cohesiveness and the permeability are the important properties of soil which can be quantitatively measured.

Soil structure and agricultural significance

Soil provides the favourable environment for plants to germinate, emergence of seedlings and development of root system. Favourable environment of soil ensures sufficient air-water regime and nutrients. Soils of bulk densities limit the uptake of nutrients due to lack or excess of water and check the root proliferation. Soil density increases at the cost of decreases of aeration resulting deficiency of oxygen. Microbiological activities also impeded at higher bulk density. Therefore, the unfavourable soil structure is considered to be an important soil fertility parameter which affects much in crop production. Soil structure affects the physical characteristics of soil such as porosity, hydraulic conductivity, infiltration, water holding capacity and erodibility. Silt-size aggregates provide poor drainability because the too small size pores.

Tilth is a soil physical condition related to cultivation of plants. The soil in good tilth indicates the condition of maximum infiltration at rainfall, sufficient moisture, adequate aeration and favourable soil temperature. Soils in good tilth are mellow, crumbly and easily worked whereas the poor tilth soils are relatively hard, cloddy and difficult to work. Dispersed clay particles represent the infavourable soil surface. The ploughing of soil consists of furrow slice, granulation, turn down the residues, weeds and green manures and loosening the soils. The success of ploughing depends on the granulation of furrow slice. The granular nature of the soil and the perfect moisture content in respect to ploughed soil determine the result of ploughing. A soil not in friable condition makes clods at ploughing.

Cultivation decreases the bulk density and increases the porosity and thereby the total volume of air in soil. Degradation of aggregates is correlated to decrease in organic matter. Rapid infiltration is important in respect to crop cultivation. Loose soil permits better infiltration. Cultivation increases the infiltration rates in crusted and compacted soil. Rain drops try to form soil crust in certain soils. Granular soils resist the depression action of rain drops. Aggregates should be stable to impact of rain drops and temporary submergence. The vertical axis of aggregates is preferred over the horizontal axis. Sand-size aggregates are better for pore distribution than angular size.

Cultivation of suitably formed soil structure is important in a certain soil-agro-climatic condition. Appropriate land use, tillage practices, rotation of crops, sub soiling, mulching, addition of organic matter, controlled irrigation, soil conservation practices, use of soil conditioners, etc., help in improving and conserving desirable structure in soils.

6.2 Soil-Water Relationship

There are usually three phases, namely, solid, liquid (water) and gas (air) in a soil mass as shown in Fig.6.3. The Fig.6.3 shows the three phases of soil in relative proportion both in volume and masses and is used to define the relationship among them.

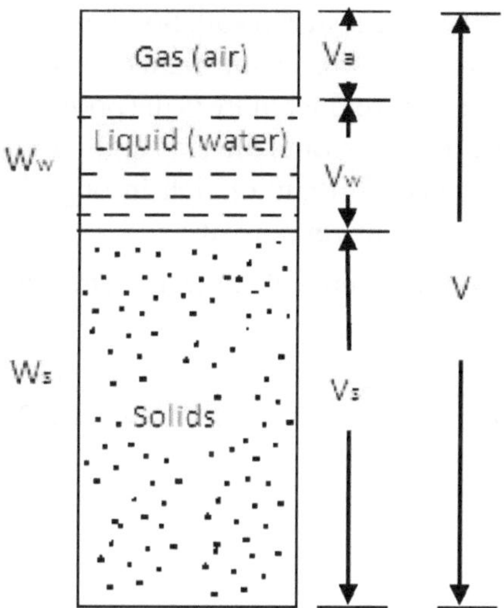

Fig. 6.3 Soil composition

Referring to Fig.6.3,

$$or, e = \frac{V - V_v}{V_s} = \frac{V}{V_s} - 1 = \frac{V}{W_s / G\gamma_w} - 1$$

V_a = volume of gas (air)

V_w = volume of liquid (water)

V_s = volume of solids

V_v = volume of voids (V_a+V_w)

V = total volume (V_v+Vs)

W_w = weight of water

W_s = weight of solids

Porosity: Any soil mass is surrounded by certain amount of voids which are termed as pore spaces. The total volume of void in unit volume of soil mass is called the porosity of soil (n). Porosity is usually expressed in percent.

$$\text{So, } n = \frac{V_v}{V} = \frac{V_a + V_w}{V_a + V_w + V_s} \tag{6.1}$$

$$\text{Again, } n = \frac{V - V_s}{V} = 1 - \frac{V_s}{V} = 1 - \frac{W_s}{\gamma_w G V} \tag{6.2}$$

Where, G = specific gravity of solid soil

γ_w = unit weight of water

The porosity of the soil is the measure of relative pore space in soil. Soil porosity depends on its texture and structure. Fine textured soils are having higher porosity than coarse textured soils. More the particles in soil are finer more is the porosity. Usually the porosity of sandy soils ranges from 35-50% and the clay soils from 40-60%. Higher porosity of a soil indicates loose soil and or rich in organic matter. Dispersion and compaction of soils provide lower porosity.

Void ratio: The ratio of the volume of voids to the volume of solids is termed as void ratio, e. Void ratio is expressed in fraction.

$$e = \frac{V_v}{V_s} \tag{6.3}$$

$$\text{Or, } e = \frac{V - V_s}{V_s} = \frac{V}{V_s} - 1 = \frac{V}{W_s / G_m} - 1$$

$$= \frac{VG\gamma_w}{W_s} - 1 \tag{6.4}$$

Let us consider an unit volume of soil whose length, breadth and height are of unit length. Therefore, the porosity of the soil, $n = \frac{V_v}{V = 1} = V_v$. The void ratio can be expressed as below.

$$e = \frac{V_v}{V_s} = \frac{V_v}{V - V_v} = \frac{n}{1 - n} \tag{6.5}$$

$$or, e - ne = n$$

$$or, n(1 + e) = e$$

$$\therefore n = \frac{e}{1 + e} \tag{6.6}$$

The porosity of a soil may very due to change in soil structure, compaction of soil, etc. In any change of volume of void the total volume of soil also changes. Whereas, in void ratio the volume of void may change without changing the volume of solids. Volume of solids can be considered as a constant quantity. This character of void ratio gives the idea of external load and void ratio. Using void ratio is convenient and preferably used in relation to foundation of engineering works.

Density of soil: It is the mass of the soil per unit volume.

Total or wet or bulk density of soil, ρ_w: : It is the mass of moist soil of unit volume of soil.

$$\rho_w = \frac{M_s + M_w}{V} = \frac{M_s + M_w}{V_a + V_w + V_s} \tag{6.7}$$

Dry density of soil, ρ_d **:** It is the dry mass of solids in unit volume of soil.

$$\rho_d = \frac{M_s}{V} = \frac{M_s}{V_a + V_w + V_s} \tag{6.8}$$

The dry bulk density of soil is expressed in g/cm³. The term dry bulk density and apparent specific gravity are often used as synonymous term. The apparent specific gravity is dimensionless. However, since 1gm of water occupy the volume of 1cm³ in normal temperature, the dry bulk density and apparent specific gravity give the same numerical value. Dry bulk density is influenced by soil texture, structure, compaction and organic matter. Coarse soils give higher bulk density. Dry bulk density is very important for physical properties of soil like permeability and water holding capacity.

Saturated bulk density, ρ_{sat} : It is the mass of unit volume of saturated soil mass.

$$\rho_{sat} = \frac{M_s + M_w}{V} = \frac{M_s + M_w}{V_a + V_w + V_s} \tag{6.9}$$

The equations for wet bulk density and saturated bulk density are same. However, the saturated bulk density refers the density only when the soil voids are completely filled up with water.

Weight of soil: It is the weight of soil per unit volume.

Bulk unit weight or wet weight of soil, γ : It is weight of the moist soil of unit volume

$$\gamma = \frac{W}{V} = \frac{W_s + W_w}{V} \tag{6.10}$$

Dry unit weight of soil, γ_d : It is weight of the solids in unit soil volume.

$$\gamma_d = \frac{W_s}{V} \tag{6.11}$$

Apparent specific gravity of soil: Apparent specific gravity and the dry unit of soil has the same numerical value in CGS.

Soil moisture content or mass wetness, *w*: It is the ratio of weight of water to weight of solids in a soil mass. It is usually expressed in percent.

$$w = \frac{W_w}{W_s} \tag{6.12}$$

The soil moisture content in root zone depth of soil at any time is very important in cultivation. In fact, for all soil water study soil moisture content is used as an important reference to explain the other characteristics. Soil moisture content may be expressed in decimal; however, it is usually expressed in percentage multiplying by 100.

Volumetric water content or mass wetness, *v*: It is the ratio of volume of water available in a total soil volume.

$$v = \frac{V_w}{V_t} = \frac{V_w}{V_a + V_w + V_s} \tag{6.13}$$

Degree of saturation, *S*: It is the ratio of volume of water present in the voids of a soil mass. It gives the idea of portion of voids filled with water.

$$S = \frac{V_w}{V_v} = \frac{V_w}{V_a + V_w} \tag{6.14}$$

Relative density or density index, ρ_r: It is defined as the ratio of the difference between the void ratio of the soil in its loosest state e_{max} and its natural void ratio *e* to the difference between the void ratio in its loosest state and densest state e_{min}

$$\rho_r = \frac{e_{\max} - e}{e_{\max} - e_{\min}} \tag{6.15}$$

The term is related to cohesionless soil and useful to understand its density under the natural process of deposition. When e is maximum, $\rho_r = 0$ and when e is minimum, $\rho_r = 1$. Thus, ρ_r varies between 0 to 1.

Functional relationship

i) Relation between the volumetric moisture content (v) and soil moisture content (*w*):

The moisture content in soil is usually measured on dry weight basis. It is often required to express it in volumetric basis. The soil moisture content is multiplied by the apparent specific gravity of the soil to represent volumetric moisture content. This may be proved as below

We have,

$$w = \frac{W_w}{W_s}, v = \frac{V_w}{V_t}, \gamma_d = \frac{Ws}{V}$$

Assuming specific gravity of water as 1, the volume of water (V_w) and weight of water (W_w) has the same numerical value in CGS unit.

$$\text{Therefore, } v = \frac{V_w}{V_t} = \frac{W_w}{W_s} \cdot \frac{W_s}{V_t} = w\gamma_d \tag{6.16}$$

ii) Relation among void ratio (e), specific gravity of soil (G), soil moisture content (w) and degree of saturation (S):

Referring to Fig.6.3 and assuming V_s as equal to unity,

$$e = \frac{V_v}{V_s} \frac{V_v}{1} = V_v$$

$$\text{Degree of saturation, } S = \frac{V_w}{V_v} = \frac{V_w}{e}$$

$$\therefore V_w = Se \tag{6.17}$$

$$\text{Again, } w = \frac{W_w}{W_s} = \frac{V_w \gamma_w}{\gamma_s V_s} = \frac{V_w \gamma_w}{\gamma_s}$$

$$G = \frac{\gamma s}{\gamma w} \ or, \gamma_s = G\gamma_w$$

$$\therefore w = \frac{V_w \gamma_w}{G\gamma_w} = \frac{V_w}{G}$$

$$or, V_w = wG \tag{6.18}$$

From **Eqs. 6.17** and 6.**18**

$$wG = Se$$

$$or, e = \frac{wG}{S} \tag{6.19}$$

iii) Relation among γ, G, e and S

$$\text{We have, wet unit weight, } \gamma = \frac{W_s + W_w}{V}$$

$$= \frac{V_s \gamma_s + V_w \gamma_w}{V}$$

From **Fig.6.3**, V_s=1, V=1+e

From **Eq.6.17**, V_w= Se

and $\gamma_s = G\gamma_w$

$$\therefore \gamma = \frac{G\gamma_w + Se\gamma_w}{1+e} \quad (6.20)$$

When the soil is dry, S = 0. The wet unit weight γ become converts to dry unit weight γ_d.

$$\therefore \rho_d = \frac{G\gamma_w}{1+e} \quad (6.21)$$

When S=1, the soil is fully saturated, and ρ_w becomes $\rho_{sat} = \frac{(G+e)\gamma_w}{1+e}$ (6.22)

iv) Relation among γ_α γ and w

Soil moisture content, $w = \frac{W_w}{W_s}$

$$\therefore 1+w = \frac{W_w}{W_s} + 1 = \frac{W_w + W_s}{W_s} = \frac{W}{W_s}$$

$$\text{Or}, W_s = \frac{W}{1+W}$$

$$\text{Dry weight}, \gamma_d = \frac{W_s}{V} = \frac{\text{W}}{(1+W)V}$$

$$= \frac{\gamma}{1+w} \quad (6.23)$$

Example 6.1 Calculate (a) void ratio, (b) dry density and (c) saturated bulk density of a soil mass if its porosity is 35% and specific gravity 2.66.

Solution:

Given, n = 35%, G=2.66

a) Void ratio, $e = \frac{n}{1-n} = \frac{0.35}{1-0.35} = 0.538$

b) Following Eq.6.21, dry bulk density, $\therefore \rho_d = \frac{G\gamma_w}{1+e}$

$= \frac{2.66 \times 1\text{g/cc}}{1+0.538}$, (Assuming unit weight of water = 1g/cm^{3}or 9.81KN/m^3)

$= 1.73\text{g}/\text{cm}^3$

(c) Following Eq.6.22, saturated bulk density, $\rho_{sat} = \dfrac{(G+e)\gamma_w}{1+e}$

$$= \frac{(2.66+0.538)1.0\,g/cm^3}{1+0.538}$$

$= 2.08 g/cm^3$

Example 6.2.A soil sample of weight 225g and volume 120cm^3 was reduced to 190g on oven drying. The specific weight of soil was 2.65. Determine (a) apparent specific gravity, (b) soil moisture content, (c) void ratio, and (d) degree of saturation.

Solution:

Weight of solid soil mass, W_s=190g

Weight of water, W_w= 225-190=35g

a) Apparent specific gravity, $\rho_d = \dfrac{W_s}{V} = \dfrac{190}{120} = 1.58$

b) Soil moisture content, $w = \dfrac{W_w}{W_s} = \dfrac{35}{190} = 0.1842 = 18.42\%$

c) Void ratio, $e = \dfrac{VG\gamma_w}{Ws} - 1$

$$= \frac{120 \times 2.65 \times 1}{190} - 1$$

= 1.67-1=0.67

d) Degree of saturation, $S = \dfrac{wG}{e}$

$$= \frac{0.1842 \times 2.65}{0.67}$$

$= 0.7285 = 72.85\%$

Example 6.3 A soil sample at 20% moisture content measures 1.95g/cm^3 wet bulk density. What is the moisture content if the wet bulk density reduces to 1.85gm/cm^3 on partial drying.

Solution:

Wet bulk density, ρ_w=1.95g/cm^3

Moisture content, w = 20%

Dry bulk density, $\rho_d = \frac{\rho_w}{1+w}$

$$= \frac{1.95}{1+0.20}$$

$$= 1.625$$

Since the dry bulk density does not change on drying, the same equation can also be used to determine the moisture content after drying.

$$\therefore \rho_d = \frac{\rho_w}{1+w}$$

$$or, 1.625 = \frac{1.85}{1+w}$$

$$or, 1+w = \frac{1.85}{1.625} = 1.1385$$

$$or, w = 0.1385 = 13.85\%$$

Example 6.4 A saturated soil sample of 38.5cm³ volume and 65.5g mass reduced to 42.4g on oven drying. Find (i) w, (ii) G, (iii) e, (iv) ρ_{sat}, (v)ρ_d

Solution:

i) Weight of water, W_w = 65.5-42.4=23.1g, W_s = 42.4g

$$w = \frac{W_w}{W_s} = \frac{23.1}{42.4} = 0.5448 = 54.48\%$$

ii) Volume of solid, V_s = 38.5-23.1=15.4

$$G = \frac{W_s}{V_s} = \frac{42.4}{15.4} 2.75$$

iii) $e = \frac{V_v}{V_s} = \frac{23.1}{15.4} = 1.5$

(iv) $\rho_{sat} = \frac{W_w + W_s}{V} = \frac{65.5}{38.5} = 1.70 g/cm^3$

(v) $\rho_d = \frac{W_s}{V} = \frac{42.4}{38.5} = 1.10 g/cm^3$

6.3 Classes of Soil Water

The pore spaces of solid matrix in soil are usually filled by air, water vapour and water. Water vapour and air in soil get displaced when water is added by rain or irrigation to dry or moist soil. When all the pore spaces are filled by water it is said to be saturated and at this state it attains the maximum water retaining capacity. The soil water can be divided into two types: (i) gravitational or free water, and (ii) held water. Gravitational or free water moves under the influence of gravity. Free water temporary retains in soil. It disappears quickly from the coarse soil. The held water remains in soil pores by the attraction of soil particles. The held water may further be divided to: (i) structural water, (ii) absorbed or hygroscopic water, and (iii) capillary water. Structural water can not be separated from the soil particles even by oven drying. This water forms the structure of the soil minerals and held in it till the structure is unbroken. Absorbed or hygroscopic water is held around the soil particles by the force of adhesion of soil particles. Adhesion is the attraction of solid surfaces for water particles. Hygroscopic water develops very thin film and is held in tension more than 31 atmospheres hence not useable by the plants. The capillary water develops comparatively thicker film around the soil particles by the process of adhesion and cohesion of water particles. Cohesion is the attraction of water molecules for each other. Capillary water retains in between 31 to1/3rd atmosphere tension and suitable for plant use. The amount of capillary water to be held in a soil depends on the structure, texture and organic matter. Fine textured, granular structure and organic matter in soil influence in increasing the capillary capacity

Soil -water potential: The total soil-water potential is defined as "the amount of work that must be done per unit quantity of pure water in order to transport reversibly and isothermally an infinitesimal quantity of water from a pool of pure water at a specified elevation at atmospheric pressure to the soil water at the point under consideration". In a simpler way the soil-water potential can be defined as the amount of work done or potential energy stored, per unit soil mass in bringing any mass, m from the reference level to the point of consideration. The soil-water potential differs from pure water as because the soil-water is subjected to different force field. These force fields result from the presence of solutes, due to external gas pressure and gravitational effect. Therefore, the total soil water potential can be represented as $\phi_s = \phi_g + \phi_p + \phi_0$

Where ϕ_s is the total potential, ϕ_g the gravitational potential, ϕ_p the pressure (or matric) potential and ϕ_0 the osmotic potential.

Gravitational potential (ϕ_g): Everybody on the earth's surface is attracted towards the center of the earth by the gravitational force which is the multiplication of mass of the body and the acceleration due to gravity. Energy is required to lift the body against this force. In a lifted body this energy is stored in the form of gravitational potential energy.

Gravitational potential can always be taken as positive or zero. This is measured at each point with respect to the reference level set within or below the soil profile such that gravitational potential always will record positive or zero potential.

Considering a mass M of water, occupying a volume V at a height z above the reference level has the gravitational energy,

$E_s = Mgz = \rho_w Vgz$

Where ρ_w is the density of water and g is acceleration due to gravity.

Thus, gravitational potential energy per unit mass,

$$\phi_g(\text{mass}) = \frac{E_s}{M} = \frac{Mgz}{M} = gz \qquad (6.24)$$

Gravitational potential per unit volume,

$$\phi_g(Volume) = \frac{E_s}{V} = \frac{\rho_w Vgz}{V} = \rho_w gz \qquad (6.25)$$

$$\text{Gravitational potential per unit weight} = \frac{\rho_w Vgz}{\rho_w Vg} = z \qquad (6.26)$$

The **Eq.6.26** expresses the height of liquid column corresponding to given pressure per unit weight as the potential energy. A pressure of 1 atmosphere is equivalent to a vertical water column of 1033cm or a mercury head of 76cm. Potential is also expressed in bars (1 bar=10^6 dynes/cm^2 or 1013cm of water column taking g=981cm/s^2).

Pressure potential, (ϕ_p): Soil-water may subjects to positive or negative pressure. Positive pressure occurs when hydrostatic pressure greater than atmospheric pressure. Negative pressure occurs when hydrostatic pressure is lower than atmospheric pressure. Negative pressure may also be said as tension, suction and capillary or matric potential. Water with a free surface is a positive pressure but the water at such a surface is at zero pressure potential. However, the negative pressure potential characterizes if the water has risen to above of this surface through capillary tube.

Osmotic potential: Osmotic potential is defined as the "amount of work that a quantity of water in an equilibrium soil water system is capable of doing when it moves to another equilibrium system identical in all respects except that there is no solution". Osmotic potential is also termed as solute potential and exists only when solutes are present in the soil water. Solutes present in soil water lower the potential energy. In fluid flow its effect is not prominent but when there is membrane solute water retards the flow, which is important phenomenon in the interaction between plant roots and soil.

Surface tension: Surface tension is the molecular phenomenon by which molecules of the particles are held together. Surface tension involves two types of molecular forces, (i) adhesion and (ii) cohesion. Adhesion is the force of attraction of the molecules of dissimilar substances. It is different in different pairs of substances, such as, gum has a greater adhesive force than water or alcohol. Cohesion is the force of attraction of the molecules of like substances. The force of attraction inversely varies with the distance between two molecules. The distance up to which this attraction works is called the molecular range. The molecular range of different solid and liquid differs but lie within the range of 10^{-7}cm measured from the center of the molecules. Cohesive force is greatest in solids, less in liquids and least or nearly negligible in gases at normal temperature and pressure. These characteristics provide a solid definite shape, liquid a free surface and a gas has neither of these.

Due to consequences of adhesion and cohesion the surface of a liquid behaves like a skin or membrane, which has a normal tendency of contract. A molecule inside a liquid attracted equally by the molecules lying within the sphere of influence, which causes no resultant force to the molecules. A molecule on the surface of the liquid is attracted downward due to the presence of fewer molecules of vapour above the liquid. For this inward pull, the surface of the liquid always tends to contact to the smallest possible area. This is the reason the soap bubble has the minimum surface area for a given volume. If a cut is made along any line on the liquid surface, immediately some force holds the separated portions together. The force, which holds the separated portions together, is proportional to the length of cut and the value per unit length is surface tension usually represented by γ. The dimensions of surface of surface tension are energy per unit area, ergs/cm^2 or dynes/cm. The increase in temperature decreases the cohesive force at the surface and inside the liquid and thereby reduces the surface tension.

The expression of surface tension may be analyzed by the consideration of forces that acts to exist a soap bubble together. The bubble is subjected to atmospheric pressure (P_a) acting on the outside of the bubble, the inside pressure (P_i) of bubble acts outward and the surface tension acts inward. At an imaginary central place, which divides the soap bubble into two equal hemispheres are held together due to the force of surface tension.

The surface tension acts throughout the length of inside and outside circular face of the hemisphere. Therefore, if σ is the surface tension for unit length, the total surface tension equals to $4\pi r\sigma$ At the state of existence of bubble,

$$(p_i - p_a)\pi r^2 = 4\pi r\sigma$$

$$or, (p_i - p_a) = \frac{4\sigma}{r} \tag{6.27}$$

Eq.6.27 shows that the difference in inside and outside pressure depends on the radius of bubble. Bigger the radius of bubble, smaller is the difference in pressure. Similarly, greater the curvature of the bubble, greater is the inside pressure.

Capillarity: The voids in the soil resemble the fine bore tube called as capillary tube. When the capillary tube is dipped in water immediately the water rises up in to it. This is the most important effect of surface tension. The height to which liquid rises is determined by the surface tension and weight of the liquid column. Referring to Fig. 6.4 the sides of the capillary tube attract the water to form a concave meniscus due to surface tension. Surface tension will occur only when the walls of the capillary have the more attraction on the liquid than the molecules of the liquid have for each other. As shown in Fig.6.4 h is the height of water raised in a capillary tube due to vertical component of surface tension.

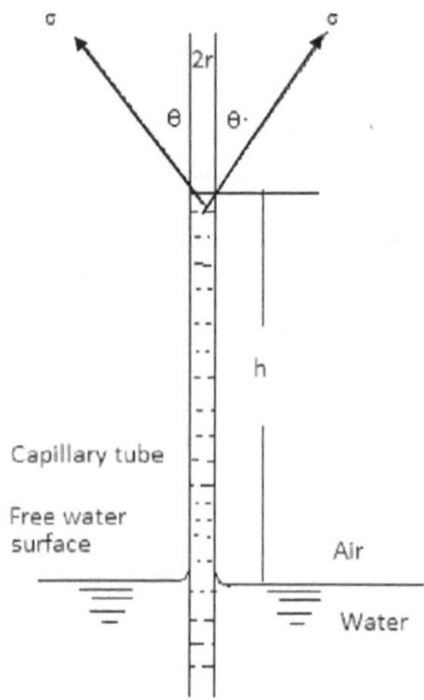

Fig. 6.4. Rise of water in capillary tube

The vertical component of surface tension is $2\pi r \sigma \cos\theta$, where r is the radius of tube (thus 2πr represents length of contact of liquid in tube), σ is the surface tension of water and θ is the angle of contact. The weight of the water column is $\pi r^2 h\rho g$, where, h is height of water column, ρ is the density of water and g is the acceleration due to gravity. Therefore,

$$\pi r^2 h\rho g = 2\pi r\sigma \cos\theta$$

$$\text{or, } h = \frac{2\sigma\cos\theta}{r\rho g} \tag{6.28}$$

In water contact angle is so small that it can be assumed to zero. The height of water rises a capillary tube depends on its diameter. Smaller the diameter, higher is the height of rise as shown in Fig 6.4. The surface tension may be expressed as dynes/cm or g/s^2.

Viscosity: Fluid d over a fixed horizontal surface considered to be the infinite number movement of fluid layers with differential velocities. The velocity of the layer with the fixed surface is stationary and the velocity of the layers increases with the distance increases up to nearly the surface of fluid. The fluid flow occurs in shear due to sliding of the adjacent layers to each other. This shear or retarding force or internal friction acts tangentially on any layer of fluid is called the viscous force. The viscous force is directly proportional to its surface area and velocity, inversely proportional to its distance from the stationary layer. The viscous force between the stationary fluid layers at some distance apart may be expressed as

$$F \propto A\frac{dv}{dx}$$

$$or, F = -\mu A\frac{dv}{dx} \tag{6.29}$$

Where F= viscous force, dynes; μ= coefficient of viscosity or dynamic or absolute viscosity; A=surface area, cm^2; dv=velocity difference between the two layers of fluid, cm/s and dx=distance between two layers of fluid, cm. The value of μ depends on the nature of fluid. Negative sign indicates the direction of viscous force is opposite to that of velocity.

For a unit surface area and the velocity gradient of unity,

$$F = \mu \tag{6.30}$$

This states that the coefficient of viscosity as the force per unit area necessary to maintain a viscosity difference of 1cm/s between two parallel layers of fluid which are 1cm apart. The unit of viscosity is named as poise in the honour of J.L.M.Poiseuille(1844). Poise is one dyne-second per square cm. It can be checked from Eq.6.29 such that

$$\mu = \frac{F}{A\,{}^{dv}\!/_{dx}} = \frac{Force}{Area \times Velocity\ gradient}$$

$$= \frac{MLT^{-2}}{(L^2)\left(\frac{L/T}{L}\right)} = \frac{MLT^{-2}}{L^2T^{-1}} = ML^{-1}T^{-1}$$

The velocity of water is greatly influenced by the temperature of water. This influence is related as

$$\mu = \frac{0.0179}{1 + 0.03368 + 0.0002217T^2}$$

where μ = *dynamic viscosity of water, dyne -s/cm*^{2}

and T= temperature of water, °C.

Fluid of lower viscosity flows more readily and called as greater fluidity. The fluidity, f, is the reciprocal of viscosity,

$$F = \frac{1}{\mu} \tag{6.31}$$

Viscosity of water decreases by 3% per 1°C rise in temperature. Liquid having high viscosity do not flow easily and are known as viscous fluid.

There is another representation of coefficient of viscosity called as kinematic viscosity, and defined as,

$$v = \frac{dynamic\ vis\ cosity\ \mu}{mass\ density\ \rho} = \frac{\mu}{\gamma / g} = \frac{\mu g}{\gamma} \tag{6.32}$$

The unit of v is cm^2/s.

Example 6.5 In a capillary tube of diameter 1.2mm water rises to 2.55cm with the angle of water contact 1.5^0. Calculate the surface tension. Assume density of water 1.0g/cm^3.

Solution:

Diameter of , capillary tube, r = 1.2 mm

Density of water, ρ=1.0g/cm^3

Angle of water contact, $\theta = 1.5^0$

We know, $h = \dfrac{2\sigma \cos\theta}{r\rho g}$

$$\therefore\ \sigma = \frac{hr\rho g}{2\cos\theta} = \frac{2.55cmx0.06cmx1g/cm^3 \times 981cm/s^2}{2 \times \cos 1.5^0}$$

$$= \frac{150.09}{2 \times 0.999}$$

$= 75.12g/s^2 = 75.12gcm/s^2\ cm$

$= 75.12$ *dynes/cm*

6.4 Infiltration

Infiltration is the process of entry of water in to soil. It occurs when water crosses the soil-air boundary to soil. It also may be defined as the 'downward movement of water in to soil'. Thus, it implies one-dimensional vertical flow of water. However, when water enters in to soil from an irrigation furrow or from a waste water trench, the lateral movement of water becomes two-or-three-dimensional

Infiltration rate is defined as the rate of entry of water in to soil at any given time. Infiltration capacity indicates the soil characteristics of maximum rate of infiltration in a specified soil condition. It signifies soil infiltrability and expresses the characteristics of how much the rain water will be absorbed by the soil and how much will follow the surface run off. Infiltration rate has the dimension of velocity (m/s). In a process of continuous infiltration, the water entry in to soil reaches to a nearly constant rate; this rate is called basic infiltration. Cumulative infiltration is the total quantity of water infiltrated in a given time.

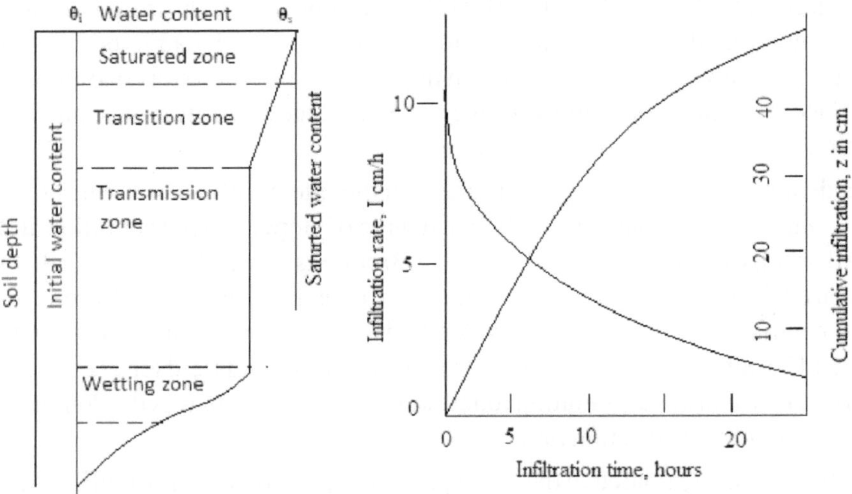

Fig. 6.5 Soil-moisture profile during ponded infiltration

Fig. 6.6 Usual characteristics of infiltration

During the process of ponded infiltration in to dry soil in a uniform profile the moisture profile may be divided in to the following zones (Fig.6.5).

1. Saturated zone: This is the zone of saturation extends up to about 1.5cm from the soil surface.
2. Transition zone. This zone is abou5cm thick located below the saturated zone. Rapid decrease in water occurs in this zone.
3. Transmission zone: This is the lengthy zone where little change of water contact occurs throughout the depth as well as with the time. In this zone, suction gradient is negligible and the movement of water takes place due to gravity.
4. Wetting zone: In wetting zone sharp decrease in water content is observed.
5. Wetting front: This is the zone of the sharp steep moisture gradient and forms the boundary between the wet and dry soil. This zone limits the moisture penetration and beyond this zone there is no visible penetration of water.

Factors affecting infiltration

The factors which mainly influence the infiltration are: antecedent (initial) moisture content, hydraulic conductivity of the soil profile, roughness of soil surface, texture, porosity, degree of swelling of soil colloids and organic matter, vegetation on soil surface, duration of rainfall or irrigation and viscosity of water. Ghosh(1975) summarized the works of antecedent moisture content on infiltration. Horton (1933) was one of the pioneer to recognize the effect of antecedent moisture on infiltration, but later he opined that the infiltration process primarily by condition at the soil surface. The combined work of Duley (1939) and Bodman and Colman (1944) emphasized that water infiltration in to soils is a dynamic process in which both the moisture potentials within the soil and soil structural conditions at the surface may change strikingly during the early stage of infiltration, causing a reduction in infiltration with elapsed time.

Antecedent moisture gives a greater effect on the rate of infiltration during the first 20 minutes of simulated rainfall than degree of slope, surface condition and rainfall-intensity factors (Neal, 1938). Breakensick and Frevert (1961) emphasized the need of accounting soil moisture content when using infiltrometer data to estimate run off on actual watersheds. Jamison and Thornton (1961) analyzed 12 years rainfall-run off data from runoff plots and field terraces under various surface and cover condition found soil moisture content was the dominant factor determining infiltration rate.

Infiltration is closely related to the pore space available for storage of infiltrating water (Hansen, 1955). On some soils, the total pore space may be reduced when soil colloids swell upon wetting. If the volume change of a soil, especially subsoil exceeds 20 percent upon wetting, then infiltration rate is reduced to extremely low values (Browning, 1939). Open and closed graded silica sand columns used by Free and Palmer (1940) to evaluate the effect of entrapped and escaped air upon infiltration rate shows that for open column infiltration rate continues to decrease until the entire column is wet as a function of diameter of the particles making up the column. In closed column, infiltration rate is governed by the pore –size and air-water movement.

Infiltration models

Many different infiltration equations have been published in literature. The most common of these are presented here in chronological order with a brief discussion of the deviations, attributes and deficiencies. The symbol I is used as the cumulative quantity of water infiltrated in time t per unit cross-section of soil surface and $i = \frac{dI}{dt}$ as the infiltrability (a flux).

The earliest equation was that of Green and Ampt propounded in 1911 (cited by Mein and Larson (1971)).

$$i = K_s \frac{L + P_w}{L} \tag{6.33}$$

This assumes that the soil surface is covered by a pool of water whose depth can be neglected and that the conductivity in the wetted zone is K_s. L is the distance from the soil surface to the wetting front and P_w is the capillary suction at the wetting front. By continuity, $(\theta_s - \theta_i) L = I$, and $(\theta_s - \theta_i)$ is the initial moisture content deficit of the soil mass or profile.

Thus, $L = \frac{I}{\theta_s - \theta_i}$

$$\therefore \frac{dI}{dt} = i = \left(\frac{\frac{I}{\theta_s - \theta_i} + P_w}{\frac{I}{\theta_s - \theta_i}} K_s \right)$$

$$= \frac{I + P_w(\theta_s - \theta_i)}{I} K_s$$

$$= K_s \left\{ 1 + \frac{(\theta_s - \theta_i)}{I} P_w \right\} \quad (6.34$$

Integrating and substituting I = 0 at t = o

$$I - (\theta_s - \theta_i) P_w l_n \left\{ \frac{I + (\theta_s - \theta_i) P_w}{(\theta_s - \theta_i) P_w} \right\} = K_s t \tag{6.35}$$

This forms the Green-Ampt law and is more convenient than Eq.6.33 for watershed modeling, as it relates the infiltration volume to the time from the start of infiltration. From it, one can obtain the volume of any given time or vice-versa. The advantage of Eq.6.35 is that all of the parameters are physical properties of the soil water system and can be measured. Estimation of P_w, however, is difficult, since the capillary suction always varies widely with moisture content.

The next equation is that of Kostiakov (1932) as cited by Kein and Larson (1971).

$$I = at^b \tag{6.36}$$

Where a & b are constants $(0 \angle b \angle 1)$.

The form of this equation is simple but the constants a and b cannot be predicted in advance. The simplicity has encouraged the use of this equation in field irrigation studies. The constants a depends on θ_i as well as on the soil type, and physical significance of a, directly depends on b and, therefore, indirectly t-range. The constant b depends on the soil type and on the range of t. This form does provide an infinite i, rather than a constant non-zero value. This could relevance for purely horizontal water desorption, wherein Philip's (1957) exact solution becomes only the first term of the series equation.

$\left[I = St^{\frac{1}{2}} + (A_2 + K_i)t + A_3 t^{\frac{3}{2}} + A_4 t^2 + \ldots.. \right]$ with a=S and b=1/2. This is good for a common time range and for initially dry fine-textured soils in which the capillary potential gradients tend to be relatively more important than the gravitational potential gradients. For wet sandy soils in which gravity becomes important, b will be closer to one. Again, the physical significance of a, is obscure and will vary with the b-value (Ghosh, 1975).

Kostiakov met these limitations by stipulating that Eq.6.36 holds good only for $i > K_s$ and for $t > (\frac{ab}{K_s})^{\frac{1}{1-b}}, i = K_s$. This procedure presupposes knowledge of K_s and is therefore difficult to know its validity.

Horton proposed the following equation in 1940 (Childs, 1969).

$$i = f_c + (f_0 - f_c)e^{-k^* t} \tag{6.37}$$

Where f_c, f_0 and k^* are the characterizing constants.

At t=o, the infiltrability is not infinite, but takes on the finite value f_c. The constant k^* which depends on the soil type and θ ditermines how quickly i will change from f_0 to f_c.

This is from is also integrable and provides I as an explicit function of t, but not vice-versa. Under certain conditions, this equation may be more cumbersome to work with since it contains three constants instead of two.

The fourth equation (Eq.6.38) has been suggested by Philip (1957):

$$I = St^{1/2} + At \tag{6.38}$$

Where S is designated as sorptivity and A is a second parameter.

Assumptions of Philip's Infiltration Equation

1. The soil is homogeneous and is of semi-infinite depth
2. The moisture content is uniform in the whole profile
3. The time, t, is 'short'

4. Value of A is relatively small compared with S
5. Moisture movement is one-dimensional (vertically downward)
6. The soil is a non-swelling type
7. There is no structural collapse at the soil surface (soil and water interface) throttling flow of water
8. There must be no rearrangement of soil particles upon wetting
9. The air movement does not influence the water movement. This condition requires water movement is analogous to water flow where consideration is given to only single phase
10. The properties of water are uniform regardless of the position occupied by water
11. An isothermal condition exists.

The **Eq.6.38** is convenient and S and A can be predicted in advance, though the method of finding their values is complex. On the basis of tests on Green-Ampt and Philip's equation, Whisler and Bouwer (1970) concluded that Green-Ampt equation was not only simpler to use, but also gave better results. They experienced considerable difficulties in predicting S and A for Eq.6.38.

According to Philip (1969), whether the time is short or not, depends on certain conditions (Ghosh, 1975). Thus a characteristic time, t_{grav}, based on S and $(K_s$-$K_i)$ of the infiltration process is introduced as:

$$t_{grav} = \left(\frac{S}{K_s - K_i} \right)^2 . \tag{6.39}$$

The order of magnitude of the t value at which the effect of gravity on the process can be expected to be as great as that of capillarity is given by t_{grav}.

Sorptivity model

According to Philip (1957,1972) and others the sorptivity S and the parameter, A, both are functions of θ. Attempts have been made by several workers to establish relationships of S and A with θ and other soil parameters. A few of these are given here.

Youngs (1968) developed the following relationships:

$$S = \sqrt{\left[2\int_0^{\alpha} (\theta - \theta_i)\lambda d\lambda \right] (\theta_s - \theta_c)^{0.5}} \tag{6.40}$$

in which $\lambda = xt^{-\frac{1}{2}}$, the Boltzman transformation used to give the moisture content θ as a function of λ.

According to Brutsaert (1968) exact solution derived from Philip's series expansion, S, is given by

$$S = \sqrt{\frac{2A\,k_s}{n}}\,(\theta_s - \theta_i)^{0.5} \tag{6.41}$$

wher n is given by $K = K_s S_n$ in which K stands for unsaturated hydraulic conductivity at moisture content, θ, and S_n is the normalized moisture content and is equal to $(\theta - \theta_i)/(\theta_s - \theta_l)$ or A' (A'-h), where A' is a constant and h is the pressure in water (height).

Talsma (1969, 1973) observed that for the initial 1 to 2 minutes of infiltration in the field (i.e. *in situ* situation), the first term of the Philip's equation (Eq.6.38) dominates the flow and thuss

$$I = St^{1/2} \tag{6.42}$$

For S, Talsma proposes the relation

$$S = \sqrt{2K_s\,(P_w + h)}\,(\theta_s - \theta_i)^{0.5} \tag{6.43}$$

Where K_s is the saturated hydraulic conductivity, P_w is the suction at the wetting front and h is the head of water in the cylinder infiltrometer.

P_w is not a strictly measureable variable. This is generally measured by tensiometers, neutron scattering method, gamma ray method, etc. Mein and Larson (1971) gave a suitable method of estimating P_w as discussed below.

Estimation of P_w

These would appear to be two alternative ways of determining P_w, from the $\varphi - \theta$ relationship or from the $\varphi - K$ relationship from the soil in question. Capillary conductivity is effectively zero at a moisture content of about 30 percent. For moisture below the level, soil moisture movement is comparatively very slow, despite the high suction in the region. Since we are interested in the moving front, it would seem that the average capillary suction, φ, is best determined from the $\varphi - K$ curve. It is proposed, therefore, that P_w be determined by

$$P_w = \frac{\int_{K_i}^{K_s} \varphi\, dK}{K_s - K_i} \tag{6.44}$$

It is more convenient to use the relative conductivities, K_r, instead of the absolute conductivity, K.

$$K_r = K/K_s \tag{6.45}$$

Substituting **Eq.6.45** in to **Eq.6.44** we have

$$P_w = \frac{\int_{K_r(initial)}^{1} \varphi dK_r}{1 - K_r(initial)} \qquad (6.46)$$

Typically the initial moisture content before an event is such that $K_r \leq 1, (\theta \leq \textit{field capacity})$, so that we can write $K_r(initial) \approx 0$ with negligible error. Then,

$$P_w = \int_0^1 \varphi dk_r \qquad (6.47)$$

or simply the area under φ - K_r curve, between $K_r = 0$ and $K_r = 1$.

Model of second parameter of Philip's equation

Young's (1968) gives a method of estimating a value of A by comparing Eq.6.38 with the infiltration equation (Eq.6.33) of Green and Ampt which refers to an idealized porous material in which the moisture profile during infiltration is rectangular in shape. This equation (Green and Ampt) which has been discussed further by Philip may be written in the form

$$I - K_s t = \frac{S^2}{2K_s} \ln\left(1 + \frac{2IK_s}{S^2}\right) \qquad (6.48)$$

By expanding this equation in powers of I and writing Eq.6.38 in the form of

$$t = \frac{2AI + S^2}{2A^2}\left\{1 - \sqrt{1 - \frac{4A^2I^2}{(2AI + S)^2}}\right\} \qquad (6.49)$$

and by expanding this in term of I, two equations may be compared at small times; the coefficient of I^2 in the two expanded expression are the same ($=1/s^2$) and those of I^3 if equated:

$$A = \frac{2}{3}K_s \qquad (6.50)$$

Philip (1957) considers Eq.6.38 to be useful enough for all but large times and to be useful in applied hydrological studies although he observes that since

$$\left(\frac{dI}{dt}\right)_{t \to \alpha} = K_s \qquad (6.51)$$

the equation must fail with these coefficients, because A has the value, $(1/3)k_s$ approximately in the example of infiltration on yellow light clay soil, he used. It has also been shown by Philip (1969) that for very large times A approaches the value of $\frac{K_s - K_i}{\theta_0 - \theta_i}$. Again it is mentioned by Philip (1969) that

$$A=A_2+K_i \tag{6.52}$$

and for which usual case with $K_i/K_s \leq 1$ it is found to be empirically and by analysis that

$$A_2 \approx \tfrac{1}{2}(K_s - K_i) \tag{6.53}$$

$$\text{Thus, } A= \tfrac{1}{2}(K_s + K_i) \tag{6.54}$$

$$\text{or,} (2A/Ks - 1) \approx K_r (\text{initial}) \tag{6.55}$$

Philip (1972) observes that for a homogeneous non-swelling soil it can be said at least that A is of the same order of magnitude as (K_s-K_i) as stated by Ghosh (1975). If one wishes to fit the data to the truncated Eq.6.38 of Philip, A will vary perhaps as little as $\tfrac{1}{3}(K_s - K_i)$, when data is fitted from t=0 to t=T with T small, to (K_s-K_i) in the limit of $T \rightarrow \alpha$.

Ghosh's Model

Ghosh (1975) studied the effect of initial soil moisture on infiltration to investigate (a) the prediction of infiltration behaviour as influenced by antecedent moisture, making use of Philip's two-term infiltration equation and, (b) the applicability of the predictive models under field condition where soil is generally non-homogeneous. Based on the study of ten soils at different antecedent moisture levels simple models to predict S (sorptivity) and A (second parameter) of the Philip's two-term infiltration equation of the time series function viz.,

$$I = St^{1/2} + At$$

were proposed for the non-idealized case of ponded water infiltrating into a heterogeneous field soil at average (not uniform throughout) initial moisture content:

$$S = \alpha(\theta_0 - \theta_i)^{\beta} \tag{6.56}$$

$$A = -S + 10^{\alpha_0 - 10^{\beta_0}} {}^{-\gamma_0(\theta_0 - \theta_i)} \tag{6.57}$$

Where S is the sorptivity in cm/ (min)$^{1/2}$, A is the second parameter of the Phill ip's equation (Eq.6.38) in cm/min, $(\theta_0 - \theta_i)$ is the initial moisture deficit of the soil mass or profile in vol / vol and $\alpha, \beta, \alpha_0, \beta_0$ and γ_0 are constants. The values of these constants depend probably on soil properties, but could not be ascertained fully in this study.

Measurement of infiltration

The techniques of measurement of infiltration in soils may be grouped in to two types: (i) Measurement by noting the difference between the rainfall/irrigation and runoff and, (ii) Measurement of entry of water into soils in controlled condition where run-off is checked (Mal, 1995).

In controlled condition infiltration is measured by using either single ring or double ring infiltrometer. The infiltrometers are usually 30 cm diameter and made of rolled steel sheet of 2-3 mm thickness. If a m.s. flat (20mm x 5mm) welded around the top outer edge of top of it makes the infiltrometer hardy, robust and capable of receiving intense hammering during the process of inserting it into the soil. The lower edge of the infiltrometer is sharpened to bevel shape for permitting easy penetration into soil. In a single ring infiltrometer divergence of flow occurs while passing though it crosses the bottom of the cylinder in the subsoil. This leads to inaccurate result of infiltration. Therefore, double ring infiltrometer is widely used where there is an additional ring around the inner ring to provide a buffer zone through where water also infiltrates to check the divergence of flow.

Double ring infiltrometer

As shown in Fig.6.7 the double ring infiltrometer is usually of 30 cm diameter inner ring, 60 cm diameter of outer ring and 25 cm height. The rings are inserted 10 cm depth in to soil in co-centric fashion during the experimentation. To provide with the marking of 10 cm level of rings ensures equal depth of penetration of the rings at every site into soils. The rings are driven into the soil by the falling weight striking the wooden plate on the rings. If the outer edge of the top of the rings is welded properly direct light hammering on the edge may not cause much damage on the edges. During the process of insertion of the rings into soil care is to taken that hammering is always done in such a way that ensure uniform vertical penetration and avoid disturbance of topsoil.

The hook gauge or point gauge is commonly used to indicate the level of water in the infiltrometer. This is fitted inside the inner ring. Sometime the common plastic scale is placed inside the inner ring to serves the purpose. However, hook or point gauge provides the accurate result. The tip of the measuring rod of the gauge is set at the desired level to which water is added. About 10 cm water level is maintained inside the cylinder, which is the usual depth of flow in border/ basin irrigation. In a properly set infiltrometer water is added into the inner ring to the level of tip of the rod. During adding of water jute or paddy straw mats are provided to avoid soil erosion. As soon as the inner ring is filled with water the outer ring is also filled with water of equal depth. Readings are taken at short interval at the beginning and it progressively increases with the advancement of time. It is to be monitored that before lowering of water more than 1 cm water should be added into the inner ring to its initial level from a source containing measured amount of water. The depth of water infiltration in any reading can be had by dividing the volume of water added by the area of inner ring. The readings are tabulated as cumulative depth of infiltration against time. These data are processed for fitting the curves or equations.

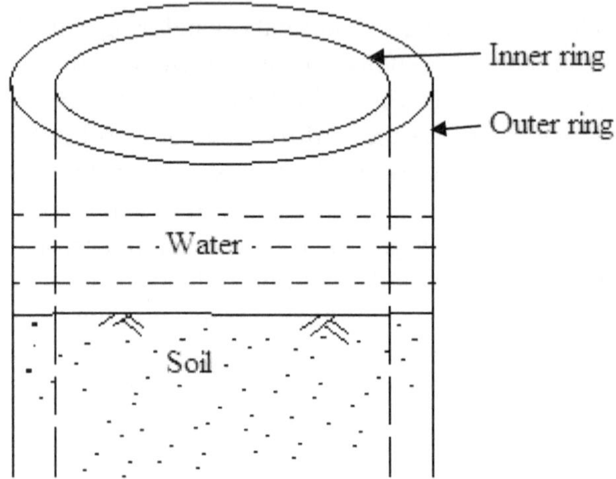

Fig. 6.7 Double ring infiltrometer

Curve fittings: The Table 6.1 containing the experimental infiltration data. These data may be processed graphically and statistically. The use of computer may faster the processing.

Table 6.2. Observed infiltration data and its processing

t, min	I, cm	logI (y)	logt (x)	logtxlogI (xy)	$(\log t)^2$
1	0.035	-1.4559	0	0	0
3	0.082	-1.0862	0.4771	- 0.5182	0.2276
8	0.182	- 0.7394	0.9031	- 0.6678	0.8256
13	0.264	- 0.5784	1.1139	- 0.6443	1.2408
23	0.424	- .3747	1.3617	- 0.5102	1.8542
43	0.703	- 0.1530	1.6334	- 0.2499	2.6680
53	0.925	- 0.0338	1.7242	- 0.058	2.9729
68	1.097	0.0402	1.8325	0.0737	3.3580
	$\sum =$	-4.3912	9.0459	-2.5747	13.1371

a) Equation of the form, $I = at^b$

Graphical: I versus t data of Table 1 are plotted in log-log graph paper and logI versus logt in plain graph paper (squared) as shown in Figs. 6.8 and 6.9. Any one of these two plots may serve the purpose. In fact the plot in squared graph paper is the substitute to plotting in log-log graph paper.

From Fig.6.8 when t = 1, I = 0.0355. But when t = 1, I = a. Therefore, a = 0.0355.

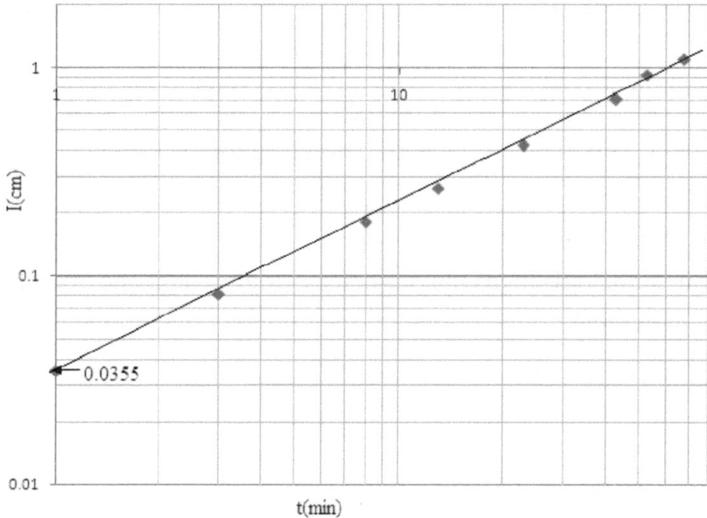

Fig. 6.8 I vs. t on log-log paper

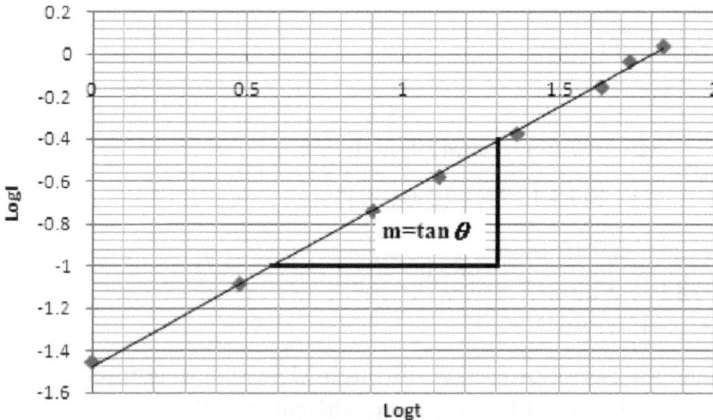

Fig. 6.9 logI vs. logt in squared paper

From Fig.6.8,

$$\tan\theta = \frac{\log(0.222) - \log(0.0355)}{\log 10} = \frac{\log\left(0.222/0.0355\right)}{\log 10} = \frac{0.796}{1} = 0.796$$

The straight line equation y = mx + c, where m is the slope if y versus x are drawn. If log is taken for the equation $I = at^b$,

$$\log I = \log a + b \log t \tag{6.58}$$

The logI and logt substitute the y and x and loga and b are c and m respectively. Therefore, the slope $\tan\theta = m = b = 0.796$.

The fitted equation is become, $I = 0.0355t^{0.796}$. (6.59)

From Fig.6.9, slope of the plot $m = b = \dfrac{-0.646-(1.0)}{1.0-0.56} = 0.8045$

The plot intercepts at -1.45 in the vertical axis. Therefore loga=-1.45.

$\therefore\ a = 0.03548$,

The fitted equation, $I = 0.03548t^{0.8045}$ (6.60)

Statistical: Using **Eq. 6.58**, logI=loga+blogt,where, y=logI, x=logt and c=loga. Using the least square method,

$$c = \log a = \frac{\Sigma x^2 \Sigma y - \Sigma x \Sigma xy}{N \Sigma x^2 - (\Sigma x)^2}$$

$$= \frac{13.1371(-4.3912) - (9.0459)(-2.5747)}{8(13.1371) - (9.0459)^2}$$

$$= \frac{-57.6876 + 23.2905}{105.0968 - 81.8283} = \frac{-34.3971}{23.1717} = 1.4844$$

$\therefore\ a = 10^{-1.4844} = 0.0328$

$$b = \frac{N \Sigma xy}{N \Sigma x^2 - (\Sigma x)^2} = \frac{8(-2.5747) - (9.0459)(-4.3912)}{8(13.1371) - (9.0459)^2}$$

$$= \frac{-20.5976 + 39.7224}{105.0968 - 81.8283} = \frac{19.1248}{23.2685} = 0.8219$$

The fitted equation, $I = 0.0328t^{0.8219}$ (6.61)

Computer: The data have been processed through MS excel in computer and the equations are taken in linear and power form with the correlation coefficient R values for comparison of the correlations. It appears that the power form gives the equation with the best correlation coefficient value as 0.998 (Fig.6.10). Therefore, the accepted equation of the available data is

$I = 0.033t^{1.28}$ (6.62)

This equation is similar to Kostiakov or Lewis equation of the form $I = at^b$.

1(cm)	t(min)
0.035	1
0.082	3
0.182	8
0.264	13
0.424	23
0.703	43
0.925	53
1.097	68

b) Equation of the form $I = at^{\alpha}+b$ (6.63)

where, I= cumulative infiltration, cm t=elapsed time, min

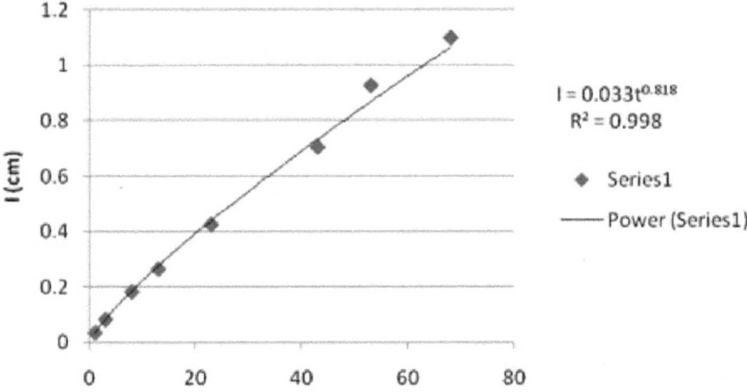

Fig. 6.10. Infiltration characteristics

The value of a, α, *b* usually varies between 0 and 1. The values of the constants a, α, *b* may be determined by plotting I against t in a plain graph paper (Murty, 2004). Select two points (t_1,I_1) and (t_2,I_2) near the extremities of the smooth curve representing the data. Find $t_3 = \sqrt{t_1 t_2}$. Get y_3 against t_3.

$$b = \frac{I_1 I_2 - I_3^2}{I_1 + I_2 - 2I_3} \tag{6.64}$$

Taking logarithm of Eq.6.63,

Log (I-b)=loga + α logt (6.65)

Put the values of b and the observed values of I and t in Eq.6.65. This will give numbers of equation.

Get two equations by adding n and N-n equations (n=1/2(N-1)). Solving these two equations give the values of a, α, *b*.

The data set in Table 6.1 are used to plot the smooth curve in plain graph paper (Fig. 6.11). From the smooth curve the two points (t_1,I_1)= (3,0.14) and (t_2,I_2) = (68,1.097) are selected on and near the extremities of the smooth curve.

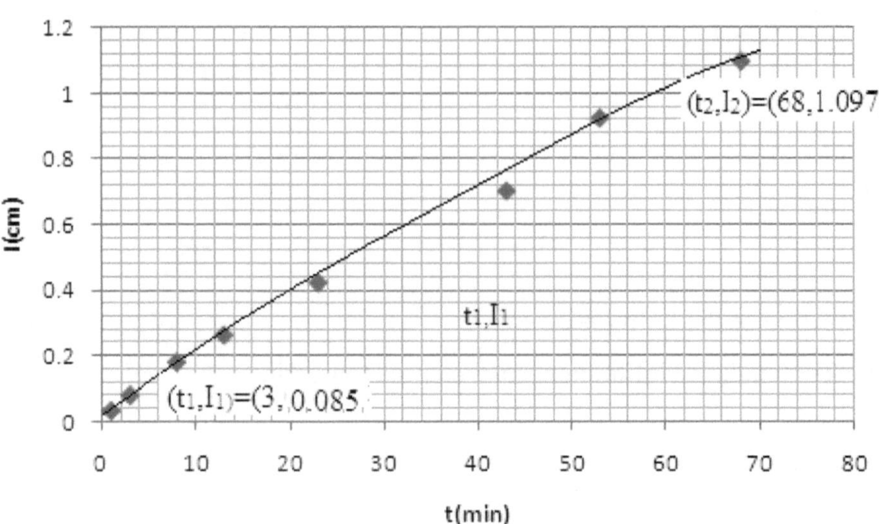

Fig. 6.11 Smooth curve of I vs. t on squared paper

Therefore, $t_3 = \sqrt{t_1 t_2}$

$= \sqrt{3 \times 68}$

$=14.2829$

The value of I_3 corresponding to t_3= 0.290

$$\therefore b = \frac{(0.085)(1.097)-(0.29)^2}{0.085+1.097-0.29\times 2}$$

$$= \frac{9.145\times 10^{-3}}{0.602} = 0.01519$$

Putting the value of b in Eq.6.65

Log (I-0.01519)=loga+α logt

Substituting the values of I and t from the data set (Table 6.1) the following equations are obtained.

-1.17516=loga+0.47712α -- (i)

-0.777777=loga+0.90309α -- (ii)

-0.610413=loga+1.11394α -- (iii)

-0.39061=loga+1.36172α -- (iv)

-0.162531=loga+1.63347α -- (v)

-0.04105=loga+1.72427α-- (vi)

0.03415=loga+1.83251α -- (vii)

Adding the first 4 and last 3 equations,

-2.94767=4loga+3.85588α --(viii)

-0.16943=3loga+5.19025α -- (ix)

Multiplying Eq.(viii) by 3 and Eq.(ix) by 4

-8.84301=12loga+11.56764α -- (x)

-0.67772=12loga+20.761α -- (xi)

Subtracting Eq.(xi) from Eq.(x)

-0.8.16529=-9.19336

$\therefore\ \alpha = 0.88817$

Again 4loga=-6.37236

$\therefore$ loga=-1.59309

or, a=0.02552

Therefore , the fitted equation is $I=0.02552t^{0.88817}+0.01519$

c) When Philip's equation is used, $I = At + St^{\frac{1}{2}}$ (6.66)

Differentiating Eq. 6.66 with respect to time

$$\frac{dI}{dI} = i = A + \frac{1}{2}St^{-\frac{1}{2}} \tag{6.67}$$

The **Eq.6.67** is like a straight line equation y=mx+c; where i=y, $t^{\frac{1}{2}} = \times, \frac{1}{2}S = m$, and c=A.

Therefore, the plot of y vs $t^{-\frac{1}{2}}$ gives the straight line. The fitted equation can be obtained by least square method.

The given data in Table 6.1 are processed and tabulated in Table 6.2.

Table 6.3 Processing of infiltration data to fit in Philip's equation

t	I	$t^{-\frac{1}{2}}$	$\left(t^{-\frac{1}{2}}\right)^2$	i	$t^{-\frac{1}{2}}.i$
		x	**(x^2)**	**(y)**	**(xy)**
1	0.035	1	1	0.035	0.035
3	0.082	0.5773	0.3333	0.0235	0.01356655
8	0.182	0.3535	0.1250	0.0200	0.00707
13	0.264	0.27735	0.0769	0.0164	0.00454854
23	0.424	0.20851	0.04348	0.0158	0.003294458
43	0.703	0.1525	0.02356	0.01405	0.002142625
53	0.925	0.1374	0.018868	0.0222	0.00305028
68	1.097	0.12127	0.014706	0.01147	0.0013909669
$\sum$=	3.71	2.8278	1.63551	0.15842	0.070063

$$c = A = \frac{\Sigma x^2 \Sigma y - \Sigma x \Sigma xy}{N \Sigma x^2 - (\Sigma x)^2}$$

$$= \frac{1.63551 \times 0.15842 - 2.8278 \times 0.070063}{8 \times 1.63551 - (2.8278)^2}$$

$$= \frac{0.2591 - 0.19812415}{13.0841 - 7.99645} = \frac{0.060975848}{5.08765} = 0.011985$$

$$m = \frac{1}{2} S = \frac{N \Sigma xy - \Sigma x \Sigma y}{N \Sigma x^2 - (\Sigma x)^2} = \frac{8x0.070063 - 2.8278x0.15842}{8x1.63551 - (2.8278)^2}$$

$$= \frac{0.560504 - 0.44798}{5.08765} = \frac{0.1125239}{5.08765} = 0.022117$$

Thus, the fitted equation is i = $0.001985+0.022117t^{\frac{-1}{2}}$

6.5 Soil Moisture Characteristics

The relationship between soil moisture content and moisture tension are expressed by soil moisture characteristics curves. The soil moisture contents for a particular tension vary in different soils depending on the soil texture, structure and some other characteristics of soils. Coarse soils can retain much less quantity of water than to fine soils at same tension. At very low-tension sandy soil retain small quantity of water whereas clay soils hold considerable quantity of water. Therefore, soil moisture tension is not necessarily the indication of availability of water for plant use.

The significance of the curves is best understood by the nature of slope of the curves. The vertical section of the curves represents the large number of pores of the same size in which a little change of tension causes a large change of water content. Horizontal section represents the situation where no pores are available of a size that will release water over the tension range concerned. The straight intermediate slope represents the pore size distribution, which uniformly drain the water with the change of tension. A slope of the curve is referred as differential water capacity, which represents the amount of water release with the change of tension. The amount of water release per unit change in tension is called the specific water capacity of the soil. Soil moisture characteristics curves provide the information of availability of water for plant use, need of water for irrigation and the amount of water held in soil before percolation starts.

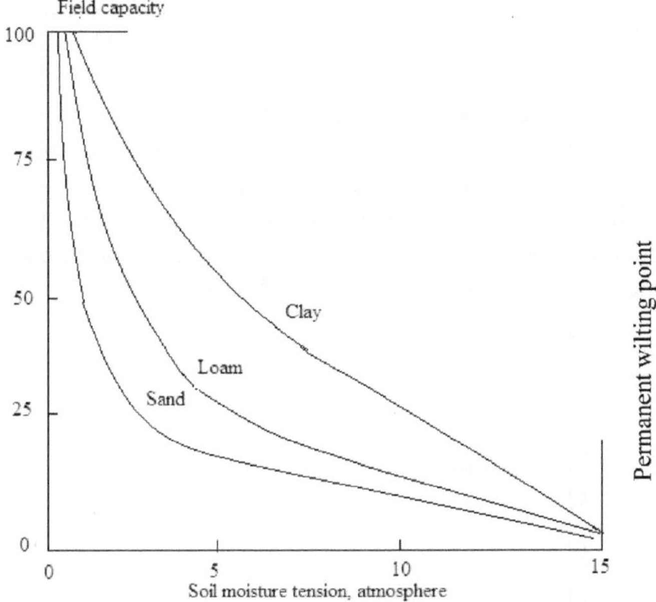

Fig. 6.12. Typical soil moisture characteristics curves for different soils. Soil moisture characteristics curves may be developed by using the tension apparatus or pressure plate apparatus.

Hysteresis: The soil moisture tension and moisture content in soil is not generally a unique character that always represents same tension at same moisture. It varies at desorption and sorption. Desorption is the drying of soil from an initially saturated sample and sorption is wetting of soil from initially dry soil. Soil suction is more in desorption than sorption at equilibrium soil wetness. Fig. 6.13 shows the typical soil moisture characteristics with hysteresis effect at equilibrium soil-moisture state.

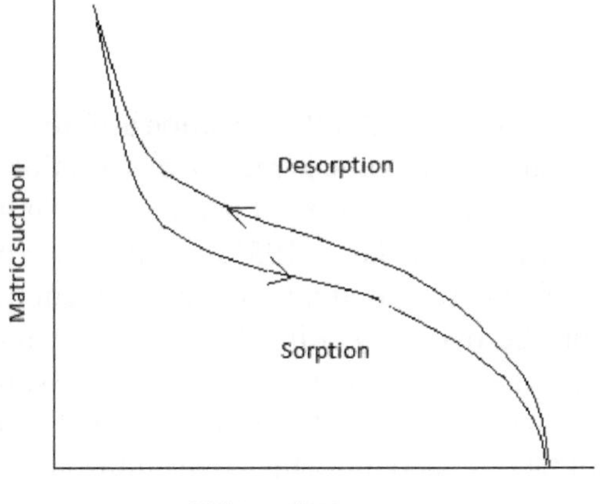

Fig. 6.13. Hysteresis curve

Tension apparatus: As shown in Fig.6.14 this consists of a sintered/ceramic plate fixed at the bottom of the glass/ceramic funnel connected with a hanging water column by flexible tube. The tube is usually transparent. However, the rubber or any opaque flexible tube with glass burette at the end of the pipe can also be used. The processed representative soil sample is placed on the ceramic plate and compacted to get the required field bulk density. The soil sample is made saturated by using the water level in the flexible tube at the level of near the soil surface for considerable time. When it is clear that saturation has attained it is to be subjected to suction head by lowering the flexible tube.

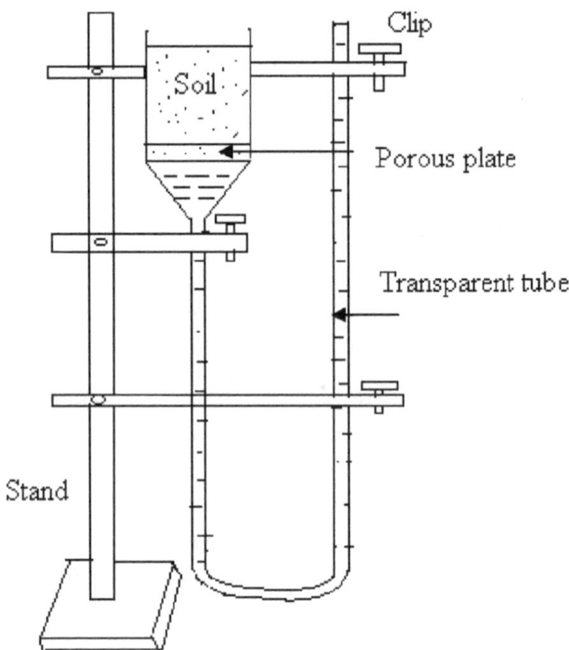

Fig.6.14. Tension apparatus

The set up is then kept untouched for a day. The next day the elevation difference between the soil sample and the water table in the tube is recorded as suction. The soil sample is taken out of the apparatus and its weight is found taking it inside a moisture box. It is then kept in oven for a day at 105^{0}C and the weight of oven dry sample is also found out. The corresponding moisture content at each suction is calculated based upon the oven dry soil. The soil moisture-tension curves are then obtained using these values. Fig.6.15 represents the soil moisture characteristics curves of the data given in Table 6.4 & 6.5. In tension apparatus theoretically suction cannot be obtained more than one atmosphere. In fact, in tension apparatus suction limit is around 0.8 atmospheres. Air enters in the tube beyond this suction.

Visser (1966) used the empirical equation to relate soil moisture tension (h) and water content

content (θ) as $h = A\frac{(E-\theta)}{e^m}$ where, E = porosity, and A, m and n are constants.

Table 6.4 Volumetric moisture content at various suction

Tension, T (cm)	Volumetric moisture content (θ), %
0	87.25
25	65.12
50	60.07
100	49.98
200	47.50
400*	39.92
800*	28.12
1000*	20.35

- **Extrapolated**

Table 6.5 Volumetric moisture content at various suction from smoothly fitted curve

Tension, T(cm)	Volumetric moisture content (θ),%
0	87.25
25	65.12
50	58
100	50
200	44.75
400	38
800	27.25
1000	20.85

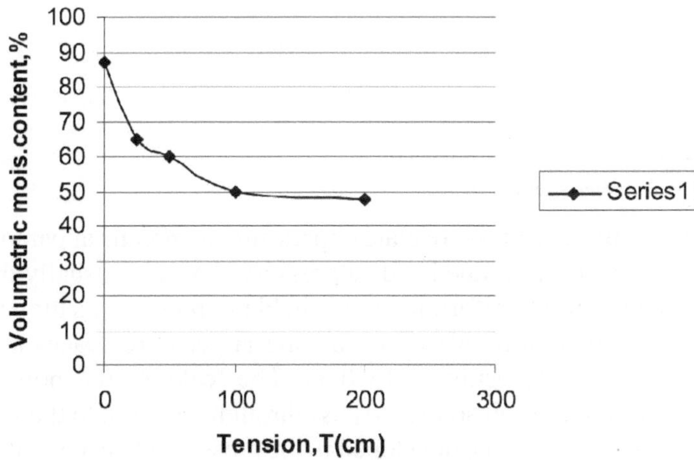

(a) Soil moisture characteristics curve with the available data

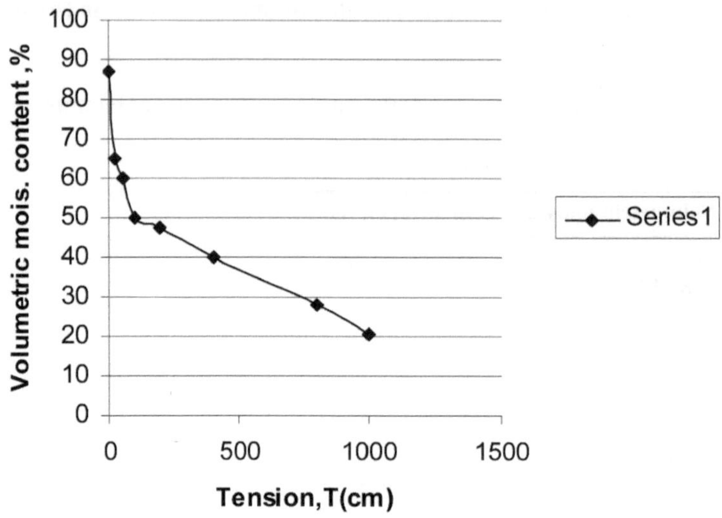

(b) Soil moisture characteristics curve with the available and extrapolated data

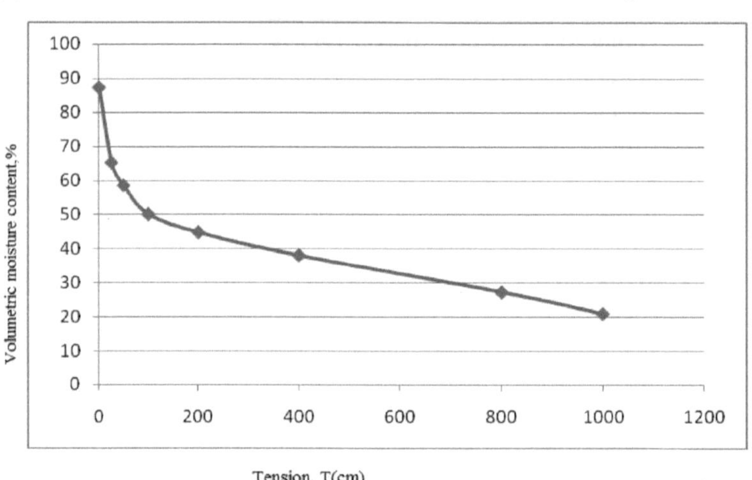

(c) Soil moisture characteristics smoothly fitted curve

Fig. 6. 15 Soil moisture characteristics curves

Pressure plate apparatus: The pressure plate or pressure membrane apparatus mainly consists of a porous membrane inside a pressure chamber. Usually the cellulose acetate provides good uniformity and suitable to pressure as high as higher than 110 bars. The ceramic plates are having larger pore spaces and effectively work within the pressure of 15 bars. The leaks of the porous membrane permit the water and soil solutes to pass through it but not to the soil matrix. The metric suction that can be developed in the pressure plate apparatus depends on the safe working pressure of the chamber and the pressure difference the membrane can admit without allowing the air bubble through the pores. Fig 6.16 shows a magnified view of pressure chamber where soil samples are in contact to cellulose membrane, which is in support to a fine mesh screen.

Pressure plate apparatus is largely used for determining water retaining characteristics of soils. Usually the processed soils are used to pressure plate apparatus. Therefore, the measurement needs special attention to use in field condition. The disturbance to soils is sometime avoided by using core samples. However, this requires more time (approximately square of the height of the sample) to get the samples at hydraulic equilibrium with the membrane. Adequate pre-wetting of samples is very important to avoid the possible errors in measurement. However, 1-2% errors are possible due to pressure and temperature fluctuation causing hysteresis effect, non-representative samples, non-attainment of out flow equilibrium, etc.

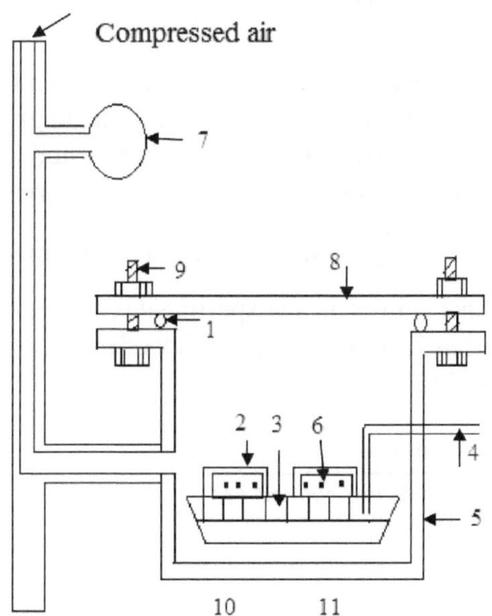

1. Pressure seal
2. Sample retaining ring
3. Ceramic porous plate
4. Outflow tube
5. Pressure vessel
6. Soil sample
7. Pressure gauge
8. Vessel lid
9. Clamping bolt
10. Rubber diaphragm
11. Pressure plate cell

Courtesy: Ghildyal & Tripathi (1987)

Fig. 6.16 Pressure plate apparatus

Determination of unsaturated hydraulic conductivity

Unlike the saturated hydraulic conductivity there is no direct method to find out the unsaturated hydraulic conductivity. It can be obtained by some indirect way with the available parameters which are saturated hydraulic conductivity and soil moisture characteristics curve. Using these parameters, Marshall and Millington Quirck have developed a method for determination of unsaturated hydraulic conductivity (Biswas, 1983).

In this method (M and MQ) the soil moisture content-tension relationship curve is divided in to M number of equal increments $(\theta_1 - \theta_{m+1})$ such that each increment of the curve will yield the same amount of moisture content, i.e., θ= constant for all increments. tentions $(T_1.....T_m)$ are obtained from the mid points at corresponding increments. The unsaturated hydraulic conductivities are then obtained using the following equation.

$$K_1 = K_0 \left(\frac{\theta_i}{\theta_0}\right)^P \frac{\sum_{J=i}^{m} (2J+1-2i)T_j^{-2}}{\sum_{J=i}^{m} (2J-1)T_j^{-2}} \tag{6.68}$$

Where,

K_i = calculated hydraulic conductivity at moisture content θ_i

K_0 = hydraulic conductivity at moisture content θ_0

m = number of increments

P = matching facto (the ratio of measured to a calculated conductivity at a particular moisture content, usually at saturation). P = 1 is found to give good result for predicting unsaturated hydraulic conductivities and i, j are indices. Using these conductivity values the relationship between the relative hydraulic conductivity K/K_0 and the tension T is obtained.

Example 6.6 Soil moisture and tension as obtained in an experiment is presented in Table 6.6. Determine the unsaturated hydraulic conductivities of this soil if the saturated hydraulic conductivity is 0.215cm/day.

Table. 6.6 Volumetric moisture content at various tension

S. No.	Tension, T (cm)	Volumetric moisture content (θ%)
1	0	87.25
2	25	65.12
3	50	60.07
4	100	49.98
5	200	47.5
*6	400	39.92
*7	800	28.12
*8	1000	20.35

*Extrapolated values (graphically)

Solution: Using the data in Table 6.6 the soil moisture characteristics curve is developed (Fig. 6.17).

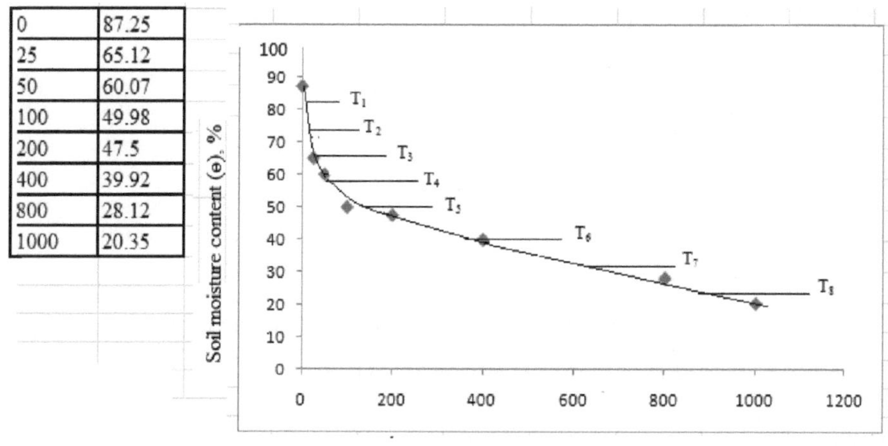

0	87.25
25	65.12
50	60.07
100	49.98
200	47.5
400	39.92
800	28.12
1000	20.35

Fig. 6.17. Soil moisture characteristics curve by using data of Table 6.6

The soil moisture content in between the first and last (87.25-20.35=66.69%) is divided in to 8 divisions to get the equal increments of 8.3625. The mid-points of the divisions represent the respective θ_i ,T_i and tabulated as below.

Tension, T(cm)	**Volumetric moisture content (θ), %**
$T_1 = 8$	83.0675
$T_2 = 15$	74.705
$T_3 = 30$	66.3424
$T_4 = 65$	57.98
$T_5 = 155$	49.6175
$T_6 = 362$	41.255
$T_7 = 630$	32.8925
$T_8 = 875$	24.53

The equation used in M and MQ method for the calculation of hydraulic conductivity (Eq. 6.68) is

$$K_i = K_0 \left(\frac{\theta i}{\theta_0}\right)^{p} \frac{\sum_{j=i}^{m}(2J+1-2i)T_J^{-2}}{\sum_{j=i}^{m}(2J-1)T_J^{-2}}$$

The hydraulic conductivity at tension T_2= 15cm of water is calculated as follows.

The volumetric moisture content, θ corresponding to T_2 is 74.705.

No. of increments, m = 8

Matching factor, P = 1

K_0= 0.215cm/day

θ_0= 87.256

T_J = tension corresponding to Jth increment

J = i-8 in numerator

1-8 in denominator

Submitting these values in **Eq. 6.68**

K_2

$$=0.215\left(\frac{74.705}{87.25}\right)$$

$$x\frac{\left(1x15^{-2}\right)+\left(3x30^{-2}\right)+\left(5x65^{-2}\right)+\left(7x55^{-2}\right)+\left(9x362^{-2}\right)+\left(11x630^{-2}\right)+\left(1.3x875^{-2}\right)}{\left(1x8^{-2}\right)+\left(3x15^{-2}\right)+\left(5x30^{-2}\right)+\left(7x65^{-2}\right)+\left(9x155^{-2}\right)+\left(11x362^{-2}\right)+\left(13x630^{-2}\right)+\left(15x875^{-2}\right)}$$

$$=0.18408x\frac{9.365x10^{-3}}{0.03668}$$

$=4.70x10^{-2}$cm/day

The values of hydraulic conductivities for the different moisture contents as calculated following Eq. 6.68 are tabulated below.

Volumetric moisture content (θ), %	Hydraulic conductivity (K), cm/day
83.0675	$2.047x10^{-2}$
74.705	$4.70x10^{-2}$
66.3424	$9.447x10^{-3}$
57.98	$1.67147x10^{-3}$
49.6175	$2.875x10^{-4}$
41.255	$6.0198x10^{-5}$
32.8925	$1.4226x10^{-5}$
24.53	$2.1524x10^{-6}$

6.6 Water Movement in Soils

Water moves in soils in liquid or vapour form. When the pore spaces are filled with water that is at saturated condition it moves under the influence of gravity. At unsaturation when the pore spaces are partially filled with water it moves under the influence of gravity and surface tension forces. Water also moves in soils through air filled pore spaces in vapour form through diffusion. The diffusion takes place only in the direction of decreasing vapour pressure.

Movement of moisture under saturated condition

Soil pores are not like the straight and smooth tubes of uniform radius so to enable to describe the flow through Poiseuille's equation but are highly irregular, tortuous and interconnected. The actual flow geometry is too complicated that it is described by the overall average macroscopic velocity over the total soil volumes. The actual flow pattern is ignored and the medium is considered as uniform to spread out the flow to entire cross section with the solid and pores alike (Hillel, 1972).

Poiseuille's law

Let the flow of fluid occurs between two plates of distance h, one is fixed and another one moving at velocity U. The velocity of fluid at stationary plate will be zero since the fluid adheres to both the plates. The velocity distributions between the plates are laminar as shown in Fig. 6.18. To maintain the movement of the plate a constant tangential force is required which is proportional to the resistance (shear) of the flow. Therefore, the shear stress (τ_s) at any point is proportional to the velocity gradient du/dy.

$$\tau_s \propto \frac{du}{dy}$$

$$or, \tau_s = -v\frac{du}{dy} \tag{6.69}$$

where, the proportionality factor v is the coefficient of viscosity. Negative sign indicates u decreases with the increasing y.

Now let us consider a coaxial fluid cylinder of length l, radius r, pressure difference between two ends of cylinder P (P_1-P_2). For maintaining constant flow velocity the frictional resistance due to shear force $2\pi y\tau_s L$ will be equal to difference of pressures between two ends. Thus,

$$P\pi r^2 = 2\pi y\tau_s L$$

$$or, \tau_s = \frac{P\pi y^2}{2\pi yL} = \frac{Py}{2L} \tag{6.70}$$

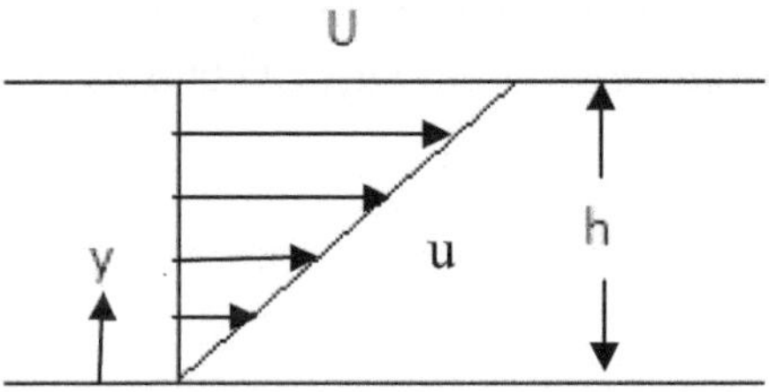

Fig. 6.18. Velocity distribution of the viscous fluid between two plates

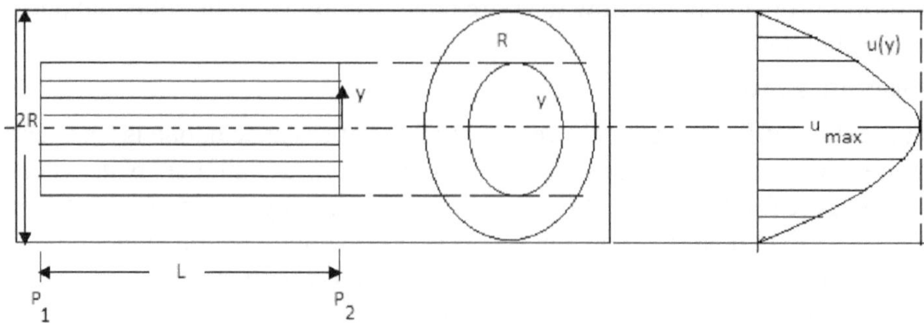

Fig. 6.19. Laminar flow through a cylindrical tube

Rearranging **Eq. 6.69** and **6.70**,

$$\frac{du}{dy}=\frac{Py}{v2L}$$

On integration, $u(y)=\frac{P}{vL}\left(C-\frac{y^2}{4}\right)$

At boundary condition when there is no slip at the wall the velocity u=0 and radius y=R, So that, $C=\frac{R^2}{4}$. Therefore,

$$u(y)=\frac{P}{4vL}\left(R^2-y^2\right) \tag{6.71}$$

Eq.6.71 indicates that the velocity is distributed parabolically over the radius with the maximum velocity when y = 0.

$$u_{max}=\frac{P}{4vL}R^2 \tag{6.72}$$

The discharge Q through the cylinder in a unit time is equal to the volume of the paraboloid made by the flow of fluid with velocities zero to u_{max}. Therefore,

$$Q=\frac{1}{2}\left(\pi R^2 x u_{max}\right)$$

$$=\frac{1}{2}\pi R^2.\frac{PR^2}{4vL}$$

$$=\frac{P\pi R^4}{8vL} \tag{6.73}$$

This equation is known as Poiseuille's law.

Usually laminar flow occurs in soil. However, radius of flow and velocity increases the flow character changes to turbulent and the mean velocity does not remain proportional to pressure drop. Pore-size distribution of a soil gives the basis of flow characteristics. In general, higher the pore size higher the flow rate, it is, therefore, rate of flow in soil will be in the order of sand>loam>clay.

Darcy's law

Henry Darcy, a French engineer, in 1856 conducted classic experiment of seepage rates in sand filters which is well known as Darcy's law. This law states that the velocity of flow of a liquid through porous media is directly proportional to the driving force on the liquid that is the hydraulic gradient and also to the property of the porous media.

$$V \infty \frac{\Delta H}{L}$$

$$or, V = K\frac{\Delta H}{L} = Ki \qquad (6.74)$$

Where, V = velocity of flow, m/s

K = proportionality constant known as hydraulic conductivity or coefficient of permeability, m/s

$\frac{\Delta H}{L}$ = hydraulic gradient, the head drop per unit distance, dimensionless

The head drop per unit distance in the direction of flow ($\frac{\Delta H}{L}$) is the hydraulic gradient, which is the driving force. The **Eq.6.74** can be generalized by replacing hydraulic head to the potential ϕ as

$$V = -\frac{Kd\phi}{L} \qquad (6.75)$$

In which L is the distance of the path of the greatest change in potential.

The negative sign arises due to flow occurs towards the decreasing head.

If a cross sectional area through which flow occurs then the discharge rate

$$Q = AV \qquad (6.76)$$

The discharge rate per unit area is $q = \frac{Q}{A}$

This is the flux density or simply the flux.

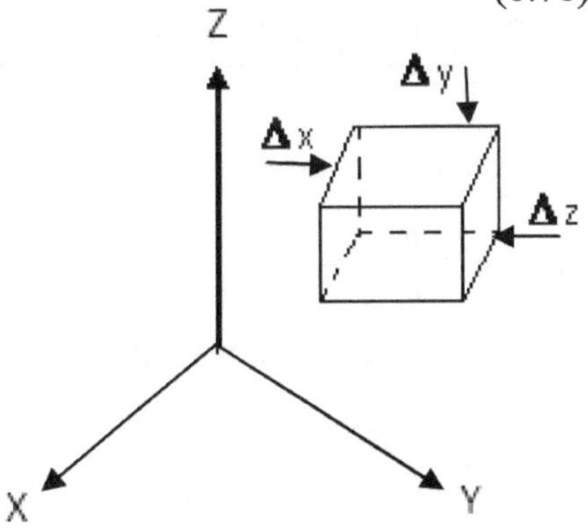

Fig.6.20. Diagram for derivation of flow equation

Richard's and Laplace's equation

The application of Darcy's law and equation of continuity to three dimensional flow of incompressible fluids in porous media results in the derivation of Richard's and Laplace's equation.

Let us consider a small rectangular parallele piped of saturated soil with xy, yz, and zx planes as shown in Fig.6.20. It is intended to obtain the expression for total flow of water in to and out of the parallele piped.

Let q_x is the quantity of water per unit area per unit time flowing in the x direction. If the rate of flow in the parallele piped in x direction is $\frac{\partial q_x}{\partial x}$ then the outflow will be $q_x + \left(\frac{\partial q_x}{\partial x}\right)\Delta x$. Therefore, the net flow in the parallele piped in x direction is

$$\text{Inflow} - \text{Outflow} = q_x.\Delta y.\Delta z - \left(q_x + \frac{\partial q_x}{\partial x}.\Delta x\right)\Delta y.\Delta z \tag{6.77}$$

If ρ is the density of water and considering the flow in x, y, z direction, the net mass of water gained or lost in the parallel piped

$$-\left[\frac{\partial(\rho q_x)}{\partial x} + \frac{\partial(\rho q_y)}{\partial y} + \frac{\partial(\rho q_z)}{\partial z}\right]\Delta x\Delta y\Delta z \tag{6.78}$$

If the soil moisture concentration factor c which is the volume of water per unit soil mass, the rate of change of water mass in parallel piped is then

$$\frac{\partial}{\partial t}(\rho c)\Delta x\Delta y\Delta z$$

Therefore, the change of water mass in the parallel piped in **Eq.6.78** can be equated as

$$-\left[\frac{\partial(\rho q_x)}{\partial x} + \frac{\partial(\rho q_y)}{\partial y} + \frac{\partial(\rho q_z)}{\partial z}\right] = \frac{\partial}{\partial t}(\rho c) \tag{6.79}$$

Darcy's law in anistropic soil can be presented as

$$q_x = -K\frac{\partial \varphi}{\partial x}, q_y = -K\frac{\partial \varphi}{\partial y}, q_z = -K\frac{\partial \varphi}{\partial z} \tag{6.80}$$

Substituting the Darcy's equation in to **Eq.6.79**

$$\frac{\partial}{\partial x}\left(\rho K\frac{\partial \varphi}{\partial x}\right) + \frac{\partial}{\partial y}\left(\rho K\frac{\partial \varphi}{\partial y}\right) + \frac{\partial}{\partial z}\left(\rho K\frac{\partial \varphi}{\partial z}\right) = \frac{\partial}{\partial t}(\rho c) \tag{6.81}$$

This is the Richard's equation for flow of fluid in porous media.

If the soil is isotropic with respect to hydraulic conductivity and hydraulic conductivity is constant, $K_x = K_y = K_z$; and if there is no loss or gain of water in the parallel piped and the water is incompressible, then $\frac{\partial}{\partial t}(\rho c) = 0$. With these assumptions the Eq.6.81 becomes

$$\frac{\partial \varphi^2}{\partial x^2} + \frac{\partial \varphi^2}{\partial y^2} + \frac{\partial \varphi^2}{\partial z^2} = 0 \tag{6.82}$$

This is the Laplace's equation.

We can also write the Laplace's equation as

$$\nabla^2 \varphi = 0 \tag{6.83}$$

And Richard's equation as

$$\frac{\partial}{\partial}(\rho c) = \nabla.(\rho K \nabla \varphi) \tag{6.84}$$

Where the symbol ∇, called 'del' is the differential operator

$$\frac{\partial}{\partial x} + \frac{\partial}{\partial y} + \frac{\partial}{\partial z}$$

and ∇^2, called 'del squared' is

$$\frac{\partial^2}{\partial x^2} + \frac{\partial^2}{\partial y^2} + \frac{\partial^2}{\partial z^2}$$

Introducing continuity equation in Darcy's law

$$\frac{\partial \theta}{\partial t} = -\nabla.q \tag{6.85}$$

$$or, \frac{\partial \theta}{\partial t} = -\nabla\left[K(\psi)\nabla H\right] \tag{6.86}$$

Hydraulic head is the combination of pressure head (or its negative or suction head) and the gravitational head (or elevation).

$$\text{So,} \frac{\partial \theta}{\partial t} = -\nabla.\left[K(\psi)\nabla(\psi - Z)\right]$$

$$= -\nabla.(K\nabla\psi) + \frac{\partial K}{\partial Z} \tag{6.87}$$

$$\text{or,} \frac{\partial}{\partial t} = -\frac{\partial}{\partial x}\left(K\frac{\partial \psi}{\partial x}\right) - \frac{\partial}{\partial y}\left(K\frac{\partial \psi}{\partial y}\right) - \frac{\partial}{\partial z}\left(K\frac{\partial \psi}{\partial Z}\right) + \frac{\partial K}{\partial Z} \tag{6.88}$$

For horizontal flow ∇Z = zero and other cases ∇Z can be assumed zero compared to strong matric suction. Therefore,

$$\frac{\partial}{\partial t} = -\nabla.\left[K(\psi)\nabla\psi\right] \tag{6.89}$$

For one dimensional horizontal flow

$$\frac{\partial\theta}{\partial t} = -\frac{\partial}{\partial x}\left(K(\psi)\frac{\partial\psi}{\partial x}\right) \tag{6.90}$$

Diffusivity

It is often advantageous to explain the flow equation by suction gradient instead of flux to the water content gradient (wetness). The matric suction gradient of a soil is $\frac{\partial\psi}{\partial t}$. It can be expanded by the chain rule to the following form

$$\frac{\partial\psi}{\partial x} = \frac{\partial\psi}{\partial\theta}\cdot\frac{\partial\theta}{\partial z} \tag{6.91}$$

Where $\frac{\partial\theta}{\partial x}$ is the wetness gradient and $\frac{\partial\psi}{\partial\theta}$ is the reciprocal to specific water capacity,

$$c(\theta) = \frac{\partial\theta}{\partial\psi} \tag{6.92}$$

This is the gradient of the soil moisture characteristics curve at any moisture content. Darcy's law can be rewritten as

$$q = K(\theta)\frac{\partial\psi}{\partial x} = -\frac{K(\theta)}{c(\theta)}\cdot\frac{\partial\theta}{\partial x} \tag{6.93}$$

Diffusivity is defined as the ratio of hydraulic conductivity $k(\theta)$ to the specific water capacity $c(\theta)$. Therefore,

$$q = -D(\theta)\frac{\partial\theta}{\partial x} \tag{6.94}$$

Using Eq.6.94 in Eq.6.90

$$\frac{\partial\theta}{\partial t} = \frac{\partial}{\partial x}\left(D(\theta)\frac{\partial\theta}{\partial x}\right) \tag{6.95}$$

Determination of diffusivity

The Eq.6.95 is the same form as the diffusion theory where concentration-dependent diffusivities are involved. Its solution depends on the knowledge of the function $D(\theta)$.

The system may be considered semi-infinite of horizontal flow with water applied at one end of a homogeneous column of porous medium. One end of a column is maintained at saturation and the column is sufficiently long such that it may be integrated as infinite length in the plus x direction where x is the horizontal coordinate to position. The boundary conditions are:

$$\theta=\theta_i \qquad x\rangle 0, t=0 \tag{6.96}$$

$$\theta=\theta_s \qquad x=0, t=0 \tag{6.97}$$

Where θ_t is the initial moisture content of the system and θ_s is the saturation moisture content.

Boltzman (1894) has shown that the solution of the **Eq.6.95** for an infinite system should contain the variable λ equal to $\frac{x}{\sqrt{t}}$. The assumption that the concentration is a function of λ can be tested by plotting x against the square root of t for a constant value of θ. For example, if it is assumed that the moisture content just behind the wetting front is essentially constant, then a plot of the position of the wetting front in the flow system described above against the square root of t is such a test. Data of this type supporting the assumption of $\frac{x}{\sqrt{t}}$ as a significant variable have been published by Kirkham and Feng (1949) using flow system of a similar nature to those to be used here.

By substituting the variable λ in to **Eq.6.95** the equation is reduced to ordinary differential equation.

$$\frac{\lambda d\theta}{2d\lambda}=\frac{d}{d\lambda}\left(D(\theta)\frac{d\theta}{d\lambda}\right) \tag{6.98}$$

Integrating this equation with respect to λ yields:

$$-\int_{\theta_i}^{\theta_x}\frac{\lambda}{2}d\theta=\int\frac{d}{d\lambda}\left(D(\theta)\frac{d\theta}{d\lambda}\right)d\lambda \tag{6.99}$$

$$=D(\theta)\left(\frac{d\theta}{d\lambda}\right)_{\theta_x}-D(\theta_i)\left(\frac{d\theta}{d\lambda}\right)_{\theta i} \tag{6.100}$$

Where the notation $\left(\frac{d\theta}{d\lambda}\right)_{\theta_x}$ indicates the value of the derivation at $\theta=\theta_x$. The last term of the right hand side of the **Eq.6.100** is zero at constant time because the variable $\left(\frac{d\theta}{d\lambda}\right)\theta_i$ is zero at value x large enough to be ahead of wetting front. Then it follows that:

$$D(\theta_x) = -\frac{1}{2}\left(\frac{d\lambda}{d\theta}\right)_{\theta_x} \int_{\theta_i}^{\theta_x} \lambda d\theta$$

or in terms of x and t at constant t;

$$D(\theta_x) = -\frac{1}{2t}\left(\frac{dx}{d\theta}\right)_{\theta_x} \int_{\theta_i}^{\theta_x} x d\theta \tag{6.101}$$

Eq.6.101 suggests the following procedure for evaluating $D(\theta)$:

a) Obtain a moisture content- distance curve from the flow system described above, i.e., θ as a function of x at a constant value of time, t.

b) From a plot of θ versus x, evaluate the integral and the derivatives in **Eq.6.101** at a series of value θ_x.

(c) Calculate D at the values of θ_x used in step (b) thereby obtaining $D(\theta)$. Therefore, Eq.6.101 can be written in the following way for estimating integral and derivative in the equation:

$$D(\theta) = -\frac{1}{2t}\left(\frac{dx}{d\theta}\right)_{\theta_x} \int_{\theta_i}^{\theta_x} \lambda t^{\frac{1}{2}} d\theta \tag{6.102}$$

or, $$D(\theta) = \frac{1}{\left(t^{\frac{1}{2}} \Delta\theta_r / \Delta x_r\right)}\left(-\frac{1}{2}\sum_{r=i}^{x} \lambda_r \Delta\theta_r\right) \tag{6.103}$$

Experiment determination

The soil water diffusivity of Memari loam was determined following the horizontal infiltration method (Crank, 1956) as stated in Eq.6.101,

$$D(\theta_x) = -\frac{1}{2t}\left(\frac{dx}{d\theta}\right)_{\theta_x} \int_{\theta_i}^{\theta_x} x d\theta$$

Where, D(θ_x) soil water deffusivity at moisture θ, cm/min

t = time required for the advancement of water up to x in infiltration column, min

$\theta = \theta i$ $\quad x \rangle 0, t = 0$

$\theta = \theta_s$ $\quad x = 0, t = 0$

The soil infiltration column consisted of 1.5cm section of tubing (ring) and 5.7cm in diameter. By multiplying by volume of one ring (38.276cm^3) with the apparent specific gravity (1.45g/cc) the weight of the soils for each ring was obtained. About 15-16 rings were attached side by side using cello tape and this set of rings was then fitted in a frame. The calculated amount of soil for one ring was then poured into the tubing and made accommodated within the ring by packing uniformly with a packer so that uniform specific gravity could be obtained. The process was repeated for other rings also and a long soil column was prepared (Fig.6.21).

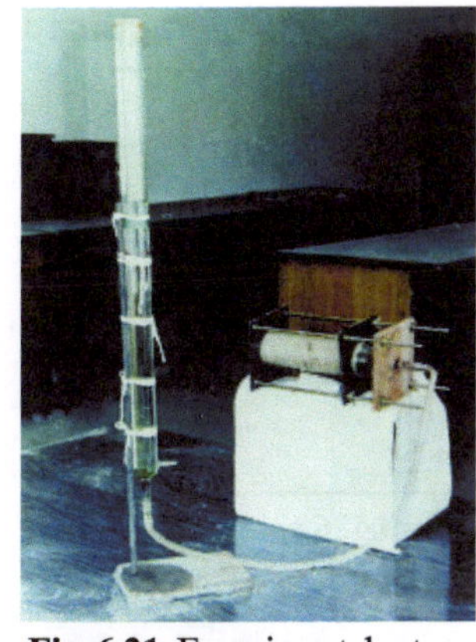

Fig. 6.21. Experimental set-up for the determination of soil-water diffusivity

The soil column was laid on the table horizontally and the water was supplied at zero head from one end of the column and the time was noted. When the water advanced 7-8 rings along the column time was noted and the water supply was cut off.

The rings of the soils up to the water front were dismantled by cutting cello tape very cautiously with the help of sharp blades passing between the rings. The soils of each ring were collected in aluminum boxes. The dry weight of the soils in aluminum boxes was taken and the differences of weight to the previous weight gave the measure of water in each ring and moisture content in percentage was determined.

The volumetric moisture content θ versus the distance x along the horizontal soil (x=0 at the saturated end) was plotted to get the smooth curves as shown in Fig.6.22. From the smooth curve the diffusivities were calculated numerically using **Eq.6.103** which is

$$D(\theta)=\frac{1}{\left(t^{\frac{1}{2}}\frac{\Delta\theta_r}{\Delta x_r}\right)}\left(-\frac{1}{2}\sum_{r=i}^{x}\lambda_r\Delta\theta_r\right)$$

$$=-\frac{1}{2t^{\frac{1}{2}}\left(\frac{\Delta\theta_r}{\Delta x_r}\right)}\left(\sum_{r=i}^{x}\lambda_r\Delta\theta_r\right) \tag{6.104}$$

The details of the $D(\theta)$ calculation is shown in **Table 6.7**.

The volumetric moisture content at different length at different ring is tabulated as below and the smooth curve as developed **(Fig.6.22)**.

Table 6.7. Volumetric moisture content at different length in soil column

X_r	θ_r
0.75	0.463
2.25	0.432
3.75	0.448
6.75	0.398
8.25	0.379
9.75	0.237
11.25	0.152

θ_s=0.486,t = 775min

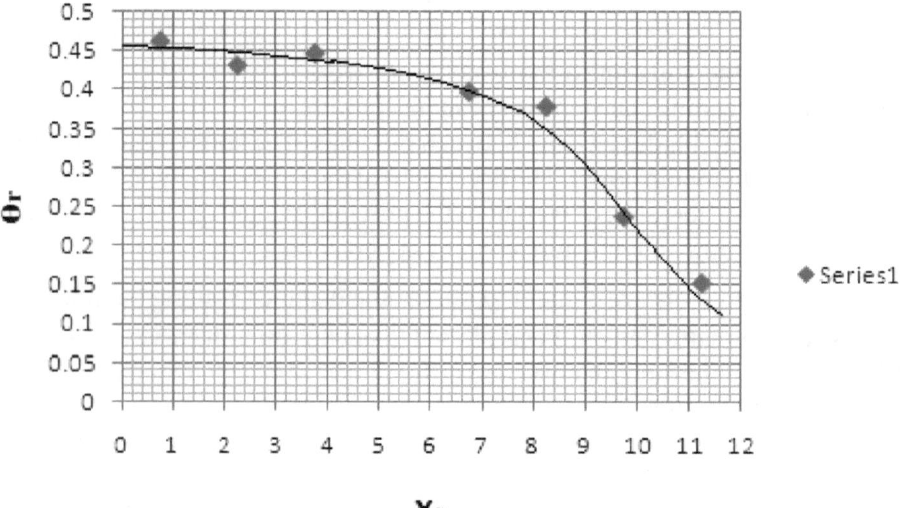

Fig. 6.22 The r-x relationship in horizontal soil column

From the smooth curve the θ at different lengths were selected considering the difference of θ is not much between the selected consecutive points (Table 6.7). The details of D(θ) calculation is shown in Table 6.8. The time of water advance in horizontal soil column was 775 minutes.

Table 6.8 Soil-water diffusivities at different moisture level of Memari loam

S.No	x_r, cm	θ_r vol/vol	$-\frac{1}{2}\lambda_r \Delta\theta_r$	$\sum_{r-i}^{x} -\frac{1}{2}\lambda_r \Delta\theta_r$	$\Delta x_r = xr - x_{r-1}$	$-t^{\frac{1}{2}} \Delta\theta_r / \Delta x_r$	$D(\theta) = \frac{1}{\left(-t^{\frac{1}{2}\Delta\theta_r/\Delta x_r}\right)} \left(-\frac{1}{2}\sum_{r=1}^{x} \lambda_r \Delta\theta_r\right)$
			cm min$^{-1/2}$	cm min$^{-1/2}$	cm	min$^{1/2}$cm^{-1}	cm^2/min
1	11.6	0.112	-2.2918x10^{-3}	-2.2918x10^{-3}	0.2	-1.5311	1.4968x10^{-3}
2	11.4	0.123	-4.7093x10^{-3}	-7.0011x10^{-3}	0.4	-1.6007	4.3738x10^{-3}
3	11.0	0.146	-5.9270x10^{-3}	-1.9929x10^{-2}	0.6	-1.3919	1.4317x10^{-2}
4	10.6	0.176	-5.3307x10^{-3}	-2.5260x10^{-2}	0.6	-1.2991	1.9444x10^{-2}
5	10.2	0.204	-2.9312x10^{-3}	-2.8191x10^{-2}	0.2	-2.2271	1.2658x10^{-2}
6	10.0	0.220	-5.7474x10^{-3}	-3.3938x10^{-2}	0.4	-2.2271	1.5239x10^{-2}
7	9.6	0.252	-5.8623x10^{-3}	-3.9800x10^{-2}	0.4	-2.3663	1.6820x10^{-2}
8	9.2	0.286	-4.9447x10^{-3}	-4.4748x10^{-2}	0.4	-2.1575	2.0741x10^{-2}
9	8.8	0.317	-5.2157x10^{-3}	-4.9964x10^{-2}	0.6	-1.5311	3.2632x10^{-2}
10	8.2	0.350	-4.4183x10^{-3}	-5.4378x10^{-2}	0.8	-1.0440	5.2080x10^{-2}
11	7.4	0.380	-1.3291x10^{-3}	-5.5707x10^{-2}	0.4	-0.6960	8.0042x10^{-2}
12	7.0	0.390	-3.0174x10^{-3}	-5.8724x10^{-2}	1.0	-0.6657	3.9094x10^{-2}
13	6.0	0.414	-2.5863x10^{-3}	-6.1310x10^{-2}	1.0	-0.6681	4.0961x10^{-2}
14	5.0	0.438	-1.7961x10^{-4}	-6.1490x10^{-2}	1.6	-0.2088	1.2838x10^{-2}
15	3.4	0.440	-4.8863x10^{-4}	-6.1979x10^{-2}	1.4	-1.1591	3.8961x10^{-2}
16	2.0	0.448	-7.1842x10^{-5}	-6.2051x10^{-2}	1.0		

Evaporation using diffusivity

A soil profile subjected to drying the depth may be considered as semi-infinite in the initial stage of drying or until about 50% of the water in the profile is evaporated (Gardner, 1959). The flow equation with the boundary conditions is:

$$\frac{\partial\theta}{\partial t}=\frac{\partial}{\partial x}\left(D(\theta)\frac{\partial\theta}{\partial x}\right) \tag{6.105}$$

$$\theta=\theta_i \qquad x\geq 0, t=0$$

$$\theta=\theta_\alpha \qquad x=0, t>0$$

Where θ is the volumetric water content, x is the distance and D is the soil-water diffusivity. Crank (1956) solved the Eq. 6.105 analytically for a semi-infinite slab with constant diffusivity and the upward flux, q, at boundary as,

$$q=(\theta_i-\theta_\alpha)(D/\pi t)^{1/2} \tag{6.106}$$

Where θ_i is the initial water content assumed constant for x=0, θ, θ_α is the water content at the boundary and assumed constant for t>0. Integration of Eq.6.106 with respect to time gives the cumulative evaporation

$$\text{CE} = 2(\theta_i-\theta_a)(\overline{D}\, t/\pi)^{1/2} \tag{6.107}$$

Where $\overline{D}$ is the weighted-mean diffusivity.

Further Crank (1956) gave the expression relating weighted-mean diffusivity to the true diffusivity. For desorption process the relation was given by the integral:

$$\overline{D}=\frac{1.85}{(\theta_i-\theta_a)^{1.85}}\int_{\theta_a}^{\theta_i} D(\theta)(\theta_i-\theta)^{0.85}\,d\theta \tag{6.108}$$

Knowing the D- θ relation numerically the value of weighted-mean diffusivity $\overline{D}$ can be calculated. Putting the $\overline{D}$ value in Eq.6.107 the cumulative evaporation for a period can be estimated when moisture is changed from θ_i to θ_α.

D- θ relation

Using the value of diffusivity D for different soil moisture content θ in Table 6.8 the curve and the equation is developed as shown in Fig 6.23.

$$\log D(\theta)=789\,\theta-2.514 \tag{6.109}$$

θ	D(θ)	logD(θ)
0.112	0.001497	-2.82484
0.123	0.004377	-2.35884
0.146	0.014317	-1.84415
0.176	0.019444	-1.71121
0.204	0.012658	-1.89763
0.22	0.015239	-1.81704
0.252	0.01682	-1.77417
0.286	0.020741	-1.68317
0.317	0.03263	-1.48638
0.35	0.05208	-1.28333
0.38	0.080040	-1.09668
0.39	0.039094	-1.40789
0.414	0.04096	-1.38764
0.438	0.012838	-1.8915
0.44	0.038961	-1.40937

Weighted mean diffusivity ($\overline{D}$)

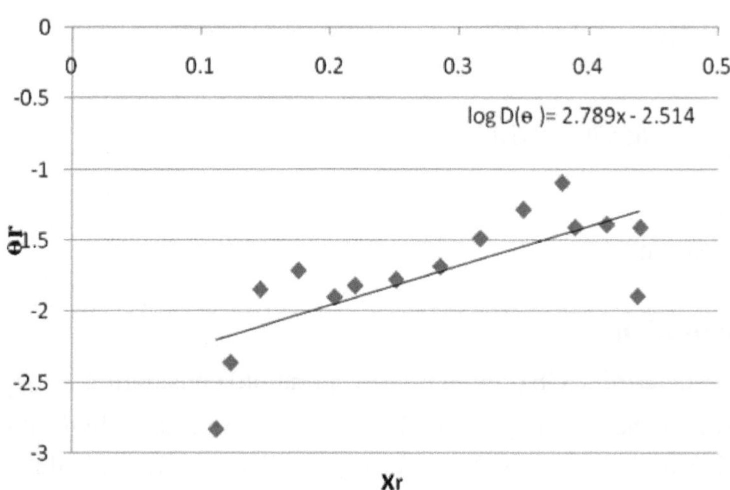

Fig.6.23 θ -x relationship

The soil moisture content at field capacity (θ_f) and air dried soil on the surface (θ_α) were 0.395 and 0.025 respectively at the experimental field. Using the **Eq.6.108** and the incremental moisture contents between θ_f and θ_α, the diffusivity for each moisture content and the weighted diffusivity is calculated.

Table 6.9. Calculation of weighted mean diffusivity

θ vol/vol	θ_f- θ vol/vol	$d\theta$ vol/vol	θ_f- $\theta^{0.85} d\theta$	D(θ) cm²/min	θ_f- $\theta^{0.85} d\theta$ D(θ)	Weighted mean diffusivity, $\overline{D}$,cm²/day
0.37	0.025	0.05	2.17379x10⁻³	0.032956	7.163975x10⁻⁴	11.9217
0.32	0.075	-	5.53060x10⁻³	0.023904	1.322063x10⁻⁴	cm²/day
0.27	0.125	-	8.537752x10⁻³	0.017339	1.480381x10⁻⁴	
0.22	0.175	-	0.01136455	0.012577	1.429329x10⁻⁴	
0.17	0.225	-	0.01407101	0.017339	2.4398058x10⁻⁴	
0.12	0.275	-	0.01668794	6.617287x10⁻³	1.1042889x10⁻⁴	
0.07	0.325	-	0.01923405	4.799875x10⁻³	9.23210511x10⁻⁵	
0.025	0.37	0.045	0.01932787	3.595216x10⁻³	6.9487869x10⁻⁵	
					1.65579319x10⁻³	

Using the **Eq.6.108**,

$$\overline{D}=\frac{1.85}{(\theta_i-\theta_\alpha)^{1.85}}\int_{\theta_\alpha}^{\theta_i} D(\theta)(\theta_i-\theta)^{0.85}\, d\theta$$

$$=\frac{1.85}{(\theta_f-\theta_a)^{1.85}}\int_{\theta_a}^{\theta_f} D(\theta)(\theta_f-\theta)^{0.85}\, d\theta$$

$$=\frac{1.85}{(0.395-0.025)^{1.85}}\times 1.65579319x10^{-3}$$

$= 5x1.65579319x10^{-3}$

$= 8.278966x10\text{-}3\ cm/\min$

$= 11.9217 cm^2/day$

Cumulative evaporation

Eq. 6.107 is used to determine CE (cumulative evaporation) considering θ_i- θ_f and θ_a- θ of the soil on the sampling day in field. The CE on any sampling date has been determined by Eq.6.107 with respect to drying date (2 days after irrigation) which is fixed for all sampling dates till it is air dried or earlier. The estimated CE thus obtained is shown in last column of Table 6.10.

Evaporation

The evaporation in between two sampling dates are obtained by subtracting the former cumulative evaporation from the later and are shown in the last but one column in Table 6.10. The day to day evaporation can be had from the curve of the time versus cumulative evaporation (Fig.6.24).

For Block A

t, days	CE, cm
26	3
28	3.237
57	6.304
59	6.474
74	7.977

For Block B

t, days	CE, cm
26	2.738
28	3
57	6.194
59	6.388
74	7.506

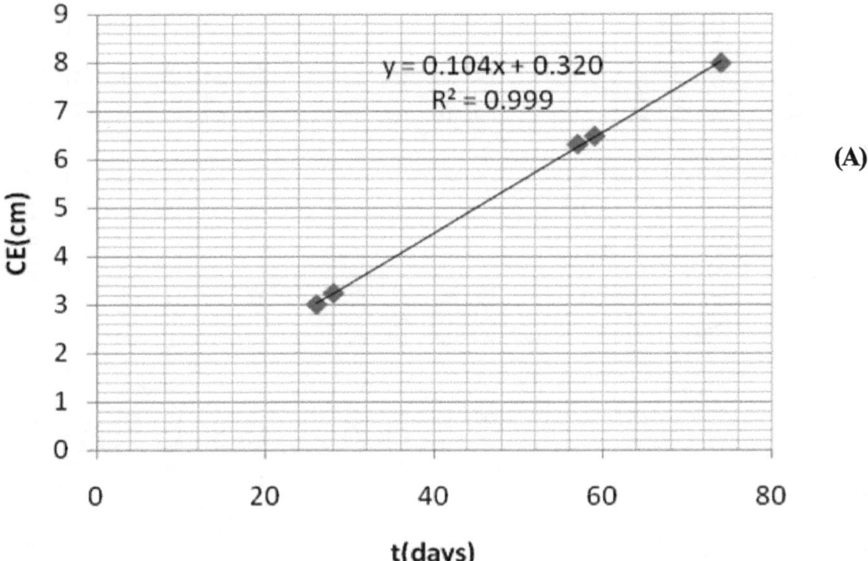

(A)

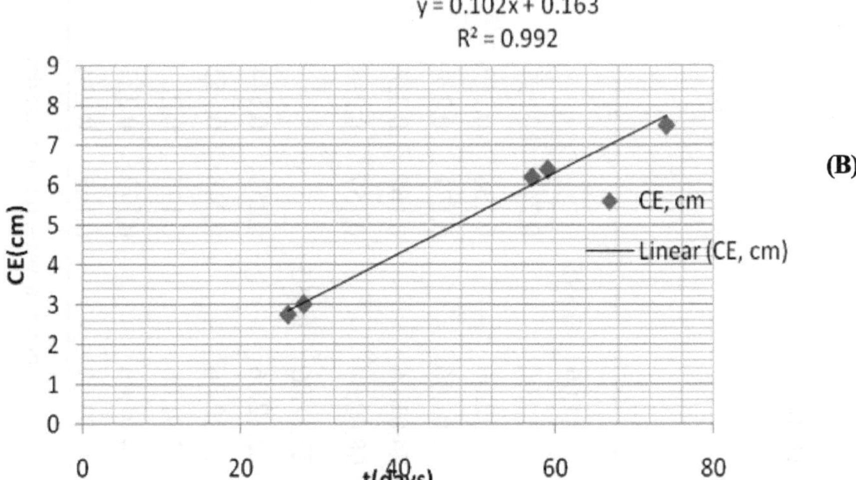

(B)

Fig. 6.24. Time vs. cumulative evaporation

Table 6.10. Calculation of evaporation using diffusivity

Plot No.	Date	θ_f	θ	$\theta_f - \theta$ vol/vol	t, days	$\left(\frac{\overline{D}t}{\pi}\right)^{1/2}$	$CE = 2(\theta_f - \theta)$ $(\overline{D}\,t/\pi)^{1/2}$	Evaporation at different time interval, mm	Evaporation from(27.11. 06-08.02.06), mm
A	27.11.06	0.395							79.77
	22.12.06	-	0.244	0.151	26	9.933	3.000	30	
	24.12.06	-	0.238	0.157	28	10.308	3.237	2.37	
	22.01.06	-	0.171	0.224	57	14.077	6.304	30.67	
	24.01.06	-	0.169	0.226	59	14.322	6.474	1.70	
	08.02.07	-	0.157	0.238	74	16.758	7.977	15.03	
B	27.11.06	0.395							75.06
	22.12.06	-	0.251	0.144	26	9.507	2.738	27.38	
	24.12.06	-	0.243	0.152	28	9.866	3	2.62	
	22.01.06	-	0.175	0.22	57	14.077	6.194	31.94	
	24.01.06	-	0.172	0.223	59	14.322	6.388	1.94	
	08.02.07	-	0.161	0.234	74	16.039	7.506	11.18	

Using the CE at different days the relations are developed for the plot A and B and are shown in **Fig. 6.24** (A&B).

Evaporation using capillary conductivity

Dalton (1802) was the first to enunciate the fundamental law of evaporation (Ghyldayal & Tripathi, 1988). The rate of evaporation from free water surface is expressed as

$$E_p = (e_0 - e_d) f(u) \tag{6.110}$$

Where,

e_0 = mean vapour pressure at the water surface,

e_d = mean vapour pressure in the air at some observational height above the water surface,

f(u) = a function whose value depends on the wind velocity.

Assuming that a layer of a thin film of vapour saturated air is formed adjacent to the water surface, under isothermal conditions; the evaporation function is analogous to Darcy's equation and can be expressed as

$$E_p = -K' \frac{de}{dz'} \tag{6.111}$$

Oswal and Dakshinamurti (1973) used Eq.6.112 to determine capillary conductivity of soil layers. The Eq.6.112 for upward flow

$$\frac{\partial \theta}{\partial t} = \frac{\partial}{\partial z}\left(\frac{\partial \psi}{\partial z} - K\right) \tag{6.112}$$

The change in moisture in time 'dt' of an elementary layer of thickness 'dz' is given by

$$\frac{\partial \theta}{\partial t} dt = \frac{\partial}{\partial z}\left(K \frac{\partial \psi}{\partial z} - K\right) dt \tag{6.113}$$

For a soil layer bounded by depths z_1 and z_2 the total change in moisture in time t is obtained by integrating above equation.

$$\overline{K} = \int_{t_1}^{t_2}\int_{z_1}^{z_2} \frac{d\theta}{dt} dt.dz = \int_{t_1}^{t_2} \overline{K}\left(\frac{\partial \psi}{\partial z} - 1\right) dt \tag{6.114}$$

$$= \frac{\int_{t_1}^{t_2}\int_{z_1}^{z_2} \frac{d\theta}{dt} dt.dz}{\left(\frac{d\psi}{dz} - 1\right) t} \tag{6.115}$$

Where $\overline{K}$ and $\overline{\psi}$ are the average values of K and ψ. By knowing the moisture θ and its corresponding tension ψ from θ-ψ relation at depths z_1 and z_2 at times t_1 and t_2, the average capillary conductivity $\overline{K}$ can be determined. Equating the $\overline{K}$ value, $q = -\overline{K} a' t \frac{d\psi}{dz}$.

For upward flow, $q = -\overline{K}a't\left(\frac{d\psi}{dz} - 1\right)$ (6.116)

Where, q=upward flow of water, or in essence E, cm/day

a' = soil surface area, cm^2

The evaporation or upward flow of water for a period t can be estimated by using Eq.6.116.

The assumption made in this method is that change in moisture content of the layer in time, $t=(t_2-t_1)$ is such that the corresponding suction difference is not large and, therefore, both ψ and K are averageable.

Soil moisture characteristics curve

Using the method of Ghosh (1993) and the pressure range of 0.1 to 10 bar the ψ-θ data sets were expressed as $\psi - \left(\frac{\theta}{\theta_0}\right)$ data sets where ψ is the suction in bar and θ_0 is the porosity.

The characteristics curve and its mathematical relation is shown in **Fig.6.25**.

ψ, bar	θ, vol/vol	θ/θ_0	$\log(\theta/\theta_0)$	$\log\psi$
0	0.4445			
0.333	0.3086	0.6943	-0.1584	-0.4775
0.9	0.1009	0.2227	-0.6523	-0.0458
2.5	0.0845	0.19	-0.7212	0.3979
5	0.08	0.178	-0.7496	0.6989
10	0.0539	0.1213	-0.9161	1

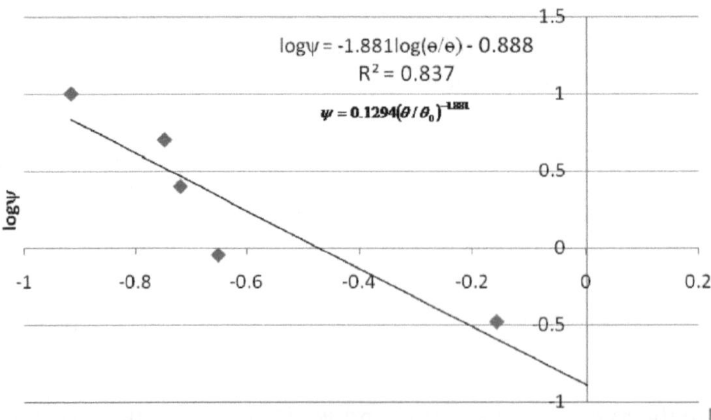

Fig.6.25 Logarithmic plot of ψ against (θ/θ_0)

Average capillary conductivity

As outlined in **Eq.6.115** the average capillary conductivities in between the top layers (each 15cm) for different suction difference are shown in Table 6.11.

Table. 6.11. Calculation of average capillary conductivity and cumulative evaporation

Plot No.	Date of sampling	Moisture (θ)	Suction, cm (0-15 cm depth)	Average suction, cm	Moisture (θ), %	Suction, cm (15-30cm depth)	Average suction, cm	$d\theta$	$\left(\frac{d\overline{\psi}}{dz}-1\right)$ cm/cm	Average conductivity (Eq.6.115) $\overline{K}$ *cm/day*	Days	Evaporation (E), (Eq.6.116), mm	ΣE, mm
A	22.12.06	23.41	440.53	-	23.72	429.81	-	-	-	-	-	-	12.8164
	24.12.06	21.79	504.17	472.35	22.48	475.44	452.63	0.0143	0.3147	0.3408	2	2.323	
	22.01.06	16.97	806.72	655.41	17.05	799.62	637.53	0.0513	0.192	0.1382	29	7.695	
	24.01.06	16.08	892.75	849.74	16.89	813.92	806.77	0.0053	1.8647	0.0213	2	0.7944	
	08.02.06	14.54	1078.88	985.81	15.75	928.26	870.09	0.0134	6.7146	0.00199	15	2.004	

1bar=1019cm of water

Steady evaporation in the presence of water table at constant depth

Darcy's law along with the continuity equation as the water flow does not only obey the Darcy's law but also the law of conservation of mass. The change in water content with time at a given position is equal to the difference between the rate of flow in to and out of the soil at the location or for flow in direction:

$$\frac{\partial\theta}{\partial t}=\frac{\partial v}{\partial z} \tag{6.117}$$

Where θ is the water content in volume, v is the velocity of flow and t is the time. Also,

$$v=-K\frac{\partial\varphi}{\partial z}(\text{Darcy's;law}) \tag{6.118}$$

Combining Eq.6.117 and Eq.6.118, we get;

$$\frac{\partial\theta}{\partial t}=-\frac{\partial}{\partial z}\left[-K\frac{\partial\varphi}{\partial z}\right] \tag{6.119}$$

For vertical flow $\varphi=\psi+z$; where $\varphi=$ total potential and z = gravitational potential.

$$\frac{\partial\theta}{\partial t}=-\frac{\partial}{\partial z}\left[-K\frac{\partial(\psi+z)}{\partial z}\right]$$

$$=\frac{\partial}{\partial z}\left[K\frac{\partial\psi}{\partial z}+1\right] \tag{6.120}$$

For a steady evaporation of water from a soil in the presence of water table

$$\frac{\partial\theta}{\partial t}=0 \tag{6.121}$$

Therefore, Eq.6.120 becomes:

$$0=\frac{\partial}{\partial z}\left[K\left(\frac{\partial\psi}{\partial z}+1\right)\right] \tag{6.122}$$

Since the movement is in upward direction.

Integrating Eq.6.122 gives:

$$q=K\left(\frac{\partial\psi}{\partial t}-1\right) \tag{6.123}$$

Where q is a constant of integration which represents the flux. Solving Eq.6.123 gives,

$$\int dz=\int\frac{d\psi}{1+\frac{q}{K}} \tag{6.124}$$

Knowing the constant distance of water table z from the ground surface, the moisture suction ψ of the top soil and the K-ψ relation, it is possible to estimate the evaporation from the soil.

The *K*-ψ equation

The soil-water diffusivity, $D(\theta)=\dfrac{K(\theta)}{C(\theta)}$

Where,

$K(\theta)$ = hydraulic conductivity of soil at moisture content θ

$C(\theta)$ = Specific water capacity, $\dfrac{d\theta}{d\psi}$ at θ

ψ = soil moisture tention at θ

Therefore, $K(\theta)=D(\theta)\dfrac{d\theta}{d\psi}$ (6.125)

The $\psi-\left(\dfrac{\theta}{\theta_0}\right)$ relation was established earlier as

$$\log\psi=-1.881\log\left(\frac{\theta}{\theta_0}\right)-0.888$$

Putting the value of $\theta_0 = 04445$ and rearranging the equation,

$\psi = 0.0282\theta^{-1.881}$ (6.126)

$\therefore \dfrac{d\psi}{d\theta}=-0.053\theta^{-2881}$ (6.127)

or, $\therefore \dfrac{d\theta}{d\psi}=-\dfrac{1}{18.87}\theta^{2.881}$ (6.128)

The D-θ relation **(Eq.6.109)** is obtained earlier. So, for a given soil moisture θ; knowing the D (θ) **(Eq.6.109)** value and the corresponding $\dfrac{d\theta}{d\psi}$ **(Eq.6.128),** the K(θ) may be determined from Eq.6.125.

Following the technique as stated ψ - θ versus θ values are obtained and shown in col. 1 and 8 of Table 6.12. The ψ - θ is shown in col.2 of the **Table 6.12.** In col.3, C(θ) = $\dfrac{d\theta}{d\psi}$ is determined and displayed. The D-θ relations are shown in col.4. In col.6 the values of $K(\theta)$ are obtained as $K(\theta)$= $C(\theta)$D(θ).

The K-ψ relation thus developed is

$K=6884\psi^{-2.10}$ (6.129)

Table 6.12 Determination of $-\psi$ relation

θ	θ-ψ	$C(\theta)$	D-θ	D-(θ)	K(θ)	$K(\theta)$	ψ	K-ψ
vol/vol	relation	$=\frac{d\theta}{d\psi}$	relation	cm²/min	$= C(\theta)$ D(θ) cm/min	cm/day	cm	relation
1	2	3	4	5	6	7	8	9
0.40		3.7824×10^{-3}	$\log D(\theta) = 2.789\,\theta - 2.514$	0.03996	1.5114×10^{-4}	0.2176	161.0454	$K = 6884.\ \psi^{-2.10}$
0.35		2.5745×10^{-3}		0.02898	7.4618×10^{-5}	0.1075	207.029	
0.3		1.6513×10^{-3}		0.02102	3.4716×10^{-5}	0.0499	276.6675	
0.25	$\psi = 0.0282\,\theta^{-1.881}$	9.7655×10^{-4}		0.01525	1.4892×10^{-5}	0.0214	389.8505	
0.20	$\frac{d\theta}{d\psi} = -\frac{1}{18.87}\theta^{2.881}$	5.1345×10^{-4}		0.01106	5.6793×10^{-6}	8.1783×10^{-3}	593.1791	
0.15		2.2415×10^{-4}		8.0232×10^{-3}	1.7984×10^{-6}	2.5897×10^{-3}	1019.0503	
0.10		6.9699×10^{-5}		5.8197×10^{-3}	4.0563×10^{-7}	5.8410×10^{-4}	2184.8584	
0.05		9.4615×10^{-6}		4.2213×10^{-3}	3.9940×10^{-8}	5.7514×10^{-5}	8047.49	

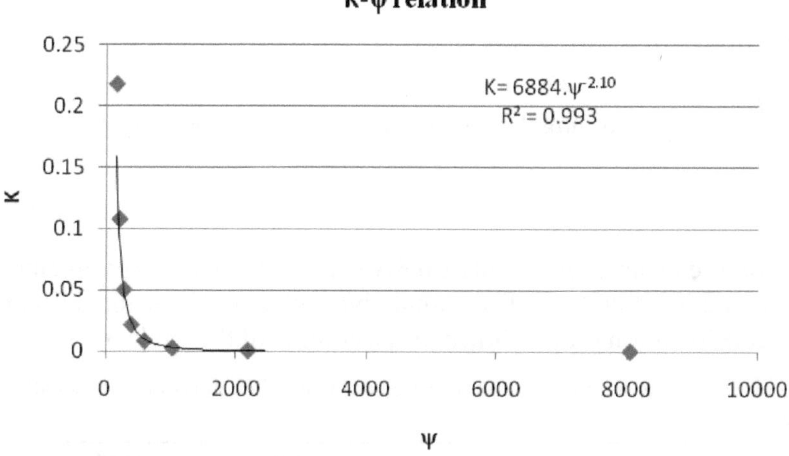

Fig.6.26 Determination of K- ψ relation

Evaporation

The upward flow or steady evaporation in the presence of water table as expressed in Eq.8.124

$$\int dz = \int \frac{d\psi}{1+\dfrac{E}{K}}$$

Where,

z = distance of ground water table from the ground surface, cm (z=Z)

Ψ = soil moisture tension at soil surface, cm of water

E = evaporation from the soil surface, cm/day

K = hydraulic conductivity, cm/day

Putting the value of $K=6884\psi^{-2.10}$ in K-ψ relation as developed

$$\int dz = \int \frac{d\psi}{1+\dfrac{E}{6884\psi^{-2.10}}}$$

$$= \int \frac{d\psi}{1+\dfrac{E\psi^{2.10}}{6884}}$$

$$= \int \left(1+\frac{E\psi 2.10}{6884}\right)^{-1} d\psi$$

$$= \int \left(1 - \frac{E\psi^{2.10}}{6884}\right) d\psi + ...$$

$$\text{or, } Z = \psi - \frac{E\psi^{3.10}}{6884x3.10} \text{ (avoiding the remaining part of the binomial expression)}$$

$$= \psi - \frac{E\psi^{3.10}}{21340.4} \tag{6.130}$$

The evaporation is calculated following **Eq.6.130** from 2 plots of experimental field and tabulated in Table 6.13. The cumulative evaporation for the days of drying under consideration is also shown in last column of the Table 6.13.

Table 6.13 Calculation of evaporation in field soils based on Darcy's law and continuity equation for Memari loam soil

Date	Average moisture(θ),	Average suction(ψ), cm	Time, days	Evaporation, mm
27.11.93-4.12.93	0.2433	407.38	7	3.115
4.12.93-8.12.93	0.2168	509.68	4	1.243
11.12.93-17.12.93	0.3091	261.55	6	
17.12.93-24.12.93	0.2394	422.95	7	
24.12.93-31.12.93	0.2177	505.72	7	
31.12.93-4.1.94	0.2052	565.22	4	

Calculation of evaporation using **Eq.6.130**

We have,

$$Z = \psi - \frac{E\psi^{3.10}}{21340.4} \quad \text{and} \quad \psi = 0.0282\,\theta^{-1.881}$$

Let us calculate the evaporation for 27.11.93-4.12.93 (Table 6.13). Using the values of Z=150cm, θ=0.2433, ψ = 0.39978bar=407.38cm of water (Eq.6.130)

$$150 = 407.38 - \frac{E(407.38)^{3.10}}{21340.4}$$

$$\text{or, } 150 - 407.38 = -\frac{E(123310222.6)}{21340.4}$$

$$-257.38 = -E5778.25$$

$$E = \frac{257.38}{5778.25} = 0.0445cm$$

$$= 0.445 mm/day$$

For 27.11.93-4.12.93=7 days

Evaporation for these 7 days = 7x0.445=3.115mm

6.7 Soil Water Uptake by Plants

Due to evaporative demand of the atmosphere plants use to transpire water in the vapor form through the stomata of the leaves. The process is called as transpiration. Plants may check the rate of transpiration by closing the stomata opening at different extent depending on the availability of soil water vis-a –vis the water potential. In doing so, the plants suffer in water stress, increased body temperature, and as a result decrease in yield. Plants draw the water from the soil between field capacity to wilting point. However, water availability less than the field capacity causes to stress to water stress to plant. This stress is different to different species of plants for same water availability. For the practical purpose a critical moisture content in between field capacity and wilting point is usually fixed through experimental trials at which crops economic yield do not suffer much.

Soil suction increases as the soil water decreases. Plants use to uptake water till the root suction is greater than the suction in the contact zone. This process of contribution of water to the plants follows the general laws of flow in unsaturated soil.

With the assumption that a typical root is infinitely long narrow cylinder of constant radius, flow through it is radial, the flow equation may be stated as below (Hillel, 1971),

$$\frac{\partial \theta}{\partial t} = \frac{1}{r}\frac{\partial}{\partial r}\left(rD\frac{\partial \theta}{\partial r}\right) \quad (6.131)$$

Where,

θ = volumetric soil wetness

D = *diffusivity*

t = time

R = radial distance from axis of the root

Gardner related the equation assuming constant flux and following initial and boundary condition,

$\theta = \theta_{0,}\ \psi = \psi_{0,}$ t=0

$$q = 2\pi ak\frac{\partial \psi}{\partial r} = 2\pi aD\frac{\partial \varphi}{\partial r}, \mathrm{r} = \mathrm{a}, \mathrm{t} > 0 \quad (6.132)$$

Where,

a = root radius

ψ = matric suction

K= hydraulic conductivity

q = rate of water uptake per unit length of root

With the assumption of K and D constant, the equation becomes,

$$\psi - \psi_0 = \Delta\psi = \frac{q}{4\pi K}\left(\ln\frac{4Dt}{r^3} - y_c\right) \tag{6.133}$$

Rose and Stern (1967) presented an equation relating soil wetness, hydraulic conductivity and rate of withdrawal of water by the root system from different soil depth zone (Hillel, 1972). Assuming the flow is vertical only for a given period of time (t_1 to t_2), the water- conservation equation is

$$\int_{t_1}^{t_2}(i - v_z - q_e)dt - \int_0^z\int_{t_1}^{t_2}\left(\frac{\partial\theta}{\partial t}\right)dzdt = \int_0^z\int_{t_1}^{t_2} r_z dzdt \tag{6.134}$$

Where,

i = rate of water supply (precipitation or irrigation)

q_e = evaporation rate from the soil surface

v_z = vertical flux of water at depth z

θ = volumetric soil wetness

r_z = the rate of decrease of soil wetness due to water uptake by root

At depth z the average rate of water uptake by the roots is

$$\bar{r}_z = \int_{t_1}^{t_2} r_z dt\,(t_2 - t_1) \tag{6.135}$$

The soil-water extraction at any small time period can be calculated by using this equation and the total (cumulative) water uptake by the roots R_z as given below

$$R_z = \int_0^z r_z dz \tag{6.136}$$

Questions and Problems

6.1. Define soil. Classify the soils based on the process of formation.

6.2. Give a brief description of Indian soils.

6.3. Define soil texture. Discuss the characteristics of major textural classes.

6.4. Define soil structure. How they are classified? Discuss the genesis of soil structure.

6.5. Discuss the soil structure and agricultural significance.

6.6. Define porosity, void ratio, wet and dry bulk density, mass wetness and degree of saturation.

6.7. Define capillarity, surface tension, dynamic and kinematic viscosity.

6.8. How the infiltration is measured? What is the advantage of using double ring infiltrometer?

6.9. Discuss the Ghosh's model of infiltration.

6.10. Derive the Richard's equation for flow of fluid in porous media.

6.11. Derive the Laplace's equation for flow of water in soils.

6.12. What is diffusivity in soil? Derive its expression, $\frac{\partial \theta}{\partial t} = \frac{\partial}{\partial x}\left(D(\theta)\frac{\partial \theta}{\partial x} \right)$.

6.13. Derive the expression, $D(\theta) = \frac{1}{t^{\frac{1}{2}} \Delta\theta_r / \Delta x_r}\left(-\frac{1}{2}\sum_{r=1}^{x} \lambda_r \Delta\theta_r \right)$ for determining the diffusivity.

6.14. Prove that volumetric moisture content in a soil is moisture content in weight multiplied by the apparent specific gravity.

6.15. A moist soil measures 1086g and 12.95% of soil moisture content. If the particle density of soil is 2.7g/cc, find (i) γ, ii) γ_d, (iii) n, (iv) e, and (iv) S_r

Ans. (i) 19.57kN/m^3 (ii) 17.33kN/m^3 (iii) 34.57% (iv) 0.528 (v) 65.93%.

6.16. A saturated soil sample of volume 77cc and mass 128.5g reduces to 81.5g on oven drying. Find (i) w,(ii) G , (iii) e, (iv) ρ_{sat} , and (iv) γ_d

Ans. (i) 57.67% (ii) 2.67, (ii) 1.564, (iii) 1.66, and (iv) 1.14g/cc

6.17. A soil sampler of 5cm diameter and 15cm length is used to collect the soil from an irrigation field. The soil sampler along with the soils weighs 675 and 635g respectively before and after the drying. Determine the (i) bulk density, (ii) dry density, (iii) degree of saturation, and (iv) soil moisture content.

Ans. (i) 1.61g/cc (ii) 1.16g/cc (iii) 23.73% (d) 11.94%

6.18. The following infiltration data were obtained in experimentation.

I(cm)	t(min)
0.03	1
0.07	4
0.15	7
0.25	15
0.43	25
0.72	45
0.92	55
1.06	70

Develop the infiltration equation of the form $I = at^{\alpha} + b$.

Ans: $I = 0.0252t^{0.881} - 0.00838$

6.19. The following data were obtained in a horizontal infiltration set up for determining the soil water diffusivity.

X_r	θ_r
0.75	0.445
2.25	0.432
3.75	0.435
6.75	0.385
8.25	0.365
9.75	0.225
11.25	0.135

Determine the diffusivities (D) at different moisture content (θ).

Ans

θ	D
0.125	3.7093×10^{-3}
0.144	6.2598×10^{-3}
0.17	7.6694×10^{-3}
0.208	1.0866×10^{-2}
0.258	1.6273×10^{-2}
0.301	2.4825×10^{-2}
0.347	4.0332×10^{-2}
0.370	5.7595×10^{-2}
0.393	1.1295×10^{-1}
0.402	1.1604×10^{-1}
0.414	1.2927×10^{-1}
0.425	1.2020×10^{-1}
0.434	2.1785×10^{-1}
0.439	4.5573×10^{-1}
0.443	1.46
0.444	

6.20. The following data of a soil were obtained by using the pressure plate apparatus. Develop the relation between them.

Tension, T(cm)	Volumetric moisture content (θ),%
3	74.55
25	59.44
50	52.13
100	44.65
200	38.87
400	32.92
800	27.75

Ans. $\theta = 99.04T^{-0.48}$, R^2=0.967

6.21. The K-ψ and ψ - θ relation of a soil is established as $K = 5887\psi^{-1.80}$ and $\psi = 0.025\theta^{-.175}$ respectively. Determine the evaporation rate at volumetric moisture content of soil 0.375 and water table at constant depth of 1.30m. Ans. 1.82mm/day

6.22. Select the appropriate answer from the following:

1. In a soil mass the volume of void and solid mass are 5 and 15cc respectively. The porosity is
   a) 25% b) 33%
   c) 50% d) 66%
2. The porosity in a soil mass is 30%. The void ratio is
   a) 0.30 b) 0.33
   c) 0.43 d) 0.70
3. The other name of bulk density of soil is
   a) Wet density b) Dry density
   c) Saturated density d) Mass wet ness
4. 100g moist soil on oven drying records 75g. The soil moisture content is
   a) 25% b) 33%
   c) 50% d) 75%
5. The unit of kinematic viscosity is
   a) cm/s^2 b) cm^2/s
   c) cm^3/s d) cm/s
6. The dimension of surface tension is
   a) MLT^{-1} b) ML^2T^{-1}
   c) $M^2L^2T^{-1}$ d) MT^{-2}
7. The porosity of a soil is 35% and specific gravity 2.66. The void ratio of the soil is
   a) 0.30 b) 0.45
   c) 0.51 d) 0.54
8. In a capillary tube of diameter 1.0mm water rises to 3.5cm with the angle of water contact 1.5^0. The surface tension in dynes/cm is
   a) 28.63 b) 57.24
   c) 85.87 d) 107.33
9. A soil sample at 30% moisture content measures $1.90 g/cm^3$. The dry bulk density in g/cm^3 is
   a) 1.46 a) 1.48
   c) 1.49 d) 1.50

10. A soil sample of weight 220g and volume 115cm^3 was reduced to 185g on oven drying. The specific weight of the soil was 2.66. The void ratio of the soil sample was

a) 0.45 b) 0.50

c) 0.55 d) 0.65

Ans.

1. a) 2. c) 3. a) 4. b) 5 b) 6. d). 7. d) 8. c)
9. a) 10. d)

6.23 Write True or False of the following statements:

1. Dry bulk density is higher than wet bulk density
2. Dry bulk density of a soil varies with the moisture content
3. A soil is said to be saturated when all the pores are filled by air, water and water.
4. The regime channel is rare
5. Regime channel theory is not applicable to artificial channel

Ans.

1. False 2. False 3. False 4. True 5. True

References

Aswa, G.L. (19930. Irrigation engineering. Wiley Eastern Ltd., New Delhi.

Biswas, R.K. (1983). Studies on Water Losses in Small Scale Irrigation. A M.Tech. dissertation submitted to IIT, Kharagpur, Dept. of Food and Agril. Engg.

Bodman, G.B. and Colman, E.A. (1944). Moisture and energy conditions during downward entry of water into soils. Soil. Sci. Soc. Amer. Proc. 8: 116-122

Boltzman, L. (1894). Zur integration dev. Diffusions gleichung bei variable diffusion koeffizienten. Annelender Physik and Chenie. 53: 959-964.

Brakensick, D.L. and Frevert, R.K. (1961). Analysis and application of infiltrometer tests. Trans. ASAE. 4: 75-77.

Browning, G.M. Volume changes of soils in relation to their infiltration rates. Soil. Sci. Soc. Amer. Proc. 4: 23-27.

Brutsaert, W.A. (1968). Solution foe vertical infiltration into dry porous medium. Water Res. Res. 4:1031-1038.

Childs, E.C. (1969). An introduction to the physical basis of soil water phenomena. John and Sons. Inc. New York and London.

Childs, E.C. and Bybordi, M. (1969). The vertical movement in water in stratified porous materials.1. Infiltration. Water Reso. Res., 5: 446-459.

Crank, J. (1956). The Mathematics of Diffussion. Oxford Univ. Press, London and New York

Duley, F.L. (1939). Surface factors affecting the rate of intake of water by soils. Soil. Sci. Soc. Amer. Proc. 4: 60-64.

Free, G.R., Browning, G.M., and Musgrave, G.W. Interrelationship of infiltration, air movement, and pore size in graded silica sand. Soil. Sci. Soc. Amer. Proc. 5: 390-398.

Ghildyal & Tripathi (1987). Soil Physics. Wiley Eastern Ltd., New Delhi.

Ghosh, R.K. (1975). Influence of Antecedent Moisture on Infiltration of Water into Soils. A Ph.D thesis submitted to Bidhan Chandra Krishi Viswavidyalaya, Kalyani, Nadia, W.B., India.

Hansen, V.E.(1955). Infiltration and soil water movement during irrigation. Soil Sci. 79: 93-105.

Hillel, D. (1972). Soil and Water Physical Principles and Processes. Academic Press, New York and London.

Horton, R.E.(1933). The role of infiltration in hydrologic cycle. Trans. Amer. Geophys. Union 14: 446-460.

Jamison, V.C. and Thorton, J.F.(1961). Water intake rates of a claypan soil from hydrograph analysis. J. Geophs. Res. 66: 1855-60.

Kirkham, D. and Feng, C.I.(1949). Some tests of the diffusivity theory and laws of capillary flow in soils. Soil Sci. 67: 29-35.

Mal, B. (1995). Soil and Water Conservation Engineering. Kalyani Publishers, New Delhi

Mein, R.G. and Larson, C.L. (1971). Modeling the infiltration component of the rainfall runoff process. Water Resources Research Center Bull. 43, Univ. of Minnesota, USA.

Neal, J.H. (1938). The effect of degree of slope and rainfall characteristics on runoff and soil erosion. Missouri Agril. Expt. Sta. Res. Bull. 280.

Philip, J.R. (1957). The theory of infiltration, 1. The infiltration equation and its solution. Soil Sci. 83: 345-357.

Philip, J.R. (1957). The theory of infiltration, 2. The profile at infinity. Soil Sci. 83: 435-448.

Philip, J.R. (1957). The theory of infiltration, 3. Moisture profile and relation to experiments.. Soil Sci. 84: 163-178.

Philip, J.R. (1957). The theory of infiltration, 4. Sorptivity and algebraic equations. Soil Sci. 84: 257-264.

Philip, J.R. (1957). The theory of infiltration, 5. The influence of initial moisture content. Soil Sci. 84: 329-339.

Punmia, B.C. (1994). Soil Mechanics and Foundation. Laxmi Publications Pvt. Ltd., New Deli.

Talsma, T. (1969). 'In Situ' measurement of sorptivity. Aust. J. Soil Res. 7: 269-276.

Whisler, F.E. and Bouwer, H. (1970) Comparison of methods for calculating vertical infiltration and drainage in soils. J. Hydrology. 10: 1-19.

7

Consumptive Use of Water

7.1 General

The major portion of water in a crop field gets lost through evaporation and transpiration. Evaporation is the process by which water is changed from the liquid state to vapor state. It takes place from the adjacent soil, water surface or from the surfaces of leaves of the plants. Transpiration is the removal of water vapour through plant body. Evapotranspiration(ET) is the sum of evaporation and transpiration. The process of evaporation and transpiration goes on simultaneously and difficult to separate them in a crop field.

The term consumptive use is used to refer evapotranspiration together with the water used in the metabolic activities of the crop plants. The actual amount of water used in metabolic activities is insignificant (less than 1% to ET). Therefore, the consumptive use and evapotranspiration are often used as synonymous term. It includes the water consumed by the plants for metabolic activities or transpiration plus the water evaporated from the land and water surfaces of the cropped area.

The factors affecting the rate of evaporation are the vapour pressure at the water surface and air above, air and water temperatures, wind speed, atmosphere pressure and size of the water body. Other factors remaining same, the rate of evaporation increases with an increase in the water temperature. Higher wind velocity causes greater scope for evaporation. Decrease in atmospheric pressure increases evaporation. The solute dissolved in water causes reduction in the rate of evaporation than that of pure water. The evaporation from soil surface occurs in same way to free water surface when the soil particles are surrounded by thin film of water and pore spaces are partially filled. However, in such situation greater resistance has to overcome for evaporation from the soil particles than while evaporating from a free water surface. The evaporation process practically stops when the moisture content in soil goes below a threshold value.

The factors influencing transpiration are availability of water to the plants, temperature and humidity of air, wind velocity, intensity and duration of sunlight, stage of plant growth, type of foliage and nature of the leaves. At growing stage of crop, high temperature and wind velocity, dry atmosphere, etc. greatly increase movement of water through the plants and thereby the rate of transpiration.

Some terminology related to evapotranspiration / consumptive use

Potential evapotranspiration (PET): It may be defined as the highest rate of evaporation (ET) by an actively growing crop or vegetation, completely covered by the plant foliage and no limitation of water in a given climate. It refers the maximum loss of water in a crop field.

Crop evapotranspiration (ET_c): It may be defined as the rate of evapotranspiration of a disease free crop growing in a field of not less than one hectare under adequate fertility and water supply so that full production potential can be achieved in prevailing environment. Evapotranspiration requirement of a crop refers the $ET_{c.}$

PET indicates highest evapotranspiration in a crop season. This evapotranspiration is used to determine highest water requirement in the crop season and thereby in designing irrigation system capacity. ET_c is the evapotranspiration at any time of the crop season. This is important in determining the requirement of water in irrigation.

Reference crop evapotranspiration (ET_0): It is the rate of evapotranspiration from an extended surface of 8 to 15cm tall green grass of uniform height, actively growing, completely shading the ground and not short of water. The ET_0 may be computed by using the empirical formulae like Blaney – Criddle, Modified Penman, Radiation and Pan Evaporation for a specified period using the mean climatic data for the period.

Actual crop evopotranspiration (ET_a): Actual crop evapotranspiration is the rate of evapotranspiration of a crop in a given soil and climatic condition. The ET_c is estimated assuming adequate water supply. There is no such assurance for ET_a and in fact very often in crop field soil water is limited. Thus, ET_a is less than or equal to ET_c.

The usual behavior of the two components of evapotranspiration viz. evaporation and transpiration are shown in Fig. 7.1. After sowing and before the germination there is only evaporation and forms the ET. During the early stage of plants when the ground cover is insignificant the rate of transpiration is low and the evaporation rate is still important in contributing to ET. As the plants grow further and shading the area by foliage the transpiration rate increases as the evaporation rate progressively decreases. The process continues till almost the entire area is covered. At this time the ET value reaches to its peak. This peak values continues with little variation till the start of maturity of the crop. The proportion of evaporation and transpiration also remain almost same.

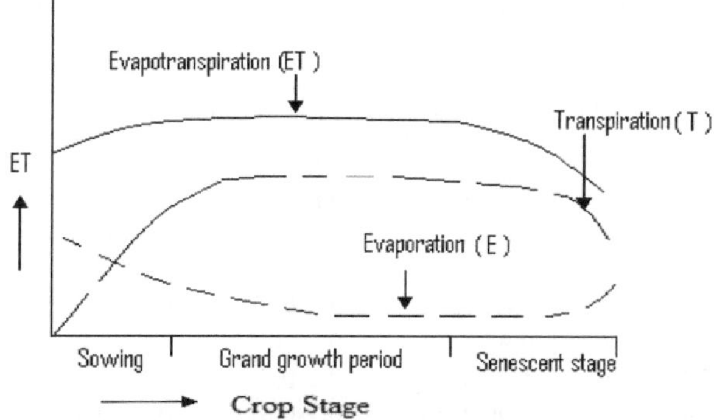

Fig. 7.1 Schematic representation of components of evapotranspiration with crop stage

Crop coefficient or crop factor (K_c): It is the ratio of crop evapotranspiration, ET_c and the reference crop evapotranspiration, ET_0, when both apply to large field under optimum growth condition.

$$K_c = \frac{ET_c}{ET_0} \tag{7.1}$$

The crop coefficient depends on crop variety, duration, growing season, stage of crop growth, depth of rooting, method of irrigation, plant population, fertilization, weed control, tillage, plant protections, etc. The crop coefficient is fixed for a given crop but the values differ at different stage of growth. It is low at the early stage of growth and increases as it approaches to grand growth stage and remains almost constant during this stage and then decline gradually (Fig.7. 2). For selecting the crop coefficient, information is required for date of sowing of the crop, the length of the growing season, the duration of the initial stage (germination to 10% ground cover), duration of the crop development stage (from 10% to 80% ground cover), the duration of the mid – season stage (from 80% ground cover to start of ripening) and duration of late season stage (from start of ripening to harvest). The crop coefficient values for important field and vegetable crops are given in Table (Table 7.1-7.5).

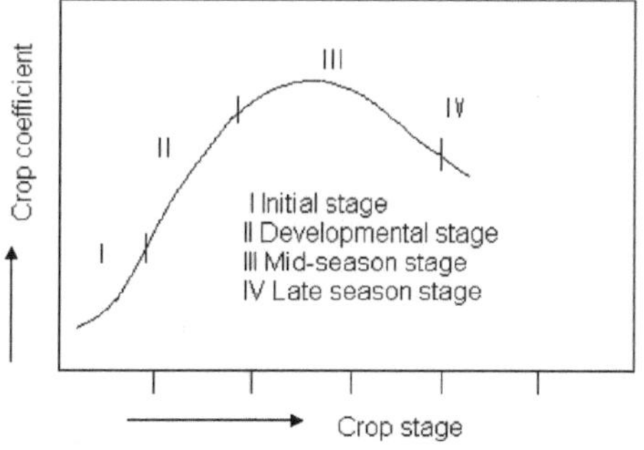

Fig 7.2 Crop coefficient at different stage of plant growth

Table. 7.1 Crop coefficient values for important field crops at different stage of development

Crop	Crop developmental stage				Total period
	Initial	Development	Mid-season	Late season	
Rice	1.10-1.15	1.1-1.5	1.1-1.3	0.95-1.05	1.05-1.2
Wheat	0.3-0.4	0.7-0.8	1.05-1.2	0.65-0.76	08.-0.9
Maize	0.3-0.5	0.7-0.85	1.05-1.2	0.8-0.95	0.75-0.9
Sorghum	0.3-0.4	0.7-0.87	1.00-1.15	0.75-0.8	0.75-0.85
Cotton	0.4-0.5	0.7-.8	1.05-1.25	0.8-0.9	0.8-0.0.9
Groundnut	0.4-0.5	0.7-0.8	0.95-1.10	0.75-0.85	0.75-0.8
Soybean	0.3-0.4	0.7-0.8	1.00-1.15	0.7-0.8	0.75-0.85
Sunflower	0.3-0.4	0.7-0.8	1.00-1.20	0.7-0.8	0.75-0.85
Sugarcane	0.4-0.5	0.7-1.0	1.0-1.2	0.75-0.8	0.85-1.05
Tobacco	0.3-0.4	0.7-0.8	1.0-1.2	0.9-1.0	0.85-0.95

Source: Reddi & Reddy(1995)

Note: First figure is for high humidity (RH min>70% and low wind velocity (U<5 m/s) conditions and the second figure refers to low humidity (RH min<20% and strong wind (U>5 m/s) conditions

Table 7.2 Crop coefficient values for selected vegetable and fruit crops

Crop	Crop developmental stage					
	Initial	Development	Mid-season	Late season	At harvest	
Cabbage	0.4-0.5	0.7-0.8	0.95-1.1	0.9-1.0	0.8-0.95	0.7-0.8
Onion	0.4-0.6	0.7-0.8	0.95-1.1	0.85-0.9	0.75-0.85	0.8-0.9
Peas	0.40.5	0.7-0.8	1.05-1.2	1.0-1.15	0.95-1.1	0.8-0.95
Potato	0.4-0.5	0.7-0.8	1.05-1.2	0.85-0.95	0.7-0.75	0.75=0.9
Tomato	0.4-0.5	0.7-0.8	1.05-1.25	0.8-0.95	0.6-0.65	0.75-0.9
Water melon	0.4-0.5	0.7-0.8	0.95-1.05	0.8-0.95	0.65-0.75	0.75-0.85
Banana	0.4-0.5	0.7-0.85	1.0-1.1	0.9-1.0	0.75-0.85	0.7-0.8
Grape	0.35-0.55	0.6-0.8	0.7-0.9	0.6-0.8	0.55-0.7	0.55-0.75
Citrus	-	-	-	-	-	0.65-0.75

Source: Reddi & Reddy (1995)

Note: The lower figure is high humidity (RH min > and low wind (U < 5m/s) and higher values for low humidity (RH min < 20%) and strong wind (U > 5 min/s) conditions.

Table 7.3 Values of crop coefficient (K_c) for some crops to estimate consumptive use from USWB pan evaporation values

Stage of crop growth (% of growth season)	Wheat (Ludhiana)	Wheat (Pune)	Maize (Ludhi-ana)	Sorghum (Albana USA)	Cotton (Pune)	Sugarcane (Hawaii)	Rice (Los Banos, Philipine)
1	2	3	4	5	6	7	8
0	0.14	0.30	0.40	0.42	0.22	0.34	1.00
5	0.17	0.4	0.42	0.44	0.22	0.37	1.02
10	0.23	0.52	0.47	0.46	0.23	0.40	1.03
15	0.33	0.62	0.54	0.48	0.24	0.44	1.05
20	0.45	0.73	0.63	0.6	0.26	0.5	1.07
25	0.6	0.84	0.75	0.52	0.35	0.6	1.09
30	0.77	0.92	0.85	0.56	0.58	0.72	1.11
35	0.81	0.96	0.96	0.59	0.8	0.86	1.13
40	0.88	1.1	1.04	0.64	0.95	0.93	1.16
45	0.9	1.1	1.07	0.71	1.03	0.98	1.18
50	0.91	1.00	1.09	0.79	1.08	1.02	1.20
55	0.9	0.91	1.1	0.92	1.08	1.07	1.21
60	89	80	1.11	1.01	1.07	1.1	1.22
65	0.86	0.65	1.10	1.07	1.05	1.13	1.22
70	0.83	0.51	1.07	1.09	1.00	1.16	1.21
75	0.80	0.4	1.04	1.09	0.93	1.19	1.19
80	076	0.3	1.00	1.05	0.85	1.20	1.16
85	0.71	0.2	0.97	0.99	0.73	1.20	1.10
90	0.65	0.12	0.89	0.91	0.62	1.19	1.03
95	0.58	0.1	0.81	0.82	0.5	1.19	0.96
100	0.51	0.1	0.7	0.7	0.4	0.89	0.86
Seasonal	0.61	0.61	0.86	0.75	0.68		1.10

Source: Water Management Division, Dept. Agri. & Irrg., Govt. of India, 1971

Table 7.4 Values of monthly crop coefficient (K_c) to compute consumptive use by Blanney-Criddle formula

Month	Sugarcane	Rice	Maize	Cotton	wheat	Berseem	vegetables	Citrus
January	0.75	-	-	-	0.50	0.50	0.50	0.50
February	0.80	-	-	-	0.70	0.55	0.55	0.55
March	0.85	-	-	-	0.75	0.60	0.60	0.55
April	0.85	0.85	0.50	0.50	0.70	0.65	0.65	0.60
May	0.90	1.00	0.60	0.60	-	0.70	0.70	0.60
June	0.95	1.15	0.70	0.75	-	0.75	0.75	0.65
July	1.00	1.30	0.80	0.90	-	0.80	0.80	0.70
August	0.95	1.25	0.80	0.85	-	0.80	0.87	0.70
September	0.90	1.10	0.60	0.75	-	0.70	0.70	0.65
October	0.85	0.9	0.50	0.55	0.70	0.60	0.60	0.60
November	0.85	-	-	0.50	0.65	0.55	0.55	0.55
December	0.75	-	-	0.50	0.60	0.50	0.50	0.55

Source: Dastane (1972) (Majumdar, 2000)

Table. 7.5. Duration of growth stage and crop coefficients (K_c) for selected garden crops

Crop	Total growing period (day)	Initial Stage		Development stage		Mid-season stage		Late season stage	
		Duration (day)	K_c	Duration (day)	K_c	Duration (day)	K_c	Duration (day)	K_c
Red amaranth	350.75	09	0.45	12	0.60	09	1.00	05	0.90
Stem amaranth	1150.75	24	0.45	36	0.70	35	1.05	20	0.95
Indian spinach	135	15	0.50	35	0.70	55	1.15	30	0.65
Spinach	45	12	0.45	18	0.60	09	1.00	06	0.90
Carrot	90	20	0.45	25	0.75	30	1.05	15	0.90
Radish	70	10	0.45	15	0.60	25	0.90	20	0.80
Sweet gourd*	140	25	0.50	45	0.75	40	1.00	30	0.80
Bottle gourd*	130	25	0.50	50	0.75	35	1.00	30	0.90
Hyacinth bean*	160	30	0.50	55	0.8	50	1.05	25	0.80
Tomato*	110	25	0.45	30	0.75	35	1.15	20	0.80
Brinjal	150	30	0.45	45	0.75	50	1.15	25	0.85
Potato	95	20	0.50	30	0.7	30	1.15	15	0.70
Garlic	100	30	0.50	35	0.7	20	1.00	15	0.85
Onion*	90	25	0.50	35	0.7	20	1.00	10	0.90
Chilli*	150	30	0.52	50	0.55	45	0.92	25	0.77
Turmeric*	285	50	0.45	90	0.9	90	1.00	55	0.75
Ginger*	250	45	0.40	85	0.88	75	1.00	45	

Source: Doorenbos & Pruitt (1977); Rashid *et al.* (1994); *Dutta (1995) and Allen *et al.* (1998)

Daily consumptive use: It is the amount of water consumptively used by the plants during a day of 24-hr. It is required to estimate the peak period demand of water by the crops for formulating the cropping pattern and decide upon the need of water supply.

Seasonal consumptive use: It is the amount of water consumptively used by the plants during the entire period of growing season of a crop. The seasonal consumptive use is important in selecting cropping pattern and sequence and to evaluate and decide upon to seasonal irrigation water requirements.

Peak period consumptive use: This is the average of the highest consumptive use rates of few days (usually 6 to 10 days) in a season. If the days are taken more the peak period consumptive use will be less leading to less cost for irrigation system but increases chances of experiencing water constraint by the plants during the peak period. Thus, the peak period consumptive use is important in designing the irrigation system capacity.

The number of days is selected on the basis of designed compromise to reduction on crop yield due to water deficit and cost of irrigation system. The peak period consumptive use occurs when the vegetation is abundant, temperature is high and crops are in flowering stage. The peak period is usually shorter in shallow rooted crops cultivated in soils of low water holding capacity.

7.2 Energy Balance

Energy balance in ET

Evaporation takes place by absorbing relatively large amount of energy either on sensible form or radiant energy. Evapotranspiration is governed by the process of energy balance at the crop field on the vegetation and soil surface. The energy received by these surfaces must be equal to energy released during the same period. The process may be expressed by the following equation.

$$R_n - G - \lambda ET - H = 0 \tag{7.2}$$

Where,

R_n = net rediation

H = sensible heat

G = soil heat

λET = latent heat flux

These terms may be positive or negative. R_n supplies energy to the surface when positive and the G, λET and H are positive when these remove energy from the surface. In Eq.7.1 the transferred vertically is only considered. The energy transferred horizontally, by advection is ignored. Similarly the other terms of energy like heat stored or released in the plant, or the energy used in metabolic activities of the plant body is ignored. The latent heat flux (λET) which represents the evapotranspiration can be estimated by the other components like net radiation (R_n), soil heat flux (G) are measured or estimated from meteorological parameters. Sensible heat (H) above is obtained by the accurate measurement of the temperature gradient. However, this is very complex and accurate measurement is difficult.

Resistance to ET

In 1948, Penman combined the energy balance with the mass transfer method and derived an equation to compute the evaporation from an open water surface from standard climatological records of sunshine, temperature, humidity and wind speed. This so-called combination method was further developed by many researchers and extended to cropped surfaces by introducing resistance factors.

The resistance factors are distinguished between aerodynamic resistance and surface resistance factors (Fig.7.3). The surface resistance parameters are combined into one parameter and called as the 'bulk' surface resistance. The surface resistance, r_s, describes the resistance of vapour flow through stomata openings, total leaf area and soil surface. The aerodynamic resistance, r_a, describes the resistance from the vegetation upward and involves friction from air flowing over vegetative surfaces.

Aerodynamic resistance (r_a)

Heat and water vapor transfer from the evaporating surface in to the air above the canopy is determined by the aerodynamic resistance as below.

$$r_a \frac{\ln\left[\frac{z_m - d}{z_{om}}\right]\ln\left[\frac{z_h - d}{z_{oh}}\right]}{k^2 u_z} \tag{7.3}$$

Where,

r_a = aerodynamic resistance, sm^{-1}

z_m = height of wind measurement, m

z_h = height of humidity measurement, m

d = zero displacement height, m

z_{om} = roughness length governing momentum transfer, m

z_{oh} = roughness length governing transfer of heat and vapor, m

K = von Karman's constant,0.41

u_z = wind speed at height z, ms^{-1}

The Eq.7.3 is restricted to where temperature, atmospheric pressure, and wind velocity distributions follow nearly adiabatic conditions (no heat exchange).

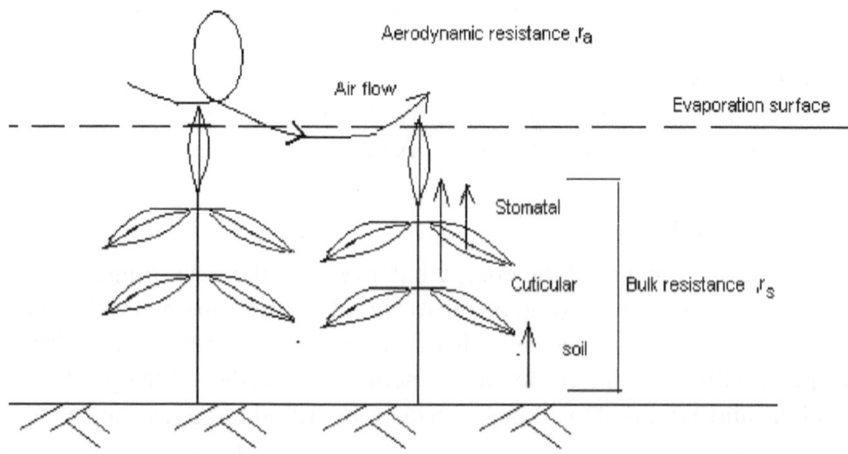

Fig.7.3. Representation of bulk surface and aerodynamic resistance to vapour flow

Several empirical equations have been developed for the estimate of d, z_m and $z_{oh.}$ The following is the acceptable approximation to a much more complex relation of the surface resistance of dense full cover vegetation (FAO, 1998).

Aerodynamic resistance for a grass reference surface

In wide range of crops the zero plane displacement height, d(m), and the roughness length governing momentum transfer, z_{om}(m), may be estimated from the crop height, h(m), as below.

$$d = 2/3h$$

$$z_{om} = 0.123h$$

$$z_{oh} = 0.1z_{om}$$

The aerodynamic resistance ra (sm^{-1}) for the grass reference surface may be expressed by Eq.7.4 assuming a constant crop height of 0.12m and a standardized height for wind speed, temperature and humidity at 2m ($z_m = z_h = 2m$).

$$r_a = \frac{\ln\left[\frac{2-2/3(0.12)}{0.123(0.12)}\right]\ln\left[\frac{2-2/3(0.12)}{(0.1)0.123(0.12)}\right]}{(0.41)^2 u_2} = \frac{208}{u_2} \tag{7.4}$$

where u_2 is the wind speed (ms^{-1}) at 2m.

$$r_s = \frac{r_l}{LAI_{active}} \tag{7.5}$$

where

r_s = bulk surface resistance (sm^{-1})

r_1 = bulk stomatal resistance of the well-illuminated leaf (sm^{-1})

LAI_{active} = active (sunlit) leaf area index [m^2 (leaf area) m^{-2} (soil surface)].

The Leaf Area Index (LAI) is the leaf area per unit area of soil below and expressed in m^2 leaf area per m^2 ground area. The active Leaf Area Index (LAI_{active}) is the index of the leaf area that actively contributes to the surface heat and vapour transfer. The LAI varies inbetween 3-5 for many mature crops. LAI usually changes at different crop stages but normally reaches its maximum before or at the flowering stage.

It is generally assumed that half of dense clipped grass actively contributes to the surface heat and vapour transfer.

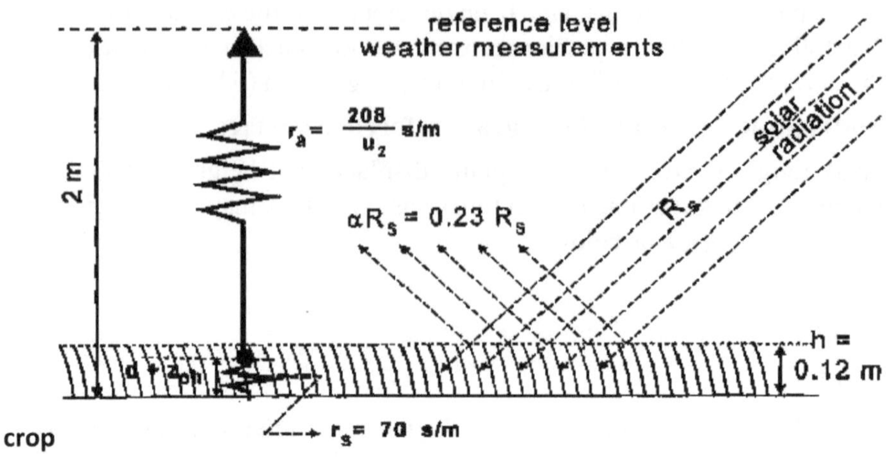

Fig. 7.4. Bulk surface resistance for grass reference

Thus $LAI_{active} = 0.5\ LAI$

For clipped grass the general equation for LAI is

LAI=24h

where h is the crop height (m).

It is further assumed that the stomatal resistance, r_1, of a single leaf has a value of about $100sm^{-1}$ under well watered conditions. Taking the crop height as 0.12m, the **Eq.7.5** for surface resistance becomes:

$$r_s = \frac{100}{0.5(24)(0.12)} \approx 70sm^{-1} \tag{7.6}$$

Reference surface

The concept of reference surface was introduced to define the unique parameters of each crop and stage of growth. In the past the water surface was proposed as reference surface for estimating ET. However, the water surface is not adequate to express the impact of aerodynamic, vegetation control and radiation characteristics of the crop field to measure ET. It is advantageous to relate a specific crop in incorporating the biological and physical process involved in ET. Grass, together with the alfalfa, is accepted worldwide as the reference surface. Since the resistance to diffusion of vapour strongly depends on crop height, ground cover, LAI and soil moisture content, the characteristic of the crop should be well defined and fixed.

The FAO Expert Consultation on Revision of FAO Methodologies for Crop Water Requirements accepted the following definition for the reference surface.

"A hypothetical reference crop with an assumed crop height of 0.12m, a fixed surface resistance of 70sm^{-1} and an albedo of 0.23".

The reference surface closely indicates a green grass surface of large area, uniform height, actively growing, completely shading the ground with adequate water.

Atmospheric pressure (P)

The pressure exerted by the earth's atmosphere is called the atmospheric pressure, P. The variation of atmospheric pressure with respect to altitude promotes evaporation, as expressed in the psychrometric constant. The effect of this is small. Using the ideal gas law and assuming a standard atmosphere of 20^{0}C, P can be calculated as,

$$p = 101.3\left[\frac{293 - 0.0065z}{293}\right]^{5.26} \tag{7.7}$$

where, P = atmospheric pressure (kPa)

z = elevation above sea level (m)

Atmospheric pressure at different altitude is shown in **Table 7.6**

Latent heat of vaporization (l)

Latent heat of vaporization (λ) is the energy required to change a unit mass of water from liquid state to vapor state at constant pressure and temperature. The value of the latent heat varies as a function of temperature. Less energy is required at higher temperature. The value of λ varies a little over normal temperature range. Therefore, it is usually taken a single value of 2.45MJkg^{-1} at an air temperature of 20^{0}C.

The latent heat flux (λET) is the rate of evapotranspiration and expressed in MJm^{-2}day^{-1}. One kg of water volume is equal to 0.001m^3. This 0.001m^3 volume of water creates a depth of 1.0 mm over an area of 1.0m^2. Therefore, evaporation at the rate of 1.0mm/m^2/day is equivalent to 2.45MJkg^{-1} and 1.0MJkg^{-1}is equivalent to 0.408 mm/m^2/day.

Let, solar energy is received in a water pond at a rate of 16MJ/m^2/day. The 75% of this energy is being used to vaporize the water. Evaporation equivalent of this energy = (16x0.75x0.408)=4.896*mm/m*^{2}/day.

Psychrometric constant (λ)

$$\gamma = \frac{C_p p}{\varepsilon\lambda} = 0.665\text{x}10^{-3}p \tag{7.8}$$

Where,

γ = psychrometric constant [kPa^{0}C^{-1}]

P = atmospheric pressure [kPa]

λ = latent heat of vaporization, 2.45 [$MJkg^{-1}$]

c_p = specific latent heat of pressure, 1.03x10^{-3} [$MJkg^{-10}C^{-1}$]

ε = ratio of molecular weight of water vapor/dry year = 0.622

Vapor pressure

Water vapor is a gas. The atmospheric pressure is the combination of air and the partial pressure of water vapor. Therefore, the water content in the air is a direct measure from the partial pressure of the water vapor. When there is air above the evaporating water surface, the water molecule escaping and returning to the water reservoir, it is understood that equilibrium is reached, and at that moment the air is said to be saturated.

Dew point temperature

It is the temperature at which the air is cooled to make the air saturated. Drier the air, larger will be the difference between air and dew point temperature. Higher water vapor in air indicates higher dew point temperature.

Table 9.6 Atmospheric pressure (P) for different altitudes (z)

Table 7.6. Atmospheric pressure (P) for different altitudes (z)

$$P = 101.3\left(\frac{293 - 0.0065z}{293}\right)5.26$$

(Eq. 7.7)

z(m)	P(kPa)	z(m)	P(kPa)	z(m)	P(kPa)	z(m)	P(kPa)
0	101.3	1000	90.0	2000	79.8	3000	70.5
50	100.7	1050	89.5	2050	79.3	3050	70.1
100	100.1	1100	89.0	2100	78.8	3100	69.6
150	99.5	1150	88.4	2150	78.3	3150	69.2
200	99.0	1200	87.9	2200	77.9	3200	68.8
250	98.4	1250	87.4	2250	77.4	3250	68.3
300	97.8	1300	86.8	2300	76.9	3300	67.9
350	97.2	1350	86.3	2350	76.4	3350	67.5
400	96.7	1400	85.8	2400	76.0	3400	67.1
450	96.1	1450	85.3.	2450	75.5	3450	66.6
500	95.5	1500	84.8	2500	75.0	3500	66.2
550	95.0	1550	84.3	2550	74.6	3550	65.8
600	94.4	1600	83.8	2600	74.1	3600	65.4
650	93.8	1650	83.3	2650	73.7	3650	65.0
700	93.3	1700	82.8	2700	73.2	3700	64.6
750	92.7	1750	82.3	2750	72.7	3750	64.1
800	92.2	1800	81.8	2800	72.3	3800	63.7
850	91.6	1850	81.3	2850	71.8	3850	63.3
900	91.1	1900	80.8	2900	71.4	3900	62.9
950	90.6	1950	80.3	2950	71.0	3950	62.5
1000	90.0	2000	79.8	3000	70.5	4000	62.1

Table 7.7 Psychometric constant (γ) for different altitudes (z)

$$\gamma = \frac{c_p p}{\varepsilon\lambda} = 0.665x^{10-3}$$

(Eq. 7.8)

z(m)	γ kPa/°C	z(m)	γ kPa/°C	z(m)	γ kPa/°C	z(m)	γ kPa/°C
0	0.067	1000	0.060	2000	0.053	3000	0.047
100	0.067	1100	0.059	2100	0.052	3100	0.046
200	0.066	1200	0.058	2200	0.052	3200	0.046
300	0.065	1300	0.058	2300	0.051	3300	0.045
400	0.064	1400	0.057	2400	0.051	3400	0.045
500	0.064	1500	0.056	2500	0.050	3500	0.044
600	0.063	1600	0.056	2600	0.049	3600	0.043
700	0.062	1700	0.055	2700	0.049	3700	0.043
800	0.061	1800	0.054	2800	0.048	3800	0.042
900	0.061	1900	0.054	2900	0.047	3900	0.042
1000	0.060	2000	0.053	3000	0.047	4000	0.041

Based on λ = 2.45 MJ kg^{-1} at 20°C. *Source:* FAO, 1998

Relative humidity

Relative humidity is the ratio of the actual (e_a) to the saturation vapor pressure (e^0) at the same temperature. It is, therefore, the ratio of amount of water it holds at ambient air to the amount of water it could hold at the same temperature is the relative humidity.

$$RH = 100\frac{e_a}{e^0} \tag{7.9}$$

The relative humidity varies much during different time of a day depending on the variation of temperature, though the actual vapor pressure might be relatively constant.

Mean saturation vapor pressure (e_a)

Since the saturation vapor pressure is related to air temperature, it is expressed as,

$$e^0(T) = 0.6108\exp\left[\frac{17.27T}{T+237.3}\right] \tag{7.10}$$

Where, e^0 (T) = saturation vapor pressure at the air temperature T(kPa)

T = air temperature, °C.

Saturation vapor pressure varies with the air temperature (Table 7.9). It is non-linear in nature. Therefore, to be more explicit to the actual situation, the saturation vapor pressure for a day, week, month, year or decade is taken as the mean saturation vapor pressure at the mean daily maximum and minimum air temperature for the period. It is expressed as,

$$e_a = \frac{e^0(T_{max}) + e^0(T_{min})}{2} \tag{7.11}$$

Using mean air temperature instead of daily minimum and maximum temperatures results in lower estimates for the mean saturation vapor pressure. The corresponding vapor pressure deficit (a parameter expressing the evaporating power of the atmosphere) will also be smaller and the result will be some under estimation of the reference crop evapotranspiration. Therefore, the mean saturation vapor pressure should be calculated as the mean between the saturation vapor pressure at both the daily maximum and minimum air temperature.

Slope of saturation vapor pressure (Δ)

The slope of saturation vapor pressure, Δ at a given temperature is given by

$$\Delta = \frac{4098\left[0.6108\exp\left(\frac{17.27T}{T+237.3}\right)\right]}{(T+237.3)^2} \tag{7.12}$$

Actual vapor pressure (e_a) derived from the dew point temperature

Dew point temperature is the temperature to which the air needs to be cooled to make the air saturated. Therefore, the actual vapor pressure (e_a) is the saturation vapor pressure at the dew point temperature (T_{dew})[^{0}C], or,

$$e_a = e^0(T_{dew}) = 0.6108\exp\left[\frac{17.27T_{dew}}{T_{dew}273.3}\right] \tag{7.13}$$

Actual vapor pressure (e_a) derived from psychrometric data

In a psychrometer the vapor pressure is measured by knowing the difference of temperature between two thermometers, the so-called dry and wet thermometer. The dry thermometer measures the temperature of the air. The bulb of the wet thermometer is kept always wet by covering constantly with a saturated wick. The evaporation from the wick requires more energy and lowers the temperature of the thermometer. The difference between the dry and wet bulb temperature is called the wet bulb depression and is the measure of air humidity. The relationship is expressed as,

$$e_a = e^0(T_{wet}) - \lambda_{psy}(T_{dry} - T_{wet}) \tag{7.14}$$

Where,

e_a = actual vapor pressure (kPa)

$e^0(T_{wet})$ = saturation vapor pressure at wet bulb (kPa)

λ_{psy} = psychrometer constant of the instrument (kPa^{0}C^{-1})

T_{dry}-T_{wet} = wet bulb depression

T_{dry} and T_{wet} = dry and wet bulb temperature respectively, (^{0}C)

λ_{psy} = $a_{psy}p$ (7.15)

a_{psy} = *a* a coefficient depends on the type of the ventilation of the wet bulb ($^0C^{-1}$)

P = atmospheric pressure (kPa)

The coefficient a_{psy} depends on the design of the psychrometer and the rate of ventilation around the bulb.

Actual vapor pressure (e_a) derived from the relative humidity data

The actual vapor pressure can be expressed by using the relative humidity data.

$$e_a = \frac{e^0\left(T_{\min}\right)\frac{RH_{\max}}{100} + e^0\frac{RH_{\min}}{100}}{2} \quad (7.16)$$

Where,

e_a = actual vapor pressure, kPa

$e^0(T_{min})$ = saturation vapor pressure at daily minimum temperature, kPa

$e^0(T_{max})$ = saturation vapor pressure at daily maximum temperature, kPa

RH_{max} = maximum relative humidity, %

RH_{min} = minimum relative humidity, %

Wind profile relationship

Wind speed is lowest at the soil surface and increases with height. For this reason to measure the wind speed the anemometers are placed at 10m height in meteorology and 2 or 3m in agro-meteorology. It is, therefore, the wind speed measured at any other height is converted to standard height of 2m by using the following equation.

$$u_2 = u_z \frac{4.87}{\ln\left(67.8z - 5.42\right)} \quad (7.17)$$

Where,

u_2 = wind speed at 2m above ground surface (m/s)

u_z = measured speed at z m above the ground surface (m/s)

z = height of measurement above ground surface (m)

7.3 Radiation

Extraterrestrial radiation (R_a)

The radiation that strikes the surface perpendicular to the sun's rays on earth's atmosphere is called solar constant. The solar constant is about 0.082MJm-

$^{2}min^{-1}$. The solar radiation which is received on the horizontal plane at the top of the earth's atmosphere is called the extraterrestrial (solar) radiation, R_a. When the sun is directly overhead the angle of incidence is zero and the extraterrestrial radiation and solar constant is same 0.082$MJm^{-2}min^{-1}$. The value of extraterrestrial radiation varies at different time of the day, as well as it varies at different latitude and longitude in different season.

Solar or short wave radiation (R_s)

Some of the extraterrestrial radiation on penetration to earth's atmosphere get scattered, reflected or absorbed by the atmospheric gases, clouds and dust. The remaining portion which reaches to the horizontal plane of the earth's surface is called the solar radiation. The solar radiation is also called as short wave radiation since the sun emits energy by means of electromagnetic wave of shortwave length in character. R_s is about 75% of extraterrestrial radiation in a cloudless day.

Relative short wave radiation (R_s/R_{so})

The relative short wave radiation is the ratio of solar radiation (R_s) to solar radiation in clear-sky (R_{so}). This ratio is useful in expressing the cloudiness of the atmosphere. Its value varies from 0.33 to maximum of 1 (clear-sky).

Relative sun shines duration (n/N)

It is the ratio of actual duration of sun shine, n, to the maximum possible duration of sun shine or day light hours, N. The value of n depends on the presence or absence of clouds. When there is no cloud the value of n is equal to N.

Albedo(α) and net solar radiation (R_{ns})

A portion of the solar radiation on earth's surface get reflected is called as the albedo. The albedo depends on the type of reflection surface and the angle of incidence. It is as high as 0.95 for freshly fallen snow and 0.05 for a wet bare soil. The portion of the solar radiation which does not get reflected is called the net solar radiation. Thus, its value is $(1-\alpha)R_s$.

Net long wave radiation (R_{nl})

The solar radiation received by the earth get converted to heat energy and loses this energy mostly through emission with wave length much longer than the sun and is called as the long wave radiation. The emitted long wave radiation either absorbed in the atmosphere or get lost in space. The absorption of long wave radiation by the atmosphere increases its temperature and as a result atmosphere radiates energy back to the earth's surface. Thus, the earth's surface both receives and emits the long wave radiation. The difference between the incoming and outgoing long wave radiation is called the net long wave radiation, R_{nl}.

Net radiation (R_n)

It is the difference between the net incoming and outgoing radiation of both short wave and long wave on earth's surface. It is normally positive during day time and negative at night. However, daily value of R_n is always almost positive excepting extreme cases at high altitude.

Soil heat flux (G)

It is the energy which is used in heating the soil. G is considered positive when it is warming the soil and negative during cooling. It is negligible compared to R_n. However, theoretically it is subtracted or added from R_n while estimating evapotranspiration.

Calculation of extraterrestrial radiation (R_a)

Extraterrestrial radiation (R_a) can be calculated by using the following equation.

$$R_a = \frac{24(60)}{\pi} G_{sc} d_r \left[\varepsilon_s \sin(\varphi)\sin(\delta) + \cos(\varphi)\cos(\delta)\sin(\omega_s) \right] \quad (7.18)$$

Where,

R_a = extraterrestrial radiation[$MJm^{-2}day^{-1}$]

G_{sc} = solar constant = 0.082$MJm^{-2}min^{-1}$

d_r = inverse relative distance earth-sun

$$= 1 + 0.33\cos\left(\frac{2\pi}{365}J\right) \quad (7.19)$$

ω_s = sun set angle

$$= \arccos\left[-\tan(\omega)\tan(\delta)\right] \quad (7.20)$$

$$\text{or, } \omega_s = \frac{\pi}{2}\arctan\left[\frac{-\tan(\varphi)\tan(\delta)}{X^{0.5}}\right] \quad (7.21)$$

$$\text{Where, } X = 1 - \left[\tan(\varphi)\right]^3\left[\tan(\delta)\right]^2 \quad (7.22)$$

and $X = 0.00001$ if $X \le 0$

φ = latitude [radian]

δ = solar decimation [radian]

$$= 0.409\sin\left(\frac{2\pi}{365}J - 1.39\right) \quad (7.23)$$

Example 7.1 Determine the atmospheric pressure and psychrometric constant at an elevation of 2000m.

Solution:

Using Eq.7.7, $P = 101.3\left[\frac{293 - 0.0065z}{293}\right]^{5.26}$

$$= 101.3\left[\frac{293 - 0.0065x2000}{293}\right]^{5.26}$$

$= 101.3x0.7876 = 79.78kPa$

Using Eq.7.8, $\gamma = 0.665x10^{-3}P$

$0.665x10^{-3}x79.78 = 0.052$

Example 7.2 Determine the saturation vapor pressure of a day of maximum and minimum temperature are 38°C and 22°C respectively.

Solution:

Using Eq.7.10, $e^0(T) = 0.6108\exp\left[\frac{17.27T}{T + 237.3}\right]$

$$e^0(38)^0 = 0.6108\exp\left[\frac{17.27x38}{38 + 237.3}\right]$$

$0.6108\exp[2.38]$

$= 6.62kPa$

$$e^0(22)^0 = 0.6108\exp\left[\frac{17.27x22}{22 + 237.3}\right]$$

$0.6108\exp[1.465] = 2.64kPa$

Example 7.3 Determine the extraterrestrial radiation for the 8 September at latitude 23°N.

Solution:

For 23°N, $\varphi = \frac{\pi}{180}x23 =$ 0.40 rad (the value is positive for the northern hemisphere)

From **Table 7.8** the number of the day in the year, J = 251 days

From **Eq.7.19**, $d_r = 1 + 0.033\cos\left(\frac{2\pi}{365}x251\right) = 1.033$ rad

From **Eq.7.23** $\delta = 0.409\sin\left(\frac{2\pi}{365}x251\text{-}1.39\right) = 0.086$ rad

From **Eq.7.20** $\omega_s =$ arccos $\left[-\tan(0.40)\tan(0.021)\right] = 1.571$ rad

$\sin(\varphi)\sin(\delta) = \sin 0.4 \sin 0.021 = 0.0082$
$\cos(\varphi)\cos(\delta) = \cos(0.4)\cos(0.021) = 0.92$
From **Eq. 7.18,**

$$R_a = \frac{24(60)}{\pi} G_{sc} d_r \left[\varepsilon_s \sin(\varphi)\sin(\delta) + \cos(\varphi)\cos(\delta)\sin(\omega_s)\right]$$
$$= \frac{24(60)}{\pi} x0.082x1.033\left[1.571x0.086 + 0.92\sin 1.571\right] = 40.97 \text{MJm}^{-2}\text{d}^{-1}$$

Equivalent evaporation = 40.97x0.408 = 16.71 mm/day

Table 7.8. Number of the day in the year (J)

Day	January	February	March*	April*	May*	June*	July*	August*	September*	October*	November*	December*
1	1	32	60	91	121	152	182	213	244	274	305	335
2	2	33	61	92	122	153	183	214	245	275	306	336
3	3	34	62	93	123	154	184	215	246	276	307	337
4	4	35	63	94	124	155	185	216	247	277	308	338
5	5	36	64	95	125	156	186	217	248	278	309	339
6	6	37	65	96	126	157	187	218	249	279	310	340
7	7	38	66	97	127	158	188	219	250	280	311	341
8	8	39	67	98	128	159	189	220	251	281	312	342
9	9	40	68	99	129	160	190	221	252	282	313	343
10	10	41	69	100	130	161	191	222	253	283	314	344
11	11	42	70	101	131	162	192	223	254	284	315	345
12	12	43	71	102	132	163	193	224	255	285	316	346
13	13	44	72	103	133	164	194	225	256	286	317	347
14	14	45	73	104	134	165	195	226	257	287	318	348
15	15	46	74	105	135	166	196	227	258	288	319	349
16	16	47	75	106	136	167	197	228	259	289	320	350
17	17	48	76	107	137	168	198	229	260	290	321	351
18	18	49	77	108	138	169	199	230	261	291	322	352
19	19	50	78	109	139	170	200	231	262	292	323	353
20	20	51	79	110	140	171	201	232	263	293	324	354
21	21	52	80	111	141	172	202	233	264	294	325	355
22	22	53	81	112	142	173	203	234	265	295	326	356
23	23	54	82	113	143	174	204	235	266	296	327	357
24	24	55	83	114	144	175	205	236	267	297	328	358
25	25	56	84	115	145	176	206	237	268	298	329	359
26	26	57	85	116	146	177,	207	238	269	299	330	360
27	27	58	86	117	147	178	208	239	270	300	331	361
28	28	59	87	118	148	179	209	240	271	301	332	362
29	29	60	88	119	149	180	210	241	272	302	333	363
30	30	-	89	120	150	181	211	242	273	303	334	364
31	31	-	90	-	151	-	212	243	-	304	-	365

* add 1 if leap year

J can be determined for each day (D) of month (M) by

J = **INTEGER** (275 M/9 - 30 + D) - 2

IF (M < 3) **THEN J = J + 2**

also, **IF** (leap year and (M > 2)) **THEN** J = J + 1

For ten-day calculations, compute J for day D = 5, 15 and 25. For monthly calculations, J at the middle of the month is approximately given by

J = **INTEGER** (30.4 M - 15)

7.4 Measurement of Evapotranspiration

The methods of measuring evapotranspiration may be grouped into (i) direct methods, (ii) pan evaporation method and (iii) empirical methods.

Direct Methods

The direct methods of measurement of evapotranspiration are: (i) lysimeter, (ii) field experimentation, (iii) soil moisture depletion studies and (iv) inflow – outflow methods. These methods are costly, laborious, time consuming, require elaborate installation and precise measurement, however, give suitable results.

Lysimeter method

Lysimeter consists of growing of crops in an isolated tank filled with either disturbed or undisturbed large soil block. The tank (lysimeter) is isolated from the surrounding but as identical as possible with the surrounding soil and vegetation and study the change of soil moisture. Lysimeters are non-weighing and weighing type. Weighing type lysimeter measures directly the change of mass and provides more accurate results than non-weighing type. In non-weighing lysimeter the ET is determined over a given period by deducting the drainage water collected at the bottom of the lysimeter from the total water input. The most accurate lysimeters may detect ET rates over time periods shorter than one hour. Other lysimeters can record the changes occur over a day or more. Lysimeter directly measures the water fluxes from the vegetative surface, therefore, provides the basis of measurement for other methods.

There are so many designs available for weighing and non – weighing type lysimeters. Among these weighing type lysimeter suggested by Pruitt and Angus (1960) and non – weighing oil drum type suggested by Gilbert and Van Bevel (1954) is proved to be more useful (Majumdar, 2000). A lysimeter should have (i) the soil moisture relationship inside the lysimeters very close to soil under natural conditions, (ii) sufficient deep to extend well below the root zone and maintain same level of moisture at the bottom as the surrounding areas, (iii) crop cultivated in the lysimeter exactly similar manner to surrounding cropped area, and (iv) the ratio of wall surface area to the enclosed lysimeter area should be minimum to avoid the influence of advection from the uncropped areas. Non-weighing type lysimeters work following the soil moisture depletion method. Using neutron scattering method monitors the soil moisture. Gravimetric method requires replicated soil sampling. Non – weighing type lysimeters are cheap and can be installed easily.

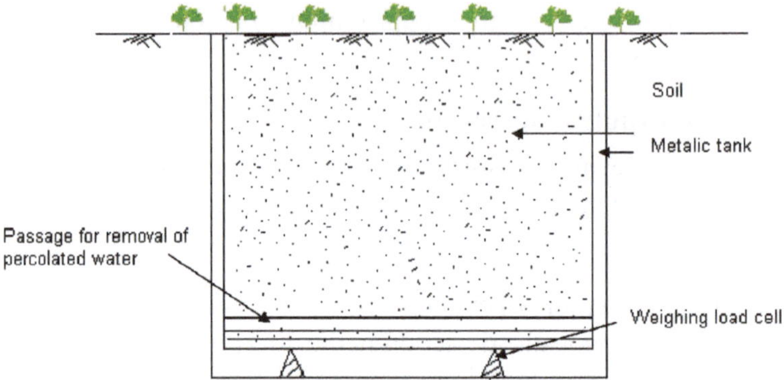

Fig. 7.5. Schematic diagram of a weighing type lysimeter

In a weighing type lysimeter the operator can study the water balance by regularly noting the water added, water retained in the soil and the losses of water in the form of evapotranspiration and deep percolation (Fig.7.5). These process requires weighing which may be made by weigh bridge with continuous recording of weight change or by floating the lysimeter inside a tank in suitable liquid (water or $Zncl_2$ solution) in which case the change in liquid displacement computes the water loss from the tank. A continuous record of such liquid displacement gives the total field water loss providing the tank is permanently buried in ground and surrounded by the large area of crop of same height. The measured amount of water is applied in the lysimeter from the supply tank. The water level in the lysimeter tank is maintained of a predetermined level. Float mechanism is provided to receive the excess water that tends to builds up in the tank. The records of overflow, deep percolation, rainfall, etc. give the calculation of evapotranspiration following the expression.

$$ET = P + IR_n + \Delta SW - (R + DP) \tag{7.24}$$

$$= ER + IR_n + \Delta SW$$

Where, P = precipitation

IR_n + net irrigation requirement of crop

Where, P= precipitation

Δ SW = soil water contribution (the difference between soil water contents at sowing and at harvest of crop).

R = surface runoff

DP = deep percolation

ER = effective rainfall

= P-(R+DP)

The lysimeter direly and most precisely measures the water losses in crop. However, there are certain limitations. The limitations are of reproduction of the physical condition such as soil texture and density, water table temperature, etc. within the lysimeter comparable to surrounding field. It is difficult to maintain the drainage likewise it occurs in natural way. The soil temperature in the lysimeters sometime also rises to large extent.

Field experimentation method

Field experiments at different level of irrigation are carried out to estimate the seasonal consumptive use of irrigated crops. The records of effective rainfall, irrigation and change of profile soil moisture provide the basis for estimating the consumptive use of crop. Among the different level of irrigation, which gives the maximum profit considering the yield and the water use, is taken as the consumptive use.

$CU=ER+IR_n \Delta SW$

$$= ER + IR_n + \sum_{i=1}^{n} \frac{M_{bi} - M_{ei}}{100} A_i D_i \qquad (7.25)$$

Where,

W_{bi}= soil moisture percentage at the beginning of the season in the ith layer.

M_{ei}= soil moisture percentage at the end of season in the ith layer of these soil.

A_{ei}= apparent specific gravity of the ith layer

D_i= layer depth of the ith layer

This method requires the accurate measurement of irrigation water supplied to the field on the basis of soil moisture depletion and excess application is avoided to prevent deep percolation. The method is satisfactory for computing seasonal water requirements but does not provide the information regarding the short period or peak use rate of the crop and deep percolation.

Soil moisture depletion studies

The soil moisture method is employed in fairly uniform soil of ground water table at such depth (at least 3m deep) that it cannot influence the soil moisture fluctuation in root zone. Soil moisture content in different layers of the root zone are measured just be- fore and after the irrigation or rainfall as early as sampling is possible and in between the two successive irrigation as frequently as possible depending on the level of accuracy desired. The irrigation is given at a predetermined level of soil moisture depletion. The soil moisture depletion between two successive soil sampling considered to be the consumptive use (CU) of that period. The water losses during the short period(s) just after irrigation (s) and soil sampling may be taken at a rate of potential evapotranspiration (PET).

The expressions for consumptive use may be expressed as

$$CU = \sum u$$

$$\text{and } u = \sum_{i=1}^{n} \frac{M_{1i} - M_{2i}}{100} A_{si} D_i \qquad (7.26)$$

Where,

u = consumptive use between the period of two successive soil sampling

M_{1i} = soil moisture percent at the time of first sampling in i^{th} layer

M_{2i} = soil moisture percent at the time of second soil sampling in i^{th} layer

A_{si} = apparent specific gravity of the soil in i^{th} layer

D_i = depth of ith layer of soil, cm

N = number of soil layer in root depth, D

PET = potential evapotranspiration

$= E_p \times K_c \times K_p$

E_p = pan evapotrnspiration

K_c = crop factor

In estimating seasonal consumptive use the inflow and outflow to subsurface water and the change of moisture in soil profile assumed negligible.

Inflow-outflow method

This method is also called as water balance method and used for estimating consumptive use for large area. It may be expressed as

$$CU = P + I + \Delta GW - R \qquad (7.27)$$

Where,

CU = consumptive use

P = seasonal or yearly consumptive use, Ha-m

I = surface water that flows in to the area, Ha-m

ΔGW change in the ground water storage, Ha-m

R = run-off from the area, Ha-m

Pan evaporimeter method

The close relationship is observed between the rate of consumptive use and the rate of evaporation from a pan evaporimeter. The evaporation from a pan is entirely dependent on atmospheric factors and independent to plant and soil factors. Therefore, pan evaporation data are very useful in estimating the short-term evapotranspiration, which are not possible by the empirical formulae employed to estimate it.

The evapotranspiration (ET_c) is estimated by multiplying the pan evaporation (E_{pan}) to the crop factor (K_c) and pan factor (K_p).

$$ET_c = E_{pan} \times K_p \times K_c \quad (7.28)$$

The crop factor varies with the stage of plant growth, extent of cover by the foliage, climate, location, etc. Evapotranspiration is required for each crop for each agro – climatic region to fix up the values of crop factors. The experiment may be set up by installing the lysimeter and pan evaporation side by side. The available data for ET_c provided by the lysimeter and evaporation from the evaporation pan will give the relation for different stage of plant growth. The values of crop factors for some selected location are given in Table 7.1-7.5.

USWB Class A Pan: The USWB Class A pan is widely used for determining the evaporation from free water surface (Fig.7.6). It consists of 120cm diameter and 25cm deep pan made of 22 gauge galvanized iron sheet with a stilling well in it, covered by wire net and exposed on the wooden frame in such a way that air may circulate freely beneath the pan. The pan and stilling well are painted white. There is a pointer in the stilling well, the tip of which indicates the level of water constantly maintained in the pan. The loss of water and thereby the lowering of water level in the pan due to evaporation may be measured by using a point gauge in the stilling well or by noting the amount of water added to bring the water to its original level. The depth of water level measured by the point gauge from the tip of the pointer is the evaporation. The volume of water added to be divided by the pan area to determine the rate of evaporation. Usually the evaporation readings are taken once every day at a fixed time. After each reading water is added to bring its level to original position.

Sunken screen pan evaporimeter

The sunken screen pan evaporimeter was developed by Sharma & Dastane (1968) which provides more close evaporation values then USWB Class-A pan evaporimeter (Majumdar, 2000). The pan evaporation to evapotranspiration ratio (ET_c/E_p) in sunken pan screen diameter was 0.95 to 1.05 in comparison to 1.2 to 1.5 in USWB class-A pan.

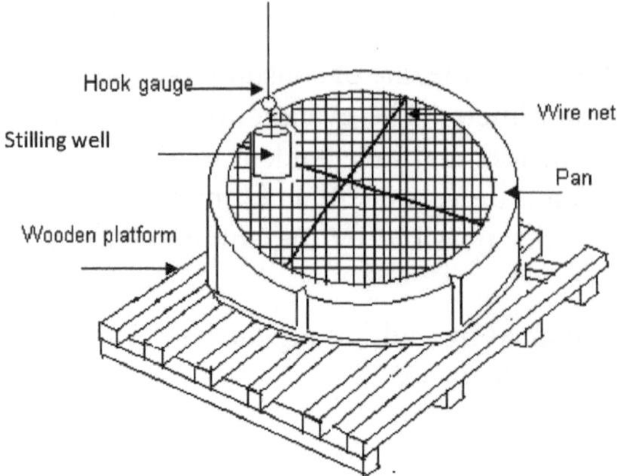

Fig. 7.6 USWB Class-A pan evaporimeter

The sunken screen pan evapoirmeter consists of 60 cm diameter and 45cm deep pan made of 20 gauge galvanized sheet and a stilling basin of diameter 15cm connected to pan by a 15cm length in tube. The pan and the stilling well are painted white and screened at the top by wire net of 6/20 meshes. The stilling well is provided by a pointer. The pan and the stilling basin are buried in the soil leaving 10cm only above the ground. Similar to USWB class A pan the rate of evaporation is determined by measuring the lowering of water level by the point gauge or by dividing the volume of water level to the marked position by the pan area (Fig.7.7).

Empirical Methods

A large number of empirical and semi-empirical equations are available to estimate E, ET_0 or ET_c from meteorological data. These methods are valid under specific agro-climatic conditions from where those were developed and thereby versatile use is restricted. American Society of Civil Engineers (ASCE) and a consortium of European Research Institutes evaluated the performances of various evapotranspiration methods using data from different lysimeter studies in Europe and summarized the comparative studies as follows (Mehta, 2006).

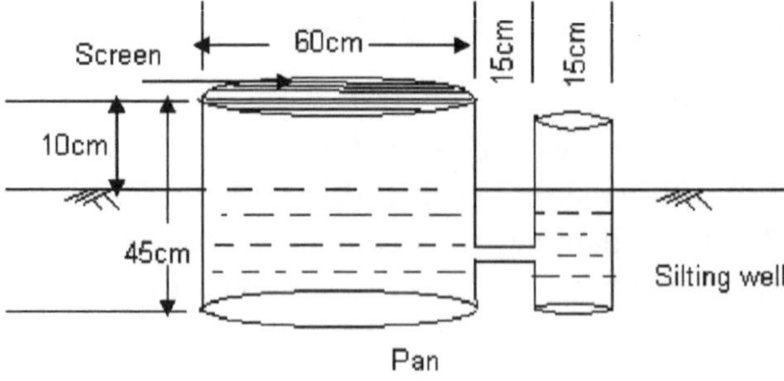

Fig. 7.7. Sunken screen pan evaporimeter

- The Penman methods may require local calibration of the wind function to achieve satisfactory results.
- The radiation methods show good results in humid climates where the aerodynamic term is relatively small, but performance in arid conditions are erratic and tend to underestimate evapotranspiration.
- Temperature methods remain empirical and require local calibration in order to achieve satisfactory results. A possible exception is the 1985 Hargreave's method which has shown reasonable ET_0 results with a global validity.
- Pan evaporation methods clearly reflect the shortcomings of predicting crop evapotranspiration from open water evaporation. The methods are susceptible to the microclimatic conditions under which the pans are opeating and the rigor of station maintenance. Their performance prove erratic
- Pan evaporation methods clearly reflect the shortcomings of predicting crop evapotranspiration from open water evaporation. The methods are susceptible
- The relatively accurate and consistent performance of the Penman-Monteith approach in both arid and humid climate has been indicated in both the ASCE and European studies.

Blaney - Criddle Method

Blaney and Criddle (1950) developed the formula for estimating consumptive use (CU) relating the mean monthly temperature, day light hours and locally developed crop coefficient. The relationship is given below:

$$\text{CU (or U)} = \sum \text{kf (or cu)} \sum \text{kf} = \sum \frac{\text{ktp}}{100} \tag{7.29}$$

Where,

CU or U	=	seasonal consumptive use for a given period, inches
u	=	monthly consumptive use, inches
t	=	mean monthly temperature, °F
p	=	monthly day light hours expressed as percentage of day light hours of the year
k	=	empirical monthly consumptive use for the growing season
f	=	monthly consumptive use factor for the growing season

$$f = \frac{tp}{100} \tag{7.30}$$

$$u = \frac{kp(1.8t + 32)}{40} \tag{7.31}$$

When, u = mean monthly consumptive use, cm

t = mean monthly temperature, ^{0}C

The monthly percentage of day light hours (p) for different latitude is given in Table 7.22 (Majumdar, 2000). Mean monthly temperature of a locality is obtained from weather station. The crop coefficients for different crops determined locally by experimentation. Unless such data is available the crop coefficient (K_c) can be used for common irrigated crops given in Table 7.1-7.5.

Blaney and Criddle formula gives sufficiently accurate estimates if locally developed crop coefficient (K_c) values are used. The formula based on the temperature and day light hours' parameter. There may be much variation in consumptive use due to variation in wind velocity and humidity at same temperature. Doorenbos and Pruitt (1975) rejected the use of coefficient (K_c) as usually done in original Blaney- Criddle approach (Michael, 1985). They recommended the following relationship factor (mm/day) in Blaney- Criddle formula excluding the crop coefficient.

$$f = P(0.46+8.13), ^{\circ}C \tag{7.32}$$

$$f = 25.4\frac{pxt}{100}, ^{\circ}F \tag{7.33}$$

Where, t = mean of daily maximum and minimum temperature in ^{0}C or ^{0}F for the month under consideration

P = the mean daily percentage of annual day time hours for a given month and latitude.

Example 7.4 Estimate the consumptive use of water (CU) for vegetables at Mohanpur (23.5^{0}N) by using Blaney-Criddle formula with the following data.

Month	Mean monthly temperature, ^{0}C
Jan	16.8
Feb	20.71
March	26.66
April	29.55
May	31.05
June	28.16
July	27.96
Aug	28.38
Sep	28.92
Oct	25.88
Nov	22.65
Dec	17.75

Solution: Estimated consumptive use of crop is shown in following tabulated form

Month	Mean monthly air temperature, (t), °C	Percent monthly day light hours (p) (by using Table 7.22)	Crop coefficient (K_c) from Table 7.4	Consumptive use (u) of the month, $\frac{kp}{40}$ (1.8t+32), cm
Jan	16.8	7.625	0.5	5.93
Feb	20.71	7.183	0.55	6.84
March	26.66	8.403	0.6	10.08
April	29.55	8.593	0.65	11.89
May	31.05	9.28	0.7	14.27
June	28.16	9.173	0.75	14.22
July	27.96	9.380	0.8	15.45
Aug	28.38	9.038	0.8	15.02
Sep	28.92	8.308	0.7	12.22
Oct	25.88	8.108	0.6	9.56
Nov	22.65	7.448	0.55	7.45
Dec	17.75	7.485	0.5	5.98

$$\therefore U = \sum_{i=1}^{12} u = 128.91 cm$$

Radiation method

The reference crop evapotranspiration (ET_0) is estimated by using the following formula:

$$ET_0 = C\,(W.R_s) \tag{7.34}$$

Where,

R_s = solar radiation in equivalent evaporation (mm/day)

W = temperature and altitude dependent weightage factor (Table 7.9)

C = adjustment factor made graphically on $W.R_s$ using estimated values of RH_{mean} and day time wind velocity (Fig.7.8)

R_s = can also be obtained from measured sunshine records as

R_s = $(0.25+0.50n/N)R_a$

Where,

R_a = extra terrestrial radiation in equivalent evaporation (mm/day)

n = actual measured bright sunshine hours

N = maximum possible sunshine hours

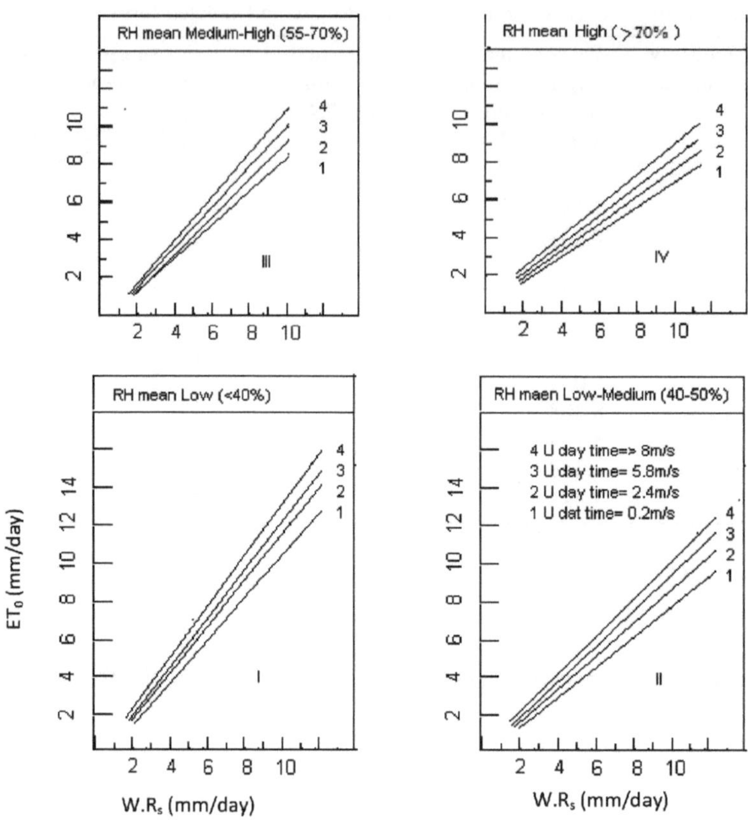

Fig.7.8 Prediction of ET_0 from $W.R_s$ for different conditions of mean relative humidity and day time wind

Example 7.5 Estimate ET_0 from the following data

Month = May

$T_{mean} = 35°C$

$n_{mean} = 8$ hours/day

Day time wind velocity=moderate (2-5 m/s)

RH_{mean} = High

Place = Mohanpur (23.5^{0}N latitude, 12.5m elevation)

Solution:

Ra for 23.5^{0}N for May = 16.375mm/day (Table 7.23)

N for 23.5^{0}N for May = 13.24 hours (Table 7.10)

$R_s = (0.25+050n/N)R_a$

$= (0.25{+}0.50x\frac{8}{13.24})16.375$

$= 9.04$ mm/day

W for 35°C & altitude 12.5m (Table 7.9) = 0.825

W.Rs = 9.04 x 0.825

= 7.458 mm

$\therefore ET_0 = 5.95$ mm/day for W.Rs = 7.458 mm (Fig.7.8)

Table 7.9 Values of weightage factor (W) for the influence of radiation on ET_0 at different temperature and altitude

Temperature	Altitude (m)					
(°C)	0	500	1000	2000	3000	4000
2	0.43	0.45	0.46	0.49	0.53	0.55
6	0.49	0.51	0.52	0.55	0.58	0.61
10	0.55	0.57	0.58	0.61	0.64	0.66
14	0.61	0.62	0.64	0.66	0.69	0.71
18	0.66	0.67	0.69	0.71	0.73	0.76
22	0.71	0.72	0.73	0.75	0.77	0.79
26	0.75	0.76	0.77	0.79	0.81	0.83
30	0.78	0.79	0.80	0.82	0.84	0.85
34	0.82	0.82	0.83	0.85	0.86	0.88
36	0.83	0.84	0.85	0.86	0.88	0.89
8	0.84	0.85	0.86	0.87	0.88	0.90
40	0.85	0.87	0.87	0.88	0.89	0.90

Source: Reddi & Reddy, 1995

Table 7.10 Mean daily duration of maximum possible sunshine hours (N) for different months in north and south latitude

North Lat	Jan	Feb	Mar	April	May	June	July	Aug	Sep	Oct	Nov	Dec
South Lat	July	Aug	Sep	Oct	Nov	Dec	Jan	Feb	Mar	April	May	June
0	12.4	12.1	12.1	12.1	12.1	12.1	12.1	12.1	12.1	12.1	12.1	12.1
5	11.8	11.9	12.0	12.2	12.3	12.4	12.3	12.3	12.1	12.0	11.9	11.8
10	11.6	11.8	12.0	12.3	12.6	12.7	12.6	12.4	12.1	11.8	11.6	11.5
15	11.3	11.6	12.0	12.5	12.8	13.0	12.9	12.6	12.2	11.8	11.4	11.2
20	11.0	11.5	12.0	12.6	13.1	13.3	13.2	12.8	12.3	11.7	11.2	10.9
25	10.7	11.3	12.0	12.7	13.3	13.7	17.5	13.0	12.3	11.6	10.9	10.6
30	10.4	11.1	12.0	12.9	13.6	14.0	13.9	13.2	12.4	11.5	10.6	10.2
35	10.1	11.0	11.9	13.1	14.0	14.5	14.3	13.5	12.4	11.3	10.3	9.8
40	9.6	10.7	11.9	13.3	14.4	15.0	14.7	13.7	12.5	11.2	10.0	9.3
50	8.5	10.1	11.8	13.8	15.4	16.3	15.9	14.5	12.7	10.8	9.1	8.1

Source: Remddi & Reddy, 1995

Thornthwaite method

Thornwaite (1948) assumed the relationship between mean monthly temperatures and mean monthly evapotranspiration and proposed the following formula (Michael, 1985):

$$e = 1.6(10t/I)^a \qquad (7.35)$$

Where,

e = unadjusted potential evapotranspiration, cm per month (month of 30 days each and 12 hours day time)

t = mean monthly temperature, °C

I = annual or seasonal heat index, the summation of all values of monthly heat indices (i) when,

$i = (t/5)^{1.514}$

a = an empirical exponent

$= 0.00000675I^3 - 0.0000771\, I^2 + 0.01792\, I + 0.49239$

The unadjusted value of e is corrected for actual day light hours and days in a month. The correction is made as follows:

$$e' = e\,(X/12)\,(Z/30)$$

$$\text{or, } \frac{XZ}{360} = \frac{e'}{e}$$

where,

e' = corrected value of PET, cm

X = actual day light hours

Z = days in a month

Thornthwaite formula does not consider the variation of crop and land use. It also does not consider the wind effect, effect of warm and cool air and lag of air temperature of a place behind the radiation. The formula is based on only temperature parameter which is sufficient to estimate the evapotrnspiration. This formula indicates that at zero temperature the evapotranspiration will cease which is not true. Thus, the Thornthwaite formula gives reasonable estimate of potential evapotranspiration in the temperate, continental climate of North America where the formula was originally developed because there the temperature and radiation are strongly correlated. A modification to the original formula was suggested by Thornthwaite and Mathur (1955) for use in other parts of the world (Majumdar, 2000).

Example 7.6 Calculate the potential evapotranspiration (PET) for Mohanpur (23.5^{0}N) from the following data by using the Thornthwaite formula.

Month	Mean monthly air temperature(t), °C
Jan	17.5
Feb	20.65
March	27.10
April	30.34
May	31.26
June	27.98
July	28.82
Aug	28.35
Sep	29.10
Oct	26.55
Nov	22.74
Dec	18.76

Solution

I = the summation of 12 month heat indices

$$= \sum i = \sum (t/5)^{1.514}$$

$$= \left(\frac{17.5}{5}\right)^{1.514} + \left(\frac{20.65}{5}\right)^{1.514} + ... = \left(\frac{18.76}{5}\right)^{1.514}$$

$= 6.66+8.56+12.92+15.33+16.04+13.56+14.18+13.83+14.39+12.53+9.91+7.40$

$= 145.31 \text{cm}$

$a = 0.000000675\, I^3 - 0.0000771\, I^2 + 0.01792\, I + 0.49239$

$= 0.000000675(145.31)^3 - 0.0000771(145.31)^2 + 0.01792(145.31) + 0.49239$

$= 2.07 - 1.628 + 2.604 + 0.49$

$= 3.53$

Month	t	10 t/I	e=1.6(10 t/I)^a	Correction factor (Table 7.24)	Corrected values of PET (Col 4x5)
1	2	3	4	5	6
Jan	17.5	1.204	3.081	0.936	2.884
Feb	20.65	1.421	5.531	0.893	4.929
March	27.1	1.865	14.442	1.03	14.875
April	30.34	2.088	21.516	1.057	22.742
May	31.26	2.151	23.897	1.144	27.338
June	27.98	1.926	16.179	1.131	18.298
July	28.82	1.983	17.934	1.161	20.821
Aug	28.35	1.951	16.933	1.117	18.914
Sep	29.1	2.003	18.58	1.02	18.952
Oct	26.55	1.827	13.429	0.993	13.335
Nov	22.74	1.565	7.776	0.916	7.123
Dec	18.76	1.291	3.942	0.919	3.623
					$\sum$ =173.844cm

Penman method

Penman (1948) proposed the equation for estimating evaporation from free water surface using the important climatic parameter such as solar radiation, temperature, vapour pressure and wind velocity. The equation is expressed as

$$E_0 = \frac{\Delta Q_n + \gamma Ea}{\Delta + \gamma} \tag{7.36}$$

Where, $E_{0=}$ evaporation from free water surface, mm/day

Δ = slope of saturation vapour pressure vs temperature curve (de_a/dt) at the mean air temperature T_a, mmHg per ^{0}C

e_a = saturation vapour pressure of the evaporating surface (e_s) in mmHg at mean air temperature Ta.[Here,e_s is considered equal to e_a by assuming zero temperature gradient between surface (s) and air temperature]

T_a = mean air temperature in ^{0}K=273+^{0}C

Q_n = net radiation (mm of water)

$= (Q_A)(1-\text{r})(0.18+0.55n/N) - \sigma Ta^4 (0.55-0.092)\sqrt{e_d})(0.10+0.90n/N)$

r = reflection coefficient of evaporation surface, 0.06 for free water surface.

Q_A=Angot's value of mean monthly extra-terrestrial radiation in mm of water/ day (Table 7.23).

n/N= ratio between actual and possible hours of bright sun shine (Table 7.10)

s = Stefan-Boltzman constant (Table 7.11 for sTa^4)

e_d =saturation vapour pressure of the atmosphere in mm of Hg at dew point temperature

$$= \frac{RH_{mean}}{100} x e_a \tag{7.37}$$

Where, RH is the mean relative humidity.

γ = psychrometric constant or the ratio of the specific heat of air to the latent heat of evaporation of water (0.49 for 0^0C and mmHg.)

E_a = an aerodynamic component in which e_s is considered equal to e_a

=0.35(e_a-e_d) (1+0.0098u_2)

u_2 =wind speed in miles/day at 2m height

= u_1(log 6.6/logh), where u_1 is the wind speed in miles/day at any other height, h in feet.

Example 7.7 Estimate the evaporation from free water surface from the following data.

Mean air temperature = 30^{0}C

Mean relative humidity = 90%

Mean sun shine hours, n = 6.5

Location = 23.5^{0}N latitude

Month = April

Wind speed at 3m height = 85 miles/day

Solution:

N for 23.5^{0}N for April = 12.67 (Table 7.10)

n/N = 6.5/12.67=0.51

r = 0.06

Q_A = 15.425 (Table 7.23)

Solving expression

1. Q_A (1-r) (0.18+0.55xn/N)

= 15.425(1-0.06) (0.18+0.55x0.51)

= 6.68

σ Ta^4 = 17.01 (Table 7.11) for Ta = 273+30 = 303^{0}K)

e_a = 32 **(Fig. 7.10)**

$$e_d = \frac{RH_{mean}}{100} \times e_a = \frac{90x32}{100} = 28.8$$

$\therefore$ $\sqrt{e_d}$ = 5.37

Solving expression

2. σ Ta^4 (0.55-0.092$\sqrt{e_d}$) (0.10+0.90n/N)

=17.01(0.55-0.092 x 5.37) (0.10+0.90 X 0.51)

=17.01 x 0.056 x 0.56

= 0.53

Q_n = Item 1-Item 2= 6.68-0.53= 6.15

Solving for E_a = 0.35 (e_a-e_d)(1+0.0098u_2)

= 0.35 (32-28.8) (1+0.0098 x 70.159)

= 0.35 x 3.2 x 1.687

= 1.89

Solving for E_a = 0.35 (e_a-e_d) (1+0.0098u_2)

= 0.35 (32-28.8) (1+0.0098 x 70.159)

= 0.35 x 3.2 x 1.687

= 1.89

Solving for $E_0 = \dfrac{\Delta Q_n + \gamma E_a}{\Delta + \gamma}$

$$= \frac{1.8x6.15 + 0.49x1.89}{1.8 + 0.49} \left(\text{For } \Delta \text{ value, see } \textbf{Fig.7.9}\right)$$

$$= \frac{11.07 + 0.926}{2.29}$$

= 5.24 mm/day

Table 7.11 Values of σT_a^4 for various temperatures when computing evapotranspiration by the Penman method (after Criddle)

Temperature (⁰K)	(mm water/day)
270	10.73
275	11.51
280	12.40
285	13.20
290	14.26
295	15.30
300	16.34
305	17.46
310	18.60
315	19.85
320	21.15
325	22.50

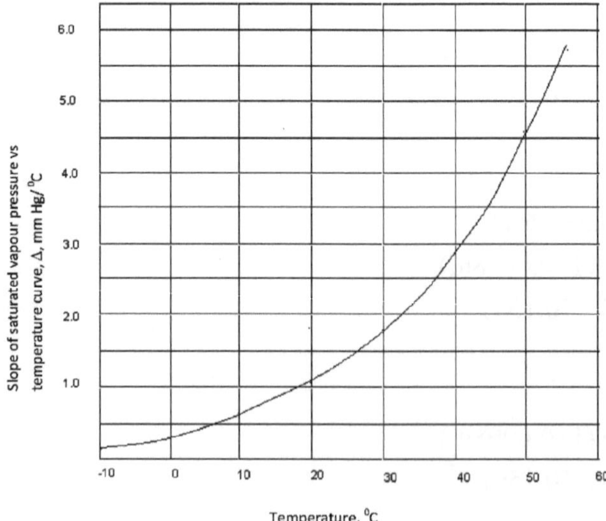

Fig.7.9 Slope of the vapour pressure vs. temperature curve

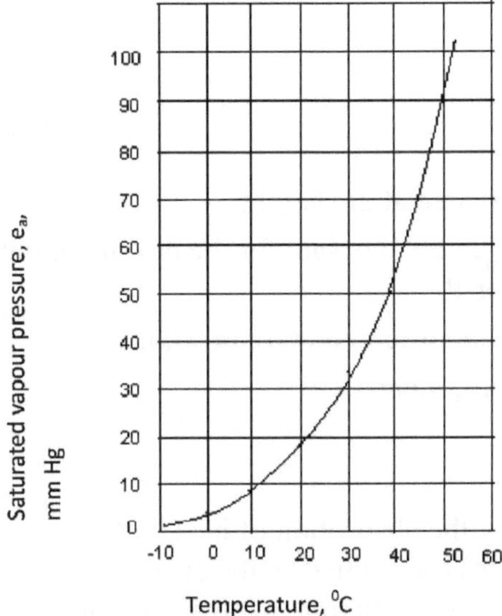

Fig. 7.10. Saturated vapour pressure versus temperature

Modified Penman Method

Modified Penman method was suggested by Doorenbes and Pruitt (1957). The method estimates fairly accurately the reference crop evapotranspiration(ET_0) as it utilizes all the meteorological parameter responsible for evapotranspiration. The method is given as

$$ET_0^* = W.R_n + (1-W)f(u).e_d - e_d) \tag{7.38}$$

Where,

ET_0^* = unadjusted reference crop evapotranspiration, mm/day

ET_0 = CET_0^*

ET_0 = reference crop evapotranspiration, mm/day

W = temperature and altitude related weightage factor for the effect of radiation on ET_0

R_n = net radiation in equivalent evaporation, mm/day

R_{ns} = the net incoming short wave solar radiation as in radiation method $= R_A(0.25+0.50n/N)(1\text{-}r)$ in which R_A is extra-terrestrial solar radiation expressed in equivalent evaporation in mm/day, n/N is the ratio between n= actual duration of bright sunshine hours and r is the reflection coefficient. The value of r is taken as 0.25 for most crops (Michael, 1985).

R_{nl}= net long wave radiation (mm/day) which is a function of temperature, vapour pressure and sunshine duration

$f(e_d) = 0.34\text{-}0.444\sqrt{e_d}$

$f(n/N) = (1\text{-}r)\ 0.25 + 0.5n/N)$

$f(U)$ = wind relation function

$= 0.27\ (1 + U/100)$

Where, U = wind velocity in km/day measured at 2m height

C = adjustment factor (ratio of $U_{day}/U_{night)}$ to compensate for the day and night weather effects for maximum RH and for R_{ns}

e_a = saturation vapour pressure at the mean air temperature in ^{0}C, mbar

e_d = mean actual vapour pressure of the air, mbar

$$ed = e_a x \frac{RH_{mean}}{100}$$

The process of evaporation is initiated by the absorption of heat by the water body due to solar radiation and the process is continued by the displacement of saturated air by the unsaturated air. Thus, solar radiation, relative humidity and wind velocity is important in evaporation process and governs the modified Penman method by a radiation term, wR_{n1}-w and aerodynamic term $f(U)(e_a\text{-}e_d)$.

Example 7.8 Calculate the crop evapotranspiration (ET_0) from the following data given below.

Month = May

$T_{mean} = 35^0C$

Wind velocity (day) = 15 km/h

Wind velocity (night) = 12 km/h

$n_{mean} = 8.5$ h

$RH_{max} = 85\%$, RHmin = 45%

Place = 25^0N and 30m altitude

Solution: e_a at 35^0C= 56.2 mbar **(Table 7.11)**

$$e_d = \frac{e_a x RH_{mean}}{100} = \frac{56.2(85+45)/2}{100} = \frac{56.2x65}{100} = 36.53mbar$$

$e_a\text{-}e_d$=56.2-36.53 =19.67 mbar

R_A= 16.45 mm/day (Table 7.23)

$R_{ns} = R_A(0.25 + 0.5\ n/N)(1\text{-}r)$, r = 0.25

= 0.75 x 16.45(0.25+0.5 x 8.5/14), N=14 (Table 7.10)

= 0.75 x 16.4 x 0.55

= 6.829 mm/day

R_{nl} = f(t)f(e_d)f(n/N)

f(t) = 14.9 (Table 7.13)

f(e_d) =0.077 (Table 7.14)

f(n/N) =0.647 for n/N = 8.5/14 = 0.607 (Table 7.15)

R_{nl}=14.9 x 0.077 x 0.647= 0.742 mm/day

R_n= R_{ns}-R_{nl}=6.829-0.742= 6.087 mm/day

W for 35°C, 30m altitude= 0.825 (Table 7.9)

f(U)=0.27 (1+U/100)

U= average of day and night wind velocity, km/day

$$= \frac{15+12}{2} \times 24 = 324 km/day = 3.75 m/s$$

f(U)= 0.27 (1+324/100)

= 1.145

C for RH_{max} of 85%, R_s of 9.08 mm/day

U_{day} / U_{night} of 1.25 and U_{day} of 4.167 m/s=1.01 (Table 7.16)

ET_0=C[WR_n+(1-W) f(U)(e_a-e_d)]

= 1.01[0.825 x 6.087 + (1-0.825)1.145 x 19.67]

= 5.07+3.986

= 9.056 mm/day

Table 7.12 Saturation vapour pressure (e_a) in mbar as a function of mean air temperature

Temperature (°C)	0	1	2	3	4	5	6	7	8	9	10
e_a (m bar)	601	6.6	7.1	7.6	8.1	8.7	9.6	10	10.7	11.5	12.3
Temperature (°C)	11	12	13	14	15	16	17	18	19	20	21
e_a (m bar)	13.1	14	15	16.1	17	18.2	19.4	20.6	22	23.4	24.9
Temperature (°C)	22	23	24	25	26	27	28	29	30	31	32
e_a (m bar)	26.4	28.1	29.8	31.7	33.6	35.7	37.8	40.1	42.4	44.9	47.6
Temperature (°C)	33	34	35	36	37	38	39	40			
e_a (m bar)	50.3	53.2	56.2	56.4	62.8	66.3	69.9	73.6			

Table 7.13 Effect of temperature (f (T)) on long wave radiation (R_{nl})

T(°C)	0	2	4	6	8	10	12	14	16	18
f(T)	11	11.4	11.7	12	12.4	12.7	13.1	13.5	13.8	14.2
T(°C)	20	22	24	26	28	30	32	34	36	
f(T)	14.6	15	15.4	15.9	16.3	16.7	11.7	11.7	18.1	

Table 7.14 Effect of vapour pressure ($f(e_d)$) on long wave radiation (R_{nl})

e_d (m bar)	6	10	12	14	16	18	20	22	24
$e(e_d)$	0.22	0.2	0.19	0.18	0.16	0.15	0.14	0.13	0.12
e_d (m bar)	28	30	32	34	36	38	40		
$e(e_d)$	0.11	0.1	0.09	0.08	0.08	0.07	0.06		

Table 7.15 Effect of the ratio of actual and maximum bright sunshine hours (f(n/N)) on long wave radiation

(R_{nl})n/N =	0	0.05	0.1	0.15	0.2	0.25	0.3	0.35	0.4
F(n/N) =	0.1	0.15	0.19	0.24	0.28	0.33	0.37	0.42	0.46
n/N =	0.45	0.5	0.55	0.6	0.65	0.7	0.75	0.8	0.85
F(n/N) =	0.51	0.55	0.6	0.64	0.69	0.73	0.78	0.82	0.98
n/N =	0.9	0.95	1.0						
F(n/N) =	0.87	0.91	1.0						

Table 7.16 Adjustment factor (C) in Penman equation

R_s (mm/day)	**RH_{max} = 30%**				**RH_{max} = 60%**				**RH_{max} = 90%**			
U day (m/s)	3	6	9	12	3	6	9	12	3	6	9	12
				U day/U night = 4.0								
0	0.86	0.90	1.00	1.00	0.96	0.98	1.05	1.05	1.02	1.06	1.10	1.10
3	0.79	0.84	0.92	0.97	0.92	1.00	1.11	1.19	0.99	1.10	1.27	1.32
6	0.68	0.77	0.87	0.93	0.85	0.96	1.11	1.19	0.94	1.10	1.26	1.33
9	0.55	0.65	0.78	0.9	0.76	0.88	1.02	1.14	0.88	1.01	1.16	1.27
				U day/U night = 3.0								
0	0.86	0.90	1.00	1.00	0.96	0.98	1.05	1.05	1.02	1.06	1.10	1.10
3	0.76	0.81	0.88	0.94	0.87	0.96	1.06	1.12	0.94	1.04	1.18	1.28
6	0.61	0.88	0.81	0.88	0.77	0.88	1.02	1.10	0.86	1.01	1.15	1.22
9	0.46	0.56	0.72	0.82	0.67	0.79	0.88	1.05	0.78	0.92	1.06	1.18
				U day/U night = 2.0								
0	0.86	0.90	1.00	1.00	0.96	0.98	1.05	1.05	1.02	1.06	1.10	1.10
3	0.69	0.76	0.85	0.92	0.83	0.91	0.99	1.05	0.89	0.98	1.10	1.14
6	0.53	0.61	0.74	0.84	0.70	0.80	0.94	1.02	0.79	0.92	1.05	1.12
9	0.37	0.48	0.65	0.76	0.59	0.70	0.84	0.95	0.71	0.81	0.96	1.06
				U day/U night = 1.0								
0	0.86	0.90	1.00	1.00	0.96	0.98	1.05	1.05	1.02	1.06	1.10	1.10
3	0.64	0.71	0.82	0.89	0.78	0.86	0.94	0.99	0.85	0.92	1.01	1.05
6	0.43	0.53	0.68	0.79	0.62	0.78	0.84	0.93	0.72	0.82	0.95	1.06
9	0.27	0.41	0.59	0.70	0.50	0.60	0.75	0.76	0.62	0.72	0.87	0.96

The following are some methods which may be used for estimating E, ET_0 or ET_c. Among these methods FAO Penman-Monteith method is recommended as the sole standard method for computation of reference crop evapotranspiration (Mehta, 2006).

Priestly-Taylor, Hargreaves, and FAO-Monteith method

1. Potential evapotranspiration (PET) by Priestly-Taylor estimate

$$PET = 1.26\frac{\Delta(R_{net} - G)}{(\Delta + \gamma)} \tag{7.39}$$

2. Reference crop evapotranspiration (ET_0) by Hargreaves method

$$ET_0 = 0.408x0.0023R_a\left(T_{mean} + 17.8\right)\sqrt{\left(T_{max} - T_{min}\right)} \quad (7.40)$$

Where, ET_0 is in mm

day, R_a is in $MJm^{-2}day^{-1}$

If $T_{mean} < 0$, set $ET_0 = 0$

3. Reference crop evapotranspiration (ET_0) by FAO Penman-Monteith

$$\text{equation } ET_0 = \frac{0.408\Delta\left(R_n - G\right) + \gamma\frac{900}{T+273}u_2\left(e_s - e_a\right)}{\Delta + \gamma\left(1 + 0.34u_2\right)} \quad (7.41)$$

Where,

R_{net} = net radiation at the crop surface ($MJm^{-2}day^{-1}$)

G = soil heat flux density ($MJm^{-2}day^{-1}$)

T = mean daily air temperature at 2m height (^0C)

u_2 = wind speed at 2m height (ms^{-1})

e_s = saturation vapour pressure (*kPa*)

e_a = actual vapour pressure (*kPa*)

e_s-e_a = saturation vapour pressure deficit (*kPa*)

γ = psychrometric constant (*kPa*)

Δ = slope vapour pressure curve (kPa^0C^{-1})

The following steps may be adopted in calculation ET_0 or ET_c by the above methods.

Step 1. Calculation of potential extraterrestrial radiation

1.1 Calculation of day of the year (J)

Table 7.17 Calculation of day of the year

Month	J
Jan	15
Feb	45
Mar	75
Apr	105
May	135
June	165
July	195
Aug	225
Sep	255
Oct	285
Nov	315
Dec	345

1.2 Calculation of declination (δ)

$$\delta = 0.4102\sin\left(\frac{2\pi}{365}(J-80)\right) \qquad (7.42)$$

Where declination is in radians

1.3 Conversion of latitude to radians (ϕ)

$$\phi = \left(\text{latitude} \times \frac{\pi}{180}\right) \qquad (7.43)$$

Where latitude is in decimal degrees.

1.4 Estimation of potential extraterrestrial radiation (R_a)

$$R_a = \frac{118}{\pi}\left\{\cos^{-1}(-\tan\delta\tan\phi)\sin\phi\sin\delta + \cos\phi\cos\delta\sin\left[\cos^{-1}(-\tan\delta\tan\phi)\right]\right\} \qquad (7.44)$$

where R_a is in MJm-[2] day-[1]

Step 2. Estimatation reference crop evapotranspiration (ET_0) (Hargreaves method)

2.1 $ET_0 = 0.408 \text{x} 0.0023 R_a \ (T_{mean} + 17.8)\sqrt{(T_{max} - T_{min})}$ (7.45)

Step 3. Estimation of net radiation (R_{net})

3.1 Estimation of downward solar (short wave) radiation (R_a)

This is the Hargreaves radiation formula

$$R_s = 0.16 R_a \sqrt{(T_{max} - T_{mean})} \qquad (7.46)$$

Where, R_s and R_a are $MJm^{-2}\ day^{-1}$

3.2 Estimation of atmospheric emissivity (ε_a)

This is the modified equation of the original to suit the units of the database.

$\varepsilon_a = (0.72 + 0.005T_{mean})(1 - 0.00.84\ \text{Cloud}) + 0.0084\ \text{Cloud}$ (7.47)

3.3 Estimation of net long wave radiation (R_L)

This is the modified equation of the original to suit the units of the database. Assume terrestrial emmisivity = 0.97

$(R_L) = (4.903 \text{x} 10^{-3})(\varepsilon_a - 0.97) + (T_{mean} + 273)^4$ (7.48)

where R_L is in $MJm^{-2}\ day^{-1}$

3.4 Estimation of net radiation (R_{net})

$R_{net} = (0.77R_s) - R_L$

where $R_{net,}$ is in $MJm^{-2}\ day^{-1}$

Step 4 Estimation of ground heat index (G)

$G_{month} = 0.07(T_{mean, month+1} - T_{mean, month-1})$

Where, $T_{mean, month+1}$ is average temperature for next month

$T_{mean, month-1}$ is average temperature for previous month.

Step 5 Estimation of potential temperature (PET)

5.1 Estimate $\frac{\Delta}{\Delta+\lambda}$ (7.49)

$$\frac{\Delta}{\Delta+\lambda} = (-0.00008xT_{mean}^{2}) + (0.0139xT_{mean}) + 0.4235$$

Step 5.2 Estimation of PET

This is the modified equation of the original to suit the units of the database and uses in Priestly-Taylor method.

$$PET = 0.408x1.26\left(\frac{\Delta}{\Delta+\lambda}\right)(R_{net} - G)$$

Where, PET is in mm/day. If T_{mean} <0, set PET = 0.

Example 7.9 The following is some information of an arbitrary chosen location. Calculate the PET by Priestly-Taylor method.

Latitude = 13.5^0

Month = June

$T_{max} = 37^0C$

$T_{min} = 27^0C$

Cloud = 10%

Step 1.1 Day of the year = 165 (**Table 7.17**)

Step 1.2 Declination, $\delta = 0.4102\sin\left(\frac{2\pi}{365}(J-80)\right)$

$= 0.4102\sin(26.68) = 0.4102x0.449 = 0.1892$ radians

Step 1.3 Latitude, $\phi = \left(\text{latitude}\, x\frac{\pi}{180}\right)$

$= 13.5x\frac{\pi}{180} = 0.2356$ radians

Step 1.4 Potential extraterrestrial radiation (R_a)

$$R_a = \frac{118}{\pi}\left\{\cos^{-1}(-\tan\delta\tan\phi)\sin\phi\sin\delta + \cos\phi\cos\delta\sin\left[\cos^{-1}(-\tan\delta\tan\phi)\right]\right\}$$

$$= \frac{118}{\pi}\left\{\begin{matrix}\cos^{-1}(-\tan(0.1842)\tan(0.2356))\sin(0.2356)\sin(0.1842) + \cos(0.2356)\cos(0.1842) \\ \sin\left[\cos^{-1}(-\tan(0.1842)\tan(0.2356))\right]\end{matrix}\right\}$$

$$=\frac{118}{\pi}\left\{\cos^{-1}\left(-1.32x10^{-05}\right)1.32x10^{-05}+0.999\sin\left[\cos^{-1}\left(1.32x10^{-05}\right)\right]\right\}$$

$$=\frac{118}{\pi}\left\{89.999x1.32x10^{-5}+0.999\sin\left[90\right]\right\}$$

$$=\frac{118}{\pi}\left\{1.188x10^{-03}+0.999x1\right\}$$

$$=\frac{118}{\pi}x1$$

= 37.56 MJ m^{-2} day^{-1}

Step 2 Reference crop evapotranspiration (ET_0)

$$ET_0=0.408\times0.0023R_a\left(T_{mean}+17.8\right)\sqrt{\left(T_{max}-T_{min}\right)}$$

$$=0.408x0.0023x37.56(32+17.8)\sqrt{(37-27)}$$

$$=0.0352x49.8x3.162$$

$$=5.54mm/\text{day}$$

Step 3.1 Downward solar radiations, $R_s=0.16R_a\sqrt{\left(T_{max}-T_{min}\right)}$

$$R_s=0.16x37.56x\sqrt{(37-27)}$$

$$=19.0\text{MJm}^{-2}\text{day}^{-1}$$

Step 3.2 Estimation of atmospheric emissivity

$$\varepsilon_a=\left(0.72+0.005T_{mean}\right)\left(1-0.0084Cloud\right)$$

$$=\left(0.72+005x32\right)\left(1-0.0084x10\right)+0.0084x10$$

$$=\left(0.72+0.16\right)\left(0.92\right)+0.084$$

$$=0.81+0.084=0.89$$

3.3 Estimation of net long wave radiation (R_L)

$$R_L=\left(4.903x10^{-9}\right)\left(\varepsilon_a-0.97\right)\left(T_{mean}+273\right)^4$$

$$=(4.903x10^{-9})(0.89-0.97)(32+273)^4$$

$$=\left(4.903x10^{-9}\right)\left(-0.08\right)\left(8.65x10^9\right)$$

$$=-3.394\,\text{MJm}^{-2}\text{day}^{-1}$$

3.4 Estimation of net radiation (R_{net})

$$R_{net}=\left(0.77R_s\right)-R_L$$

$$=\left(0.77x19\right)+3.394=18.024\,\text{MJm}^{-2}\text{day}^{-1}$$

Step 4 Estimation of ground heat index (G)

$$G_{month} = 0.07\left(T_{mean,month+1} - T_{mean,month-1}\right)$$

$$= 0.07\,MJm^{-2}day^{-1}$$

Step 5.1 Estimation of $\frac{\Delta}{\Delta+\lambda}$

$$\frac{\Delta}{\Delta+\lambda} = \left(-0.00008xT_{mean}{}^{2}\right) + \left(0.0139xT_{mean}\right) + 0.4235$$

$$= \left(-0.00008x32^{2}\right) + \left(00139x32\right) + 0.4235$$

$$= -0.08192 + 0.4448 + 0.4235$$

$$= 0.786\,MJm^{-2}day^{-1}$$

Step 5.2 Estimation of PET

$$PET = 0.408x1.26\left(\frac{\Delta}{\Delta+\lambda}\right)\left(R_{net} - G\right)$$

$$= 0.408x1.26(0.786)(18.023 - 0.07)$$

$$= 0.514(0.786)(17.953)$$

$$= 7.253\,mm/day$$

Steps to be taken in calculating ET_0 by FAO Penman-Monteith method

Parameters

T_{max}	°C	
T_{min}	°C	T_{mean} - $(T_{max} + T_{min})/2$
	°C	
T_{mean}	°C	Δ (Table 7.18)
	kPa/°C	
Altitude	m	Δ (Table 7.7)
	kPa/°C	
u_2	m/s	$(1 + 0.34\,u_2)$
$\Delta/[\Delta+\gamma(1 + 0.34\,u_2)]$		
$\lambda/[\Delta+\gamma(1 + 0.34\,u_2)]$		
$[900/(T_{mean} + 273)]\,u_2$		
Vapor pressure deficit		
T_{max}	°C	$e°(T_{max})$ (Table 7.10)
	kPa	
T_{min}	°C	$e°(T_{min})$ (Table 7.19)
	kPa	
Saturation vapor pressure	$e_s = [(e°(T_{max}) + e°(T_{min})]/2$	
	kPa	

e_a derived from dewpoint temperature:

T_{dew}	°C	$e_a = e°(T_{dew})$ (Table 7.19)
	kPa	

OR e_a derived from maximum and minimum relative humidity:

RH_{max}	%	$e°(T_{min})\ RH_{max}/100$
	kPa	
RH_{min}	%	$e°(T_{max})\ RH_{min}/100$
	kPa	
	e_a: (average)	
	kPa	

OR e_a derived from maximum relative humidity: (recommended if there are errors in RH_{min})

RH_{max}	%	$e_a = e°(T_{min})\ RH_{max}/100$
	kPa	

OR e_a derived from mean relative humidity: (less recommended due to non-linearities)

RH_{mean}	%	$e_a = e_s\ RH_{mean}/100$
	kPa	

Vapour pressure deficit (e_s - e_a)

kPa

Radiation

Latitude		
Day		R_a (Table 7.17)
	MJ m^{-2} d^{-1}	
Month		N (Table 7.10)
	hours	
n	hours	n/N

If no R_s data available: $R_s = (0.25 + 0.50\ n/N)\ R_a$

MJ m^{-2} d^{-1}

$R_{so} = [0.75 + 2\ (\text{Altitude})/100000]\ R_a$

MJ m^{-2} d^{-1}

R_s/R_{so}

$R_{ns} = 0.77\ R_s$

MJ m^{-2} d^{-1}

T_{max}		$\sigma T_{max\ K}4$ (Table 7.20)
	MJ m^{-2} d^{-1}	
T_{min}		$\sigma T_{min\ K}4$ (Table 7.20)
	MJ m^{-2} d^{-1}	

$(\sigma T_{max,\ K}4 + \sigma T_{min,\ K}4)/2$

MJ m^{-2} d^{-1}

e_a	kPa	$(0.34+0.14\ \sqrt{e_a})$
R_s/R_{so}		$(1.35\ R_s/R_{so} - 0.35)$

$R_{nl} = (\sigma T_{max\ K}4 + \sigma T_{min,\ K}4/2(0.34-0.14\sqrt{e_a})(1.35R_sR_{so}-0.35)$

$R_n = R_{ns} - R_{nl}$

T_{month} °C G_{day} (assume)

0

$T_{month-1}$ °C $G_{month} = 0.14\ (T_{month} - T_{month-1})$

$R_n - G$

MJ $m^{-2}\ d^{-1}$

$0.408\ (R_n - G)$

mm/day

Grass reference evapotranspiration

$$\left[\frac{\Delta}{\Delta+\gamma(1+0.34u_2)}\right]\left[0.408(R_n - G)\right]$$

mm/day

$$\left[\frac{\gamma}{\Delta+\gamma(1+0.34u_2)}\right]\left[\frac{900}{T+273}\right]u_2\left[(e_s - e_a)\right]$$

mm/day

$$ET_o = \frac{0.408\Delta(R_n - G)+\gamma\frac{900}{T+273}u_2(e_s - e_a)}{\Delta+\gamma(1+0.34u_2)}$$

mm/day

Example 7.10 Calculate the ET_0 by FAO Penman-Monteith method by using the following data of an arbitrary chosen location.

Latitude = 12.625°N

Month = July

Monthly average daily maximum temperature (T_{max}) = 35.2°C

Monthly average daily minimum temperature (T_{min}) = 25.4°C

Monthly average daily vapor pressure (e_a) = 2.90kP_a

Monthly average daily wind speed measured at 2m height (u_2) = 2.5m/s

Monthly average sunshine duration (n) = 8.2 hours/day

Mean monthly average temperature for July ($T_{month,)}$ = 30.3°C

Mean monthly average temperature for June ($T_{month-1}$) = 29.3.2°C

Solution:

The calculation is made following the steps as stated earlier.

$T_{mean} = (T_{max} + T_{min})/2 = (35.2+25.4)/2=30.3$°C

Δ = 0.2466 *kPa/°C* (Table 7.18)

P = 101.3 kPa/°C (Table 7.6) for altitude 2m

$\gamma = 0.067$ (Table 7.7) for altitude 2m

$(1+ 0.3\ u_2) = (1+0.34x2.5) = 1.85$

$\Delta/[\Delta+\gamma(1 + 0.34\ u_2)] = 0.2466/[0.2466+0.067(1.85)]=0.2466/0.37055=0.665$

$\gamma/[\Delta+\gamma(1 + 0.34\ u_2)] = 0.067/0.3705=0.1808$

$[900/(T_{mean} + 273)]\ u_2= [900/(30.3+273)2.5] = 2250/303.3=7.418$

Vapour pressure deficit

$e^0\ (T_{max}=35.2^0) = 5.6868kPa$ (Table 7.19)

$e^0\ (T_{min}=24.4^0C) = 3.244kPa$

Saturation vapour pressure

$e_s = [(e°(T_{max}) + e°(T_{min})]/2$

$= [5.6868+3.244]/2=4.4654kPa$

$e_a=2.90kP_a$ (Given)

Vapour pressure deficit = $(e_s\text{-}e_a) = (4.465\text{-}2.90) = 1.565$kPa

Radiation (for month July)

J = 195 (for July) (Table 7.17)

Latitude = 12.625^{0}N

$R_a= 40.7004$ MJm^{-2} day^{-1} (Table 7.21)

Day length, N = 12.7575 hours (Table 7.10)

n/N=8.2/12.7575 =0.64277

$R_s = (0.25 + 0.50\ n/N)\ R_a =$

$(0.25+0.50x0.6427(40.7004 = 0.5714x40.7004 = 23.254MJm^{-2}\ day^{-1}$

$R_{so} = [0.75 + 2\ (\text{Altitude})/100000]\ R_a$

$= (0.75+2(2)/\ 100000)40.7004 = 30.5269\text{MJm}^{-2}\ \text{day}^{-1}$

$R_s/R_{so}= 23.254/30.5269 = 0.7618$

$R_{ns} = 0.77\ R_s = 0.77x23.254 = 17.905MJm^{-2}\ day^{-1}$

$\sigma\ T_{max},\ _K4 = 44.326MJm^{-2}\ day^{-1}$ (Table 7.20)

$\sigma\ T_{min},\ _K4 = 38.7604MJm^{-2}\ day^{-1}$

$(\sigma\ T_{max},\ _K4\ ^+\ \sigma\ T_{min,\ K}4)/2 = (44.326 + 38.7604)/2 = 41.5432MJm^{-2}\ day^{-1}$

For $e_a = 2.90kPa$

$(0.34\text{-}0.14\sqrt{e_a}\ ) = (0.34\text{-}0.2384) = 0.1016$

For $R_s/R_{so}= 0.7618$

$(1.35\ R_s/R_{so}\text{-}0.35) = (1.35x0.7618\text{-}0.35) = (1.0284\text{-}0.35) = 0.6784$

$Rnl = (\sigma T_{max, K}4 + \sigma T_{min, K}4)/2(0.34\text{-}0.14\sqrt{e_a})(1.35R_s/R_{so}\text{-}0.35$

$= (41.5432)(0.1016)(0.6784) = 2.8633\ MJm^{-2}\ day^{-1}$

$R_n = R_{ns} - R_{nl} = 17.905\text{-}2.8633 = 15.0417\ MJm^{-2}\ day^{-1}$

$G_{month} = 0.14\ (T_{month} - T_{month-1})$

$= 0.14\ (30.3\text{-}29.3) = 0.14$

$(R_n\text{-}G) = 15.15.0417\text{-}0.14 = 14.9017\ MJm^{-2}day^{-1}$

$0.408(R_n\text{-}G) = 0.408 \times 14.9017 = 6.08$ mm/day

Grass reference evapotranspiration

$$\left[\frac{\Delta}{\Delta+\gamma(1+0.34u_2)}\right]\left[0.408(R_n-G)\right]$$

$$=(0.665)(0.408x14.9017)=4.043 mm/day$$

$$\left[\frac{\gamma}{\Delta+\gamma(1+0.34u_2)}\right]\left[\frac{900}{T+273}\right]u_2\left[(e_s-e_a)\right]$$

$$=(0.1808)(7.418)(1.565)=2.099 mm/day$$

$$ET_o=\frac{0.408\Delta(R_n-G)+\gamma\frac{900}{T+273}u_2(e_s-e_a)}{\Delta+\gamma(1+0.34u_2)}$$

$$=4.043+2.099=6.1419\cong 6.14 mm\backslash day$$

For daily data

The ET_0 calculation by FAO Penman-Monteith method for 24-hours requires the following meteorological data.

1. Air temperature: Maximum (T_{max}) and minimum(T_{min}) daily air temperature
2. Air humidity: Mean daily actual vapour pressure (e_a) derived from psychrometric, dew point temperature or relative humidity data.
3. Wind speed: Daily average wind speed for 24 hours measured at 2m height (u_2).
4. Radiation: Net radiation (R_n) is measured or computed from solar and long wave radiation or from the actual duration of sunshine (n). The extraterrestrial radiation (R_a) and day light hours (N) may be computed using Eq.7.18 and 7.21. The soil heat flux (G) beneath the reference grass surface is negligible and ignored for calculating ET_0 for a day.

Table 7.18 Slope of vapor pressure curve (Δ) for different temperatures (T)

$$\Delta = \frac{4098\left[0.6108\exp\left(\frac{17.27T}{T+237.3}\right)\right]}{(T+237.3)^2} \quad \textbf{(Eq. 7.12)}$$

T °C	Δ kPa/°C	T °C	Δ kPa/°C	T °C	Δ kPa/°C	T °C	Δ kPa/°C
1	0.047	13	0.098	25.0	0.189	37.0	0.342
1.5	0.049	13.5	0.101	25.5	0.194	37.5	0.350
2.0	0.050	14.0	0.104	26.0	0.199	38.0	0.358
2.5	0.052	14.5	0.107	26.5	0.204	38.5	0.367
3.0	0.054	15.0	0.110	27.0	0.209	39.0	0.375
3.5	0.055	15.5	0.113	27.5	0.215	39.5	0.384
4.0	0.057	16.0	0.116	28.0	0.220	40.0	0.393
4.5	0.059	16.5	0.119	28.5	0.226	40.5	0.402
5.0	0.061	17.0	0.123	29.0	0.231	41.0	0.412
5.5	0.063	17.5	0.126	29.5	0.237	41.5	0.421
6.0	0.065	18.0	0.130	30.0	0.243	42.0	0.431
6.5	0.067	18.5	0.133	30.5	0.249	42.5	0.441
7.0	0.069	19.0	0.137	31.0	0.256	43.0	0.451
7.5	0.071	19.5	0.141	31.5	0.262	43.5	0.461
8.0	0.073	20.0	0.145	32.0	0.269	44.0	0.471
8.5	0.075	20.5	0.149	32.5	0.275	44.5	0.482
9.0	0.078	21.0	0.153	33.0	0.282	45.0	0.493
9.5	0.080	21.5	0.157	33.5	0.289	45.5	0.504
10.0	0.082	22.0	0.161	34.0	0.296	46.0	0.515
10.5	0.085	22.5	0.165	34.5	0.303	46.5	0.526
11.0	0.087	23.0	0.170	35.0	0.311	47.0	0.538
11.5	0.090	23.5	0.174	35.5	0.318	47.5	0.550
12.0	0.092	24.0	0.179	36.0	0.326	48.0	0.562
12.5	0.095	24.5	0.184	36.5	0.334	48.5	0.574

Table 7.19 Saturation vapour pressure (e°(T)) for different temperatures (T)

$$e^{\circ}(T) = 0.6108\exp\left[\frac{17.27T}{T+237.3}\right]$$

(Eq. 7.10)

T °C	e_s kPa	T °C	e°(T) kPa	T °C	e°(T) kPa	T °C	e_s kPa
1.0	0.657	13.0	1.498	25.0	3.168	37.0	6.275
1.5	0.681	13.5	1.547	25.5	3.263	37.5	6.448
2.0	0.706	14.0	1.599	26.0	3.361	38.0	6.625
2.5	0.731	14.5	1.651	26.5	3.462	38.5	6.806
3.0	0.758	15.0	1.705	27.0	3.565	39.0	6.991
3.5	0.785	15.5	1.761	27.5	3.671	39.5	7.181
4.0	0.813	16.0	1.818	28.0	3.780	40.0	7.376
4.5	0.842	16.5	1.877	28.5	3.891	40.5	7.574
5.0	0.872	17.0	1.938	29.0	4.006	41.0	7.778
5.5	0.903	17.5	2.000	29.5	4.123	41.5	7.986

6.0	0.935	18.0	2.064	30.0	4.243	42.0	8.199
6.5	0.968	18.5	2.130	30.5	4.366	42.5	8.417
7.0	1.002	19.0	2.197	31.0	4.493	43.0	8.640
7.5	1.037	19.5	2.267	31.5	4.622	43.5	8.867
8.0	1.073	20.0	2.338	32.0	4.755	44.0	9.101
8.5	1.11	20.5	2.412	32.5	4.891	44.5	9.339
9.0	1.148	21.0	2.487	33.0	5.030	45.0	9.582
9.5	1.187	21.5	2.564	33.5	5.173	45.5	9.832
10.0	1.228	22.0	2.644	34.0	5.319	46.0	10.086
10.5	1.27	22.5	2.726	34.5	5.469	46.5	10.347
11.0	1.313	23.0	2.809	35.0	5.623	47.0	10.613
11.5	1.357	23.5	2.896	35.5	5.78	47.5	10.885
12.0	1.403	24.0	2.984	36.0	5.941	48.0	11.163
12.5	1.449	24.5	3.075	36.5	6.106	48.5	11.447

Table 7.20 $\sigma T_K 4$ (Stefan-Boltzmann law) at different temperatures (T)

With = 4.903 10^{-9} MJ K^{-4} m^{-2} day^{-1} and T_K = T[°C] + 273.16

T(°C)	$\sigma T_K 4$ (MJ m^{-2} d^{-1})	T(°C)	$\sigma T_K 4$	T(°C)	$\sigma T_K 4$ (MJ m^{-2} d^{-1})
1.0	27.70	17.0	34.75	33.0	43.08
1.5	27.90	17.5	34.99	33.5	43.36
2.0	28.11	18.0	35.24	34.0	43.64
2.5	28.31	18.5	35.48	34.5	43.93
3.0	28.52	19.0	35.72	35.0	44.21
3.5	28.72	19.5	35.97	35.5	44.50
4.0	28.93	20.0	36.21	36.0	44.79
4.5	29.14	20.5	36.46	36.5	45.08
5.0	29.35	21.0	36.71	37.0	45.37
5.5	29.56	21.5	36.96	37.5	45.67
6.0	29.78	22.0	37.21	38.0	45.96
6.5	29.99	22.5	37.47	38.5	46.26
7.0	30.21	23.0	37.72	39.0	46.56
7.5	30.42	23.5	37.98	39.5	46.85
8.0	30.64	24.0	38.23	40.0	47.15
8.5	30.86	24.5	38.49	40.5	47.46
9.0	31.08	25.0	38.75	41.0	47.76
9.5	31.30	25.5	39.01	41.5	48.06
10.0	31.52	26.0	39.27	42.0	48.37
10.5	31.74	26.5	39.53	42.5	48.68
11.0	31.97	27.0	39.80	43.0	48.99
11.5	32.19	27.5	40.06	43.5	49.30
12.0	32.42	28.0	40.33	44.0	49.61
12.5	32.65	28.5	40.60	44.5	49.92
13.0	32.88	29.0	40.87	45.0	50.24
13.5	33.11	29.5	41.14	45.5	50.56
14.0	33.34	30.0	41.41	46.0	50.87
14.5	33.57	30.5	41.69	46.5	51.19
15.0	33.81	31.0	41.96	47.0	51.51
15.5	34.04	31.5	42.24	47.5	51.84
16.0	34.28	32.0	42.52	48.0	52.16
16.5	34,52	32.5	42.80	48.5	52.49

Example 7.11 Calculate the ET_0 on July 7 from the following meteorological information of a location at 23.5°N and at a height of 12.5 m above sea level.

Maximum air temperature (T_{max}) = 35.2°C

Minimum air temperature (T_{min}) = 25.4°C

Maximum relative humidity (RH_{max}) = 92%

Minimum relative humidity (RH_{min}) = 78%

Wind speed measured at 10m height = 10 km/h

Actual hours of sunshine hours (n) = 9.2 hours

Solution:

Wind speed at 10m height = 10 km/h or u_2 = 2.78m/s

At standard height, with z = 10m, u_2 = 2.079m/s

Eq. 7.17,

$$u_2 = u_z \frac{4.87}{\ln(67.8z - 5.42)}$$

$$= 2.78\frac{4.87}{\ln(67.8x10 - 5.42)} = 2.78x\frac{4.87}{6.511} = 2.079 = 2.08 m/s$$

Parameters

For altitude = 12.5m

P = 101.27 KP_a

Eq.7.7,

$$P = 101.3\left(\frac{293 - 0.0065z}{293}\right)^{5.26}$$

$$= 101.3\left(\frac{293 - 0.0065x12.5}{293}\right)^{5.26} = 101.3\left(\frac{293 - 0.08125}{293}\right)^{5.26} = 101.3x0.999$$

$= 101.27\ kPa$

T_{mean} = (35.2+25.4)/2 = 30.3°C

For T_{mean} = 30.3°C

Δ = 0.152 $kPa/^0C$

Eq. 7.12,

$$\Delta = \frac{4098\left[0.6108\exp\left(\frac{17.27T}{T + 237.3}\right)\right]}{(T + 237.3)^2}$$

$$= \frac{4098\left[0.6108\exp\left(\frac{17.27x30.3}{(30.3+273.3)}\right)\right]}{(30.3+273.3)^2}$$

$$= \frac{4098\left[0.6108\exp\left(\frac{523.281}{303.6}\right)\right]}{(303.6)^2}$$

$$= \frac{4098\lfloor 0.61085.60446 \rfloor}{(92172.96)} = 4098x3.7139x10^{-05} = 0.152kPa/^0C$$

For P = 101.27$kPa/^0C$

Eq.7.8,

$$\gamma = \frac{c_p p}{\varepsilon\lambda} = 0.665 \text{x} 10^{-3} p$$

$= 0.665x101.27x10^{-3} = 0.0673kPa/^0C$

$(1 + 0.34\, u_2) = (1+0.34x2.08) = 1.707$

$\Delta/[\Delta + \gamma\,(1+0.34\,u_2)] = 0.152/[0.152+0.067(1.707)] = 0.152/0.266 = 0.5706$

$\Delta/[\Delta + \gamma\,(1+0.34\,u_2)] = 0.067/0.266 = 0.2519$

$[900/(T_{mean} + 273)]\, u_2 = [900/(30.3+273)]2.08 = 1872/303.3 = 6.172$

Vapour pressure deficit

For $T_{max} = 35.2^0C$, $e^0\,(T_{max})$ = **(Eq.7.10)**

Eq.7.10,

$$e^o(T) = 0.6108\exp\left[\frac{17.27T}{T+237.3}\right]$$

$\therefore e^0\,(T\text{max})$

$$= 0.6108\exp\left[\frac{17.27x35.2}{35.2+237.3}\right]$$

$= 0.6108 \exp 2.2308 = 0.6108x9.307 = 5.6849kPa$

$$\& e^0(T_{min}) = 0.6108\exp\left[\frac{17.27x25.4}{25.4+373.3}\right]$$

$$= 0.6108\exp\frac{438.658}{262.7} = 0.6108 \exp 1.6698 = 0.6108x5.311 = 3.224kpa$$

$e_s = [(e^o(T_{max}) + e^o(T_{min})]/2$

= (5.6849+3.224/2 = 4.4545kPa

Given RH_{max} = 92%

RH_{min} =78%

Eq.7.16,

$$e_a = \frac{e^o(T_{min})\frac{RH_{max}}{100} + e^o(T_{max})\frac{RH_{min}}{100}}{2}$$

$$= \frac{3.224\frac{92}{100} + 5.6849\frac{78}{100}}{2}$$

$$= \frac{2.966 + 3.4179}{2} = 3.70kPa$$

Vapor pressure deficit (e_s-e_a) = 4.4545-3.70=0.7545kPa

Radiation

For July 7, J = 188 (Table 7.8)

For latitude 23.5°C

R_a=39.825 MJm^{-2} day^{-1} (Table 7.21)

N = 13.85 hours (Table 7.10)

$n/N = 9.2/13.275 = 0.693$

$R_s = [0.25+0.509n/N]\ R_a$

$= [0.25+0.3465]39.825 = 23.756\ MJm^{-2}\ day^{-1}$

$R_{so} = (0.75+2x10^{-5}z)R_a$

$R_{so} = (0.75+2x10^{-5}x12.5)39.825 = 29.879\ MJm^{-2}\ day^{-1}$

$R_s / R_{so} = 23.756 / 29.879 = 0.795$

$R_{ns} = (1\text{-}a)R_{s,}\ a = 0.23$

$= (1\text{-}0.23)23.756 = 18.292\ MJm^{-2}\ day^{-1}$

$T_{max} = 35.2^0\ C$

$T_{max,\ K}4 = 35.2 + 273.16 = 308.36^0\ K$

$T_{max',\ K}4 = 38.958\ MJm^{-2}\ day^{-1}$ **(Table 7.20)**

$T_{min} = 25.4^0$C

$T_{min,\ K}4 = 25.4 + 273.16 = 298.56^0K$

$\sigma T_{max,\ K}4 = 38.958\ MJm^{-2}\ day^{-1}$ (Table 7.20)

$\left(0.34 - 0.14\sqrt{e_a}\right) = \left(0.34 - 0.14\sqrt{3.70}\right) = 0.07kPa$

$(1.35R_s/R_{so}\text{- }0.35) = 1.35x0.795 - 0.35 = 0.7233$

$$R_{nl} = \left[\frac{\sigma T_{max,K^4}\sigma T_{min,K^4}}{2}\right]\left(0.34-0.14\sqrt{e_a}\right)\left(1.35R_s/R_{so}-0.35\right)$$

$= 41.642x(0.0707x0.6275) = 1.84432\ MJm^{-2}\ day^{-1}$

$R_n = R_{ns}\text{- }R_{nl}$

$= 18.292\text{-}1.8443 = 16.4477\ MJm^{-2}\ day^{-1}$

G=0, soil heat flux for a day is ignored

$R_n\text{- }G = 16.4477\text{-}0 = 16.4477\ MJm^{-2}\ day^{-1}$

Grass reference evapotranspiration

$$\left[\frac{\Delta}{\Delta+\gamma(1+0.34u_2)}\right]\left[0.408(R_n-G)\right]$$

$= (0.5706)(0.408x16.4477) = 3.829\ mm/day$

$$\left[\frac{\gamma}{\Delta+\gamma(1+0.34u_2)}\right]\left[\frac{900}{T+273}\right]u_2\left[(e_s-e_a)\right]$$

$(0.2519)(2.967)(2.08)(0.7545) = 1.1729\ mm/day$

$$ET_0\,\frac{0.408\Delta(R_n-G)+\gamma\frac{900}{T+273}u_2(e_s-e_a)}{\Delta+\gamma(1+0.34u_2)}$$

$= 3.829+1.1729 = 5.00 mm/day$

Example 7.12 Three geometrically identical Lysimeters (A, B and C) in a paddy field have uniform initial pending depths of 12cm. After a week, the recorded water depths in A, B and C are 10.4 cm, 8.6 cm, and 7.0 cm, respectively. Rainfall received during the week is 15mm, and the crop coefficient of the paddy is 0.94. Lysimeter 'A' has a closed bottom with no plant. Lysimeter 'B' has an open bottom with no plant. Lysimeter 'C' has an open bottom with paddy plants. Neglecting the boundary effects and groundwater contribution in the Lysimeter, the weekly potential evapotranspiration will be....cm. (GATE, 2018)

Solution:

Initial pending depth and rainfall to all lysimeter = 12 cm + 15 mm = 13.5 cm

Evaporation from lysimeter A =13.5-10.4 = 3.1cm

Percolation from lysimeter B = 13.5-8.6-3.1 = 1.8 cm

Evapotranspiration from lysimeter C = 13.5-1.8-7.0 = 4.7 cm

Potential evapotranspiration = Evapotranspiration / Crop coefficient = 4.7/0.94 = 5.0 cm

Table 7.21 Daily extraterrestrial radiation (R_a) for different latitudes for the 15th day of the month

$$Ra = \frac{24(60)}{\pi} G_{sc} d_r \left[\omega_s \sin(\varphi)\sin(\delta) + \cos(\varphi)\cos(\delta)\sin(\omega_s)\right]$$

(values in MJ m^{-2} day^{-1})

Northern Hemisphere												Lat	Southern Hemisphere											
Jan	Feb	Mar	Apr	May	Jun	Jul	Aug	Sep	Oct	Nov	Dec	deg	Jan	Feb	Mar	Apr	May	Jun	Jul	Aug	Sep	Oct	Nov	Dec
0.0	2.6	10,4	23.0	35.2	42.5	39.4	28.0	19.9	4.9	0.1	0.0	70	41.4	28.6	15.8	4.9	0.2	0.0	0.0	2.2	10.7	23.5	37.3	45.3
0.1	3.7	11.7	23.9	35.3	42.0	38.9	28.6	16.1	6.0	0.7	0.0	68	41.0	29.3	16.9	6.0	0.8	0.0	0.0	3.2	11.9	24.4	37.4	44.7
0.6	4.8	12.9	24.8	35.6	41.4	38.8	29.3	17.3	7.2	1.5	0.1	66	40.9	30.0	18.1	7.2	1.5	0.1	0.5	4.2	13.1	25.4	37.6	44.1
1.4	5.9	14.5	25.8	35.9	41.2	38.8	30.0	18.4	8.5	2.4	0.6	64	41.0	30.8	19.3	8.4	2.4	0.6	1.2	5.3	14.4	26.3	38.6	43.9
2.3	7.1	15.4	26.6	36.3	41.2	39.0	30.6	18.5	9.7	3.4	1.3	62	41.2	51.5	20.4	9.6	3.4	1.2	2.0	6.4	15.5	27.2	38.3	43.9
3.3	8.3	16.6	27.5	36.6	41.2	39.2	31.3	20.6	10.9	4.9	2.2	60	41.5	32.3	21.5	10.8	4.4	2.0	2.9	7.6	16.7	28.1	38.7	43.9
4.3	9.6	17.7	26.4	37.0	41.3	29.4	32.0	21.7	12.1	5.5	3.1	58	41.2	33.0	22.6	12.0	5.5	2.9	3.9	8.7	17.9	28.9	39.1	44.0
5.4	10.8	18.9	29.2	37.4	41.4	39.6	32.6	22.7	13.3	6.7	4.2	56	42.0	33.7	23.6	13.8	6.6	3.9	4.0	9.9	19.0	29.8	39.5	44.1
6.5	12.0	20.0	30.0	37.8	41.5	39.8	33.2	23.7	14.5	7.8	5.2	54	42.2	34.3	24.6	14.4	7.7	4.9	6.0	11.1	20.1	30.6	39.9	44.3
7.7	13.2	21.1	30.8	38.2	41.6	40.1	33.8	24.7	15.7	9.0	6.4	52	42.5	35.0	25.6	15.6	8.8	6.0	7.1	12.2	21.2	31.4	40.2	44.4
8.9	14.4	22.2	31.5	38.5	41.7	40.2	34.4	25.7	16.9	10.2	7.5	50	42.7	35.6	26.6	16.7	10.0	7.1	8.2	13.4	22.2	32.1	40.6	44.5
10.1	15.7	23.3	32.2	38.8	41.8	40.4	34.9	26.6	18.1	11.4	8.7	48	42.9	36.2	27.5	17.9	11.1	8.2	9.3	14.6	23.3	32.8	40.9	44.5
11.3	16.9	24.3	32.9	39.1	41.9	40.6	35.4	27.5	19.2	12.6	9.9	46	43.0	36.7	28.4	19.0	12.3	9.3	10.4	15.7	24.3	33.5	41.1	44.6
12.5	18.0	25.3	33.5	39.3	41.9	40.7	35.9	28.4	20.3	13.9	11.1	44	43.2	37.2	29.3	20.1	13.5	10.5	11.6	16.8	25.2	34.1	41.4	44.6
13.8	19.2	26.3	34.1	39.5	41.9	40.8	36.3	29.2	21.4	15.1	12.4	42	43.3	37.7	30.1	21.2	14.6	11.6	12.8	18.0	26.2	34.7	41.6	44.6
15.0	20.4	27.2	34.7	39.7	41.9	40.8	36.7	30.0	22.5	16.3	13.6	40	43.4	38.1	30.9	22.3	15.8	12.8	13.9	19.1	27.1	35.3	41.8	44.6
16.2	21.5	38.1	35.2	39.9	41.8	40.8	37.0	30.7	23.6	17.5	14.8	38	43.4	38.5	31.7	23.3	16.9	13.9	15.1	20.2	28.0	35.8	41.9	44.5
17.5	22.6	29.0	35.7	40.0	41.7	40.8	37.4	31.5	24.6	18.7	16.1	36	43.4	38.9	32.4	24.3	18.1	15.1	16.2	21.2	28.8	36.3	42.0	44.4
18.7	23.7	29.9	36.1	40.0	41.6	40.8	37.6	32.1	25.6	19.9	17.3	34	43.4	39.2	33.0	25.3	19.2	16.2	17.4	22.3	29.6	36.7	42.0	44.3
19.9	24.8	30.7	36.5	40.0	41.4	40.7	37.9	32.8	26.6	21.1	18.5	32	43.3	39.4	33.7	26.3	20.3	17.4	18.5	23.3	30.4	37.1	42.0	44.1
21.1	25.8	31.4	36.8	40.0	41.2	40.6	38.0	33.4	27.6	22.2	19.8	30	43.1	39.6	34.3	27.2	21.4	18.5	19.6	24.3	31.1	37.5	42.0	43.9

22.3	26.8	32.2	37.1	40.0	40.9	40.4	38.3	33.9	28.5	23.3	21.0	28	43.0	39.8	34.8	28.1	22.5	19.7	20.7	25.3	31.8	37.8	41.9	43.6
23.4	27.8	32.8	37.4	39.9	40.6	40.2	38.3	34.5	29.3	24.5	22.2	26	42.8	39.9	35.3	29.0	23.5	20.8	21.8	26.3	32.5	38.0	41.8	43.3
24.6	28.8	33.5	37.6	39.7	40.3	39.9	38.3	34.9	30.2	25.5	23.3	24	42.5	40.0	35.8	29.8	24.6	21.9	22.9	27.2	33.1	38.3	41.7	43.0
25.7	29.7	34.1	37.8	39.5	40.0	39.6	38.4	35.4	31.0	26.6	24.5	22	42.2	40.1	36.3	30.6	25.6	23.0	24.0	28.1	33.7	38.4	41.4	42.6
26.8	30.6	34.7	37.9	39.3	39.5	39.3	38.3	35.8	31.8	27.7	25.6	20	41.9	40.1	36.6	31.3	26.6	24.1	25.0	28.9	34.2	38.6	41.2	42.1
27.9	31.5	35.2	38.0	39.0	39.1	38.9	38.2	36.1	32.5	28.7	26.8	18	41.5	40.0	37.0	32.1	27.5	25.1	26.0	29.8	34.7	38.7	40.9	41.7
28.9	32.3	35.7	38.1	38.7	38.6	38.5	38.1	36.4	33.2	29.6	27.9	16	41.1	39.9	37.2	32.8	28.5	26.2	27.0	30.6	35.2	38.7	40.6	41.2
29.9	33.1	36.1	38.1	38.4	38.1	38.1	38	36.7	33.9	30.6	28.9	14	40.6	39.7	37.5	33.4	29.4	27.2	27.9	31.3	35.6	38.1	40.2	40.6
30.9	33.8	36.5	38.0	38.0	37.6	37.6	37.8	36.9	34.5	31.5	30.0	12	40.1	39.6	37.7	34.0	30.2	28.1	28.9	32.1	36.0	38.6	39.8	40.0
31.9	34.5	36.9	37.9	37.6	37.0	37.1	37.5	37.1	35.1	32.4	31.0	10	39.5	39.3	37.8	34.6	31.1	29.1	29.8	32.8	36.3	38.5	39.3	39.4
32.8	35.2	37.2	37.8	37.1	36.3	36.5	37.2	37.2	35.6	33.3	32.0	8	38.9	39.0	37.9	35.1	31.9	30.0	30.7	33.4	36.6	38.4	38.8	38.7
33.7	35.8	37.4	37.6	36.6	35.7	35.9	36.9	37.3	36.1	34.1	32.9	6	38.3	38.7	38.0	35.6	32.7	30.9	31.5	34.0	36.8	38.2	38.2	38.0
34.6	36.4	37.6	37.4	36.0	35.0	35.3	36.5	37.3	36.6	34.9	33.9	4.0	37.6	38.3	38.0	36.0	33.4	31.8	32.3	34.6	37.0	38.0	37.6	37.2
35.4	37.0	37.8	37.1	35.4	34.2	34.4	36.1	37.3	37	35.6	34.8	2	36.9	33.9	38.0	36.4	34.1	32.6	33.1	35.2	37.1	37.7	37.0	36.4
36.2	37.5	37.9	36.8	34.6	33.4	33.9	36.7	37.2	37.4	36.3	35.6	0	36.2	37.5	37.9	36.8	34.8	33.4	33.9	35.7	37.2	37.4	36.3	35.6

Values for R_a on the 15th day of the month provide a good estimate (error < 1 %) of R_a averaged over all days within the month. Only for high latitudes greater than 55° (N or S) during winter months deviations may be more than 1%.

Table 7.22 Monthly percentage of day light hours of the year (p) for different latitude

Latitude (degree)	Jan	Feb	march	April	May	June	July	Aug	Sep	Oct	Nov	Dec
1	2	3	4	5	6	7	8	9	10	11	12	13
					NORTH							
60	4.67	5.70	8.05	9.66	11.72	12.39	12.33	10.72	8.57	7.00	5.04	4.15
59	4.81	5.78	8.05	9.60	11.61	12.23	12.21	10.6	8.56	7.07	5.09	4.31
58	4.99	5.85	8.06	9.55	11.44	12	12.00	10.56	8.56	7.13	5.13	4.55
57	5.14	5.93	8.07	9.51	11.32	11.77	11.87	10.47	8.54	7.19	5.27	4.69
56	5.29	6.00	8.10	9.45	11.2	11.67	11.69	10.4	8.52	7.25	5.54	4.89
55	5.39	6.06	8.12	9.41	11.11	11.53	11.59	10.32	8.51	7.30	5.62	5.01
54	5.53	6.12	8.15	9.36	11.0	11.40	11.43	10.27	8.50	7.33	5.74	5.17
53	5.64	6.19	8.16	9.32	10.88	11.31	11.34	10.19	8.50	7.38	5.83	5.31
52	5.75	6.23	8.17	9.28	10.81	11.13	11.22	10.15	8.49	7.40	5.94	5.43
50	5.98	6.32	8.25	9.25	10.69	10.93	10.99	10.00	8.44	7.43	6.07	5.65
48	6.13	6.42	8.22	9.15	10.50	10.72	10.83	9.92	8.45	7.56	6.24	5.86
46	6.13	6.42	8.22	9.15	10.50	10.72	10.83	9.92	8.45	7.56	6.24	5.86
44	6.45	6.59	8.25	9.04	10.22	1.038	10.5	9.73	8.43	7.67	6.51	6.23
42	6.60	6.66	8.28	8.97	10.10	10.21	10.37	9.64	8.42	7.73	6.63	6.39
40	6.73	6.73	8.30	8.92	9.99	10.08	10.34	9.56	8.41	7.78	6.73	6.53
38	6.87	6.79	8.34	8.90	9.92	9.95	10.1	9.47	8.38	7.80	6.82	6.66
36	6.99	6.86	8.35	8.85	9.31	9.83	9.99	9.40	8.36	7.85	6.92	6.79
34	7.10	6.91	8.36	8.80	9.72	9.70	9.88	9.33	8.36	7.90	7.07	6.92
32	7.20	6.97	8.37	8.72	9.63	9.60	9.77	9.28	8.34	7.93	7.11	7.05
30	7.30	7.03	8.38	8.72	9.53	9.49	9.67	9.22	8.34	7.99	7.19	7.14
28	7.40	7.02	8.39	8.68	9.46	9.38	9.58	9.16	8.32	8.02	7.27	7.27
26	7.49	7.12	8.40	8.64	9.37	9.30	9.49	9.1	8.32	8.06	7.36	7.35
24	7.58	7.17	8.40	8.6 0	9.30	9.19	9.41	9.05	8.31	8.10	7.43	7.46
22	7.76	7.22	8.41	8.57	9.22	9.12	9.31	9.00	8.30	7.13	7.5	7.56
20	7.73	7.26	8.20	8.52	9.14	9.02	9.25	8.95	8.30	8.19	7.58	7.88
18	7.88	7.26	8.40	8.46	9.06	8.99	9.20	8.81	8.29	8.24	7.67	7.89
16	7.94	8.42	8.42	8.45	8.98	8.98	9.07	8.80	8.28	8.24	7.72	7.9
14	7.08	8.43	8.43	8.44	8.90	8.73	8.99	8.79	8.28	8.28	7.85	8.04
12	8.08	8.44	8.44	8.43	8.84	8.64	8.90	8.78	8.27	8.28	7.85	8.05

10	8.11	8.44	8.44	8.43	8.91	8.57	8.84	8.74	8.26	8.29	7.89	8.08
8	8.13	8.45	8.45	8.39	8.75	8.51	8.77	8.70	8.25	8.31	7.89	8.11
6	8.19	8.45	8.45	8.39	8.73	8.48	8.75	8.69	8.25	8.41	7.95	8.19
4	8.20	8.46	8.46	8.33	8.65	8.4	8.67	8.63	8.21	8.43	7.95	8.2
2	8.43	8.47	8.47	8.22	8.51	8.25	8.52	8.5	8.2	8.46	8.16	8.42
0	8.49	8.49	8.49	8.22	8.49	8.22	8.49	8.49	8.19	8.49	8.22	8.49
					SOUTH							
0	8.49	7.67	8.49	8.22	8.49	8.22	8.49	8.49	8.19	8.49	8.22	8.49
2	8.55	7.71	8.49	8.19	8.44	8.17	8.43	8.44	8.19	8.52	8.27	8.55
4	8.64	7.76	8.5	8.17	8.39	8.08	8.2	8.41	8.19	8.56	8.33	8.65
6.00	8.71	7.81	8.5	8.12	8.30	8.00	8.19	8.37	8.18	8.59	8.38	8.74
8.00	8.79	8.84	8.51	8.11	8.24	7.91	8.13	8.32	8.18	8.62	8.47	8.84
10.0	8.85	7.86	8.52	8.09	8.18	7.84	8.11	8.28	8.18	8.65	8.52	8.90
12.0	8.91	7.91	8.53	8.06	8.15	7.79	8.08	8.28	8.17	8.67	8.58	8.95
14.0	8.97	7.97	8.54	8.03	8.07	7.70	7.08	8.19	8.16	8.69	8.65	9.01
16.0	9.09	8.02	8.56	7.98	7.96	7.57	7.94	8.14	8.14	8.76	8.72	9.17
18.0	9.18	8.06	8.57	7.93	7.99	7.50	7.88	8.90	8.14	8.8	8.8	9.24
20.0	9.25	8.09	8.58	7.92	7.83	7.41	7.73	8.05	8.13	8.83	8.85	9.32
22	9.36	8.12	8.58	7.89	7.74	7.3	7.76	8.03	8.13	8.86	8.9	9.38
24	9.44	8.17	8.59	7.87	7.60	7.24	7.58	7.99	8.12	8.89	8.96	9.47
26	9.52	8.28	8.60	7.81	7.56	7.07	7.49	7.87	8.11	8.94	9.10	9.16
28	9.61	8.31	8.61	7.79	7.49	6.99	7.40	7.85	8.1	8.97	9.19	9.74
30	9.69	8.33	8.63	7.75	7.43	6.94	7.3	7.80	8.09	9	9.24	9.80
32	9.76	8.36	8.64	7.70	7.39	6.85	7.2	7.73	8.08	9.04	9.31	9.87
34	9.88	8.41	8.65	7.68	7.3	6.73	7.1	7.69	8.06	9.07	9.38	9.99
36	10.06	8.53	8.67	7.61	7.10	6.59	6.99	7.59	8.06	9.15	9.51	10.21
38	10.14	8.61	8.68	7.59	7.03	6.46	6.87	7.51	8.05	9.19	9.60	10.34
40	10.24	8.65	8.7	7.54	6.96	6.33	6.73	7.46	8.04	9.23	9.69	10.42
42	10.39	8.72	8.71	7.49	6.85	6.20	6.6	7.39	8.01	9.27	9.79	10.57
44	10.52	8.81	8.72	7.44	6.73	6.04	6.45	7.3	8.00	9.34	9.91	10.72
46	10.68	8.88	8.73	7.39	6.61	5.87	6.3	7.21	7.98	9.41	10.03	10.90
48	10.85	8.98	8.76	7.32	6.45	5.69	6.13	7.12	7.96	9.47	10.17	11.09
50	11.03	9.06	8.77	7.25	6.31	5.48	5.98	7.03	7.95	9.53	10.32	11.30

Source: U.S.D.A, A.R.S. (1962)

Table 7.23 Extra-terrestrial radiation Q_A (or R_{AE}) Ra expressed in equivalent evaporation in mm/day

Latitude (degree)	Jan	Feb	march	April	May	June	July	Aug	Sep	Oct	Nov	Dec
					NORTH							
50	3.8	6.1	9.4	12.7	15.8	17.1	16.4	14.1	10.9	7.4	4.5	3.3
48	4.3	6.6	9.8	13.0	15.9	17.2	16.5	14.3	11.2	7.8	5.0	3.7
46	4.9	7.1	10.2	13.3	16	17.2	16.6	14.5	11.5	8.3	5.5	4.3
44	5.3	7.6	10.6	13.7	16.1	17.2	16.6	14.7	11.9	8.7	6.0	4.7
42	5.9	8.1	11.0	14.0	16.2	17.3	16.7	15.0	12.2	9.1	6.5	5.2
40	6.4	8.6	11.4	14.3	16.4	17.3	16.7	15.2	12.5	9.6	7.0	5.7
38	6.9	9.0	11.8	14.5	16.4	17.2	16.7	15.3	12.8	70.0	7.5	6.1
36	7.4	9.4	12.1	14.7	16.4	17.2	16.7	15.4	13.1	10.6	8.0	6.6
34	7.9	9.8	12.4	14.8	16.5	17.1	16.8	15.5	13.4	10.8	8.5	7.2
32	8.3	10.2	12.8	15.0	16.5	17.0	16.8	15.6	13.6	11.2	9.0	7.8
30	8.8	10.7	13.1	15.2	16.5	17.0	16.8	15.7	13.9	11.6	9.5	8.3
28	9.3	11.1	13.4	15.3	16.5	16.8	16.7	15.7	14.1	12.0	9.9	8.8
26	9.8	11.5	13.7	15.3	16.4	16.7	16.6	15.7	14.3	12.3	10.3	9.3
24	10.2	11.9	13.9	15.4	16.4	16.6	16.6	15.8	14.5	12.6	10.7	9.7
22	10.7	12.3	14.2	15.5	16.3	16.4	16.4	15.8	14.6	13.0	11.0	10.2
20	11.2	12.7	14.4	15.6	16.3	16.4	16.4	15.9	14.8	13.3	11.6	10.7
18	11.6	13	14.6	15.6	16.1	16.1	16.1	15.8	14.9	13.6	12.0	11.1
16	12	13.3	14.7	15.6	16.0	15.9	15.9	15.7	15.0	13.9	12.4	11.6
14	12.4	13.6	14.9	15.7	15.8	15.7	15.7	15.7	15.1	14.1	12.8	12
12	12.8	13.9	15.1	15.7	15.7	15.5	15.5	15.6	15.2	14.4	12.3	12.5
10	13.2	14.2	15.3	15.7	15.5	15.3	15.3	15.5	15.3	14.7	12.6	12.9
8	13.6	14.5	15.3	15.6	15.3	15.0	15.1	15.4	15.3	14.8	13.9	13.3
6	13.9	14.8	15.4	15.4	15.1	14.7	14.9	15.2	15.3	15.0	14.2	13.7
4	14.3	15	15.5	15.5	14.9	14.4	14.6	15.1	15.3	15.1	14.5	14.1
2	14.7	15.3	15.6	15.3	14.6	14.2	14.3	14.9	15.3	15.3	14.8	14.4
0	15	15.5	15.7	15.3	14.4	13.9	14.1	14.8	15.3	15.4	15.1	14.8

SOUTH												
50	17.5	14.7	10.9	7.0	4.2	3.1	3.5	5.5	8.9	12.9	16.5	18.2
48	17.6	14.9	11.2	7.5	4.7	3.5	4.0	6.0	9.3	13.2	16.6	18.2
46	17.7	15.1	11.5	7.9	5.2	4.0	4.4	6.5	9.7	13.4	16.7	18.3
44	17.8	15.3	11.9	8.4	5.7	4.4	4.9	6.9	10.2	13.7	16.7	18.3
42	17.8	15.5	12.2	8.8	6.1	4.9	5.4	7.4	10.6	14	16.8	18.3
40	17.9	15.7	12.5	9.2	6.6	5.3	5.9	7.9	11	14.2	16.9	18.3
38	17.9	15.8	12.8	9.6	7.1	5.8	6.3	8.3	11.4	14.4	17	18.3
36	17.9	16	13.2	10.1	7.5	6.3	6.8	8.8	11.7	14.6	17	18.2
34	17.8	16.2	13.8	10.9	8.5	7.3	7.7	9.6	12.4	15.4	17.2	18.1
32	17.8	16.2	13.8	10.9	8.5	7.3	7.7	9.6	12.4	15.4	17.2	18.1
30	17.8	16.4	14	11.3	8.9	7.9	8	10.1	12.7	15.3	17.3	18.1
28	17.7	16.4	14.3	11.6	9.3	8.2	8.6	10.4	13	15.4	17.2	17.9
26	17.6	16.4	14.4	12	9.7	8.7	9.1	10.9	13.2	15.5	17.2	17.8
24	17.5	16.5	14.6	12.3	10.2	9.1	9.5	11.2	13.4	15.6	17.1	17.7
22	17.4	16.5	14.8	12.6	10.6	9.6	10	11.6	13.7	15.7	17	17.5
20	17.3	16.5	15.0	13.0	11.0	10	10.4	12	13.9	15.8	16.8	17.1
18	17.1	16.5	15.1	13.2	11.4	10.4	10.8	12.3	14.1	15.8	16.8	17.1
16	16.9	16.4	15.2	13.5	11.7	10.8	11.2	12.6	14.3	15.8	16.7	16.8
14	16.9	16.4	15.2	13.5	11.7	10.8	11.2	12.6	14.3	15.8	16.7	16.8
12	16.6	16.3	15.4	14.0	12.5	11.6	12.0	13.2	14.7	15.8	16.4	16.5
10	16.4	16.3	15.5	14.2	12.8	12.0	12.4	13.5	14.8	15.9	16.2	16.2
8	16.1	16.1	15.5	14.4	13.1	12.4	12.7	13.7	14.9	15.8	16	16.0
6	15.8	16	15.6	14.7	13.4	12.8	13.1	14	15.0	15.7	15.8	15.7
4	15.5	15.8	15.6	14.9	13.8	13.2	13.4	14.3	15.1	15.6	15.5	15.4
2	15.3	15.7	15.7	15.1	14.1	13.5	13.7	14.5	15.2	15.5	15.3	15.1
0	15	15.5	15.7	15.3	14.4	13.9	14.1	14.8	15.3	15.4	15.1	14.8

Table 7.24 Correction factor for length and number of days in a month for use in the Thornthwaite formula (mean possible sunlight expressed in units of 12 hours each) (Thornwaite, 1948)

Latitude (degree)	Jan	Feb	march	April	May	June	July	Aug	Sep	Oct	Nov	Dec
					NORTH							
0	1.04	0.94	1.04	1.01	1.04	1.01	1.04	1.04	1.01	1.04	1.01	1.04
5	1.02	0.93	1.03	1.02	1.06	1.03	1.06	1.05	1.01	1.03	0.99	1.02
10	1.00	0.91	1.03	1.03	1.08	1.06	1.08	1.07	1.02	1.02	0.98	0.99
15	0.97	0.91	1.03	1.04	1.11	1.08	1.12	1.08	1.02	1.01	0.95	0.97
20	0.95	0.9	1.03	1.05	1.13	1.11	1.14	1.11	1.02	1.00	0.93	0.94
25	0.93	0.89	1.03	1.06	1.15	1.14	1.17	1.12	1.02	0.99	0.91	0.91
26	0.92	0.88	1.03	1.06	1.15	1.15	1.17	1.12	1.02	0.99	0.91	0.91
27	0.92	0.88	1.03	1.07	1.16	1.15	1.18	1.13	1.02	0.99	0.90	0.90
28	0.91	0.88	1.03	1.07	1.16	1.16	1.18	1.13	1.02	0.98	0.90	0.90
29	0.90	0.87	1.03	1.07	1.17	1.16	1.19	1.13	1.03	0.98	0.90	0.89
30	0.90	0.87	1.03	1.08	1.18	1.17	1.20	1.14	1.03	0.98	0.89	0.88
31	0.90	0.87	1.03	1.08	1.18	1.18	1.20	1.14	1.03	0.98	0.89	0.88
32	0.89	0.86	1.03	1.08	1.19	1.19	1.21	1.15	1.03	0.98	0.88	0.87
33	0.88	0.86	1.03	1.09	1.19	1.20	1.22	1.15	1.03	0.97	0.88	0.86
34	0.88	0.85	1.03	1.09	1.20	1.20	1.22	1.16	1.03	0.97	0.87	0.86
35	0.87	0.85	1.03	1.09	1.21	1.21	1.23	1.16	1.03	0.97	0.86	0.85
36	0.87	0.85	1.03	1.10	1.21	1.22	1.24	1.16	1.03	0.97	0.86	0.84
37	0.86	0.84	1.03	1.10	1.22	1.23	1.25	1.17	1.03	0.97	0.85	0.83
38	0.85	0.84	1.03	1.10	1.23	1.24	1.25	1.17	1.04	0.96	0.84	0.83
39	0.85	0.84	1.03	1.11	1.23	1.24	1.26	1.18	1.04	0.96	0.84	0.82
40	0.84	2.83	1.03	1.11	1.24	1.25	1.27	1.18	1.04	0.96	0.83	0.81
41	0.83	0.83	1.03	1.11	1.25	1.26	1.27	1.19	1.04	0.96	0.82	0.80
42	0.82	0.83	1.03	1.12	1.26	1.27	1.28	1.19	1.04	0.95	0.82	0.79
43	0.81	0.82	1.02	1.12	1.26	1.28	1.29	1.20	1.04	0.95	0.81	0.77
44	0.81	0.82	1.02	1.13	1.27	1.29	1.3	1.20	1.04	0.95	0.8	0.76

45	0.80	0.81	1.02	1.13	1.28	1.29	1.31	1.21	1.04	0.94	0.79	0.75
46	0.79	0.81	1.02	1.13	1.29	1.31	1.32	1.22	1.04	0.94	0.79	0.74
47	0.77	0.80	1.02	1.14	1.30	1.32	1.33	1.22	1.04	0.93	0.78	0.73
48	0.76	0.80	1.02	1.14	1.31	1.33	1.34	1.23	1.05	0.93	0.77	0.72
49	0.75	0.79	1.02	1.14	1.32	1.34	1.35	1.24	1.05	0.93	0.76	0.71
50	0.74	0.78	1.02	1.15	1.33	1.36	1.37	1.25	1.06	0.92	0.76	0.70
50	1.06	0.95	1.04	1.00	1.02	0.99	1.02	1.03	1.00	1.05	1.03	1.06
10	1.08	0.97	1.05	0.99	1.01	0.96	1.00	1.01	1.00	1.06	1.05	1.10
15	1.12	0.98	1.05	0.98	0.98	0.94	0.97	1.00	1.00	1.07	1.07	1.12
20	1.14	1.00	1.05	0.97	0.96	0.91	0.95	0.99	1.00	1.08	1.09	1.15
25	1.17	1.01	1.05	0.96	0.94	0.88	0.93	0.98	1.00	1.1	1.11	1.18
30	1.20	1.03	1.06	0.95	0.92	0.85	0.9	0.96	1.00	1.12	1.14	1.21
35	1.23	1.04	1.06	0.94	0.89	0.82	0.87	0.94	1.00	1.13	1.17	1.25
40	1.27	1.06	1.07	0.93	0.86	0.78	0.84	0.92	1.00	1.15	1.2	1.29
42	1.28	1.07	1.07	0.92	0.85	0.76	0.82	0.92	1.00	1.16	1.22	1.31
44	1.3	1.08	1.07	0.92	0.83	0.74	0.81	0.91	0.99	1.17	1.23	1.33
46	1.32	1.1	1.07	0.91	0.82	0.72	0.79	0.9	0.99	1.17	1.25	1.35
48	1.34	1.11	1.08	0.9	0.8	0.7	0.76	0.89	0.99	1.18	1.27	1.37
50	1.3	1.12	1.08	0.89	0.77	0.67	0.74	0.88	0.99	1.1	1.21	1.41

Questions & Problems

7.1 Define evaporation, transpiration, evapotranspiration and consumptive use. Discuss the factors influence the evaporation and transpiration.

7.2 Define potential evapotranspiration, crop evapotranspiration, reference crop evapotranspiration, actual evapotranspiration and crop coefficient. Discuss the factors on which the crop coefficient value ofion a crop depends?

7.3 Define daily, seasonal and peak period consumptive use. Discuss the principle of selecting peak period consumptive use.

7.4 Explain with the expression the energy balance in evapotranspiration (ET) and aerodynamic resistance (r_a).

7.5 Define reference surface crop, latent heat of vaporization (λ), psychrometric constant (γ) and relative humidity (RH).

7.6 Define extraterrestrial radiation, solar radiation, solar shortwave radiation, relative shortwave radiation, relative sunshine duration and albedo.

7.7 Discuss the mean saturation vapor pressure and slope of the saturation vapor pressure.

7.8 Describe the principle of measurement of actual vapor pressure derived from psychrometric and relative humidity data.

7.9 What is a lysimeter? Describe the principle of working of a weighing type lysimeter with necessary sketch.

7.10 Describe the method of estimating evapotranspiration by field experimentation and soil moisture depletion studies.

7.11 Describe with sketches USWB Class A pan and Sunken screen pan for measuring evaporation.

7.12 Describe the method of estimating consumptive use by Blaney-Criddle and Radiation method.

7.13 Determine the atmospheric pressure and psychrometric constant at an elevation of 1500m.

Ans. P=84.78kPa, $\gamma = 0.056$

7.14 Determine the saturation vapor pressure of a day for maximum and minimum temperatures are 35°C and 20°C respectively.

Ans. e^0 (35°C) =5.62kPa, e^0 (20°C) =3.98kPa

7.15 Calculate the extraterrestrial radiation (R_a) for 15th August at 20°N.

Ans. 15.77 mm/day

7.16 Estimate the consumptive use of crop for cotton at latitude 23°N by using Blaney-Criddle method with the following data.

Month	Mean monthly air temperature,(t), °C
April	29.55
May	31.05
June	28.16
July	27.96
Aug	28.38
Sep	28.92
Oct	25.88
Nov	22.65
Dec	17.75

Ans. 103.53cm

7.17 Estimate ET_0 from following data

Month = June

T_{mean} = 33 °C

n_{mean} = 8 hours/day

Day time wind velocity=moderate (2-5 m/s)

RH_{mean} = High

Place = Mohanpur (23.5°N latitude, 12.5m elevation)

Ans. 5.82 mm/day

7.18 Calculate the PET by Priestly-Taylor method from the following data.

Latitude = 12.5°

Month = June

T_{max} = 36°C

T_{min} = 24°C

Cloud = 10%

Ans. 7.09 *mm / day*

7.19 Calculate the ET_0 by FAO Penman-Monteith method by using the following data of an arbitrary chosen location.

Latitude = 15°N

Month=July

Monthly average daily maximum temperature (T_{max}) = 35°C

Monthly average daily minimum temperature (T_{min}) = 25°C

Monthly average daily vapor pressure (e_a)=2.90kP$_a$

Monthly average daily wind speed measured at 2m height (u_2) = 2.5m/s

Monthly average sunshine duration (n) = 8.0 hours/day

Mean monthly average temperature for July ($T_{month)}$ = 30⁰C
Mean monthly average temperature for June ($T_{month-1}$) = 29⁰C
Potential extraterrestrial radiation (Ra)
Ans.6.0 mm/day

7.20 Select the appropriate answer from the following multiple-choice questions.

1. Potential evapotranspiration is more or equal to evapotranspiration
   a) Crop evapotranspiration
   b) Actual crop evapotranspiration
   c) Unadjusted
   d) Pan evaporation
2. At the early stage of crop maximum contribution to evapotranspiration is
   a) Transpiration b) Evaporation
   c) Percolation d) Leaching
3. Peak period consumptive consumptive use is the average of the highest consumptive use rates of few days usually
   a) 2-10 days b) 6-10 days
   c) 5-15 days d) 10-15 days
4. Irrigation system capacity depends on
   a) Daily consumptive use rate
   b) Seasonal consumptive use rate
   c) Peak period consumptive use rate
   d) Mid-season consumptive use rate
5. 1.0MJm^{-2} day^{-1}is equivalent to
   a) 0.104 mm/m²/day b) 0.208 mm/m²/day
   c) 0.306 mm/m²/day d) 0.408 mm/m²/day
6. Hypothetical crop in the process of estimating evapotranspiration has a fixed surface resistance of
   a) 30s/m b) 50s/m
   c) 60s/m d) 70s/m
7. Latent heat of 5kg water at an air temperature of 20⁰C is
   a) 1.55 MJm^{-2} day^{-1} b) 2.45 MJm^{-2} day^{-1}
   c) 4.10 MJm^{-2} day^{-1} d) 12.25 MJm^{-2} day^{-1}
8. Psychrometric constant at atmospheric pressure 102.5kPa is
   a) 0.066 b) 0.067
   c) 0.068 d) 0.069

Ans:

1. c) 2. b) 3. b) 4. c) 5. d) 6. d) 7. d) 8. c)

7.21 State True or False of the following statements.

1. Crop factor and crop coefficient is synonymous.
2. Plant population is a factor of crop coefficient.
3. Crop factor is always less than the reference crop evapotranspiration.
4. The duration of initial stage of a crop is germination to 15% of ground cover.
5. Soil heat flux is considered negligible.
6. The specific heat at constant pressure on average atmospheric condition is 1.013×10^{-3} $MJm^{-2} day^{-}$.
7. Relative humidity varies much more than the actual vapor pressure in different time of a day.
8. Actual vapor pressure is the saturated vapor pressure at the dew point temperature.
9. The Penman methods require local calibration of the wind to achieve satisfactory results.
10. The radiation methods show good results in arid conditions.

Ans.

1. True 2. True 3. False 4. False 5. True 6. True 7. True
8. True 9. True 10. False

References

Allen, R. G, Pereira, L.S., Raes, D., Smith, M. (1998). "Crop evapotranspiration: Guidelines for computing crop water requirements" FAO Irrigation and drainage paper 56, Rome, Italy.

Blaney, H.F. & Criddle,W.D.(1957). Determining water requirements in irrigated areas from climatological and irrigation data. U.S.Dept. Agr. SCS-TP 96.

Doorenbos J. and Pruitt W.O. (1977). Guidelines for predicting crop water requirement (Revised). FAO Irrigation and Drainage Paper 24. Rome. 143 p.

Majumdar,D.K. (2000). Irrigation Water Management Principles and Practices. Printice Hall of India Pvt.Ltd., New Delhi. p.152,160.

Mehta, V.K. (2006). Estimating evaporation from weather data. Argham/Cornell University.

Michael, A.M. (1985). Irrigation Theory and Practice. Vikas Publishing House Pvt. Ltd., New Delhi.p.523

Rashid, M.H.U., M.S.Islam & D.N.Sharma (1994). Irrigation information package for planning irrigation for diversified crops. Canadian Executing Agency (CEA/Crop Diversification Programme (CDP).

Reddy, G.H.S. & Reddy,T. Yellamanda (1995). Efficient Use of Irrigation Water. Kalyani Publishers,New Delhi. P.104-105.

8

Groundwater Hydraulics & Wells

Hydraulics of wells means the laws that are applicable to understand the flow of water through the formation and certain aquifer properties that determine the discharge of a well. Darcy's law is used in studying groundwater flow though Darcy's law is not valid under certain cases.

8.1 Water Bearing Formation

The earth's crust is called the lithosphere which consists of hard rocks and disintegrated rock materials. The outer parts of the earth's crust are porous and containing partially and fully filled up with water. The surface strata of the earth is partially filled with water and is called as zone of aeration and the zone below this which is completely filled with water is called the zone of saturation. The zone of aeration is divided into (i) the soil water zone (ii) the intermediate zone and (iii) the capillary fringe (Fig.8.1).

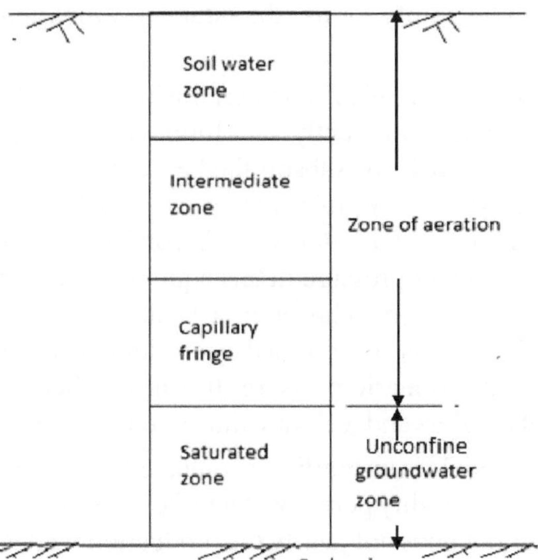

Fig.8.1. Classification of subsurface water

Soil water zone: The zone lying with the ground surface and contains the major root system of the vegetation from which water is lost to the atmosphere as evaporation or evapotranspiration.

Capillary fringe: This is the zone in which water is held up by capillary action. The zone is extended from the saturated zone to the height of the capillary rise.

Intermediate zone: This is the the zone in between the capillary and soil water zone.

Porosity: The portion of the rocks or soil which is not occupied by solid materials is called voids, interstices or pore space. The porosity is a measure of these interstices. This portion is filled up with air or water. It is assumed as the ratio of the volume of interstices (V_v) to the total volume (V) and expressed as percentage.

$$n = \frac{Vv}{V} x100 \quad (8.1)$$

n = porosity (%)

V_v = Volume of water required to fill or saturate the pore spaces (cm^3 or m^3)

V = total volume of rocks or soils (cm^3 or m^3)

Qualitatively the porosity greater than 20% is considered as large, between 5-10% as medium and less than 5% as small. The pore spaces in soils or rocks filled up with water does not always indicate the availability of water for withdrawal due to different physical properties. Clay has more spaces and water than the coarse soil but readily available water in coarse soil is more due to large pores. On the basis of groundwater availability for use the saturated formation are classified as follows.

Aquifer: An aquifer is a saturated formation of earth material which not only contains sufficient water but also can yield significantly. This formation has high water permeability. Among all kinds of rocks or subsoil the best aquifers are unconsolidated deposits of sands and gravels. Aquifer may be classified as the confined and unconfined aquifer. In confined aquifer water is confined by an overlying impermeable layer. Water is under pressure in this aquifer and water will rise above the bottom of overlying layers. This aquifer is also called as artesian aquifer. In unconfined aquifer water is not confined by any impermeable layer. Water in this aquifer is under atmospheric pressure. It is also called as water table aquifer. Different type of aquifers and wells are illustrated in Fig.8.2.

Aquitard: This is a formation in which yield is insignificant compared to aquifer. It is semi pervious in nature and does not readily permit water to flow through it. It may store large quantities of water but does not justify the production well is placed in it. An aquitard is a leaky confining bed which slowly transmits water to or from an adjacent aquifer. Clay, loams and shales are typical aquitards.

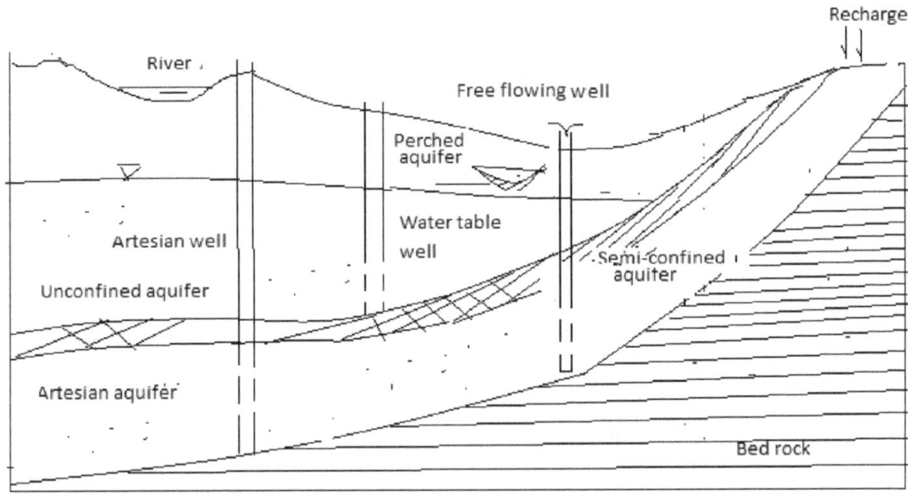

Fig.8.2 Different type of aquifers and wells

Aquiclude: This is a geologic formation practically impermeable to flow of water. Aquiclude may be of high porosity containing huge amount of water but movement is restricted. Clay layer is good example of aquiclude.

Aquifuge: This is a geologic formation which is neither porous nor permeable or interconnected openings to transmit water. Rock formation without any fracture is aquifuge.

Perched aquifer: Perched aquifer is a special type of unconfined aquifer which occurs when a groundwater body is separated from the main groundwater by a comparatively impermeable layer of small areal extent.

Semi-confined or leaky aquifer: It is the completely saturated over or below of which there is an impervious layer. When piezometric head decreases due to pumping from the above or below the impervious layer the water flow takes place vertically to the aquifer from which pumping occurs.

Effluent and influent stream: Effluent stream has the bed lower than the groundwater table. During the period of low flow the stream water level may go down to groundwater level. In such situation the groundwater contributes to the flow of stream. Such streams are called as effluent streams. Opposite to this when bed of the stream is above the groundwater table the stream water percolates to groundwater. Such streams are called as influent streams (Fig.8.3).

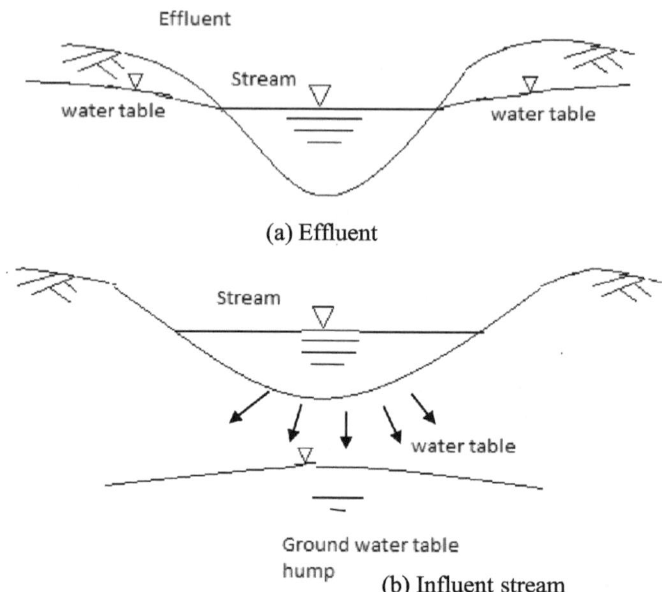

Fig.8.3 Effluent and influent stream

8.2 Aquifer Properties

Aquifer properties are the characteristics or capacity of release the water in the pores and the ability of flow.

Storage in aquifer: In unconfined aquifer water released from the aquifer causes voids in the aquifer but in confined aquifer it remains saturated. The water released or taken in to storage is due to change in aquifer compressibility and water density.

Stress in aquifer: The vertical force per unit area in a confined aquifer is sum of the weight per unit area of the geologic materials and water called as intergranular stress σ_z and pore water stress (p), respectively. Thus,

Total stress = $\sigma_z + p$ (8.2)

If the water is withdrawn, the piezometric head will decrease but the total stress will remain constant. This indicates that intergranular stress increases as proportional to pore water decreases. Thus

$d\sigma_z = \sigma_z + dp$

or, $d\sigma_z + dp = 0$ (8.3)

The increased intergranular stress causes to compaction of the skeleton by reducing thickness of the aquifer and pore volume.

Specific storage and storage coefficients: The volume of water release from per unit area in confined aquifer changes due to aquifer compaction and reduced water density. If water is considered incompressible similar to unconfined aquifer, the mass of a small saturated element is

$$M = \rho n\Delta x\Delta y\Delta z \tag{8.4}$$

Assuming that change of deformation of the aquifer takes place only in vertical direction, the change of mass

$$dM = \rho d(n\Delta z) + n\Delta z dp \tag{8.5}$$

The quantity $dM_1 = \rho d(n\Delta z)$ is the contribution of change in mass (M) to change of pore volume and $dM_2 = n\Delta z dp$ due to change in density of water.

The change in pore volume per unit change in intergranular stress is the pore-volume compressibility (α_p) and defined as

$$\alpha_p = -\frac{1}{n\Delta z}\frac{d(n\Delta z)}{d\sigma_z} = \frac{1}{n\Delta z}\frac{d(n\Delta z)}{dp}$$

$$\therefore dM_1 = \rho\alpha_p n\Delta z dp \tag{8.6}$$

Let V_w be the volume of water. The compressibility of water may be stated as

$$\beta = -\frac{1}{Vw}\frac{dVw}{dp} \tag{8.7}$$

Since the mass is constant,

$$M = \rho V_w = C$$

$$or,\ V_w d\rho + \rho dV_w = 0$$

$$or, d\rho = -\frac{\rho dV_w}{V_w}$$

$$or, \frac{dV_w}{V_w} = -\frac{d\rho}{\rho}$$

From **Eq.8.7**

$$\frac{dV_w}{V_w} = -\beta d\rho$$

$$or, \beta d\rho = \frac{d\rho}{\rho}$$

$$or, d\rho = \rho\beta d\rho \tag{8.8}$$

Therefore, the change of mass per unit area due to change in water compressibility

$$dM_2 = n\Delta z d\rho = n\Delta z \rho \beta dp \tag{8.9}$$

The total change in mass for unit volume of aquifer is

$$\frac{\Delta M}{\Delta x \Delta y \Delta z} = dM_1 + dM_2$$

$$= \rho \alpha_p n \Delta z d_p + n\Delta z \rho \beta dp$$

$$= n\rho\left(\alpha_p + \beta\right) dp \tag{8.10}$$

Hydrologists have more interest in volume of water than mass. If Eq.8.10 is divided by ρ it is become

$$\frac{d\nabla w}{\Delta x \Delta y \Delta z} = n\left(\alpha_{p+} + \beta\right) dp \tag{8.11}$$

In which is the change in water in the element ΔxΔyΔz for both water and aquifer compressibility.

The pressure in the aquifer is not measured directly but it is observed as the pressure head, $h_p = \frac{p}{pg}$ at any point in the aquifer. The change in head

$$dp = \rho g dh_p \tag{8.12}$$

From **Eq.8.11 & 8.12**

$$\frac{d\nabla w}{\Delta x \Delta y \Delta z} = n\rho g(\alpha_p + \beta)$$

$$or, \frac{1}{\Delta x \Delta y \Delta z} \frac{d\nabla w}{dh_p} = n\rho g(\alpha_p + \beta)$$

$$or, S_s = n\rho g(\alpha_p + \beta) \tag{8.13}$$

S_s is called as specific storage. Specific storage by definition is the volume of water released from storage per unit volume of aquifer per unit decline in pressure head. When the specific storage is multiplied by the thickness of the aquifer it is called storage coefficient (S). If the thickness of the aquifer is B, storage coefficient

$$S = S_s B \tag{8.14}$$

Storage coefficient is the volume of water released from a column of unit area per unit decline of pressure head.

Example 8.1 The water and pore volume compressibility of a confined aquifer are 4.8×10^{-11} cm^2/dyne & 4.35×10^{-11} cm^2/dyne, respectively. Estimate specific storage and storage coefficient of the aquifer for the thickness of the aquifer 30m and porosity 0.35.

Solution

Using **Eq.8.13**

Specific storage, $S_s = n\rho g(\alpha_p + \beta)$

$$= \frac{980 dynes}{cm^3}(0.35)(4.35x10^{-11} + 4.8x10^{-11})\frac{cm^2}{dyne}$$

$= 343x9.15x10^{-11}$

$= 3.14x10^{-8}/cm$

Storage coefficient, $S = BS_s$

$= 30mx3.14x10^{-8}$/cm

$= 9.42x10^{-5}$

Example 8.2 The storage coefficient of a 40m thick confined aquifer is $3.5x10^{-4}$ as determined by a pumping test at a location. Estimate the volume of water recovered per km^2 if the pressure head in the aquifer is reduced by 30m and the average volume per km^2 is $2.5x10^7 m^3$.

Solution

$$\text{We have, } S_s = \frac{S}{B} = \frac{3.5x10^{-4}}{40m} = 8.75x10^{-5} / m$$

Volume recovered per km^2 = $8.75x10^{-5}/mx2.5x10^7 m^3 x30m$

$= 65625m^3$

Permeability: It is the measure of ability of transmits water through the soil. The flow of water takes place through the voids in soil. The voids are interconnected. The flow does not take place in straight way but in a zigzag way though the velocity is considered to be in a straight line at an effective velocity. It is measured by the term coefficient of permeability or hydraulic conductivity (K). A medium will have a coefficient of permeability unit length per unit time if unit volume of water is transmitted per unit time through unit cross-sectional area. The unit of it is length per unit time (L/T). K values are given in Table 8.1 for some common aquifer materials.

Table 8.1 Coefficient of permeability of some common aquifer materials

Material	K(cm/s)	K_0(darcys)
A. Granular materials		
1. Clean gravel	1-100	10^3-10^5
2. Clean coarse sand	0.010-0.01	1—10^3
3. Mixed sand	0.005-0.01	10-May
4. Fine sand	0.001-0.05	Jan-50
5. Silty sand	$1x10^{-4}$-$2x10^{-3}$	0.1-2
6. Silt	$1x10^{-5}$-$5x10^{-4}$	0.01-0.3
7. Clay	6-Oct	3-Oct
B. Consolidated materials		
1. Sand stone	10^{-6}-10^{-3}	10^{-3}-1.0
2. Carbonate rock with secondary porosity	10^{-5}-10^{-3}	10^{-3}-1.0
3. Shale	10-Oct	7-Oct
4. Fractured and weathered rock (aquifer)	10^{-6}-10^{-3}	10^{-3}-1.0

Source: Subramania, 2013

Specific yield (S_y) & specific retention (S_r): Specific yield is the ratio of actual volume of water that can be derived by gravity or can be extracted from unit volume of aquifer. The porosity gives a measure of total pore volume and so the maximum volume of water in the soil medium. The portion of the total water total volume which remains in the soil pores due to molecular attraction and surface tension after the removal of water by the force of gravity is called the specific retention (S_r)

Porosity, $n = S_y + S_r$ (8.15)

The prosody and specific yield of some common aquifer materials are given in Table 8.2.

Table 8.2 Porosity and specific yield of different soil materials

Materials	Porosity (%)	Specific yield (%)
Clay	45-55	10-Jan
Sand	30-40	30-Oct
Gravel	30-40	15-30
Sand stone	20-Oct	15-May
Shale	10-Jan	0.5-5
Limestone	10-Jan	0.5-5

Source; V.V.N.Murty

The coefficient of permeability reflects the combined effect of porous media and fluid properties.

Coefficient of permeability increases to greater size of particles, void ratio, degree of saturation and decreases to higher specific surface area, velocity of water and entrapped gases.

The coefficient of permeability can be expressed in the following equations

$$K = cd_m^{\,2}\frac{\gamma}{\mu} \tag{8.16}$$

$$K = \frac{1}{2}\frac{e^2}{1+e}\frac{\gamma}{\mu}\frac{1}{S^2} \tag{8.17}$$

Where,

d_m = Mean particle size of the porous medium

γ = Unit weight of water

μ = Dynamic viscosity of water

c = A shape factor which depends on porosity, packing, shape of gravel and grain size distribution of the porous medium

S = specific surface

From Eq.8.16, $K = cd_m^{\,2}\frac{\rho g}{\mu}$

$$= cd_m^{\,2}\frac{g}{\frac{\mu}{\rho}}$$

$$\text{or, } K_s \propto \frac{1}{\vartheta_s} \tag{8.18}$$

Where, $\vartheta = \frac{\mu}{\rho}$ = kinematic viscosity

When the standard value of coefficient of permeability (K_s) is measured in the laboratoty at the temperature 20°C

$$K_S \propto \frac{1}{\vartheta_S} \tag{8.19}$$

At any other temperature t, coefficient of permeability (K_i) is

$$K_i \propto \frac{1}{\vartheta_i} \tag{8.20}$$

$$\text{From Eq.8.19 \& 8.20, } K_S = K_i\frac{\vartheta_i}{\vartheta_S} \tag{8.21}$$

Where ϑ_S & ϑ_i are kinematic viscosity at 20°C and t°C respectively.

Intrinsic permeability: Coefficient of permeability is the combined properties of the soil medium and the fluid properties. The Eq.8.16 can be rewritten as

$$K = K_0\frac{\gamma}{\mu} \tag{8.22}$$

where $K_0 = cd_m^{\ 2}$

The parameter K_0 is the function of medium only and is called as intrinsic permeability. The dimension of K_0 is L^2. It is also expressed in darcy. One darcy is equal to $9.87 \times 10^{-13} m^2$.

Transmissibility: It is the multiplication of hydraulic conductivity (K) of the aquifer with its thickness B for unit width of aquifer.

$$T = KB \tag{8.23}$$

The unit of T is L^2/T.

Hydraulic resistance: Hydraulic resistance or the reciprocal leakage is the property of a semi-confined aquifer. It is the character which indicates the resistance of flow through semi-impervious layer. It is defined by the ratio of thickness of saturated zone (B) to the hydraulic conductivity (K) of the semi-impervious layer i.e. B/K and designated by c.

$$C = \frac{F^2}{KB} \tag{8.24}$$

Where,

c = hydraulic resistance (days)

F = leakage factor, m

K = hydraulic conductivity, m/day

B = thickness of horizontal saturated zone between the semi-impervious and impermeable layer, m

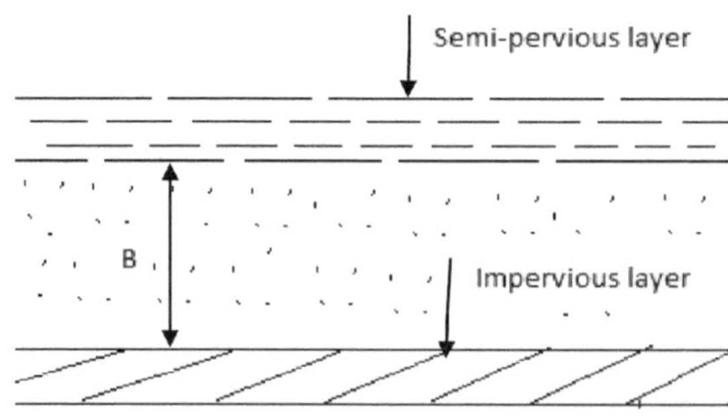

Fig.8.4 Semi-impervious aquifer

The value of c usually varies between 100 to 10, 00,000 minutes.

Leakage factor: Leakage factor determines the origin of withdrawn of water from below or above the semi-impervious layer. It may be expressed as

$$b = \sqrt{KBc} \tag{8.25}$$

Higher value of leakage factor (b) is an indication of high resistance of impervious layer to flow compared to the aquifer. The dimension of b is length, L.

Surface tension: Surface tension is a molecular phenomenon of forces. These are two types (i) adhesion or adhesive forces and (ii) cohesion or cohesive forces. Adhesion is the force of attraction between molecules of unlike substances. It is different in different pair of substances. Attractive forces between two molecules rapidly decrease with the distance between them. Adhesive force is more in solid, less in liquids and the least in gases at ordinary temperature and pressure.

The surface of a liquid behaves like a skin or membrane due to influence of adhesive and cohesive force. If this membrane is cut a force is applied to bring the separated portion together. This force is called surface tension and is proportional to length of the cut. Surface tension may be represented by the symbol σ and is defined by the force in dynes acting along the surface of a liquid right angle to any line of 1cm length. Thus, the unit of surface tension is dynes/cm. Surface tension of water at different condition is given in Table 8.3.

Table 8.3 Surface tension of water at different temperature, vapour pressure in saturated air and water density

Temperature, ^{0}C	Vapour pressure of water (mm of Hg)	Mass of water vapour in saturated air (g/m^3)	Surface tension of water (dynes/cm)
-20	0.776	0.892	-
-10	1.65	2.154	-
0	4.579	4.835	75.6
4	6.101	6.32	75
5	6.543	6.761	74.9
10	9.209	9.33	74.2
15	12.788	12.712	73.5
20	17.535	17.118	72.7
25	23.756	22.796	72
30	31.824	30.039	71.2
40	55.324	50.5	69.6
50	92.51	-	67.9
75	289.1	-	63.5
100	760	-	58.9

1 atmosphere=1036 cm of water or 76.39 cm of mercury

1 bar=10^6 dynes/cm^2=1023 cm of water

1 millibar=1/1000 bar

Bulk pore velocity: The velocity of flow in porous media calculates by Darcy's law is not actual, it is apparent. The flow takes place through the irregular path of porous media and the velocity varies point to point. The bulk pore velocity represents the actual velocity of travel and expressed in

$$V_{\propto} = \frac{V}{n} \tag{8.26}$$

Where, n = porosity

V_a = pore velocity

Viscosity: In an open channel flow takes place over a stationary surface. If the depth of flow is divided in to a number of layers, it appears that the bottom layers of the adjacent layers try to retard the flow of the above layers. Thus the relative motion of all the layers is to some extent lost due to internal friction. A continuous force is required to apply for flow in the channel. In the open channel this force is the gravity. The influence of retardness of flow by the lower layer to the immediately upper layer is called viscosity of fluid. Fluid with large viscosity retards the motion due to its strong intermolecular force. Opposite to it, a fluid with low viscosity flows easily. Gases have negligible viscosity. Viscosity of liquid decreases with the increase of temperature but the viscosity increases with temperature in gases.

When two layers of fluid in flow at some distance apart the viscous force may be expressed as

$$F = -\eta A \frac{dv}{dx} \tag{8.27}$$

Where,

F = viscous force, dynes

η = coefficient of viscosity of the fluid

A = surface area, cm^2

dv = difference in velocity between the adjacent layers of fluid, cm/s

dx = distance between the adjacent layers, cm

For unit surface area and velocity gradient

$$F = -\eta$$

This enables the coefficient of viscosity to define as the tangential force per unit area required to maintain unit difference of velocity between two adjacent layers of fluid in motion unit distance apart. The area in cm^2, velocity in cm/s and distance in cm, the corresponding unit force for viscosity is 1dyne-s/cm^2 and is called poise after the name of J. L. M. Poisseuille (1844). This value of dynamic viscosity for different temperature of water may be obtained by the following equation.

$$\mu = \frac{0.0179}{0 + 0.03368T + 0.000221T^2}$$

Where, μ = dynamic viscosity of water dyne-s/cm^2

T=temperature in ^{0}C.

One-hundredth of poise is called centipoises. The viscosity of water at 20^{0}C is almost 1 centipoise. Dynamic and kinematic viscosity is related as below

$$\vartheta = \frac{\mu}{\rho}$$

Where,

ϑ = kinematic viscosity of water, cm^2/s

ρ = mass density of water, g/cm^3

Permeameter: Soil water permeameter is used to measure the coefficient of permeability. Permeameter may be constant head or falling head. In the permeameter water percolates through the soil of cross-sectional area A and length L. If a constant head of H is maintained in the permeameter as shown in Fig.8.5(a), the discharge is

$$Q = AV = AK\frac{H}{L}$$

$$or, K = \frac{Q}{A}.\frac{L}{H} \tag{8.28}$$

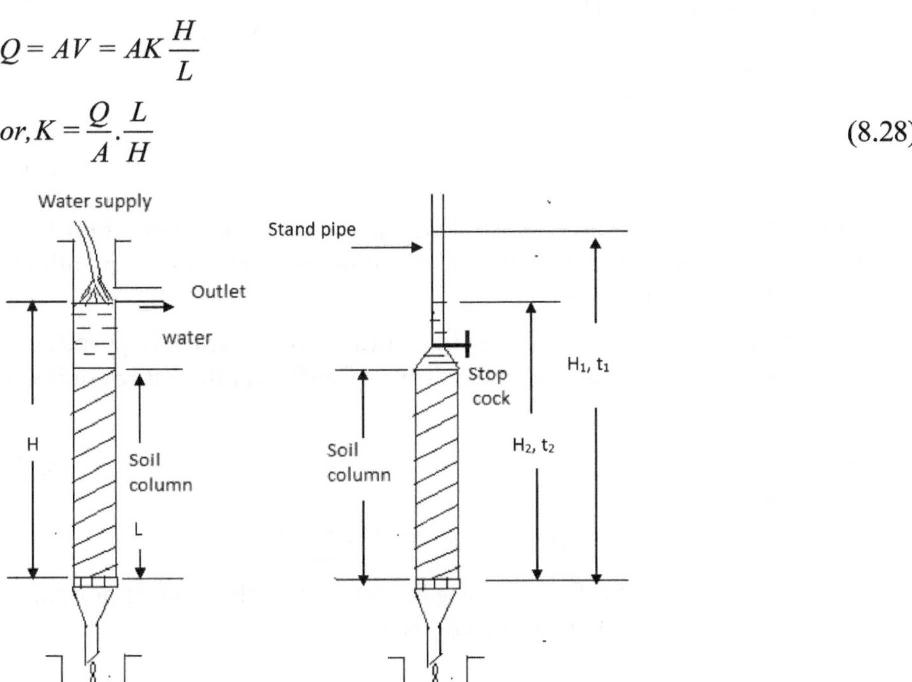

Fig.8.5 Soil permeameter

In a variable head permeameter as shown in Fig.8.5(b), let

H = height of water in the stand pipe at any time, t

H = H_1 when $t=t_1$

H = H_2 when $t=t_2$

a = cross-sectional are of the stand pipe

u = velocity of flow in stand pipe = $-\frac{dH}{dt}$

u_1 = velocity of flow in soil column = $Ki = K\frac{H}{L}$

The discharge Q through the stand pipe and through the soil column is same. Therefore,

$Q = au = Au_1$

$$\text{or, } -a\frac{dH}{dt} = AK\frac{H}{L}$$

$$\text{or, } -aL\frac{dH}{H} = AKdt$$

Integrating between H_1 to H_2 and t_1 to t_2

$$aL\ln\frac{H_1}{H_2} = AK\left(t_2 - t_1\right)$$

$$\therefore K = \frac{aL}{A\left(t_2 - t_1\right)} ln\frac{H_1}{H_2} \tag{8.29}$$

Stratification: An aquifer may be stratified and each stratification may have different permeability. There may be two situations of flow. Flow is parallel to the stratification or perpendicular to it.

i) Flow parallel to stratification (Fig. 8.6(a)): When flow is parallel to stratification the equivalent permeability of flow K_e for the entire aquifer is

$$K_e = \frac{\sum_1^n K_i B_i}{\sum_1^n B_i} \tag{8.30}$$

The transmissibility of the aquifer, $T = K_e \sum B_i = \sum_i^n K_i B_i$

ii) Flow is normal to stratification (Fig.8.6(b)): when the flow is normal to stratification the equivalent permeability K_e is

$$K_e = \frac{\sum_1^n L_i}{\sum_1^n \frac{L_i}{K_i}} \tag{8.31}$$

Where, $L = \sum_1^n L_i$ = length of the aquifer

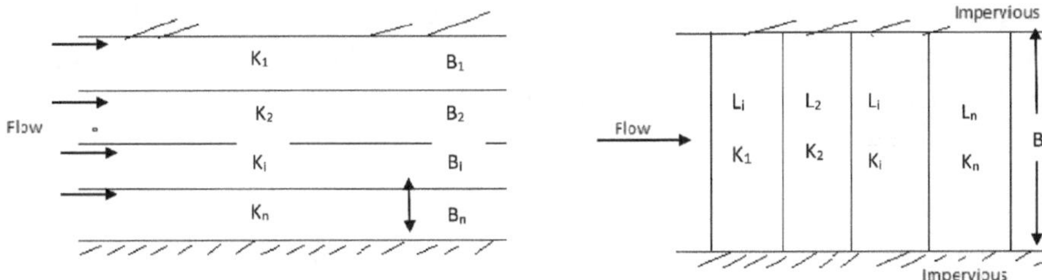

Fig. 8.6 Flow through stratification

Example 8.3. A tracer records 12 hour to travel one observation well to another at 65m apart and water level difference of 0.4m. Find the coefficient of permeability, intrinsic permeability and the Reynolds number of flow if the aquifer has the characteristics of mean particle size diameter 0.18 mm, soil porosity 0.25 and viscosity of water 0.01cm²/s.

Solution:

Tracer velocity, $V_t = \frac{60m}{12h} = \frac{60x100\text{cm}}{12x3600s} = \frac{0.15cm}{s}$

Apparent velocity, $V = V_t xn = 0.15x0.25 = 0.0376$ cm/s

Coefficient of permeability, $K = \frac{V}{i} = \frac{0.0376cm/s}{\frac{0.4m}{65m}} = 6.11cm/s$

Intrinsic permeability, $K_0 = \frac{Kv}{g} = \frac{6.11cm/sx0.01cm^2/s}{981cm/s^2} = 6.23x10^{-5}cm^2$

Example 8.4 The oil flows at the rate of 1l/s in a pipe of 40mm diameter. Calculate the flow of oil in the pipe and the Reynolds number if density of oil is 920kg/m³ and the dynamic viscosity of the oil is 40mPas.

Solution:

$$\text{Velocity of flow, } V = \frac{Q}{A} = \frac{1l/s}{\pi\frac{D^2}{4}} = \frac{1x10^{-3}m^3/s}{\pi\frac{(\frac{40}{1000})^2}{4}m^2} = 0.79m/s$$

$$\text{Reynolds number, } R_e = \frac{\rho VD}{\vartheta}$$

$$= \frac{920kg/m^3x0.79\text{m}/sx40x10^{-3}m}{\frac{40mx1kg}{m^2}.s}$$

$$= 726.8$$

Example 8.5 Calculate the velocity of flow of water through a circular pipe of diameter 400mm. The density of water is 1g/cm³ and the dynamic viscosity of water 1 centipoise, Reynolds number 2000.

Solution:

$$R_e = \frac{\rho VD}{\vartheta}$$

$$or, 2000 = \frac{1g/cm^3 x V x \left(\frac{400}{10}\right) cm}{1 dynes\text{-}s/cm^2}$$

$$\therefore V = \frac{2000}{1.0x4.0x100} = 5cm/s$$

Example 8.6 In a constant head permeameter of diameter 10cm, soil column length 15cm, water height 20cm collects 500cm³ water in 70 seconds. Calculate the permeability of the soil.

Solution:

Area of cross-section of soil column, $A = \pi\frac{D^2}{4} = \pi\frac{10^2}{4} = 78.54cm^2$

Hydraulic gradient, $\frac{H}{L} = \frac{12}{20} = 0.6$

Discharge, $Q = \frac{Volume\ of\ water\ collected}{time} = \frac{500cm^3}{70s} = 7.14cm/s$

Example 8.7 In a variable head permeameter of stand pipe diameter 5mm recorded 0.5m drop of water from the initial head of 1.2m in 2.5 hours. The soil column was 180mm long and diameter 100mm. What is the coefficient of permeability of soil?

Solution:

Cross-sectional area of stand pipe, $a = \frac{\pi D^2}{4} = \frac{\pi(0.5)^2}{4} = 0.196cm^2$

Length of soil column, L=180 mm=18 cm

Cross-sectional area of soil column, $A = \frac{\pi D^2}{4} = \frac{\pi(100/10)^2}{4} = 78.54cm^2$

Time $(t_2\text{-}t_1)$=2.5h=2.5x3600=9000s

From Eq.8.29, $K = \frac{aL}{A(t_2 - t_1)} ln\frac{H_1}{H_2} = \frac{0.196x18}{18.54x9000} ln\frac{120}{70} = 1.4x10^{-5} cm/s$

Example 8.8 The permeability of a stratified soil deposits are $7.5x10^{-4}$, $3.6x10^{-4}$ and $6.2x10^{-4}$cm/s and thickness of the layers are 5.6, 4.5 and 8.0m respectively. Calculate the average permeability of the soil perpendicular to the deposit.

Solution:

Thickness of the layers, L_1, L_2, L3=5.6m, 4.5m, 8.0m

Permeability of the layers K_1, K_2, K_3= $7.5x10^{-4}$cm/s, $3.6x10^{-4}$cm/s, $6.2x10^{-4}$cm/s

Using Eq.8.31 the average permeability

$$K_e = \frac{\sum_1^n Li}{\sum_1^n Ki} = \frac{5.6m+,4.5m+8.0m}{\frac{5.6m}{7.5x10^{-4} cm/s}+\frac{4.5m}{3.6x10^{-4} cm/s}+\frac{8.0m}{6.2x10^{-4} cm/s}}$$

$$= \frac{1810cm}{\frac{560cm}{7.5x10^{-4} cm/s}+\frac{450cm}{3.6x10^{-4} cm/s}+\frac{800ccm}{6.2x10^{-4} cm/s}}$$

$$= \frac{1810cm}{(74.67+125+129.03)10^4 s} = \frac{1810cm}{328.7x10^4} = 5.51x10^{-4} \text{cm/s}$$

Example 8.9 A saturated soil sample of volume 2 liter weighing 4.5kg was measured 4.02kg when the water in it was completely drained out. On drying the weight of the soil sample was found 3.98kg. Assuming unit weight of water 1000kg/m^3, find the specific yield and particle density of the solids.

Solution:

Weight of drain out water = 4.50-4.02=0.48kg

Volume of drain out water=0.48kg/1000kg/m^3=$4.8x10^{-4}$m^3

$$\text{Specific yield, } S_y = \frac{\text{Volume of water drained out}}{\text{Volume of the soil sample}} = \frac{4.8x10^{-4} m^3}{2x10^{-3} m^3} = 0.24$$

Actual weight of water in the sample=4.50-3.98=0.52g

Actual volume of water in the sample=0.52kg/1000kg/m^3=$5.2x10^{-4}$m^3

Volume of solids in the sample=$2x10^{-3}$-$5.2x10^{-4}$=$1.48x10^{-3}$m^3

$$\text{Density of solids} = \frac{\text{Weight of the soilds}}{\text{Volume of the soilds}} = \frac{3.98\text{kg}}{1.48x10^{-3} \text{m}^3} = 2.69x10^3 \text{kg}/m^3$$

Example 8.10 A pipe of diameter 1.0m in a reservoir got clogged for some length by sediment. The sediments were of different types with the coefficient of permeability of 10m/day at the upstream 150m length, 25m/day at the middle 100m length and 2.0m/day at the downstream unknown length. Estimate this unknown length of the pipe if there was a head difference of 20m between the upstream and downstream of clogged length of pipe and with the seepage flow of 0.5m³/day.

Solution:

Let x be the unknown length of the clogged pipe.

Equivalent coefficient of permeability, $K_e = \dfrac{\sum_1^n L_i}{\sum_1^n \dfrac{L_i}{K_i}}$

$$= \frac{150+100+x}{\frac{150}{10}+\frac{100}{25}+\frac{x}{2}} = \frac{250}{15+4+0.5x} = \frac{250+x}{19+0.5x}$$

$$Q = AV = \frac{\pi D^2}{4} K_e i = \frac{\pi 1.0^2}{4} \times \frac{250+x}{19+0.5x} \times \frac{20}{250+x}$$

$$or, 0.5m^3/day = \frac{0.785x20}{19+0.5x} = \frac{15.7}{19+0.5x}$$

$or,\ 9.5 + 0.25x = 15.7$

$$\therefore\ x = \frac{6.2}{0.25} = 24.8m$$

Example 8.11 A watershed of 100km² is underlain by an unconfined aquifer having hydraulic conductivity of 15m/day and specific yield 0f 0.20. If 30 million m³ of water is pumped from this aquifer through uniformly distributed wells, the average drop of water table over the watershed in meter will be (GATE, 2017)

a) 1.50 b) 0.75

c) 0.06 d) 6.00

Solution:

Watershed area, A=100km²=100x1000²m²

Volume of water pumped =30million m³=30x10⁶m³

Net depth of water pumped $= \dfrac{100\times1000\times1000}{30\times10^6} = 0.3m$

Average drop of water $\dfrac{0.3m}{sp.yield} = \dfrac{0.3}{0.2} = 1.5m$

Example 8.12 An unconfined aquifer covering an area of 50 ha has a hydraulic conductivity of 20m/day and specific yield 12%. After a significant rainfall event, the water table rises from 17m to 14.5m below the ground level. Assuming no obstruction and outflow of groundwater during the recharge period, the amount of groundwater recharge contributed by the rainfall in m^3 is ..... (GATE, 2015)

Solution:

Area of aquifer, A=50ha

Hydraulic conductivity, K=20m/day

Specific yield, S_y=12%

Water table rise=17m-14.5=2.5m

Net depth of water rise=2.5mx0.12=0.3m

Amount of ground water recharge=5hax0.3m=50x10000x0.3=150000m^3

8.3 Equation of Motion

Let us consider velocity of flow u, v, w in the Cartesian coordinate direction x, y, z respectively through an aquifer elemental prism as shown in Fig.8.6. The equation of continuity of fluid flow or rate of change of mass flow through three pairs of faces is

$$\frac{\partial}{\partial t}(\Delta m) = -\left[\frac{\partial}{\partial x}(\rho\mu) + \frac{\partial}{\partial y}(\rho v) + \frac{\partial}{\partial z}(\rho z)\right] \quad (8.32)$$

Using Eq.8.10 for specific storage (S_s)

$$\frac{dM}{\Delta x \Delta y \Delta z} = n\rho(\alpha_p + \beta)dp$$

$$or, \frac{dM}{\Delta x \Delta y \Delta z dh} = ng(\alpha_p + \beta), p = \rho gh \;\&\; dp = \rho gdh$$

$$or, \frac{dM}{\Delta x \Delta y \Delta z} = S_s \rho dh \quad (8.33)$$

The term $\frac{dM}{\Delta x \Delta y \Delta z}$ is the change of mass for volume. This can be taken as change of mass (Δm) for unit volume

$$\therefore \Delta m = S_s \rho dh \quad (8.34)$$

$$\rho u + \frac{\partial}{\partial x}(\rho u)\Delta x$$

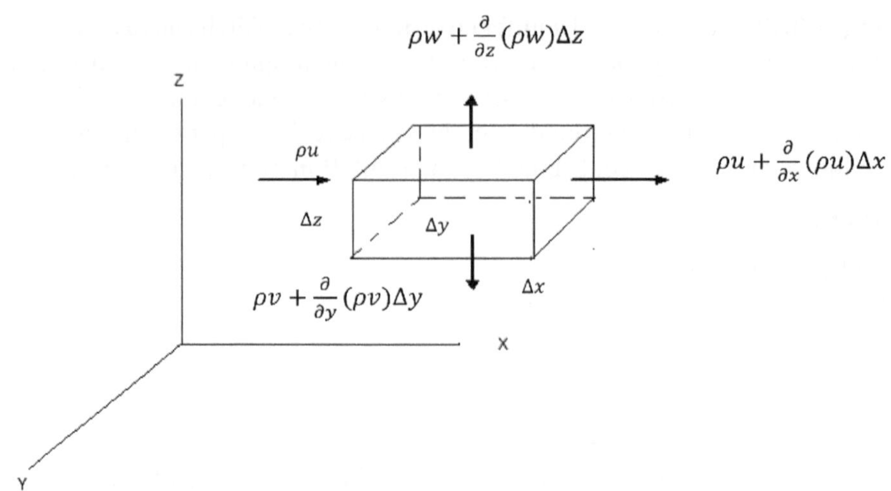

Fig. 8.7 Ground water flow through an elemental prism

Differentiating Eq.8.34 with respect to time and taking the limit of volume ΔV approaches to zero

$$\frac{\partial}{\partial t}(\Delta m) = S_s \rho \frac{dh}{dt} \tag{8.35}$$

Assuming the aquifer is isotropic with permeability K, the Darcy's equation for x, y, z directions are

$$u = -K\frac{\partial h}{\partial x^2}, v = -K\frac{\partial h}{\partial y^2}, w = -\frac{\partial h}{\partial z} \tag{8.36}$$

Using **Eq.8.35** the various terms of right hand side of Eq.8.32 becomes

$$\frac{\partial}{\partial x}(\rho\mu) = \rho\frac{\partial u}{\partial x} + u\frac{\partial \rho}{\partial x} = -K\rho\frac{\partial^2 h}{\partial x} - K\frac{\partial h}{\partial x}\rho^2\beta g\frac{\partial h}{\partial x}$$

$$= -K\rho\frac{\partial^2 h}{\partial x^2} - K\rho^2\beta g\left(\frac{\partial h}{\partial x}\right)^2$$

Similarly, $\frac{\partial}{\partial y}(\rho v) = \rho\frac{\partial v}{\partial y} + v\frac{\partial \rho}{\partial y} = -K\rho\frac{\partial^2 h}{\partial y^2} - K\rho^2\beta g\left(\frac{\partial h}{\partial y}\right)^2$

$$\frac{\partial}{\partial y}(\rho w) = \rho\frac{\partial w}{\partial z} + w\frac{\partial \rho}{\partial z} = -K\rho\frac{\partial^2 h}{\partial z^2} - K\frac{\partial h}{\partial z}\left(\frac{\rho\beta dp}{\partial z}\right),$$

We know, $h = \frac{p}{\rho g} + z$

$$or, \frac{\partial h}{\partial z} = \frac{\mathrm{dp}}{\rho g dz} + 1$$

$$or, dp = \left(\frac{\partial h}{\partial z} - 1\right)\rho g dz$$

$$\therefore \frac{\partial}{\partial y}(\rho w) = -K\rho\frac{\partial^2 h}{\partial z^2} - K\frac{\partial h}{\partial z}\left(\frac{\rho\beta\left(\frac{\partial h}{\partial z} - 1\right)\rho g dz}{\partial z}\right)$$

$$= -K\rho\frac{\partial^2 h}{\partial z^2} - K\rho^2\beta g\left[\left(\frac{\partial h}{\partial z}\right)^2 - \frac{\partial h}{\partial z}\right]$$

Using Eq.8.35 and the above, the Eq.8.32 can be written as

$$K\rho\left[\frac{\partial^2 \mathrm{h}}{\partial \mathrm{x\,x}^2} + \frac{\partial^2 h}{\partial \mathrm{y}^2} + \frac{\partial^2 h}{\partial \mathrm{y\,z}^2}\right] + K\rho^2\beta g\left[\left(\frac{\partial h}{\partial \mathrm{x}}\right)^2 + \left(\frac{\partial h}{\partial \mathrm{y}}\right)^2 + \left(\frac{\partial h}{\partial \mathrm{y}}\right)^2 - \frac{\partial h}{\partial z}\right] = S_s\rho\frac{\partial h}{\partial t} \quad (8.37)$$

The second term of left hand side negligible especially when $\frac{\partial h}{\partial x} \ll 1$ Thus Eq.8.38 is rearranged as

$$\frac{\partial^2 h}{\partial x^2} + \frac{\partial^2 h}{\partial y^2} + \frac{\partial^2 h}{\partial z^2} = \frac{S_s}{K}\rho\frac{\partial h}{\partial t} \quad (8.38)$$

Since $S_s B$ = S, KB = T, *and* $\nabla^2 \mathrm{h} = \left(\frac{\partial^2 h}{\partial x^2} + \frac{\partial^2 h}{\partial y^2} + \frac{\partial^2 h}{\partial z^2}\right)$, Eq.8.38 can be rewritten as

$$\nabla^2 h = \frac{S}{T}\frac{\partial h}{\partial t} \quad (8.39)$$

Eq.8.39 is the basic differential equation for groundwater flow in a homogeneous isotropic confined aquifer.

In a steady state flow condition the term $\frac{\partial h}{\partial t}$ will not exist and Eq.8.39 converts to

$$\nabla^2 h = 0 \quad (8.40)$$

The Eq.8.40 is known as Laplace equation. This is the fundamental equation used in analyzing the flow problems.

(b) Unconfined flow

The flow through the unconfined aquifer is analyzed with the assumptions of Dupuit (1863) which states that (i) the water table or free surface is slightly inclined and the streamlines may be assumed to be horizontal in all sections and equipotentials are vertical, (ii) slope of free surface is equal to hydraulic gradient to any depth.

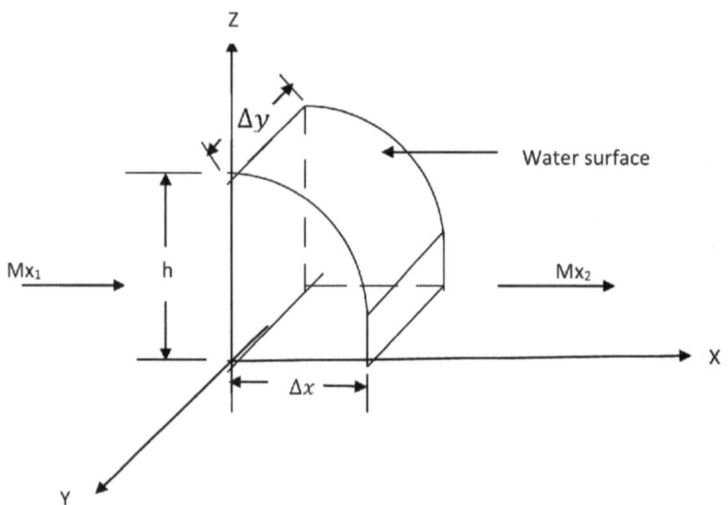

Fig. 8.8 Ground water flow in unconfined aquifer

Let us consider a prism of fluid with the boundary of water surface as shown in **Fig. 8.8**. Mass of water entering to this prism to x direction, $Mx_1 = \rho V_x h \Delta y$

Mass of water leaving the prism to x direction,

$$Mx_2 = \rho V_x h\Delta y + \frac{\partial}{\partial x}(\rho V_x h\Delta y)\Delta x$$

Net mass of water change as the prism at inflow and outflow in x direction

$$Mx_1 - Mx_2 = \Delta M_x = -\frac{\partial}{\partial x}(\rho V_x h\Delta y)\Delta x$$

Similarly the mass of water change due to flow in y direction

$$\Delta M_y = -\frac{\partial}{\partial y}(\rho V_y h\Delta x)\Delta y$$

There is no inflow or outflow in the z direction. For steady incompressible flow there should not be any change of mass in the prism. Therefore,

$$\Delta M_x + \Delta M_y = 0 \tag{8.41}$$

Assembling ΔM_x and ΔM_y

$$-\frac{\partial}{\partial x}(\rho V_x h\Delta y)\Delta x - \frac{\partial}{\partial y}\left(\rho V_y h\Delta x\right)\Delta y = 0$$

$$or, \frac{\partial}{\partial x}(V_x h) + \frac{\partial}{\partial y}\left(V_y h\right) = 0 \tag{8.42}$$

Using Darcy's law, $V_x A = -K\frac{\partial h}{\partial x^2}, V_y = -K\frac{\partial h}{\partial y}$

Eq.8.42 becomes

$$\frac{\partial}{\partial x}\left(-Kh\frac{\partial h}{\partial x}\right) + \frac{\partial}{\partial y}\left(-Kh\frac{\partial h}{\partial y}\right) = 0$$

$$\text{Or,} \frac{\partial}{\partial x}\left(h\frac{\partial h}{\partial x}\right) + \frac{\partial}{\partial y}\left(h\frac{\partial h}{\partial y}\right) = 0$$

$$\text{Or,} \frac{1}{2}\left[\frac{\partial(h^2)}{\partial x}\right] + \frac{1}{2}\left[\frac{\partial(h^2)}{\partial y}\right] = 0$$

$$\text{Or,} \frac{\partial^2 h^2}{\partial x^2} + \frac{\partial^2 h^2}{\partial y^2} = 0$$

$$\text{Or, } \Delta^2 h^2 = 0 \tag{8.43}$$

Eq.8.43 is the Laplace equation in h^2 for flow in unconfined aquifer.

(c) Unconfined flow with recharge

Let R be the steady recharge rate from the ground surface to the aquifer. In consideration to Fig.8.9 there will an additional contribution of flow, say ΔM_2 through the elemental prism of surface area $\Delta z\Delta y$ in the vertical (z) direction. In the steady incompressible flow the continuity equation for the flow is

$$\Delta M_x + \Delta M_y + \Delta M_2 = 0$$

$$\text{i.e.} -\frac{\partial}{\partial x}(\rho V_x h\Delta y)\Delta x - \frac{\partial}{\partial y}(\rho V_y h\Delta x)\Delta y + \rho R\Delta x\Delta y = 0$$

$$\text{or,} -\frac{\partial}{\partial x}\left[\rho\left(-K\frac{\partial h}{\partial x}\right)h\Delta x\Delta y\right] - \frac{\partial}{\partial y}\left[\rho\left(-K\frac{\partial h}{\partial y}\right)h\Delta x\Delta y\right] = 0$$

$$or, \frac{\partial}{\partial x}\left(h\frac{\partial h}{\partial x}\right) + \frac{\partial}{\partial y}\left(h\frac{\partial h}{\partial y}\right) = -\frac{R}{K}$$

$$or, \frac{1}{2}\frac{\partial}{\partial x}\left(\frac{\partial h^2}{\partial x}\right)+\frac{1}{2}\frac{\partial}{\partial y}\left(\text{h}\frac{\partial h^2}{\partial y}\right)=-\frac{R}{K}$$

$$or, \frac{\partial^2 h^2}{\partial x^2}+\frac{\partial^2 h^2}{\partial y^2}=-\frac{2R}{K} \qquad (8.44)$$

Eq.8.44 is the basic differential equation for unconfined flow with the constant recharge to ground water.

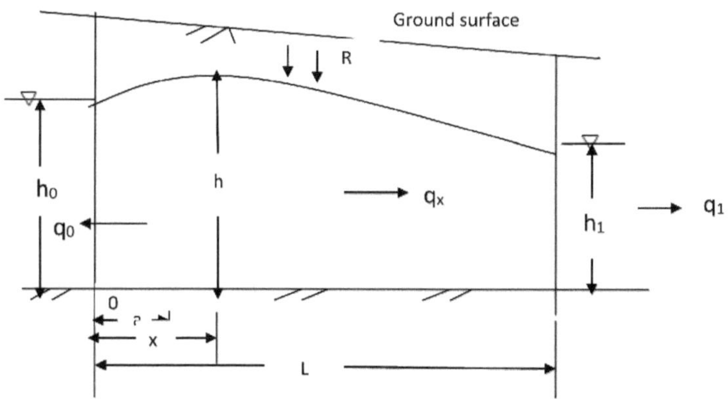

Fig. 8.9 Flow in unconfined aquifer with recharge

i) One dimensional flow with recharge

As shown in Fig.8.9 in an unconfined aquifer there is a constant rate of recharge to the ground water in the aquifer section with a difference of head.

From Eq.8.44 for one dimensional horizontal flow

$$\frac{\partial^2 h^2}{\partial \text{x}^2}=-\frac{2R}{K} \qquad (8.45)$$

Integrating with respect to x

$$h^2=-\frac{R}{K}x^2+C_1x+C_2 \qquad (8.46)$$

where C_1 and C_2 are constants of integration. The boundary condition:

i) At x=0, $h=h_0$

Putting x=0 in Eq.8.46, $C_2=h_0^2$

ii) At x=L, $h=h_1$, Eq.8.46 becomes

$$h_1^2=-\frac{R}{K}L^2+C_1L+C_2h_0^2$$

$$C_1 = \frac{h_1^2 - h_0^2 + \frac{R}{K}L^2 + C_1 L}{L}$$

Putting the values of C_1 and C_2 in Eq.8.46

$$h^2 = -\frac{R}{K}x^2 - \frac{\left(h_0^2 - h_1^2 - \frac{R}{K}L^2\right)}{L}x + h_0^2 \tag{8.47}$$

Eq.8.47 is the equation of ellipse. Due to recharge head of water h will be in general above h_0. It will be maximum at x=a, and will decline thereafter to maximum fall of h_1 at x=L.

Differentiating Eq.8.47 with respect to x

$$2h\frac{dh}{dx} = -\frac{R}{K}2x - \frac{\left(h_0^2 - h_1^2 - \frac{R}{K}L^2\right)}{L}$$

$$or, h\frac{\partial h}{\partial x} = -\frac{R}{K}x - \frac{\left(h_0^2 - h_1^2 - \frac{R}{K}L^2\right)}{2L} \tag{8.48}$$

When, $\frac{dh}{dx}$ =0, x = a

$$or, 0 = -\frac{R}{K}2a - \frac{\left(h_0^2 - h_1^2 - \frac{R}{K}L^2\right)}{L}$$

$$\therefore \frac{R}{K}2a = \frac{R}{K}\frac{L^2}{L} - \frac{\left(h_0^2 - h_1^2\right)}{L}$$

$$or, a = \frac{L}{2} - \frac{K}{R}\frac{\left(h_0^2 - h_1^2\right)}{2L} \tag{8.49}$$

The location at the length x=x is called the water divide. The flow of water left of the water divide will be towards upstream water body and right of the divide towards downstream water body. The discharge of per unit width of aquifer at any length x is

$$q_x = -Kh\frac{\partial h}{\partial x}$$

Using Eq.8.48

$$q_x = -K\left[-\frac{R}{K}x - \frac{\left(h_0^2 - h_1^2 - \frac{R}{K}L^2\right)}{2L}\right]$$

$$= R\left(x - \frac{1}{2}\right) + \frac{K}{2L}\left(h_0^2 - h_1^2\right) \tag{8.50}$$

The discharge q_x will vary depending on h at different length of x. At the upstream when x=0

$$q_x = q_0 = -\frac{RL}{2} + \frac{K}{2L}\left(h_0^2 - h_1^2\right) \tag{8.51}$$

At the downstream when x=L

$$q_x = q_1 = \frac{RL}{2} + \frac{K}{2L}\left(h_0^2 - h_1^2\right)$$

$$= RL + q_0 \tag{8.52}$$

ii) One dimensional flow without recharge

In a steady one dimensional flow without recharge (R=0) in an unconfined aquifer (Fig.8.10) the Eq.8.47

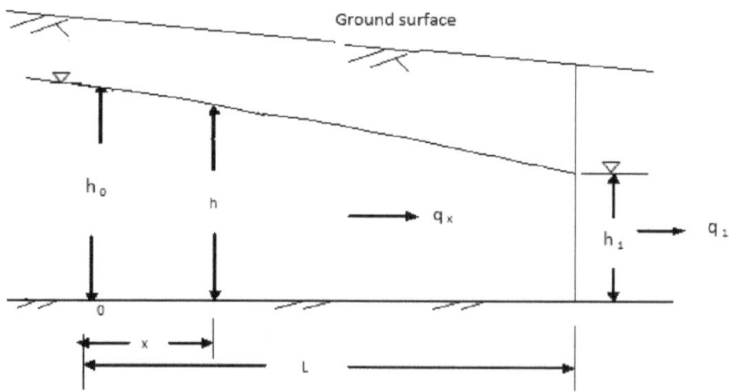

Fig.8.10 Flow in unconfined aquifer without recharge

$$h^2 = -\frac{R}{K}x^2 - \frac{\left(h_0^2 - h_1^2 - \frac{R}{K}L^2\right)}{L}x + h_0^2$$

$$or,\ h^2 - h_0^2 = \frac{\left(h_1^2 - h_0^2\right)}{L}x \tag{8.53}$$

This is an equation of parabola. Differentiating Eq.8.53 with respect to x

$$2h\frac{\partial h}{\partial x}=\frac{\left(h_1^2-h_0^2\right)}{L}$$

The discharge per unit width of aquifer

$$q=-Kh\frac{\partial h}{\partial x}=\frac{\left(h_0^2-h_1^2\right)}{2L}K \tag{8.54}$$

Example 8.13 In an unconfined aquifer ground water flow water table heights are 15m and 10m at upstream and downstream respectively over the horizontal impervious layer at 2 kilometer distance apart. Estimate the per unit width aquifer discharge to downstream with no recharge and with a recharge of 0.0025m/day assuming hydraulic conductivity of the aquifer 30m/day.

Solution:

h_0=15m, h_1=10m, L=2Km=2000m, R=0.0025m/day, K=30m/day

When there is no recharge the discharge per unit width of aquifer (Eq.8.52)

$$q=\frac{\left(h_0^2-h_1^2\right)}{2L}K$$

$$=\frac{(15^2-10^2)}{2x2000}x30=\frac{125x30}{4000}=0.94m^3\ /\ day$$

When there is recharge, the discharge per unit width of aquifer (Eq.8.52),

$q_x = RL+q_0$

Using **Eq.8.53**, $q_0=-\frac{RL}{2}+\frac{K}{2L}\left(h_0^2-h_1^2\right)$

$$=-\frac{0.0025x2000}{2}+\frac{39}{2x2000}\left(15^2-10^2\right)$$

= - 2.5 + 0.94 = 1.56m^3/day

$q_x = q_1$ = RL + q_0 = 0.0025x2000 - 1.56

= 5.0 - 1.56 = 3.44m^3/day

Example 8.14 Using Example 8.13 find the location and height of the water table divide.

Solution:

Using Eq.8.49 the location of the water table divide

$$a=\frac{L}{2}-\frac{K}{R}\frac{\left(h_0^2-h_1^2\right)}{2L}$$

$$=\frac{2000}{2}-\frac{30}{0.0025}\frac{\left(15^2-10^2\right)}{2x2000}$$

$$= 1000 - 375 = 625m$$

Using **Eq.8.47** the height of water table divide

$$h^2 = h_m^2 = -\frac{R}{K}x^2 - \frac{\left(h_0^2 - h_1^2 - \frac{R}{K}L^2\right)}{L}x + h_0^2$$

$$= -\frac{0.0025}{30}(625)^2 - \frac{\left(15^2 - 10^2 - \frac{0.0025}{30}x2000^2\right)}{2000}x625 + 15^2$$

$$= -32.55 - \frac{(125 - 333.33)}{2000}x625 + 255$$

$$= -\,32.55 + 65.10 + 225$$

$$= 257.55$$

$$\therefore h_m = 16.05m$$

Example 8.15 Water is recharged through a well of 200mm diameter penetrating up to the base of the aquifer. The hydraulic conductivity of the aquifer is 19m/day. During recharge the water level in the well is 32m from the base of the aquifer. A constant height of 27m above the same base is obtained at a distance of 200m from the well.

a) If the aquifer is confined with a distance of 20m, the theoretical recharge rate in liter per second is

i)	18.2	ii)	25.2
iii)	28.8	iv)	30.5

(b) If the aquifer is unconfined, the theoretical recharge rate in liter per second is

i)	10.5	ii)	17.3
iii)	26.8	iv)	30.5

(GATE, 2009)

Solution:

a)

Diameter of the well, D = 200 mm = 0.2m

Hydraulic conductivity, K = 19m/day = $2.2x10^{-4}$m/s

Water level in the well, h_1 = 32m

Water level at 200m distance, $h_2 = 27m$

Distance between h_1 & $h_2 = 200m$

Thickness of aquifer, B = 20m

For confined aquifer,

$$Q = \frac{2\pi KB(h_l - h_2)}{ln\frac{r_1}{r_2}} = \frac{2\pi x 2.2x10^{-4} x 20(32-27)}{ln\frac{200}{0.1}} = \frac{0.138}{7.6} = 0.01818m^3/s$$

= 18.18l/s

b) For unconfined aquifer,

$$Q = \frac{\pi K(h_1^2 - h_2^2)}{ln\frac{r_1}{r_2}} = \frac{\pi 2.2x10^{-4}(32^2 - 27^2)}{ln\frac{200}{0.1}} = \frac{\pi 2.2x10^{-4} x295}{7.6} = 0.0268m^3/s$$

=26.8l/s

8.4 Flow in Wells

A water well is a hydraulic hole made in the earth down to a supply of water for the purpose of take out this ground water to the ground surface. A well made by driving a tube in to the water bearing stratum is called the tube well. The flow of water to the well depends on aquifer properties, design of well and methods of construction and developing the wells.

When pumping of water takes place in an aquifer water table for unconfined aquifer and piezometric surface for confined aquifer starts to drop. At the beginning of pumping the water level near the pump drops faster and the drop is extended outward with time. The flow is unsteady at the beginning and slowly attains steady state condition in an aquifer of uniform shape and texture. The radial flow of water through the aquifer forms a conical shape of water table. This conical shape is called cone of depression and radial extent is called radius of influence. The shape of the cone of depression depends on the rate of pumping, time of pumping, aquifer properties and recharge in the zone of influence of the well. The depth of water at any radial distance from the previous static level is called the drawdown. Drawdown is maximum at the well and zero at the outer limits of depression.

Flow in confined aquifer

The flow through confined aquifer was proposed by Theim and the equation derived is referred as Theim equation. Fig.8.1 shows completely penetrating well drawing water from a confined aquifer. The thickness of the aquifer is B and the piezometric head was H before pumping. At the steady state pumping the piezometric head at the well is h_w and the drawdown s_w.

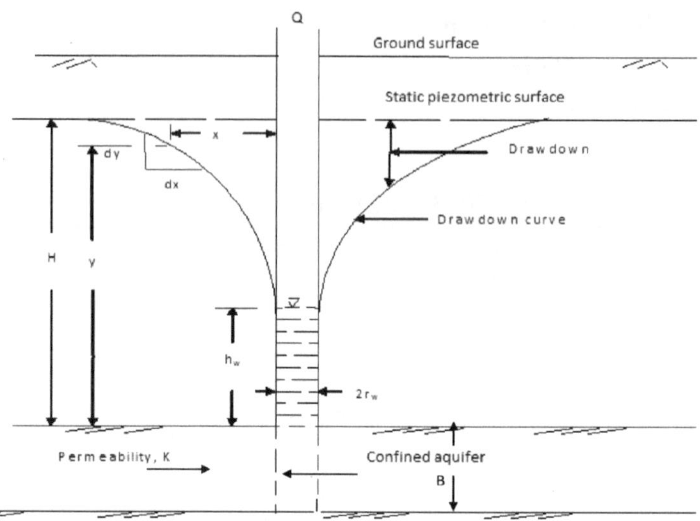

Fig.8.11 Flow to a confined aquifer

For analysis of this flow, the origin of coordinate is assumed to be at the point O located at the center of the well at its bottom. Let K be the hydraulic conductivity of the aquifer. If the piezometric head is h at radial distance of r, the velocity of flow following Darcy's law is

$$V_x = K\frac{dh}{dr} \tag{8.55}$$

The cylindrical surface through which this velocity occurs is Therefore, the discharge through the well is

$$Q = AV = 2\pi rBK\frac{dh}{dr}$$

$$or, \frac{Q}{2\pi KT}\frac{dr}{r} = dh$$

Integrating between r_1 and r_2 for the piezometric heads h_1 and h_2 respectively

$$\frac{Q}{2\pi KT}ln\frac{r_2}{r_1} = h_2 - h_1$$

$$or, Q = \frac{2\pi KB(h_2 - h_1)}{ln\frac{r_2}{r_1}} \tag{8.56}$$

The Eq.8.56 is the steady flow equation in a confined aquifer.

When the drawdown s_1 &s_2 are known

s_1=H-h_1, s_2=H-h_2, KB=T

h_2-h_1=H-s_2-H+s_1=s_1-s_2

The **Eq.8.56** can be rewritten as

$$Q = \frac{2\pi T\left(s_1 - s_2\right)}{ln\frac{r_2}{r_1}} \tag{8.57}$$

At the furthest end of radius of influence s=0, r_2=R and h_2=H and at the well wall r_1=r_w, h_1=h_w, s_1=s_w. The Eq.8.57, can be rewritten as

$$Q = \frac{2\pi T s_w}{ln\frac{R}{r_w}} \tag{8.58}$$

Unconfined aquifer

As shown in Fig.8.12, steady flow occurs in a well from a completely penetrating unconfined aquifer. The stream lines of the free surface are not strictly horizontal or straight. It is more horizontal at the bottom. However, Dupuit's assumptions are assumed to be valid for analysis of flow. The assumptions state that (i) the stream lines are to be assumed horizontal and equipotentials are vertical for small inclination of the free surface, and (ii) the hydraulic gradient is equal to the slope of the free surface and does not vary with this depth.

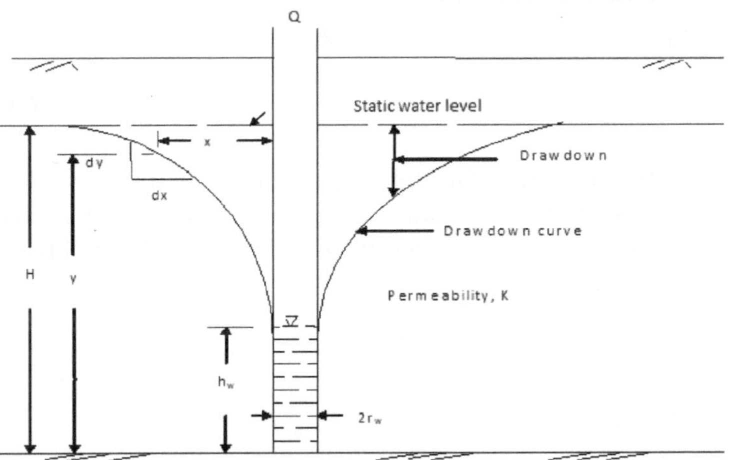

Fig.8.12 Flow in an unconfined aquifer

Similar to confined aquifer the origin of coordinate be at O located at the center of the well at the bottom and P be any point on the drawdown curve whose coordinate is (r,h). Let r_w be the radius of the well.

The discharge through the well

Q=AV

$$= 2\pi rhK\frac{dh}{dr}$$

$$= 2\pi rKh\frac{dh}{dr}$$

$$or, Q\frac{dr}{r} = 2\pi Khdh$$

Integrating between the radial distances r_1 &r_2 and with respect to depth h_1, h_2

$$Q\int_{r_1}^{r_2}\frac{dr}{r} = 2\pi K\int_{h1}^{h2} hdh$$

$$or, ln\frac{r_2}{r_1} = 2\pi K\left(\frac{h_2^{\ 2}}{2} - \frac{h_1^{\ 2}}{2} = \pi K\left(h_2^{\ 2} - h_1^{\ 2}\right)\right)$$

$$or, Q = \pi K\frac{\left(h_2^{\ 2} - h_1^{\ 2}\right)}{ln\frac{r_2}{r_1}} \tag{8.59}$$

This is the equation of equilibrium for a well in unconfined aquifer. At the extreme point of zone of influence the radial distances are R, r_w and depths H, h_w respectively. Therefore, Eq.8.59 converts to

$$or, Q = \pi K\frac{\left(H^2 - h_w^{\ 2}\right)}{ln\frac{R}{r_w}} \tag{8.60}$$

Example 8.16 What is the discharge of the tube well of diameter 20cm installed in a confined aquifer of thickness 10m and permeability 30m/day? At steady state pumping the drawdown reaches at the well is 2.5m and the radius of influence 500m.

Solution:

Radius of well, r_w=0.10m

Aquifer thickness, B=10m

Permeability, K=30m/day

Radius of influence, R=500m

Drawdown at well, s_w=2.5m

Transmissibility, T=KB=30m/dayx10m/day=300m²/day

Using **Eq.8.58**,

$$Q = \frac{2\pi T s_w}{ln\frac{R}{r_w}}$$

$$= \frac{2\pi x 300 x 2.5}{ln\frac{500}{0.10}}$$

$$= \frac{4712.4}{8.52} = \frac{553.28 m^3}{day} = 6.40 kl/s$$

Example 8.17 A 30cm well fully penetrates in a confined aquifer of 25m thickness pumping at the rate of 15l/s. The drawdowns in the observation wells are 2.5 and 2.0m at 15m and 40m respectively from the well. Determine the transmissibility of the aquifer and drawdown in the well.

Solution

Radius of well, $r_w = 0.15m$

Aquifer thickness, B = 25m

Discharge, Q = 15l/s =1296m^3/day

Drawdowns of observation wells, s_1, s_2 = 2.5m, 2.0m

Distances of observation wells, r_1, r_2 = 15m, 40m

Using Eq.8.57,

$$Q = \frac{2\pi T(s_1 - s_2)}{ln\frac{r_2}{r_1}}$$

$$or, 1296 = \frac{2\pi T(2.5 - 2.0)}{ln\frac{40}{15}} = \frac{\pi T}{0.98}$$

$$\therefore T = \frac{1296 x 0.98}{\pi} = 404.62 m^2/day$$

Further using Eq.8.57 for s_1=s_w and r_1=0.15m

$$Q = \frac{2\pi T(s_w - s_2)}{ln\frac{r_2}{r_1}}$$

$$or,\ 1296 = \frac{2\pi x 404.62(s_w - 2.0)}{ln\frac{40}{0.15}}$$

$$or,\ 1296 = \frac{2542.31(s_w - 2.0)}{5.59}$$

$$or,\ s_w = \frac{7239.46 + 5084.62}{2542.31} = \frac{12324.08}{2542.31} = 4.85m$$

Example 8.18 A 200 mm well fully penetrates a confined aquifer. After a long period of pumping at a rate of 1400l/min, the drawdowns in the observation wells located at 25m and 40m from the pumping well are found 2.6m and 1.9m, respectively. The transmissivity of the aquifer in m^2/day is

a) 19 b) 198
c) 206 d) 215

(GATE, 2012)

Solution

Diameter of well, D=200mm=0.2m

Discharge of the well, Q=1400l/min=0.0233m^3/s

Drawdowns of the observation wells, s_1, s_2=2.6m, 1.9m

Location of observation wells, r_1, r_2=25m, 40m

Transmissivity =hydraulic conductivity(K) x thickness of aquifer(B)=KB

Discharge of the well,

$$Q = \frac{2\pi KB(s_1 - s_2)}{\ln \frac{r_2}{r_1}}$$

$$0.0233 = \frac{2\pi KB(2.6\text{-}1.9)}{\ln \frac{40}{25}}$$

$$0.0233 = \frac{4.398KB}{0.47}$$

$$= 2.15.13m^2/day$$

Example 8.19 A 30cm well fully penetrates in an unconfined aquifer of saturated thickness 30m discharges at a steady rate of 1500m^3/day. The observation wells at 50 and 100m from the well shows the drop of water 3.0m and 2.0m respectively. Compute the permeability of the aquifer and drawdown at the well.

Solution:

Radius of well, r_w = 0.15m

Aquifer thickness, B = 30m

Discharge, Q =1500 m^3/day

Drawdowns of observation wells, s_1, s_2 = 3.0m, 2.0m

Distances of observation wells, r_1, r_2 = 50m, 100m

$\therefore\ h_1 = 30\text{-}3 = 27m,\ h_2 = 30\text{-}2 = 28m$

Using Eq.8.59,

$$Q = \pi K \frac{\left(h_2^{\,2} - h_1^{\,2}\right)}{ln\dfrac{r_2}{r_1}}$$

$$or, 1500 = \pi K \frac{\left(28^2 - 27^2\right)}{ln\dfrac{100}{50}}$$

$$= \pi K \frac{55}{0.69} = 250.42K$$

$$\therefore\ K = \frac{1500}{250.42} = 5.99m/day$$

Using **Eq.8.59** for the heads of water at the radial distances of r_w and 30m,

$$Q = \pi \text{K} \frac{\left(h_1^{\,2} \text{-} h_w^{\,2}\right)}{ln\dfrac{r_1}{r_w}} = \pi K \frac{\left(27^2 \text{-} h_w^{\,2}\right)}{ln\dfrac{50}{0.15}} = 3.24\left(27^2 - h_w^{\,2}\right)$$

$$or, 27^2 - h_w^{\,2} = \frac{1500}{3.24} = 462.96$$

$or,\ h_w^{\,2} = 729\text{-}462.96 = 266.04$

$\therefore\ hw = 16.31m$

Example 8.20 A 30cm well penetrates 40m below the static water table discharges at the rate of 1850m^3/day. The drawdowns at 5 and 50m from the well are 2.5 and 1.0m respectively. What is the transmissibility of the aquifer?

Solution:

Radius of well, r_w = 0.15m

Aquifer thickness, B = 40m

Discharge, Q =1850m^3/day

Drawdowns of observation wells, s_1, s_2 = 2.5m, 1.0m

Distances of observation wells, r_1, r_2 = 5m, 50m

$\therefore\ h_1 = 40\text{-}2.5 = 37.5m,\ h_2 = 40\text{-}1 = 39m$

From Eq.8.59

$$Q = \pi K \frac{\left(h_2^{\ 2} - h_1^{\ 2}\right)}{ln \frac{(r_2)}{r_1}}$$

$$or, 1850 = \pi K \frac{\left(39^2 - 37.5^2\right)}{ln \frac{50}{5}} = \pi K \frac{114.75K}{2.30} = 156.74K$$

$$\therefore\ K = \frac{1850}{156.74} = 11.80 m/day$$

Transmissibility, T=KB=11.80x40=472m³/day

Example 8.21 A 40cm diameter tube well is constructed in a 10m thick confined aquifer having hydraulic conductivity of 25m/day. The piezometric surface is observed to be 40m high from the impervious stratum at the radius of influence of 500m. The drawdown in the tube well is 30m. If the thickness of the aquifer is doubled and diameter of tube well is reduced to half, keeping all other parameters and conditions same, the change in discharge from the well is increased by 83.72%, decreased by 83.72%, increased by 82.28%, decreased by 82.28%. Take $\pi = 3.14$ (GATE, 2020).

Solution:

Radius of well, r_w=20 cm

Thickness of aquifer, T=10m

Hydraulic conductivity, K=25m/day

Radius of influence, r_1=500m

Piezometric surface at 500m radius of influence, h_1=40m

Drawdown of the tube well, h_w=30m

Discharge of the tube well,

$$Q_1 = 2\pi KB \frac{\left(h_l - h_w\right)}{\ln\left(\frac{r_1}{r_w}\right)}$$

$$= 2x3.14x25m/\text{day}x10m \frac{(40 - 30m)}{\ln\left(\frac{500m}{0.2m}\right)}$$

$$2x3.14x25m/\text{day}x10mx\frac{10m}{7.82}$$

$= 2007.67\text{m}^3/\text{day}$

When the thickness of the aquifer is doubled and diameter of the tube well is half,

$$Q_2 = 2\pi KB\frac{(h_1 - h_w)}{\ln\left(\frac{r_1}{r_w}\right)}$$

$$= 2x3.14x25m/dayx20m\,\frac{(40-30)m}{\ln\left(\frac{500m}{0.1m}\right)}$$

$$= 2x3.14x50m/dayx20mx\frac{10}{8.52}$$

$= 3685.45\text{m}^3/\,day$

Change of discharge

$$= \frac{Q_2 - Q_1}{Q_1}x100 = \frac{3685.45 - 2007.67}{2007.45}x100 = 83.58\% \approx 83.72\%\,increased$$

Example 8.22 The drawdown at the well face of 30cm diameter fully penetrating well in a confined aquifer is 3m. The radius of influence in the well is 3km. For the same discharge and aquifer condition, the drawdown at 300m away from the center of the well is ....m. (GATE, 2018)

Solution:

Head of water at well and 3km away are h_1 & h_2 respectively.

$\therefore h_2 - h_1 = 3.0\text{m}$

$$Discharge, Q = 2\pi KB\frac{(h_2 - h_1)}{\ln\left(\frac{r_2}{r_1}\right)}$$

$$= 2\pi KB\frac{3m}{\ln\frac{3000}{0.15}} = \frac{6\pi KB}{\ln 20000} = 1.90\,KB$$

$$Again\, discharge, Q = 2\pi KB\frac{(s_1 - s_2)}{\ln\left(\frac{r_2}{r_2}\right)}$$

$$= 2\pi KB\frac{(3m-s_2)}{\ln\frac{300}{0.15}} = \frac{2\pi KB(3m-s_2)}{7.60} = 0.83KB(3m-s_2)$$

$\therefore$ $1.90KB = 0.83\ KB\ (3m\text{-}s_2)$

or, $3 - s_2 = 1.90/0.83 = 2.29$

$\therefore$ $s_2 = 3 - 2.29 = 0.71m$

Example 8.23 150mm diameter well is driven fully in a confined aquifer of thickness 12m. When pumping is done at a constant rate of 150l/min for 10h,the steady-state draw-down found to be 2.20m and 0.02m at distances of 14m and 50m, respectively from the center of the well. The hydraulic conductivity of the aquifer in m/day is : (GATE, 2014)

Solution:

Diameter of the well, D=150mm=0.15m

Thickness of the aquifer, T=12m

Rate of discharge of the well, Q=150l/min=150/(1000x60)=$2.5\times10^{-3}m^3/s$

Time of operation of the well, t=10h

Drawdown, s_1=2.20m

Drawdown, s_2=0.02m

Drawdown s_1 at radius, r_1=14m

Drawdown s_2 at radius, r_2=50m

Hydraulic conductivity=K

$$We\,have, Q = \frac{2\pi KB(s_1 - s_2)}{\ln\frac{r_2}{r_1}}$$

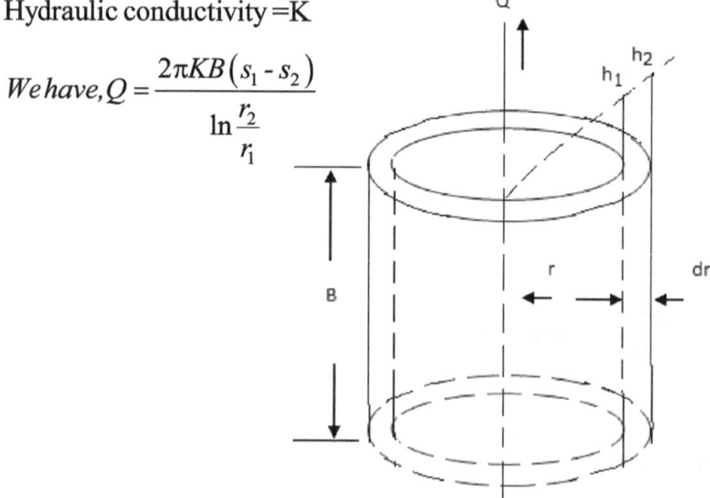

Fig.8.13 Confined aquifer flow in cylindrical sections

$$or, 2.5x10^{-3} = \frac{2\pi Kx12(2.2-0.02)}{\ln\frac{50}{14}}$$

$$= \frac{164.36K}{1.27}$$

$$\therefore K = \frac{1.27x2.5x10^{-3}}{164.36}$$

$= 21.936\text{x}10^{-5} m/s$

$= 1.67\ m/day$

Unsteady flow in confined aquifer

At the beginning of pumping in the well of a confined aquifer cone of depression forms to the piezometric surface. The cone of depression expands gradually with respect to time to attain the state of equilibrium. The flow condition in the time period in between the start and attainment of equilibrium is the unsteady flow.

As shown in Fig.8.13, there are two cylindrical sections in a confined aquifer of thickness B and radii r, r+dr at piezometric heads h_1 and h_2 respectively. A well in the center of the sections discharging at a constant rate Q. The rate of change of volume of water within the cylindrical sections may be expressed as

$$\frac{\partial v}{\partial t} = -\frac{\partial Q}{\partial r}dr = 2\pi r S\frac{\partial h}{\partial t}\text{dr} \quad \text{..... (1)} \tag{8.61}$$

Where, $\frac{\partial v}{\partial t}$ = change in volume of water between the cylindrical sections of heads h_1 and h_2 with respect to time

$\frac{\partial Q}{\partial r}$ = change in flow rate between h_1 and h_2 with distance (r)

$\frac{\partial h}{\partial t}$ = change in piezometric head between h_1 and h_2 with respect to time

S= storage coefficient

Using Darcy's law the discharge in the well

$$Q = -2\pi Tr\frac{\partial h}{\partial r} \tag{8.62}$$

Where, T=transmissibility of the aquifer =KB

Differentiating **Eq. 8.62** with respect to r

$$\frac{\partial Q}{\partial r} = -2\pi T \frac{\partial}{\partial r}\left(r\frac{\partial h}{\partial r}\right)$$

$$= -2\pi T\left(\frac{\partial r}{\partial r}\frac{\partial h}{\partial r} + \frac{\partial^2 h}{\partial r^2}\right)$$

$$= -2\pi T\left(\frac{\partial h}{\partial r} + r\frac{\partial^2 h}{\partial r^2}\right) \quad (8.63)$$

Combining **Eqs.8.61 & 8.63**

$$2\pi T\left(\frac{\partial h}{\partial r} + \frac{\partial^2 h}{\partial r^2}\right) = 2\pi r S\frac{\partial h}{\partial t}$$

$$or, \left(\frac{\partial h}{\partial r} + \frac{\partial^2 h}{\partial r^2}\right) = \frac{\mathrm{s}}{T} r \frac{\partial h}{\partial t}$$

$$or, \frac{1}{\mathrm{r}}\frac{\partial h}{\partial r} + \frac{\partial^2 h}{\partial r^2} = \frac{S}{T}\frac{\partial h}{\partial t} \quad (8.64)$$

The Eq.8.64 is the partial differential equation for unsteady radial flow.

The is developed the solution of Eq.8.64. In between the critical conditions of beginning of pumping ($h=h_0$, t=0) and after considerable period of pumping $(\mathrm{h} \to 0, \mathrm{r} \to \alpha, \mathrm{t} \geq 0)$ the flow condition is expressed as

$$\text{Drawdown}, s = h_0 - h = \frac{\mathrm{Q}}{4\pi T}\int_{\mathrm{u}}^{\alpha}\frac{\mathrm{e}^{-u}}{u}du = \frac{Q}{4\pi T}W(u) \quad (8.65)$$

$$\text{Where}, u = \frac{r^2 S}{4Tt}, S = \frac{4Ttu}{r^2} \quad (8.66)$$

Eq.8.65 is called as the Theis equation or nonequilibrium equation. W(u) is known as the Theis well function. This equation can be expanded as

$$S = \frac{Q}{4\pi T}\left[-0.5772 - lnu + u - \frac{u^2}{2.2!} + \frac{u^2}{3.3!}....\right] \quad (8.67)$$

For small values of u (≤ 0.01) (Jacob (1946, 1950) showed that first two terms of the series w(u) are adequate. Applying this assumption in **Eq.8.67**

$$S = \frac{Q}{4\pi T}\left[-0.5772 - ln\frac{r^2 S}{4\pi t}\right] \quad (8.68)$$

$$or, s = \frac{Q}{4\pi T}ln\left(\frac{2.2Tt}{r^2 S}\right) \quad (8.69)$$

When s_1 and s_2 are drawdown at time t_1 and t_2

$$S_2 - S_1 = \frac{Q}{4\pi T} ln\frac{t_2}{t_1} \tag{8.70}$$

If drawdown (s) is plotted against time (t) on a semi-log paper it will be a straight line for large values of t. Slope of this line will give the value of storage coefficient S. When the drawdown s=0 in Eq.8.69

$$\frac{2.25Tt_0}{r^2 S} = 1$$

$$or, S = \frac{2.25Tt_0}{r^2} \tag{8.71}$$

Where, t_0 is the time corresponding to 'zero' drawdown as obtained by extra plotting the straight line portion of the semi-log curve.

Example 8.24 A well in a confined aquifer of 30m thickness, permeability 25m/day, storage coefficient 0.004 and discharging at the rate of 2000m³/day. Compute the drawdown at 50 and 75m distance from the well after 15 hours of pumping.

Solution:

B=30m, K=25m/day, S=0.004, Q=2000m^3/day, t=15 hours

$\therefore T = KB = 25x30 = 750m^2$/day

At distance, r = 50m

$$u = \frac{r^2 S}{4Tt} = \frac{50^2 x0.004}{4x750x15/24} = 0.0053$$

For u=0.0053<0.01, using W(u) value to two significant digits

W(u) = - 0.5772 - ln (0.0053) + 0.0053

= - 0.5772 = 5.24 + 00.053 = 4.668

$$\text{Drawdown, } s_{50} = \frac{Q}{4\pi T} W(u)$$

$$= \frac{2000x4.668}{4\pi x750} = 0.99m$$

At distance, r = 100m

$$u = \frac{r^2 S}{4Tt} = \frac{100^2 x0.004}{4x750x15/24} = 0.0213$$

u=0.0213>using w(u) value to four significant digits

$$W(u) = -0.5772 - \ln(0.0213) + 0.0213 - \frac{(0.0213)^2}{2.2!} + \frac{(0.0213)^3}{3.3!}$$

= - 0.5772 + 3.849 + 0.0213 - 1.134 x 10^{-4} + 5.368$x$$10^{-7}$

=3.29m

Questions and Problems

8.1 Define zone of aeration. How do you divide it?

8.2 Define aquifer, aquitard, aquiclude, aquifuge.

8.3 Distinguish perched and leaky aquifer. Define specific yield, specific retention, bulk pore velocity and viscosity.

8.4 Discuss with sketch the effluent and influent stream. Define intrinsic permeability, transmissibility, hydraulic resistance and leakage factor.

8.5 Define integral stress, pore water stress, specific storage and storage coefficient.

8.6 Derive the relation for specific storage of ground water with pore-volume compressibility and compressibility of water.

8.7 What is permeability of soil? How the permeability is calculated through stratified soil?

8.8 Derive the basic differential equation for ground water flow in confined aquifer. How it converts to Laplace equation?

8.9 Derive the basic differential equation for flow in unconfined aquifer.

8.10 Derive the expressions for flow in unconfined aquifer with constant recharge to ground water.

8.11 Derive the expressions for flow in unconfined aquifer without any recharge to ground water.

8.12 Derive the expression for discharge through well in a confined aquifer.

8.13 What is water well? Derive the expression for discharge through well in an unconfined aquifer.

8.14 Derive the partial differential equation for unsteady radial flow.

8.15 The storage coefficient of a 30m thick aquifer is 8.5×10^{-5}. If the water compressibility of the aquifer is 4.8×10^{-11} $cm^2/dyne$, what is the pore-volume compressibility? Assume porosity as 0.33.

Ans. 4.5×10^{-9} $cm^2/dyne$

8.16 The volume of water recorded per km^2 of a confined aquifer is $5.5 \times 10^4 m^3$ at a pressure head of 25m. Find the specific storage if the average pressure per km^2 of aquifer is $2.75 \times 10^7 m^3$. Ans. $6.05 \times 10^{-4}/m$

8.17 An unconfined aquifer extends over an area of $1 km^2$ and has hydraulic conductivity, total porosity and specific retention of 20m/day, 30% and 10% respectively. After pumping some groundwater from this aquifer, the water table dropped to a depth of 20m from the ground level. If the water table was initially at 14.5m below the ground level, the change in groundwater storage in million cubic meters would be - (GATE, 2013)

Ans. 1.1

8.18 Calculate the mean velocity of flow in a pipe of diameter 20mm if the Reynolds number is 4,000 and the kinematic viscosity of water is $1.02 x 10^{-2} cm^2/s$.

Ans.20.4cm/s

8.19 The water pressure recorded at entrance and other end of a U-shaped manometer are 300kPa and 240kPa respectively. What is the height of the mercury column if the mercury and water densities are $13.6 x 10^3 kg/m^3$ and $1.0 x 10^3 kg/m^3$ respectively.

Ans. 0.48m

8.20 Calculate the permeability of a soil under test in a constant head permeameter of diameter 10cm, hydraulic gradient 0.75 and collects $600 cm^3$ of water in 60 seconds.

Ans.0.17cm

8.21 In a variable head permeameter test the soil sample column was 5.0 cm diameter and 20 cm length. The head of water drops from 100 cm to 50 cm in 40 minutes in the stand pipe of diameter 1.0cm. What is the coefficient of permeability of soil?

Ans.$2.31 x 10^{-4}$ cm/s

8.22 In an aquifer of layered soils of thickness 3.1, 2.9 and 3.5m has the permeability of $4.7 x 10^{-3}$, $5.9 x 10^{-3}$ and $8.9 x 10^{-2} cm/s$ respectively. Calculate the average permeability of the aquifer normal to the soil layers.

Ans.$7.98 x 10^{-3} cm/s$

8.23 An aquifer extended to $20 Km^2$ area has dropped the water table by 2.5m. The porosity and specific retention of the aquifer is 25 and 10% respectively. What is the specific yield of the aquifer and change of storage due to drop of water of water table?

Ans. 15%, $3.0 x 10^5 m^3$

8.24 A drainage pipe of diameter1.2m got clogged by sediment for some portion of its length. The coefficient of permeability were 20m/day, 5m/day and unknown for the upstream 100m, middle 125m and downstream 50m length respectively of the clogged pipe. Estimate the unknown coefficient of permeability of the downstream section if the head difference between the two sides of clogged pipe was 30m and the pipe flow of $0.5 m^3/day$.

Ans: 1.32m/day

8.25 Rainfall occurs at the rate of 5mm/day in the strip of land in between the streams 1.0Km apart. The water levels in the streams are 15 & 12m from the horizontal impervious layer. Compute the (i) discharges in the canals through per meter width of aquifer and (ii) location and height of water divide. Assume aquifer permeability 30m/day.

Ans. (i) 1.285 & 3.715m^3/day (ii) 257 & 15.36m

8.26 A 30cm diameter tube well fully penetrates in a confined aquifer of thickness 15m discharges at the rate of 500m^3/day. The piezometric head drops at well 2.5m and the radius of influence 750m. Compute the transmissibility and permeability of aquifer.

Ans. T=193.65m^2/day, K=12.91m/day

8.27 A 30cm diameter well fully penetrates in a confined aquifer of thickness 15m, permeability 25m/day and discharge 1500m^3/day. Compute the radius of influence if piezometric head drops at well 4.5m.

Ans. 176.18m

8.28 The discharge from a 20cm installed in an unconfined aquifer of saturated thickness 45m, permeability 10m/day, discharges at the rate of 1500m^3/day. The drawdown at well is 4.5m. What is the radius of influence?

Ans.315.96m

8.29 A well is to be installed in an unconfined aquifer of saturated thickness 50m and permeability 5m/day. Expected discharge rate 880m^3/day, radius of influence 500m and maximum drawdown at the well 5.0m. Determine the diameter of the well to be selected.

Ans. 20cm

8.30 A well in a confined aquifer of thickness 30m has the permeability of 25m/day, storage coefficient 0.0035 and discharge rate 1500m^3/day. Compute the drawdowns at 20 and 100m from the well after 18 hours of pumping.

Ans.1.08m, 3.59m

8.31 Select the appropriate answer from the following multiple choice questions

1. Darcy is equal to (a) 9.17x10^{-7}m^2 (b) 9.87x10^{-11}m^2 (c) 7.97x10^{-9} cm^2 (d) 9.87x10^{-9}cm^2
2. Unit of kinematic viscosity is (a) cm/s (b) cm/s^2 (c) cm^2/s (d) cm^3/s
3. Transmissibility has the dimension of (a) L/T (b) L^2/T (c) L/T^2 (d) L^2/T^2
4. Hydraulic resistance has the dimension of (a) T (b) L (c) L/T^2 (d) L^2/T
5. 1 bar pressure is equal to (a) 10^{4}dynes/cm^2 (b) 10^5 dynes/cm^2 (c) 10^6 dynes/cm^2 (d) 10^8 dynes/cm^2
6. Cohesive force is more in (a) solid (b) liquid (c) gas (d) semi-liquid
7. A sandy clay unit is an example of (a) aquifer (b) aquitard (c) aquiclue (d) aquifuge
8. Leakage factor has the dimension of (a) T (b) L (c) L/T^2 (d) L^2/T

9. Dupuit-Forchheimer assumptions are used for analyzing ground water flow in
   a) Transmission only
   b) Storage coefficient only
   c) Unconfined aquifers
   d) Both confined and unconfined aquifers

(GATE, 2015)

10. An unsteady time-drawdown pumping test was conducted in na confined aquifer and the drawdown was measured with time in an observation well located at a certain distance away from the pumping well. Using these time-drawdown data, we can determine
   a) Transmissivity only
   b) Storage coefficient only
   c) Specific storage only
   d) Both transmissivity and storage coefficient

(GATE, 2015)

Ans.

1. d) 2. c) 3. b) 4. a) 5. c) 6. a) 7. b) 8. b)
9. c) 10. d)

8.32 Write True or False of the following statements

1. Zone of aeration is situated with water
2. The portion of the soil not occupied by the solids is said to be interstices
3. Granite and sandstone are disintegrated rock materials
4. Confined aquifer is also called as water table aquifer
5. Leaky aquifer is also called as semi-confined aquifer
6. Perched aquifer is a special case of unconfined aquifer
7. Influent stream receives the ground water flow
8. Henry Darcy is a German hydraulic engineer
9. Laboratory standard value of coefficient of permeability is taken for that pure water at 20^0C
10. Darcy law determines the actual velocity of flow through the pores
11. Actual velocity of flow in soil is much less than the apparent velocity
12. Coefficient of permeability of soil is inversely proportional to kinematic viscosity of fluid.
13. The viscosity of water at 20^0C is almost 1 centipoise

14. Intrinsic permeability is a function of medium only
15. Hydraulic resistance is a property of semi-impervious layer of the semi-confined aquifer
16. Leakage factor is inversely proportional to hydraulic resistance
17. Storage coefficient is also known as storavity
18. Well function is directly proportional to transmissibility of the aquifer
19. Discharge in confined aquifer is more if drawdown is more
20. Water saturated depth is not important for discharge of a well in unconfined aquifer

Ans.

1.	False	2.	True	3.	True	4.	False	5.	True	6.	True	7.	False
8.	False	9.	True	10.	False	11.	False	12.	True	13.	True	14.	True
15.	True	16.	False	17.	True	18.	False	19.	True	20.	False		

References

Michael, A.M. (2008). Irrigation Theory and Practices. Vikas Publishing House Pvt. Ltd., New Delhi.p.132-137

Murty, V.V.N.Murty (2004). Land and Water Management Engineering. Kalyani Publishers, New Delhi.p.105-136

Raghunath, H.M. (1987). Ground Water. Wiley Eastern Limited, New Delhi. P.84-93

Subramanya, K (2013). Engineering Hydrology. McGraw Hill Education (India)

9

Lining Irrigation Channels

9.1 Introduction

Water is a precious commodity. Necessity of conservation of water supply is increasingly felt all over the world as the demand of it continues to increase when the sources of supply are increasingly scarce. The conveyance and proportional distribution of water is one of the most important part of any irrigation project. Water available in the stream, reservoirs or wells is carried to the field of use through canals. The economy of any irrigation project largely depends on how efficiently the water is being transported and distributed in respect to water losses and cost.

Water losses in irrigation channels are mainly due to evaporation from the water surface and seepage from the bed and side of the channels. Seepage losses some time become the single important factor which causes unavailability of water particularly in downstream of the channel in canal irrigation system and excess pumping of water in tube wells and river lift irrigation system.

Lining irrigation channels means to cover the channel earthen surface by any substance, which can considerably reduce the seepage loss. The concrete, brick, stone, natural clay of low permeability, bituminous material, plastic compounds, etc., are used for channel lining.

Advantages of lining

The following are the advantages of lining on an irrigation channel:

i) Reduction in seepage losses resulting in a saving of water, which can be utilized for irrigating additional areas. Thus duty of canal water is increased.

ii) Reduction in percolation to the ground water reservoir causes to prevention of water logging unless the percolated water is harvested for further use of it.

iii) Reduction of maintenance costs and possible breaching due to increased stability of the channel.

iv) Prevention and reduction of weed growth due to unfavourable hard surface of the channel.

v) Lining increases the velocity of flow and decreases silting in the channel.

vi) Saving of land due to low cross-sectional area is required for same discharge at higher velocity of flow in lined channel.

vii) Improvement of operational efficiency.

Disadvantages of lining

i) High initial cost.

ii) Repair works are problem some

iii) Permanent in nature and difficult to change the position of outlets.

iv) Berms are not usually provided which is hazardous for pedestrian and vehicular traffic.

9.2 Types of Lining

The various lining materials can be grouped as (i) Exposed surface linings, and (ii) Buried membrane linings, or (i) Conventional linings, and (ii) Non-conventional linings.

Exposed surface linings

These are the type of lining placed upon the sub grade of the canal exposed to wear, erosion and deterioration effect of flowing water, operation and maintenance equipment, disturbance of the public and cattle and other hazards. The exposed surface lining includes cement concrete, bricks or tiles, asphalted concrete, soil-cement and earth, etc.

i) Cement concrete lining

These are made of selected aggregates, Portland cement and water. The cement concrete lining is most extensively used in USA where it is usually placed and finished by mechanical means but in India it is generally done by manually.

A good concrete lining depends largely on the condition of the foundation on which the lining is made particularly when the lining is unreinforced. Before lining the earth channel is made definite geometric shape and the bottom and sides of it are compacted properly so that any possibility of cracking due to settlement of the sub grade is avoided. Drainage facilities are provided for the bank to avoid the backpressure of percolated water in the soils of low permeability.

The small channels are often lined by hand placing of cement concrete to a thickness of 4cm on the slopes and 5.0 to 6.5cm on the bottom of the channels. The bank of the channel will be such that it will be self-supporting of earth pressure and most commonly the side slopes are 1.5:1 to 1.25:1 for average soils. Curing is very important during construction of lined channels. Immediately before placing the concrete the sub grade should be suitably moistened to a depth of 30cm in sandy soil and 15cm in other soils so that the lining does not rapidly dry. Even in a good compacted sub grade there is the chance of movement of sub grade to cause cracking of concrete. Some additional thickness or reinforcement is therefore provided in case of heavy filling or weak soils. The reinforcements are provided to the extent of 0.25 to 0.30 percent of the cross-sectional area of concrete. However, steel reinforcement is suggested to omit because it increases 10 to 15 percent additional expenditure excepting special cases where it requires for safety purposes. Cement concrete lining gives long service and minimum repair and maintenance costs. It is one of the most conventional types of lining which has sufficiently been used in India and other parts of the world. It can withstand high flow velocities of up to 2.5m/s because of its greater resistance to erosion and therefore more preferable than other lining.

The disadvantages of cement concrete are its susceptibility in developing cracks due to drying, temperature changes, shrinkage and settlement of sub grade materials. The cracks thus developed cause appreciable leakage through the channel though sometime these cracks are considerably controlled by the deposition of silts and swelling of concrete by absorption of water. For the purpose of minimizing these cracks special contraction joints are provided at a regular interval in the concrete preferably at fifty times the lining thickness.

Murthy (1988) observed that cement concrete lining eliminates weed growth and thereby improves flow characteristics and last long. Even burrowing animals cannot penetrate concrete. Koll et al (1993) studied the RCC trapezoidal channel of 0.3m bottom width, depth 0.50m, with side slope 1.5:1, cement, sand, gravel mixture (1:3:4) with 0.25% reinforcement. A buffer 0.075m thick layers of sand and murum was provided on the base and the sides were properly compacted before construction. The performances of the channel were the light seepage loss, moderate repair and maintenance charges and cost of construction Rs148/m length.

Prefabricated concrete sections

Prefabricated or precast concrete sections are popularly used now days in lining small irrigation channels. These are constructed by using the definite shape moulds usually semi-elliptical or half-round. The prefabricated sections require minimum land area in field, permit cultivation very close to it and can be used in field very fast. Thus, this practice saves time and land. It is also cheaper than usual concrete lining. The steel reinforcement used in it also can be replaced by low cost reinforcement. Biswas (1999) replaced costly reinforcement and stone chips by bamboo sticks and blast furnace slag chips respectively. He designed and fabricated the moulds and tested for its performances. The details are described below for the better understanding of the learners and benefit of the possible users.

Bamboo-reinforced precast concrete with local sand and blast furnace slag as aggregates

a) Half round channel section

i) Mould for precast concrete section

The mould consisted of two half-circle sections of which one suitably matches inside the other with annular gap of 2.54cm in between them. The outer and inner diameters were 53.36cm and 48.26cm respectively for 50.8cm diameter concrete channel section (Fig. 9.1). The moulds were also made for 36.4cm and 30cm diameter sections. The length of each type of section was 152.4cm.

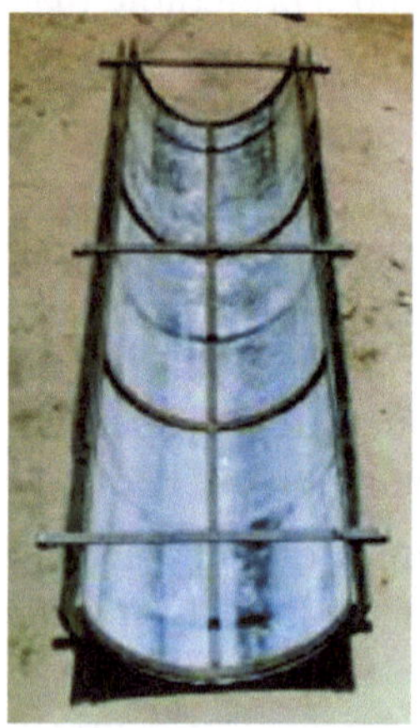

Fig. 9.1 A complete set of mould

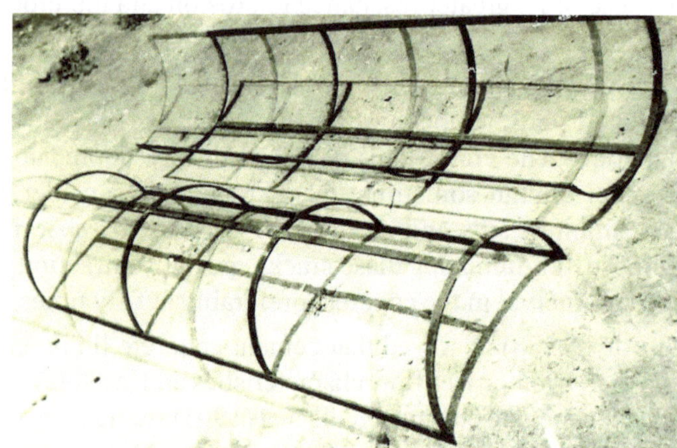

Fig. 9.2 Steel reinforcement

Galvanized iron (GI) sheet of 20 SWG was rolled to half circle to cover the section structure from outside for the inner section and inside for the outer section. After shaping it properly it was riveted on the flat bars (Fig.9.2). When the two sections form a set there was a designed gap of 2.54cm in-between the sections throughout the inside area.

ii) Bamboo reinforcement

The bamboo sticks of 5mm x 10mm cross-section and 150cm length with spacing of 10cm x 10cm or 15cm x 15cm were used to fabricate the concrete sections of 50cm and 36.4cm diameters. After binding the reinforcements by wires following the desired spacing they became the net structure of required length and breadth to fit in different diameter moulds (Fig. 9.3).

Fig. 9.3 Bamboo reinforcement in reinforced precast section

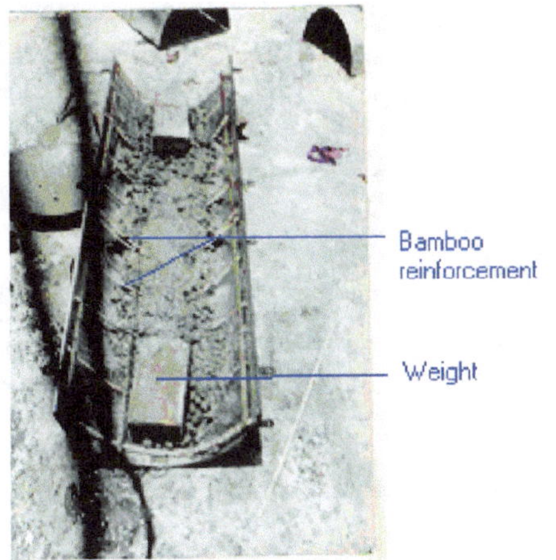

Fig. 9.4 Progress of casting the half round bamboo

iii) Aggregate and mixture

The waste of ferro-alloy factories usually called as 'slag' which were locally available at a cheap rate were used as the raw material for the coarse aggregate as substitute of costly conventional stone-chips or brick-bats. The slags were broken to 25mm down pieces and used in the mixture of 1:2:4 and 1:3:6 cement: sand: aggregate ratio.

The bamboo-sticks after binding them with wire at specified spacings were submerged in water for a few hours before these were used. After the submergence these bamboo reinforcements acquired the characteristics of bending smoothly inside the mould with little pressure and kept fixed by putting some loads (viz. brick, stone, etc.). The primary contents of the mixture was pushed in the reinforcement structure at the bottom in such an amount and way that it helped the reinforcements in maintaining the position nearly at the middle of the ultimate thickness of the section. With the addition of mixture, sufficient load was developed when the imported load on the reinforcements were removed. Further addition of mixture covered the large area of the reinforcement net. The shovel or flat bar of comfortable length were used for suitably packing and compacting the mixture within reinforcement structures. The GI sheet made side covers were then placed at both the sides to check the mixture material to slide out through end's gap. At this time the inner section of mould was placed over the outer section and was shacked and pressed on the mixture to set properly and maintaining the desired gap between the sections. This was done by binding together the two sections of the mould set by wires in the available grooves. After filling and compacting the mixture through the annular gap to the desired level it was then left for a day (Fig. 9.5).

Fig. 9.5 The section in the mould

Fig. 9.6 The sections are in the process of curing

On the next day, the inner mould section was lifted to expose the concrete mixture to atmosphere and watering was started. The concrete section was ready to remove from the mould after one another day. By lifting the mould vertically and given a slight tilt the concrete section got detached. This was then kept aside for watering for another few days (Fig. 9.6).

v) Laying of precast sections to the irrigation channels

The location and direction of the irrigation channels were fixed following the existing irrigation channels in the tube well commands. The undulation along the existing irrigation channels was determined by leveling operations at 5m intervals. Avoiding the zig-zag direction, the straight and desired slope channels were planned. Necessary earth filling was done to get the flat soil formation of appropriate slope throughout the proposed channel length.

The precast sections were carried on the formation surface i.e., bed of the channel and placed one after another. Both the outer side of the sections was filled up sufficient soils so that the sections got required support and remained fixed in the position. In general, at 20m intervals or at the points where field situation demanded for facilitation the irrigation on concrete section was skipped for construction of water diversion box. The two faces of the properly positioned sections were jointed by concrete mix and where there were required plastering by cement mortar was done to the inside sections for better look and performances. Thus, the continuous channels of desired lengths were made (Fig. 9.7).

Fig. 9.7. Channels are made by jointing the sections

b) Trapezoidal and triangular channels made of flat slabs

i) Bamboo reinforcements

The flat slabs for trapezoidal and triangular channels were made for 150cm length. The bamboo reinforcements used for all type were 5mm x 10mm in cross-section and 1 Nos. in crosswise with 5, 4 and 3 Nos. in lengthwise for 40cm, 30cm and 20cm wide slabs respectively. The bamboo reinforcements were fastened tightly by using 26 SWG wires to get the complete net of reinforcements. For 40cm, 30cm and 20cm wide slabs the thickness wire 5cm, 5cm or 4cm and 4cm or 3cm respectively.

ii) Construction of slabs

For casting the flat slabs of desired length and breadth the support of wooden pieces from all the directions were given to get the required block for casting (Fig. 9.8). The blocks were made on the leveled ground. The reinforcement net were placed on the polythene paper inside the block in such a way that it maintained a gap of about 1cm from the ground when concrete mixture was poured in it. The concrete mixture after pouring inside the blocks was compacted properly so that the shape and thickness of the slabs were ensured. It took another one or two days for watering and dismantling the wooden supports and lifted from the polythene to store in safe place.

(iii) Use of slabs in making the lined channels

Similar to the half round precast sections, the direction of the field channels were fixed following the existing irrigation channels. The slabs were placed one by one straight in a line on the channel beds (Fig.9.9). Each bottom piece was jointed by two other pieces at the sides by using the cement mortar forming different side slopes of the trapezoidal channels. Both outsides of the channels were provided with the support of soils immediately after it took the shape. The soil support and time to time dressing it properly were required to avoid the risk of breaking of the slabs (Fig.9.9).

Fig. 9.8 Flats slabs are cast by wooden support made of reinforced precast slabs is in progress

Fig. 9.9 The trapezoidal channel

(c) Triangular section

By jointing two slabs pieces together at the bottom made the triangular channels. Considering the requirement of the channel capacity at field condition only the larger breadth slabs (40cm) were used for making the triangular channels (Fig.9.11).

Hydraulic performances and economy of the precast concrete channels

The performance of the channels was tested for their discharges, velocity, seepage losses, roughness coefficient, etc. the seepage losses were measured by ponding water for about a day (24 hours). The net loss through the unit-wetted area was calculated by deducting the evaporation loss during the period. The cost of fabrication for precast bamboo-reinforced sections or slabs, carrying to the sites and laying them with proper foundation support and earthing up, etc., have been considered in calculating the cost of per metre length of the channel.

Fig.9.10 The trapezoidal channel made of bamboo-reinforced precast slabs

Fig. 9.11 Triangular channel made of bamboo-reinforced precast slabs

Similarly, the cost involved in constructing the brick lined channel of same capacity has been considered for giving a comparison to cost of different type of channels. The seepage loss through 50cm and 36.4cm diameter channels were found very much close to each other (8.86 to 11.21 $cm^3/cm^2/day$) and average of which was 10.86 $cm^3/cm^2/day$ (Table 9.1). The average value of roughness coefficient (n) was found 0.0372which might be considered on high side. This was due to good number of concrete mix joints could not be made smooth and leveled for practical reasons. The water loss through the earthen channel was 130.72 cm^3/cm^2/day (Biswas et al, 1995). Thus, there was a reduction of 91.68% seepage loss through half round precast concrete channel. The roughness coefficients and seepage losses of the channels made of 40cm, 30cm and 20cm wide slabs varied from 0.0103 to 0.0964 and 5.33 to 12.63 $cm^3/cm^2/day$ with the average of 0.036 and 8.55 $cm^3/cm^2/day$ respectively (Table 9.2). The reduction of seepage loss was 93.45%. The average seepage loss and roughness coefficient were 15.76 $cm^3/cm^2/day$ and 0.0074 respectively in triangular shaped channel. Therefore, the roughness coefficient was on low side but the reduction of seepage loss was also comparatively lower (87.93%).

The cost of construction varied from Rs.95 to 150/m length of the channel for the channels of cross-sectional area 520cm^2 to 1696cm^2 respectively (Table 9.3). For similar cross-sectional area the estimated cost of brick lined channel varied from Rs.210 to 650/m length. Thus, there was saving of cost of construction between 54.76 to 67.94% with average of 64.02%. Alternately, the construction costs would be 2.21 to 3.12 times more for brick lined channels compared to bamboo-reinforced precast channels of different shape.

Table 9.1 Hydraulic characteristics of half round precast concrete channel

Channel No.	Dia. of channel ,cm	Bed slope (S),%	X-sectional area of channel , cm^2	Depth of flow, cm	Velocity cm/s	Discharge l/s	Maximum discharge capacity l/s	Wetted x-section -al area at flow cm^2	Wetted perimeter cm	Cement: Sand: Aggregate	Seepage loss m/day/ cm	Roughness coefficient (n)
1	50	0.36	981.74	12	29.97	10.86	29.42	362.36	53.49	1:02:04	8.36	0.033
2	50	0.356	981.74	11.8	29.48	10.43	29.84	353.84	50.7	1:02:04	10.67	0.034
3	36.4	0.271	520.31	14.2	26.54	9.98	13.81	375.91	49.11	1:02:04	11.21	0.035
4	36.4	0.151	520.31	15.4	22.01	9.22	11.45	418.79	51.55	1:02:04	8.24	0.033
5	36.4	0.817	320.31	8.65	31.9	8.96	16.59	280.87	37.07	1:02:04	10.33	0.051

Table 9.2 Hydraulic characteristics of different capacity and geometric channels made of flat slabs

Channel No.	Breadth (b)& depth (d) of channels, cm	Breadth (b)& depth (d) of channel slabs, cm	Side Slope, Z	Bed slope (S), %	X-section -al area of channel, cm^2	Depth of flow, cm	Velocity cm/s	Discharge l/s	Maximum discharge capacity l/s	Wetted x-section -al area at flow, cm^2	Cement: Sand: Aggre-gate	Seepage loss m/day/ cm^2	Rough-ness coeffi-cient (n)
Trapezoidal													
1	43,32	Side=40,5 Bottom=-do-	0.3125	0.0627	9.6	24.84	10.98	42.12	441	-	1:02:04	3.73	0.0176
2	30,32	Side=40,4 Bottom=-do-	0.577	0.213	11.6	25.4	10.82	29.71	425.6	-	1:02:04	12.63	0.038
3	30,26	Side=30,4 Bottom=-do-	0.4423	0.369	9.4	33.15	10.64	35.76	321.08	-	1:03:06	7.7	0.0291

4	23,25	Side=30,4 Bottom= 20,4	0.49	0.169	12.1	30.31	10.61	26.7	350.04	-	1:03:06	5.33	0.0247
5	34,26	Side=40,5 Bottom= 30,4	0.2205	0.3606	20	12.2	7.42	12.6	604.55	-	1:02:04	10.02	0.0964
6	24,23	Side=30,5 Bottom= 20,4	0.608	0.183	15.3	14.2	7.24	12.4	509.52	-	1:02:04	11.9	0.0103
7	0,26	Side=40,5	1.1	0.0319	15	28.92	7.16	21.48	247.5	-	1:02:04	15.79	0.0077
8	0,25	Side=40,5	1.16	0.0978	13.5	36.07	7.07	26.15	196.04	-	1:02:04	14.36	0.0071

Table 9.3 Cost of construction of different type of channels

S.No	**Type of channel**	**X-section of the channel, cm²**	**Cost of construction Rs./m length**		**Saving in cost (%)**
			Channel Channel under study	**Brick lined channel**	
1	Bamboo-reinforced precast half round channel	981	125	390	67.94
2	-do-	520	95	210	54.76
3	Bamboo-reinforced precast trapezoidal channel	1696	150	650	61.53
4	-do-	1175	135	415	67.46
5	-do-	880	130	375	65.33
6	Bamboo-reinforced precast triangular channel	750	120	365	67.12

ii) Shortcrete lining

Shortcrete is the mortar consisting a thorough mixture of cement, sand and water that is forced under pressure through a nozzle to the surface of the channel by means of compressed air. Shortcrete consists of usual mix of cement and sand at 1:4.5(with maximum sand size of 0.5cm). Shortcrete lining can be placed in an irregular surface thus avoiding the requirement of dressing the sub grade. It is particularly suitable for rock cuts and cracked and leaky surfaces of old concrete lining where structures are otherwise sound and still effective. The lining may be constructed with or without reinforcement (in the form of wire mesh or expanded metal). The reinforcement increases the longevity of lining specially when used on earth sub grade. Depending on the maintenance and temperature fluctuation the shortcrete may provide satisfactory service for a period of 20 years. However, it is less permanent in nature than concrete lining. The usual thickness of shortcrete is 3.5cm. The cost of shortcrete is more than concrete of equal thickness. Therefore, use of shortcrete is only location specific.

iii) Brick lining

The brick lining was first used in Haveli canal in Punjab. It consisted of two layers of 30.5cm x 14.7cm x 5.35cm tiles in cement mortar of 1.25cm of 1:3 cement sand mortar witched in between. The lining failed in some reaches within a year of backfill and insufficient free board resulting in saturation of backfill by the canal water itself. It was considered that the lining had never been considered for resisting pressure developing behind the lining for a particular thickness and there was nothing wrong with the specification of the lining itself. After the experience in Haveli canal the brick linings have been used on Sorda canal, Bhakra canal and the Rajasthan canal and so on taking in to full consideration of defects noticed on Haveli.

Brick or tile lining is cheaper than concrete lining and is preferred for long lengths in plains. Brick can be burnt at the site of the channel which can reduce the cost of transportation and ultimately the cost of construction compared to the concrete lining where stones and sand are transported over the long distance, maintenance cost in brick lining is low and the consumption of sand is low and can be laid by semi skilled labor without the use of any machinery (Fig.9.12). The advantage of brick lining over concrete lining is summarized as below.

i) Ordinary mason without any special training can lay brick lining.
ii) The thickness remains uniform unlike in concrete when the sub grade is not perfectly plane.
iii) Serious control on work is not required. In concrete lining, grading of aggregates, proportioning, water cement ratio, etc. are carefully controlled.
iv) Necessity of expansion joints is avoided. Coefficient of expansion of brick is nearly half of concrete. Shrinkage is practically eliminated.

v) Reduces the transport cost to a great extent because brick kiln can be established near the canal site. Stones are carried from far away for concrete lining. Low consumption of cement in brick lining also reduces the cost.

vi) Rounded section in brick can easily be had without the use of any form.

Similar to concrete lining the sub grade should be properly dressed and moistened before laying the lower layer of bricks or tiles over the layer of mortar. This lower layer of brick or tile should be wetted thoroughly before laying plaster on it and at least two days should be elapsed before placing the second layer.

Taley et al (1986) compared soil established mortar faced tiles, bricks and 400-gauge polythene as irrigation channel lining materials with unlined earth channels. The coefficient of roughness expressed as Manning's 'n' was least (0.007) for polythene lining followed by brick lining (0.0144) and soil established mortar faced tiles (0.0145) and greatest for unlined earth channel (0.0298). Seepage was greatest (0.1681 m^3/m^2/day) through the unlined earth channel followed by the brick lining (0.1472 m^3/m^2/day) and soil established mortar faced tiles (0.1214 m^3/m^2/day) and least in polythene laid channels (0.0432 m^3/m^2/day). Polythene lining was the cheapest material followed by soil brick lining was more durable and less susceptible to damage than tiles and polythene. Dwivedi et al (1999) found the seepage rate, average saving of water, and cost of channel lining for rectangular channel of brick lay in cement sand mortar as 35.78 l/m^2/day, 88.70% and Rs120.24/m^2, respectively.

Fig. 9.12 Brick lined channel

iv) Stone and concrete block lining

This type of lining is practiced in the area where suitable materials, such as sandstone or basalt are available in abundance. Dressed stones are very costly and considered for lining in short reaches while good-wearing surfaces is required for decoration purposes. Undressed stones are used in laying set in mortar. Seepage losses are much in stone lining when stones are not mortared properly. For the purpose of controlling the erosion only in steep slopes non-mortared stones are used as lining. The thickness of stone lining is generally 30 to 45cm.

Precast concrete block lining is similar to brick lining. The concrete blocks are manufactured centrally at control condition and carried to the sites where it is cheaper than brick lining. The concrete block lining is simple in repair and maintenance.

v) Burnt clay tile lining

Concrete, brick, stone, etc. are long established and conventional channel lining materials which are durable and considerably reduce the seepage loss. However, due to high cost, their adoption has remained limited to small parts of command areas under government-sponsored programmes. If efficient lining could be done with cheaper material and can be used with limited skill, it may be popular to the farmers.

Several workers in the past have attempted to identify suitable lining material for irrigation channels. Ahmad and Yadav (1986) found clay based lining materials to be suitable in terms of seepage reduction and possibility of in-situ adoption with less labour requirement. Bithu (1987) used and clay tiles for canal ling in desert region and concluded that the latter was better due to favorable thermal properties, which do not promote crack formation under temperature stress. Khair (1988) recommended soil cement tiles for lining irrigation channels in developing countries. Biswas (1997) conducted the experiment to develop village potter friendly clay tile fabrication technique and farmer-friendly layout procedure. These are described below.

Half round burnt clay tiles

Moulds and fabrication of half round burnt clay tiles

The moulds were formed by shaping the soil in half round of 40cm diameter and 60cm length over the leveled open ground or the wooden mould of same shape and dimension (Fig. 9.13 & 9.14). The moulds of higher diameters and lengths were also tried but were found unsuitable as the tiles either collapsed or broke during handling and drying.

Fig. 9.13 Earthen mould for half round burnt clay tiles

Fig. 9.14 Wooden mould for half round burnt clay tiles

Different type of soils in the locality were tested for its performance in keeping proper shape of the tiles by resisting the cracking and bending upon sun drying. Soils with high percentage of silt and clay worked well. The soil collected from dry pond bed was mixed with water and worked thoroughly to bring it to an appropriate consistence. A layer of this soil of about 1cm thickness was put around the mould by hand thumping. The desired shape and thickness was made by shaping it by the knife. About six to seven hours of sun drying was required when the properly shaped semi-dried tiles could be removed from the mould for further drying (Fig.9.15).

Fig. 9.15 Semi-dried clay tiles are put in drying

Fig. 9.16 Clay tiles are burnt in potter's kiln

The completely air dried tiles were burnt in artisan kilns of local type (Fig.9.16). About 5-10% tiles were cracked or partially broken during burning, most of which were made useable by joining the cracked or broken pieces by cement mortar.

Laying of the tiles to form the channel

The tiles were placed one after another on the properly leveled channel bed giving 0.45% slope following the existing slope of the field. These were then jointed by cement mortar at the faces to get a continuous channel. After jointing the tiles the required earth filling was done from both sides of the sections for proper support and setting in position. In general, at 20m intervals or at the required places one water diversion box was constructed to divert water to any side from that point.

Fig.9.17 Half round burnt clay tiles made channel in progress

Fig.9.18 Half round burnt clay tiles made channel in operation

The constructed channel was used round the year and the performance of the channel was compared to earthen channel in respect of seepage loss (Table 9.4&9.5).

Table 9.4 Seepage through burnt clay tiles

S. No.	Days after construction	Seepage rate, $m^3/m^2/day$ Dry surrounding		Wet surrounding		Average seepage rate, $m^3/m^2/day$
		Static	**Flowing**	**Static**	**Flowing**	
1	One week	0.023	0.027	0.02	0.024	0.024
2	9 months	0.011	0.011	0.01	0.01	0.01
3	1 year 3 months	0.013	0.014	0.011	0.011	0.012
	Average	0.016	0.017	0.014	0.015	0.015

Table 9.5 Seepage through earthen channel

S. No.	Days after construction Static	Seepage rate, $m^3/m^2/day$ Dry surrounding Flowing	Static	Wet surrounding Flowing		Average seepage rate, $m^3/m^2/day$
1	One week	1.08	1.19	1.03	1.14	1.11
2	9 months	1.148	1.52	1.33	1.43	1.44
3	1 year 3 months	1.13	1.22	1.1	1.16	1.15
	Average	1.23	1.31	1.15	1.24	1.23

Revealing the Table 9.4 it may be stated that the seepage loss practically stabilizes with time from a slightly higher initial value in the clay tile channel. This is attributed to 'ageing' during which the pores in the channel material get gradually sealed due to accumulation of finer particles flown in due to wind action. The overall average of the seepage rates in burnt clay tiles was found to be 0.015 $m^3/m^2/day$. it is much better compared to brick lining (0.1472 $m^3/m^2/day$), soil established mortar faced tiles (0.212 $m^3/m^2/day$) as reported by Taley and Kohale (1986), and better than LDPE lining over laid on the sides with cement pointed bricks and with 15cm soil cover on the bottom (rectangular section, 0.02904 $m^3/m^2/day$), less than cement pointed brick lining (1:1 slope, 0.0234 $m^3/m^2/day$), but higher than cement pointed brick underlined with LDPE (2:1 slope, 0.01368 $m^3/m^2/day$ as found by Tewari *et al.*, 1990.

Burnt clay flat tiles for making trapezoidal channels

Plane surface flat tiles

The flat tiles of 30cm and 20cm breadth and 60cm length of plane surfaces. The 60cm length and 30cm breadth of the tiles could not be obtained due to excess bending in drying process. At field trials the trapezoidal made of three pieces of 20cm breadth tiles found to be inadequate for carrying the discharge through the channels. Considering the field requirements practicality in making the tiles, the tiles were made in quantity for 40cm length, 25cm breadth and thickness of 2.2 cm (Fig.9.19).

By stretching the rope the bottom pieces were laid one by one straight on the bed of the channel. Each bottom piece was jointed by two other pieces at the sides forming different side slopes of the channels. The tiles were jointed each other by cement mortar. The both outside of the channels were provided with soil support and dressed properly time-to-time to avoid the risk of dismantling the joints and thereby collapsing the channel (Fig.9.20).

Fig.9.19 Plane surface flat tiles

Fig.9.20 Channel made of plane surface flat tiles

Roof tiles

The locally available roof tiles of size 45cm x 24cm x 1.75cm have been successfully used in constructing the channels. The surfaces of these tiles are usually uneven and rough due to its articulation, which are needed for its suitable placement to house roofs. With a little effort and practice by using the mason's kornick one side of these tiles can be had plane by removing the articulations. The other sides remain as such and provide strength to the tiles. The rough surface was used as outer surface when used in channels. Similar to the plane surface tiles these were used in three pieces to form the trapezoidal shape by jointing through cement mortar. (Fig. 9.19)

Fig.9. 21 The trapezoidal channel made of roof tiles in progress

v) Asphalt concrete lining

Asphalt is a petroleum product. It is used as a lining material with the sand and gravel similar to cement in concrete. The hot mix asphalt concrete is prepared, placed and compacted at high temperature. Asphalt concrete is well adopted in smaller canals where placement can be accomplished by using slip form similar to those used for cement concrete lining.

Asphalted concrete lining is cheaper than cement concrete lining since it can use low-grade aggregates. The aggregates, which are found unsuitable for cement concrete lining, can be suitably used in asphalted concrete. Actually the difference in cost of asphalt and cement and the suitability of local aggregates for asphalted concrete determine its selection over the cement concrete.

The maximum permissible velocity is 1.5m/sec in asphalted concrete lining. It is susceptible to weed growth, poor resistance to external hydrostatic pressure and the danger of sliding during hot weather. Sterilization of sub grade against weed growth is an essential part of this lining. The longevity of asphalted concrete lining ranges between 10 to 20 years according to Chowla (1992). Other disadvantages of asphalted concrete lining are danger of weed growth and are not adequately resistance to erosion and hence are not much durable.

vi) Soil - cement

Soil-cement lining, as the name implies is made up of a mixture of cement and soil. Soil-cement as lining is recommended in only to the soils where soils are sandy in nature, climate is mild and the other suitable lining materials are not available. The quality of lining greatly depends on the grade of the soils. Though most of the local soils are considered fit for use as lining material but better results are obtained by the soils with a maximum size of 20mm and the fine particles within 10 to 35% passing through the Indian Standard Sieve No.8.

The soil-cement mix is prepared by cement and soil mixed dry thoroughly and adding water to its optimum moisture content. The material is placed on sub grade compacted to required thickness commonly applied are 7.5 to 15cm. It is cured for at least 7 days with wet earth or straw.

The soil-cement lining is poor in weather resistant, does not permit higher velocity of flow, longevity is short and maintenance cost is high. Therefore, it is not generally recommended in a newly excavated canal where high velocity is proposed for increasing capacity by reducing excavation cost. The soil-cement lining is primarily used for the purpose of reducing the seepage losses. It is considered for minor canals of low velocities.

Kale et al (1988) found seepage losses at the rate of 0.85 and 0.75 l/m^2/h for the lined channels of cement + soil plaster (2:10) and cement + soil plaster (3:1), respectively. Hegade et al (1981) found seepage losses as 109.90 l/m^2/day in channels lined with soil cement plaster (5:1). Ranade et al (1993) found seepage rate of 9.5 l/m^2/h in soil-cement (8:1) paste of 5cm thick.

vii) Earth lining

The lining made of locally or at a reasonable distance available natural or processed soil is termed as earth lining. The selected soil should be of low permeability, high stability against disintegration and good resistance to erosion.

The stability of the soil materials may be improved by using bentonite, asphalt, emulsions, resins, cement, lime, etc. However, using the additives will increase the cost of lining. The proportional combination of gravel- sand-clay mixture considered being best earth material for lining.

Earth is used in the form of a silt layer in the channel either loosely or compacted to provide a more impermeable layer over the underlying permeable strata. With the advancement of soil mechanics and improvement of earth-moving equipments and techniques the earth lining has emerged as a common lining in some countries.

Earth lining considered to be suitable in the canals passing through sandy or gravelly materials and filling reaches to prevent seepage where other exposed linings are expensive and may develop cracks due to damage by weathering action, erosion and cracks on drying. Earth lining may be classified as (i) thin compacted earth lining, (ii) thick compacted earth lining, (iii) locally placed earth lining, (iv) stabilized soil, and (v) bentonite soil lining.

Thin compacted earth lining: Thin lining has very limited use because of the risk of damage due to erosion and its use confined to straight continuously running channel with less problem of weed growth. Shally (1965) observed that in some parts of Bengal and Madras the seepage losses from channels built in heavy clay soil are so low that lining is hardly necessary or justified. Misra et al (1990) used lateritic blocks in lining the channels. It reduced the infiltration rate from 43.82 to 12.0cm^3/cm^2/day and benefit: cost of this lining was found 1.7:1.

The thick compacted earth lining is made of compacted successive layers of suitable soils each not more than 15cm thick and the total thickness vary from 60 to 90cm depending on the size of the canal. The thick compacted earth lining is cheapest if the suitable soils are available by excavation of the canal or in reasonable distance.

Loosely placed earth lining: This is the loose, uncompacted cover of selected clayey soils on the bed and sides of the canal to provide an earth lining of 30cm thick. The serviceable life of this earth lining is relatively short. It is used where initial cost is the primary consideration and temporary seepage need to be checked. The performance of the lining depends largely on the fineness of the soil, impermeable character at loose state and stability against erosion.

Stabilized soil lining: If the soils are found low in permeability than the requirements for stable sides, erosion resistance and water tightness then the quality of the soil may be improved by using certain materials viz. asphalts, resins, chemicals, cement, petrochemicals, swelling clays, etc. Addition of these materials increases the performance of the lining to a large extent and also the cost of lining which discourages the use of stabilized soil lining.

Bentonite soil lining: The bentonite is a natural earth material characterized by high water absorption and swelling properties. The bentonite is impervious on swelling and found maximum use of it due to its high impermeability. It is very cheap if it can be obtained from local deposits. Sodium bentonits have higher swelling characteristics than calcium bentonites due to larger ionic radius of sodium (INCID, 2000).There are three methods for using the bentonite for lining the canal. These are (i) bentonite-water mixture, (ii) bentonite-soil mixture and (iii) bentonite membrane lining. In bentonite water mixture method a pond is made on the canal bed with the help of bunds where water is pumped and lump free bentonite slurry is made by using suitable dispersing agent and mixing arrangement. The pond water is allowed to stand for about two days when the slurry flow into the soil pores and seal it. This method is found effective in high permeable materials. In bentonite soil-mixture type bentonite is added 5-25% to the soil to provide a 5-10cm thick-finished layer of lining. Additional soil may be used to protect the lined layer where velocity of flow is high. In bentonite-membrane lining 3-5cm thick bentonite layer is laid on the canal sub grade and protected by 15-30cm thick soil cover or by brick or tiles in cement mortar. This type of lining is suitable in continuously running canals.

Buried membrane canal

A buried membrane consists of thin and impervious water barrier covered by a protective layer. The protective cover saves the lining from damage due to exposure high water velocity and movement of maintenance equipments and animals. The commonly used buried membrane linings are:

i) Sprayed-in-site asphalted membrane lining
ii) Prefabricated asphalted membrane lining
iii) Plastic film and synthetic rubber membrane lining
iv) Bentonite and clay membrane lining

i) Sprayed-in-situ asphalted membrane lining

This is made of high-softening-point asphalt sprayed in-situ at a high temperature (175^{0}C-210^{0}C) on a prepared sub grade to form 5-8mm thick waterproof barrier. A protective cover is given over the lining by the earth or gravel. It works satisfactorily if the thickness of the asphalted layer is not proved too thin to restrict the penetration of weeds rupture from movement of covering materials.

ii) Prefabricated asphalted membrane lining

This is the membrane made of asphalt-coated felts, fibre glass mats, asbestos etc. This type of lining is used in small channel and in short reaches where sprayed-in-situ asphalted lining is costlier due to its requirement of equipment and trained personnel. The membrane has the thickness of 3-6mm and is made of in rolls of standard size. Similar to spray-in-situ type a protective cover is given on the lining. The performance of lining mainly depends on the selection of an adequately water tight, durable and low cost material.

iii) Plastic film and synthetic rubber membrane lining

The plastic film and rubber membrane are popular and extensively used now days. The most commonly used materials are plastisized polyvinyl chloride (PVC), polyethylene (PE) and butyl rubber. The polythene considered to be most economical for buried flexible membrane lining. The plastic film and synthetic rubber membrane are very light, easy to transport and handle and can be manufactured at any size. However, these materials very susceptible to mechanical damage when exposed and therefore it is covered by soil soon after it is placed on sub grade. These are not good in the places where chances of damage of the flexible linings are more from roots of excessive weed growth.

iv) Bentonite and clay membrane lining

Bentonite has got the highly absorbent power and property in swelling of water. High quality bentonite contains up to 90 percent colloidal size particles and high water sealing power. It is spreaded 3-5cm over the sub grade and covered by 15-20cm thick soil or gravel.

9.3 Selection of Lining Materials

The lining of irrigation canal is very simple structure from the engineering point of view. However, it involves huge labor and materials. So, once the necessity of lining is decided it should be the best one in a given set of conditions. The following points need to be considered in selecting the type of lining.

i) Soil properties

In a soil of high clay or gypsum content the concrete or brick or any such type of rigid lining is not suggested because it can get damaged due to swelling of clays or gypsum. The thick compacted earth lining or a buried membrane, which are flexible in nature, may be more suitable. For hard surface lining of a canal if some unwanted soils are found which may be removed and replaced by sand or other suitable materials or canal direction may be altered. The best one is selected on the economic and practical feasibility. The cement concrete or soil cement lining may be adopted where suitable and sufficient sands and gravels are available. If the available soils are suitable for compaction earth lining may be considered. A membrane may be considered with a good cover material where the soils do not fall in the above categories. However, in all the cases capacity of the lining in resisting the erosion will be taken in to consideration.

ii) Water tightness

The purpose of lining is to provide suitable watertight barriers. The water tightness is measured by the permeability of lining. A lining has to fulfill these criteria. The seepage loss in a lined channel should not be more than 10% of the unlined channel.

Where the value of water is high, the aim should be maximum reduction of seepage losses. Most probably the thin plastic, asphalted or rubber sheeting overlying by concrete lining provides the best impermeable lining.

iii) Topography

The earth linings and buried membrane linings are suitable for mild or non-sloping land. Hard surfaces can be adopted in higher slopes. This is due to consideration of permissible velocity of flow.

iv) Durability

In general durability of lining depends on the quality of materials, accuracy of construction, operation and maintenance of the canal, etc. However, following are the factors, which affects much in durability of the linings.

a) Movement due to moisture and thermal changes

The lining and other base materials are subjected to alternate wetting and drying. Accordingly it goes on expansion and contraction. The lining should be low in coefficient of expansion in respect to thermal changes. The thermal coefficient of cement concrete and brick are $5.5 \times 10^{-6}\ ^{0}F$ and $3.0 \times 10^{-6}\ ^{0}F$ respectively.

b) Chemical attack

The material to be selected should be less susceptible to available chemicals in the soil and groundwater. The most destructive salts are sulphates or sodium and magnesium, which are commonly used in alkaline soils and groundwater. The lining materials should be free from these salts and its damaging effect can be much controlled by using 30/40 grade of maxphalt coating at the lower layer of lining.

c) Structural durability

The lining should be strong enough to resist the pressure due to different settlement of sub grade, head of water in channel, high sub-soil water table, saturation of back fill by rain, etc.

d) Hydraulic efficiency

It is measured by the coefficient of rugosity of the channel surface. Higher coefficient of rugosity reduces the velocity of flow. The characteristics of providing smooth surface of any lining material are preferred in selecting a lining material.

e) Land value

In the area where the land value is high the lining materials that require large cross-section of the channel are usually avoided. Hard surface steep side slope is preferred and accordingly the material is selected.

f) Operation and maintenance

With the advance of time the lining may get damaged. It should be capable of being repaired easily and economically. The channels go on full supply of water, empty or frequent change of water level. A hard surface channel is better in this situation. In regard to weed control, repair or silt removal the buried membrane lining has little advantage over unlined channels.

g) Availability of construction materials and labor and machinery

The materials, which are locally available either in the site or within the reasonable distance, are more acceptable on the economic point of view. The hired materials from long distance may increase the lining cost to such extent that the lining project is become uneconomical. If the high transport cost is unavoidable, after due consideration to all other aspects lining by soil-cement, prefabricated concrete or possibility of manufacturing bricks in the site may be explored.

Labor in some areas is surplus and inexpensive. Labor extensive lining may be thought for these areas. The cement concrete lining is more suitable for machine installation but stone and brick are more labor consuming.

h) Cost benefit

Whichever the lining is used it has to be economically satisfactory. Thorough study is required in respect of cost of construction and maintenance, saving of water, etc. and ultimately the cost: benefit of the lining material, which provides high benefit/cost ratio should be used.

9.4 Economics of Canal Lining

Lining is justified only when annual benefit of lining exceeds the annual expenditure due to construction of lined canal. The method of calculation of annual benefit and cost are discussed below.

Let C = Cost of construction of lined channel, Rs/m^2.

s &S = Seepage losses in the unlined and lined canal respectively, m^3/m^2/day.

p &P = Wetted perimeter in unlined and lined sections respectively, m^2.

T = Maximum perimeter of lining, m

d = Number of running days of canal, days/year

W = Value of water saved, Rs/m^3

L = Length of canal, m

M = Annual saving in rupees in operation and maintenance due to lining.

B = Annual estimated value of other benefits in rupees. The other benefits may include the benefits due to prevention of water logging, reduced risk of breaching, reduce change of health hazards, increased rate of land value, etc.

X = Rate of interest on investment, percent/year.

i) Annual value of water lost by seepage from unlined section =pLsdW rupees

Annual value of water lost by seepage from lined section =PLSdW.

Annual saving in value of water otherwise lost by seepage= (pLsdW-PLSdW)

= LdW(ps-PS) rupees

Total annual benefits out of lining the canal= LdW(ps-PS) +B+M

ii) Additional expenditure due to lining the unlined channel=TLC

Depreciation

It is expected that the cost of the lining should be recovered from the benefits of lining within the useful life of lining. The determination of depreciation is required in estimating the cost of service of any material or establishment. Depreciation is the loss in value and service capacity of the material due to different reasons with the passage of time. Longer the service life of the material in hours, days or years, the lower the annual rate and cost. There are different methods in calculating the depreciation. For the requirement of cost of use of service of the material or establishment, the straight-line and compound method of estimating depreciation are more useful than other methods.

Straight-line method

The straight-line method reduces the value of the material by an equal amount each year during its useful life. The annual depreciation is

$$D = \frac{C - S}{L} \tag{9.1}$$

where,

C = original purchase price

S = salvage, trade-in or resale value at the end of the service life

L = service life

Compound- interest or sinking fund method

Compound interest method of computing depreciation provides for the payment of an equal amount in to a sinking fund, which if invested at compound interest, together with the trade-in value, would replace the machine at the end of its service life. With this method the annual depreciation charge is the sum of the equal annual amount paid in to the sinking fund and the compound interest owned by the total amount in this fund. The annual charge for depreciation therefore increases with the age of the material, as interest earned by the sinking fund grows larger.

Let a is the amount required to be deposited each year to meet up the depreciation for the period of useful life of lining (Y) and rate of interest on investment, (x%). Then, a can be determined following the principle of compound interest method of calculating the depreciation of any material.

Year	1	2	3	4	5…….	Y-1	Y
Sinking fund	a	a	a	a	a	a	a
Sinking fund+ Interest after end of the year	a_1	a_2	a_3	a_4	a_5	a_{Y-1}	a_Y
Depreciation	a_1	a_1+a_2	$a_1+a_2+a_3$		……..		

Sinking find of amount a at the end of 1 year is a_1. If rate of interest is x, then,

$a_1= a(1+x)$

Similarly, $a_2= a(1+x)(1+x)$, after 2 complete year

$a_2=a(1+x)^2$

$a_3= a(1+x)^3$,, 3 ,, ,,

$a_y= a(1+x)^y$,, Y ,, ,,

So, the additional expenditure on construction for lining, TLC = $S+a(1+x)+a(1+x)^2+.....a(1+x)^{Y-1}+a(1+x)^Y$

Let, $1+x = m$

$TLC = S+ am+am^2+am^3+...........am^{y-1}+am^y$

Let, $m^Y+m^{y-1}+........+m^3+m^2+m=Y$

$Y = m(m^{Y-1}+m^{y-2}+.....+m^2+m+1)$

$= m(Y-m^Y+1)$

$= mY-m^{Y+1}+m$

$Y(1-m)=m(1-m^Y)$

$$\text{Or, } Y = \frac{m(1-m^Y)}{(1-m)} = m(m^Y-1)/(m-1)$$

$TLC = S+am(m^y-1)/(m-1)$

$= S+a(1+x)\{(1+x)^Y-1)\}/(1+x-1)$

$$= S+a(1+x)/x\{(1+x)^Y-1\} \quad (9.2)$$

Let the value at the end of nth year is P_n.

Therefore, $P_n = TLC - D_n$

Where D_n is the depreciation up to n^{th} year similar to **Eq.9.2**. The depreciation amount up to n^{th} year is

$D_n = a(1+x)/x\{(1+x)^n - 1\}$

$\therefore P_n = S + a(1+x)/x\{(1+x)^y - 1\} - a(1+x)/x\{(1+x)^n - 1\}$

$= S + a(1+x)/x\{(1+x)^y - (1+x)^n\}$

The depreciation amount in a particular year, say the nth year

$= a(1+x)/x\{(1+x)^n - 1\} - a(1+x)/x\{(1+x)^{n-1} - 1\}$ or $a(1+x)^n$ (9.3)

Example 9.1 A brick lined channel of 500m length has been constructed at a cost of Rs1, 00,000/-. Determine the amount to be deposited/year to meet up the depreciation of the channel. Assume 12 percent rate of interest and useful life of the channel as 20 years. Determine the depreciation in 10th year of channel following compound interest method.

Solution:

We have, TLC =Rs1, 00,000/-

$x = 0.12$

$Y = 20$ years

$1{,}00{,}000 a(1+0.12)/0.12\{(1+0.12)^{20} - 1\}$

$= a \times 9.33 \times 5.65$

$\therefore$ a = Rs1, 239.16

Depreciation amount in 10th year of the channel

$= a(1+x)^n$

$= 1239.16(1+0.12)^{10}$

= Rs.3848.46

Example 9.2 A trapezoidal shape concrete lined channel of 500m was constructed to accommodate 10 cumecs of flow. Work out the economics of lining from the following data.

Bottom of the channel=2m

Depth of water in channel=2m

Depth of channel=2.4m

Side slope of channel=1:1

Wetted perimeter of unlined channel=10m

Seepage through unlined channel=1.2m^3/m^2/day

Seepage through the lined channel=0.01/m^3/m^2/day

Cost of water = Rs.0.2/m^3

Annual maintenance cost of the unlined channel = Rs5.00/10m^2

Cost of channel lining = Rs500/m^2

Assume additional data suitably, if required.

Solution:

Wetted perimeter of the lined channel, $P = B + 2d\sqrt{Z^2 + 1}$

$= 2 + 2x2\sqrt{1+1} = 7.66m$

Perimeter of the lined channel, $P = B + 2D\sqrt{Z^2 + 1}$

$= 2 + 2x2.4\sqrt{1+1} = 8.79m$

Wetted surface of the lined channel = 7.66x500 = 3830m^2

Surface area of the lined channel = 8.79x500 = 4395m^2

Wetted surface of the unlined channel=10x500=5000m^2

Seepage loss through lined channel

= 3830m^{2}x0.01m^3/m^2/day

=38.3m^3/day

Seepage loss through unlined channel

=1.2m^3/m^2/day x 5000m^2

=6000m^3/day

Saving of water due to lining

=6000-38.3=5961.7m^3/day

Let us assume that the channel is running for 250 days/year

Annual saving of water =5961.7x250=1490425m^3

Annual benefit for saving of water due to lining=1490425x0.20

= Rs2, 98,085/-

Annual maintenance cost for unlined channel=Rs.5.00/10m^{2}x5000m^2

= Rs. 2500.00

Assuming 40% of this amount is saved in lined channel.

Annual saving in maintenance charge=2500x40/100=Rs.1000.00

Cost of lining = Rs.500/ m^2x4395m^2 = Rs.2197500/-

Assuming 10% rate of interest on investment, average annual interest

$= \text{Rs.}2197500 / 2x\frac{10}{100} = \text{Rs.}109875/-$

Assuming 30 years of useful life of lining.

Depreciation/year = Rs. 2197500/30 = Rs.73250

Total annual cost = Rs.109875+Rs.73250 = 183125

Total annual benefit = Rs.298085+Rs1000-=Rs.299085

Benefit-cost ratio = Rs.299085/Rs.183125-= 1.63

Example 9.3 A field irrigation channel of a spout in a deep tube well (DTW) command area was lined by using bamboo-reinforced precast concrete slabs forming in trapezoidal shape. Determine the benefit: cost ratio of lining from the following details of DTW and channel.

i) DTW

a) Cost of DTW irrigation system (P) = Rs.850000

b) Yearly operating time, T_1 = 1320 h

c) Yearly operator's wage = Rs70, 000

d) Yearly repair and maintenance cost = Rs50, 000

e) Yearly power charge @ Rs. 3.00/unit = Rs.47520

f) Expected life of DTW = 30 years

g) Rate of interest on investment, r_1 = 12%

h) Discharge of DTW, Q_1 = 40 l/s

i) Salvage value, S_1 = 10%

j) Discharge of the spout = 10l/s

ii) Trapezoidal channel (x-sectional area=881.3cm^2) made of bamboo-reinforced precast slabs.

1. Length of channel = 250m
2. Cost of construction of channel @ Rs130/m, P_2 = Rs. 32500
3. Expected life of channel = 15 years
4. Interest on investment, r_2 = 12%
5. Salvage value, S_2 = Nil
6. Repair and maintenance = Rs. 1000/year
7. Operating period, T_2 = 550 h/year
8. Repair and maintenance of unlined channel = Rs. 2000/year
9. Saving of water due to lining = 25% of spout source

Solution:

Calculation for cost of unit volume of water of DTW

i) Depreciation $= \frac{P_1 - S_1}{L_1} = \frac{8,50,000 - 8,5000}{30} = Rs.25500 / yr$

ii) Interest $=\frac{P_1 - S_1}{2} x r_1 = \frac{850000+85000}{2} x 0.12 = 56100/yr$

iii) Repair and maintenance=Rs.50, 000/yr

iv) Power charge = Rs.47420

v) Operator's wage = Rs.70000/yr

Total of (i) to (v) = Rs.249120

vi) Discharge = $Q_1 x T_1$ = 40x1320x3600 = 190080 l/yr

vii) Cost of water = 249120/190080=Rs.1.31/m^3

Calculation of cost of lining

i) Depreciation $=\frac{P_2 - S_2}{C_2} = 32500/15 = \text{Rs.}2167/\text{yr}$

ii) Interest $= \frac{P_2 + S_2}{2} xr_2 = \frac{32500x12}{2x100} = \text{Rs.}1950/\text{yr}$

Total of (i) and (ii) = Rs.4117/yr

Saving of water due to lining

i) Yearly discharge through the spout = $Q_2 x T_2$

= 10l/sx550h

= 10x550x3600 h=19800m^3

ii) Saving of water $= \frac{19800x25}{100} = 4950m^3$

Benefit of lining

i) Saving in repair and maintenance=2000-1000=Rs.1000/yr

ii) Other benefit= Rs.750/yr (say)

iii) Benefit due to water saving = Volume of water saved x Cost of unit volume of water

= 4950x1.31=Rs.6485/yr

Total benefit = Rs.8235/yr

So, benefit: cost of lining=8235/4117=2:1

Example 9.4 The non-conventional earthen roof tiles were used for lining a field irrigation channel in a shallow tube well command at Nalkuri village in Nadia district of West Bengal. Work out the economics of lining from the following data of STW and channel.

STW

1. Cost of STW = Rs.50000

2. Yearly operating time = 2000 h
3. Yearly repair and maintenance cost = Rs.1250
4. Expected service life = 30 years
5. Discharge of STW = 6.5l/s
6. Salvage value = 10%
7. Rate of interest on investment = 12%
8. Operator's wage = Rs.12/h

Earthen roof tile trapezoidal channel (730 to 815cm^2 x-sectional area)

1. Length of the channel = 250m
2. Cost of construction of the channel @ Rs85/m = Rs.21250
3. Expected life of channel = 15 years
4. Interest on investment = 12%
5. Salvage value = Nil
6. Yearly repair and maintenance cost of lined channel= Rs.1000
7. Operating period = 1500h/year
8. Yearly repair and maintenance cost of unlined channel = Rs.2000
9. Saving of water = 25%
10. Yearly consumption of power = 7000 units
11. Cost of power =Rs.3/unit

Solution:

Cost of unit volume of water

i) Depreciation = $\dfrac{50000-50000x0.1}{30} = Rs.1500/year$

ii) Interest = $\dfrac{50000+5000}{2}x12/100 = Rs.3300/year$

iii) Repair and maintenance per year = Rs.1500

iv) Operator's wage = 2000x12=Rs. 24,000/*year*

v) Power charge = Rs.21000/*year*

Total of (i) to (v) = Rs.51300/*year*

vi) Discharge/year = 6.5*x*3600*x*2000 = 46800000*l* = 46800m^3

vii) Cost of water = $\dfrac{Rs.51300}{46800m^3} = Rs.1096/m^3$

Cost of channel

i) Depreciation = $\frac{Rs.21250/-}{15 years} = Rs.1416.67/year$

ii) Interest = $\frac{21250}{2} x12/100 = Rs.1275/year$

Total of (i) and (ii) = Rs.2691.67/year

Benefit of channel

i) Saving in repair and maintenance = Rs.1000/year

ii) Other benefit =Rs.750/year (say)

iii) Benefit due to water saving

$$\frac{46800m^3 x25}{100} xRs.1096/m^3 = Rs.2823.2/year$$

Total of (i) to (iii) = Rs.14573.2/year

Benefit: Cost of the channel = 14573.2:2691.67 = 5.41:1

Priority of Water Courses in Lining

About 50-60% of delivered water gets lost through conveyance system in a canal irrigation system and almost half of this occurs in field irrigation channels. Lining is more or less the only method to check the high seepage loss through the irrigation channels. Due to fund constraint the watercourses can be fixed on piece meal basis by fixing priorities among the outlets.

Seepage loss (S) per unit time for a given section of a channel is the function of length of the section (L), its wetted perimeter (P) and soil type. Therefore,

S= f(L ,P, soil type) (9.4)

The soil type in a channel section generally does not change and the wetted perimeter is fairly uniform. Therefore, seepage losses in volume per unit time of a channel section can be represented by

$S=aL^b$ (9.5)

Where a and b are empirical constants whose values depend on soil type. The product of S and time, t, the channel sections runs gives the seepage index (SI) as

$SI = atL^b$ (9.6)

The seepage index gives the basis of working out lining priorities.

The priorities are to be decided for (i) the length of the field channel to be lined and (ii) priorities between different canal outfits.

Length of channel to be lined: The channel is divided into sections, which run individually. Determine the SI value for all the channel sections. Get the cumulative SI values in a separate column. Get the percent of SI value to total SI value up to total channel length in another column. By comparing this percent of SI column with the column of total length one can select the channel section which to be lined for saving pre-decided (say 80%) seepage.

Outlet priority: The priority among the outlet may be determined using following formula

$$P_r = \frac{\sum_{i=1}^{n}\sum_{j=1}^{m}(SI)i,j}{\sum_{j=1}^{m}(L_0)} \tag{9.7}$$

Where,

P_r = Priority value of an individual outlet

L_0 = Optimum length up to n^{th} section of channel to be lined

M = Number of channels at the outlet priority values in descending order determine the rank of priority.

Example 9.5 Calculate the seepage index (SI) and percent to total SI from the data of a DTW inlet and fix up the channel length to be lined for 80% saving of seepage loss.

Channel section No.	Length of channel section, m	Time for each section runs, h	Seepage loss, $m^3/m^2/h$
A	50	25.2	0.18
B	65	24.5	0.2
C	120	11.2	0.15
D	150	5.5	0.22
E	70	4.0	0.15
F	60	3.5	0.16
G	35	1.5	0.2

Solution:

Channel section. No	Length of channel section, m	Cumulative length of channel, m	Time for each section runs, h	Seepage loss, $m^3/m^2/h$	Seepage Index(SI) (Col.4xCol.5)	Cumulative Seepage Index	Percent to total SI (Col.7÷Col.6)
1	2	3	4	5	6	7	8
A	50	50	25.5	0.18	4.59	4.59	33.16
B	65	115	24.5	0.20	4.9	9.49	68.57
C	120	235	11.2	0.15	1.68	11.47	80.71
D	150	385	5.5	11.22	1.21	12.38	89.45
E	70	455	4.0	0.15	0.6	12.98	93.79
F	60	515	3.5	0.16	0.56	13.54	97.83
G	35	550	1.5	0.20	0.30	13.84	100.00

By revealing the Col.8 it appears that for 80.71% saving of seepage loss of the channel section to be lined are A, B & C of total length. For exact 80% saving of seepage loss, the length of channel section C to be taken

$$= 120 - \frac{120}{(80.71 - 68.57)} x0.71 = 120 - 7.02 = 112.98\text{m}.$$

So, the total length at 80% saving of seepage=115+112.98=227.98m.

Example 9.6 The channel lengths to be lined at 80% saving of seepage loss and cumulative seepage index at different outlet of a canal irrigation system were as below. Determine the priority values and priority ranking of the outlets.

Outlet No.	Name of the channel	Length to be lined (L_0)	Seepage Index
A	A_1	327	20.8
	A_2	1200	90.5
	A_3	2500	250
B	B_1	725	25.5
	B_2	400	36.4
	B_3	900	400.2
	B_4	300	126.25
C	C_1	350	75.8
	C_2	475	112
	C_3	380	90.33

Solution:

Outlet No.	Name of the channel	Length to be lined (L_0)	Cumulative Seepage Index	Priority value, $\sum \text{Col.4} \div \text{Col.3}$	Priority ranking
1	2	3	4	5	6
A	A_1	327	20.8	0.09	3
	A_2	1200	90.5		
	A_3	2500	250		
		$\sum 2325$	$\sum = 361.30$		
	B_1	725	25.5	0.25	1
B	B_2	400	36.4		
	B_3	900	400.2		
	B_4	300	126.25		
		$\sum = 2325$	$\sum = 588.35$		
C	C_1	350	75.8	0.23	2
	C_2	475	112		
	C_3	380	90.33		
		$\sum = 1205$	$\sum = 278.13$		

9.5 Measurement of Seepage

Seepage through channels may be measured in non-flow and flow condition following (i) Ponding and (ii) Inflow-outflow method, respectively.

Ponding method

The seepage through the channels is determined by ponding water in it for a specified period. The difference in volume of water during the period of ponding is the volume of water which losses through seepage. The seepage volume is corrected by deducting the evaporation losses occur from the average surface area of water for the period. The rate of seepage is determined by dividing the seepage volume by the average wetted area of the channel.

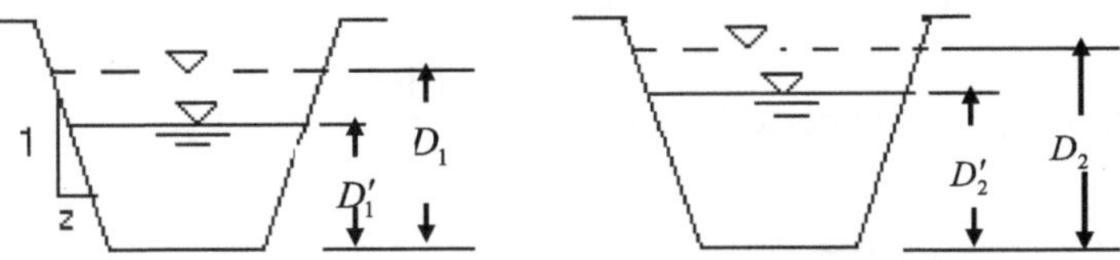

(a) Up stream channel section (b) Down stream channel section

Fig. 9.22 Upstream & downstream wetted sections of channel during experimentation

Following are the description of computation of seepage loss through ponding water for different shaped channel.

Trapezoidal channel

The bed width of the channel = B in m

The length of channel = L in m

The depth of water at u/s of the channel at time T_1= D_1 in m

The depth of water at d/s of the channel at time T_1= D_2 in m

The depth of water at u/s of the channel at time T_2= D'_1 in m

The depth of water at d/s of the channel at time T_2= D'_2 in m

The side slope of the channel = z: 1

Wetted x-sectional area of the channel at u/s at T_1, $A_1 = BD_1 + zD_1^2$

Wetted x-sectional area of the channel at d/s at T_1, $A_1 = BD_2 + zD_2^2$

Wetted x-sectional area of the channel at u/s at T_2, $A'_1 = BD'_1 + zD'^2_1$

Wetted x-sectional area of the channel at d/s at T_2, $A'_2 = BD'_2 + zD'^2_2$

Volume of water in the channel at T_1, $V_1 = \dfrac{A_1 + A_2}{2} xL$

Volume of water in the channel at T_2, $V_2 = \dfrac{A'_1 + A'_2}{2} xL$

The seepage losses through the channel for the period (T_2-T_1) including evaporation losses, $V_s = V_1 - V_2$

The evaporation of water = E in m/day

Wetted perimeter of trapezoidal channel at u/s at $T_1 = P_1$

Wetted perimeter of the channel at d/s at $T_1 = P_2$

Wetted perimeter of the channel at u/s at $T_2 = P'_1$

Wetted perimeter of the channel at d/s at $T_2 = P'_2$

Average wetted perimeter at $T_1 = \frac{P_1 + P_2}{2} = P_1$

Average wetted perimeter at $T_1 = \frac{P'_1 + P'_2}{2} = P_2$

Average wetted perimeter for the period $(T_2\text{-}T_1) = \frac{P_1 + P_2}{2}$

Average wetted area for the period, $A_w = \frac{P_1 + P_2}{2} \times L$

The average top width of water surface at $T_2, W_1 = \frac{(B + 2zD'_1) + (B + 2zD'_2)}{2}$

The average top width of water surface at $T_2, W_2 = \frac{(B + 2zD'_1) + (B + 2zD'_2)}{2}$

The average top width of water surface for the period of $(T_2\text{-}T_1)$, $W = \frac{W_1 + W_2}{2}$

The average surface area of water for consideration to evaporation loss, $A_e = WxL$

The volume of water lost through evaporation, $V_e = (W \times L)E\frac{(T_2 - T_1)}{24}$

The net seepage through the channel, $V_n = V_s\text{-}V_e$

The seepage rate, $S_r = \frac{V_a}{A_w}$ for the period $(T_1\text{-}T_2)$

$$= \frac{V_n x24}{A_w(T_2 - T_1)} \text{ m}^3/\text{m}^2/\text{day} \tag{9.8}$$

Example 9.7 Following data were obtained in ponding water during 10.15 hour to 9.00 hour of next day in a bamboo reinforced precast concrete slab made = trapezoidal shape channel at Mohanpur Instructional Farm in Nadia District of West Bengal. Compute the seepage losses through the channel. The depth of water at upstream and downstream of the channel was found 0.175m, 0.053m and 0.22m, 0.098m at the beginning and end of the period of study respectively. The width of channel bottom and side slope is 23cm and 0.49: 1 respectively. Assume evaporation rate as 2.8 mm/day.

Solution:

Wetted x-sectional area of the channel at u/s at 10.15 h, $A_1 = BD_1 + zl$

$= 0.23x0.175 + 0.49(0.175)^2$

$= 0.04 + 0.055 = 0.005m^2$

Wetted x-sectional area of the channel at d/s at 10.15h, $A_2 = BD_2 + zD_2^2$

$= 0.23x0.22 + 0.49(0.22)^2$

$= 5.06x10^{-2} + 2.37x10^{-2}$

$= 0.0743m^2$

Wetted x-sectional area of the channel at u/s at 9.00h $A'_1 = BD'_1 + zD_1'^2$

$0.23x0.053 + 0.49(0.053)^2$

$= 1.22x10^{-2} + 1.377x10^{-3}$

$= 1.356x10^{-2}m^2$

Wetted x-sectional area of the channel at d/s at 9.00h $A'_2 = BD'_2 + zD_2'^2$

$= 0.23x0.098 + 0.43x(0.098)^2$

$= 2.254x10^{-2} + 4.71x10^{-2}m^2$

$= 2.72x10^{-2}m^2$

Volume of water in the channel at 10.15h,

$$V_1 = \frac{A_1 + A_2}{2} xL$$

$$= \frac{0.095 + 0.0743}{2} x17.1$$

$= 1.447m^3$

Volume of water in the channel at 9.00h

$$V_2 = \frac{A'_1 + A'_2}{2} xL$$

$$= \frac{1.356x10^{-2} + 2.72x10^{-2}}{2} x17.1$$

$= 0.35m^3$

The seepage through the channel for the period including evaporation,

$V_s = V_1 - V_2$

$= 1.447 - 0.35 = 1.097m^3$

Wetted perimeter of the channel at d/s at 8-15 hr, $P_1 = B + 2D\sqrt{z^2 + 1}$

$= 0.23 + 2 \text{ X } 0.175 \sqrt{(0.49)^2 + 1}$

$= 0.23 + 0.39 = 0.62m$

Wetted perimeter of the channel at d/s at 8.15h

$P_2 = B + 2D \sqrt{z^2 + 1}$

$= 0.23 + 2 \text{ x } 0.175 \sqrt{(0.49)^2 + 1}$

$= 0.23 + 0.49 = 0.72m$

Wetted perimeter of the channel at u/s at 9.00h

$P^1_1 = B + ZD \sqrt{z^2 + 1}$

$=0.23+2x0.053\sqrt{(0.49)^2+1}$

$=0.23+0.1118=0.35m$

Wetted perimeter of the channel at d/s at 9.00h

$P^1_2=B+2D\sqrt{z^2+1}$

$=0.23+2x0.098\sqrt{(0.49)^2+1}$

$=0.45m$

Average wetted perimeter at 8.15h

$$P_1=\frac{P_1+P_2}{2}=\frac{0.62+0.72}{2}=0.67m$$

Average wetted perimeter at 9.00h

$$P_2=\frac{P_1+P_2}{2}=\frac{0.35+0.45}{2}=0.40m$$

Average perimeter for period of study,

$$P=\frac{P_1+P_2}{2}=\frac{0.67+0.40}{2}=0.535m$$

Average wetted area for the period of study, $A_w = PxL = 0.535x17.1 = 9.15m^2$

Average top width of water surface at 10.15h,

$$W_1=\frac{(B+2zD_1)+(B+2zD_2)}{2}$$

$= B+z\ (D_1+D_2)$

$=0.23+0.49(0.175+0.22)$

$=0.424m$

Average top width of water surface at 9.00h, $W_2 = B + z(D'_1 + D'_2)$

$= B+z\ (D_1+D_2)$

$=0.23+0.49(0.053+0.098)$

$=0.304m$

Average top width of water surface during the period of study, $W=\frac{W_1+W_2}{2}$

$$=\frac{0.424+0.302}{2}=0.364m$$

Average top width of water for consideration the evaporation loss, $A_e=WxL=0.364x17.1=6.224m^2$

Period of study 8.15h to 9.00h of next day = 24*h* 45 min = 24.75*h*

Volume of water lost through evaporation, $V_e = A_e x E \frac{24.75}{24}$

$= 6.224x0.0028x \frac{24.75}{24} = 1.795x10^{-4} m^3$

The net seepage loss through the channel

$V_n = V_s - V_e = 1.097 - 1.65x10^{-2} = 1.08m^3$

$$\text{Seepage rate,Sr} = \frac{V}{A_w x \frac{24.75}{24}} = \frac{0.783}{9.15x \frac{24.75}{24}} = 1.145x10^{-1} m^3/m^2/day$$

Half round channel

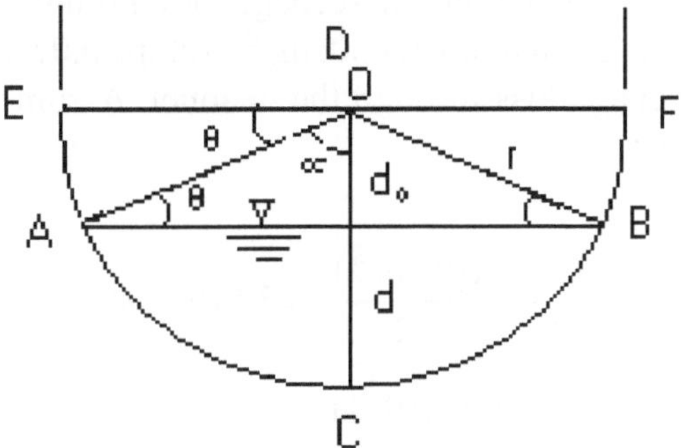

Fig. 9.23 Half round channel section

The diameter of the channel = D

The radius of the channel, r = D/2

The maximum x-sectional area of the channel = $\frac{\pi \Delta^2}{4x2}$

The depth of water in the channel = d

The wetted x-sectional area of the channel at depth of d (Fig. 9.23).

A=Area ABC = Half circle-AreaOAB-AreaOAEx2

$$= \frac{\pi r^2}{2} - d_0\sqrt{r^2 - d_0^2} - \frac{\pi r^2}{360}.\theta.2, \text{where}, d_0 = r - d$$

$$= \frac{\pi r^2}{2} - d_0\sqrt{r^2 - d_0^2} - \frac{\pi r^2}{180^0} Sin^{-1}\left(\frac{d_0}{r}\right) \tag{9.9}$$

The wetted perimeter at water depth d,

P=Perimeter of half circle-AE-BF

$$= \pi r - \frac{2\pi r}{360^0}.\theta - \frac{2\pi r}{360^0}.\theta$$

$$= \pi r - \frac{\pi r\theta}{180^0} - \frac{\pi r\theta}{180^0} = \pi r - \frac{\pi r\theta}{90^0}$$

$$= \pi r - \frac{\pi r\left(90^0 - \alpha\right)}{90^0} = \pi r - \pi r + \frac{\pi r\alpha}{90^0}$$

$$= \frac{\pi r\alpha}{90^0} \qquad (9.10)$$

Top width of water at depth d, AB=2 $\sqrt{r2 - d_0^2}$

Example 9.8 The depth of water recorded at u/s of the channel 10.5cm and 3.1cm at 7.15h and 21.30h respectively in studying the seepage loss through a burnt clay tile made field irrigation channel of diameter 30cm, bed slope 0.07% and length 58.2m. Determine the seepage loss through the channel. Assume evaporation at the rate of 2.8mm/day.

Solution:

Maximum x-sectional area of channel = $\frac{\pi r^2}{2} = \frac{\pi(0.30/2)^2}{2} = 3.5x10^{-2} m^2$

Drop of channel bed at 58.2m = $\frac{0.07}{100} x58.2 = 4.07x10^{-2} m$

∴ Depth of water in the channel

d_1= 10.50 cm at u/s at 7.15h

d_2= 14.57 cm at d/s ,, ,,

d'_1= 3.10 cm at u/s at 21.30h

d'_2= 7.17 cm at d/s ,, ,,

d_0= 15-10.5 = 4.5*cm* at u/s at 7.15h

d_0 = 4.5-4.07 = 0.43*cm* at d/s ,, ,,

d_0= 15-3.1 = 11.90*cm* at u/s at 21.30h

d_0= 11.90.4.07 = 7.830*cm* at u/s at 21.30h

Wetted x-sectional area of the channel (Eq.9.9)

$$A = \frac{\pi^2}{2} - d_0\sqrt{r^2 - d_0^2} - \frac{\pi r^2}{180^0} Sin^{-1}\left(\frac{d_0}{r}\right)$$

$$\therefore A_1 = \frac{\pi x 15^2}{2} - 4.5\sqrt{15^2 - 4.5^2} - \frac{\pi x 15^2}{180^0} Sin^{-1}\left(\frac{4.5}{15}\right)$$

$= 353.429\text{-}87.87\text{-}3.927 \times 17.457$

$= 220.495 cm^2 at\ u/s\ at\ 7.15h$

$$\therefore\ A_2 = \frac{\pi x 15^2}{2} - 0.43\sqrt{15^2 - (0.43)^2} - \frac{\pi x 15^2}{180^0} Sin^{-1}\left(\frac{0.43}{15}\right)$$

$= 353.429\text{-}6.447\text{-}3.927 \times 0.0287$

$= 346.869\ cm^2\ at\ u/s\ at\ 7.15h$

$$\therefore\ A'_1 = \frac{\pi x 15^2}{2} - 11.9\sqrt{15^2 - 11.9^2} - \frac{\pi x 15^2}{180^0} Sin^{-1}\left(\frac{11.9}{15}\right), d_0 = 15\text{-}3.10 = 11.9cm$$

$= 353.429\text{-}108.668\text{-}206.149$

$= 38.60cm^2\ at\ u/s\ at\ 21.30h$

$$\therefore\ A'_2 = \frac{\pi x 15^2}{2} - 7.83\sqrt{15^2 - 8.83^2} - \frac{\pi x 15^2}{180} Sin^{-1}\left(\frac{7.83}{15}\right), d_0 = 15\text{-}7.17 = 7.83cm$$

$= 353.429\text{-}100.178\text{-}123.568$

$= 129.68cm^2\ at\ u/s\ at\ 21.30h$

Volume of water in the channel at 7.15h, $V_1 = \dfrac{A_1 + A_2}{2} xL$

$$= \frac{220.485 + 346.869}{2} x58.2x100 = 1651000\ cm^3$$

$= 165m^3$

Volume of water in the channel at 21.30h, $\dfrac{A'_1 + A'_2}{2} xL$

$$= \frac{38.60 + 129.68}{2} x58.2x100 = 489694.8.4cm^3$$

$= 0.489\ m^3$

The seepage and the evaporation loss through the channel for the period 7.15h to 21.30h,

$V_s = V_1\text{-}V_2 = 1.65\text{-}0.489 = 1.161m^3$

Wetted perimeter in the channel,

$$P_1 = \frac{\pi ra}{90^0} = \frac{\pi}{90^0} x15xCos^{-1}\left(\frac{d_0}{r}\right)$$

$$= \frac{\pi}{90^0} x15xCos^{-1}\left(\frac{4.5}{15}\right)$$

$= 37.98\ cm$ at u/s at 7.15h

$$P_2 = \frac{\pi ra}{90^0} = \frac{\pi}{90^0} x15xCos^{-1}\left(\frac{d_0}{r}\right)$$

$$= \frac{\pi}{90^0} x15xCos^{-1}\left(\frac{0.43}{15}\right)$$

$= 46.26$ *cm at* d/s *at* 7.15h

$$P'_1 = \frac{\pi r\alpha}{90^0} = \frac{\pi}{90^0} x15xCos^{-1}\left(\frac{d_0}{r}\right)$$

$$= \frac{\pi}{90^0} x15xCos^{-1}\left(\frac{11.9}{15}\right)$$

$= 19.64$ *cm* at u/s at 21.30h

$$P'_2 = \frac{\pi r\alpha}{90^0} = \frac{\pi}{90^0} x15xCos^{-1}\left(\frac{d_0}{r}\right)$$

$$= \frac{\pi}{90^0} x15xCos^{-1}\left(\frac{7.83}{15}\right)$$

$= 30.65cm$ at d/s at 21.30h

Average wetted perimeter of the channel at 7.15h

$$= \frac{P_1 + P_2}{2} = \frac{37.98 + 46.26}{2} = 42.12cm$$

Average wetted perimeter of the channel at 21.30h,

$$= \frac{P_1 + P'_2}{2} = \frac{19.64 + 30.65}{2} = 25.15\,cm$$

Average wetted perimeter of the channel during 7.15h to 21.30h,

$$P_w = \frac{42.12 + 25.15}{2} = 33.635cm$$

Width of the water surface

$$W_1 = \frac{2\sqrt{\left(r^2 - {d_{01}}^2\right)} + 2\sqrt{\left(r^2 - {d_{02}}^2\right)}}{2}$$

$$= \frac{2\sqrt{\left(15^2 - 6.5^2\right)} + 2\sqrt{\left(15^2 - 2.43^2\right)}}{2}$$

$$= \frac{27.03 + 29.60}{2} = 28.31cm$$

$$W_2 = \frac{2\sqrt{\left(r^2 - {d_{01}}^2\right)} + 2\sqrt{\left(r^2 - {d_{01}}^2\right)}}{2}$$

$$= \frac{2\sqrt{\left(15^2 - 11.9^2\right)} + 2\sqrt{\left(15^2 - 7.8^2\right)}}{2}$$

$= 9.13 + 12.79 = 21.92$ cm

Average top width of water surface during 7.15h to 21.30h i.e., for 14.25h,

$$W = \frac{W_1 + W_2}{2} = \frac{28.31 + 21.92}{2} = 25.12cm.$$

The volume of water loss through evaporation during the period,

$V_e = WxLxE$, E = rate of evaporation = 0.28 mm/day

= 25.12 cm x 58.2 x 100 cm x 0.28 mm/day x 14.25h/24

$= 24303.1\text{cm}^3$

$= 0.0243\text{m}^3$

The seepage loss through the channel during the period

$V_n = V_s - V_e = 1.161 - 0.0243 = 1.1367\text{m}^3$

$$\text{Seepage rate} = \frac{V_n}{P_w x L} x \frac{24h}{14.25h}$$

$$= \frac{1.136x24}{\frac{33.535}{100}x58.2x14.25}$$

$= 8.608\ x\ 10^{-2} m^3/m^2/day$

Inflow-outflow method

To measure the loss of water in flowing condition, the channel is supplied with a constant flow of water by regulating the discharge at the upstream of the channel. The difference of discharge between the upstream and downstream of the channel at any time is the rate of seepage through the wetted area of the channel. The discharge through the downstream may be measured by using flumes or weirs. Average wetted area is obtained by using the average depth of flow and channel geometry. For an irregular shape earthen channel, the wetted area is determined by knowing the weighted perimeter of as many as possible channel sections. For nay section the depth of water at close intervals are to be taken. The average of these depths for different section is assumed to be the depth of water of the channel. Similarly the average of the widths of the channel at different sections is the width of the channel. Thus, the irregular shape channel is converted to a regular channel.

Seepage in varying condition

Seepage losses through the channel mainly depend on the type of lining in lined channel and the type of soil in the earthen channel. However, the surrounding soil condition and the age of use of the channel particularly in lined channel influence seepage losses. Seepage is more in dry than wet surrounding. Flowing channel records more seepage than ponded channel. Lined channel used to show slight higher seepage rate at the early age. It takes some time to be properly stabilized in respect to seepage through it.

Biswas & Mallick (1997) reported the seepage losses through the half-round burnt clay tile and earthen channels for flowing and non-flowing condition in dry and wet surroundings.They found the seepage loss more (0.023m^3/m^2/day) at one week of channel use than 9 months of use (0.011 m^3/m^2/day) for non-flowing and dry surrounding. This is attributed to ageing during which the pores in the channel material got gradually sealed by the accumulation of finer particles flown due to wind action.

Water losses through earthen irrigation channels

Only a portion of water delivered to farmland is beneficially available to the crops and one third of water diverted is lost in conveyance and distribution system through seepage, tail water losses and evaporation (Khair et al., 1984). Sur (1989) reported conveyance loss and efficiency in unlined field channel in the command area of Bhaini distributory in South-West Punjab. The overall losses were 24.2% from the channel.

Lenka (1985) reported that about 45% of the total water is lost in conveyance from the reservoir to the field (17% in main channel, 8% in distributaries and 20% in water courses). About 30% of the rest in the field and it is lost due to seepage and deep percolation (16.5%). Thus 61.62% of water is lost. About 1.5% is lost due to evaporation in all these conveyance courses. About 1% is lost due to weeds. The seepage losses in earthen channels were measured 1.296 $m^3/m^2/day$, which are about 17.82% of the DTW command supply for New Alluvial Agro-climatic Zone of West Bengal (Biswas, 1995). Dwivedi et al. (1999) found average rate of seepage as $3.578 x 10^{-2} m^3/m^2/day$ in unlined compacted earthen channel.

Example 9.9 Study was conducted to determine the seepage loss in trapezoidal shape earthen roof tile made channel of length 100m at wet surrounding. A constant discharge rate of 12 l/s was applied at the head end of the channel. Necessary arrangement was made to measure the discharge at the tail end of channel by a Cipoletti weir of 10cm crest. The weir recorded 10 cm head of water. The depth of water at the head, tail and middle point of the channel were 15, 16 & 15.8 cm respectively. The bed width of the channel was 12cm with side slope of 0.5:1. Determine the seepage rate in the channel.

Solution: Average depth of water in the channel,

$$D = \frac{15+16+15.8}{3} = 15.6cm$$

Wetted perimeter of the channel,

$$P_w = B + 2D\sqrt{Z^2+1}$$

$$= 10 + 2x15.5\sqrt{(0.5)^2+1}$$

$$= 10 + 34.88 = 44.88cm$$

Wetted area of the channel,

$A_w = P_w x$ *Length of channel*

$= 44.88$ *cm x* $100m$ = *44.88m²*

Discharge at the tail end measured by the Cipoletti weir,

$Q_2 = 0.0186LH^{3/2}$

$= 0.0186x10x16^{3/2}$

$= 11.904l/s$

Loss of water in the channel, $Q_s = Q_2 = 12\text{-}11.90 = 0.096l/s = 9.6x10^{-5} m^3/s$

Seepage rate $\frac{Q_s}{A_w} = \frac{9.6x10^{-5}}{44.88} = 2.136x10^{-6} m^3/m^3/s$

Questions and Problems

9.1 What do you mean by lining irrigation channel? What are the advantages and disadvantages of it?

9.2. Classify the lining materials. Discuss the brick and cement concrete as the lining materials.

9.3 Describe the process of making the reinforced half round concrete channel section.

9.4 Describe the process of making the half round burnt clay tiles. Discuss its performance in controlling seepage losses in irrigation channels.

9.5 An earthen tile lined channel was constructed at a cost of Rs17, 000/-. The rate of interest on investment is 10%, the expected longevity of the channel is 15 years and trade-in-value is nil. Determine the (i) depreciation per year by straight-line method and the (ii) sinking fund and depreciation of the 5th year if calculated on compound interest method.

Ans: (i) Rs.1, 133.33 (ii) Rs. 486.41 & Rs.783.37

9.6 A rectangular brick lined channel was constructed to accommodate flow of $3.5m^3/s$. The channel runs for 1200 hours in a year. Work out the economics of lining from the following data.

Bottom of the channel = 0.75m

Depth of water in the channel = 1.5m

Length of channel = 500m

Longevity of the channel = 25 years

Salvage value = Nil

Rate of interest on investment = 10%

Wetted perimeter of unlined channel = 7.0m

Seepage through the unlined channel = $1.1m^3/m^2$/day

Seepage through the lined channel = $0.015/m^3/m^2$/day

Cost of water = Rs.$0.2/m^3$

Annual maintenance cost of the unlined channel = Rs. $4.00/10m^2$

Cost of channel lining = Rs. 300/m

Assume additional data suitably, if required.

Ans: Benefit: Cost = 2.85:1

9.7 A field irrigation channel in a spout of a deep tube well (DTW) command area was lined by using half round earthen tiles (33cm diameter). Determine the benefit: cost ratio of lining from the following details of DTW and channel.

i) DTW

1. Cost of DTW irrigation system (P) = Rs. 8,00,000/-
2. Yearly operating time, T_1 = 1250 hrs.
3. Yearly operator's wage = Rs. 15,000/-
4. Yearly repair and maintenance cost = Rs. 45,000/-
5. Yearly power charge = Rs. 45,000/-
6. Expected life of DTW = 30 years
7. Rate of interest on investment = 10%
8. Discharge of DTW = 50 l/s
9. Salvage value = 10%
10. Discharge of the spout = 10 l/sec

ii) Earthen tile channel

1. Length of channel = 250m
2. Cost of construction of channel = Rs. 70/m
3. Expected life of channel = 15 years
4. Interest on investment = 10%

5. Salvage value = Nil
6. Annual repair and maintenance of lined channel = Rs. 1,000/-
7. Annual repair and maintenance of unlined channel = Rs. 1,550/-
8. Other benefit of lined channel = Rs. 500/year
9. Operating period = 650 h/year
10. Repair and maintenance of unlined channel = Rs. 2,000/yr
11. Saving of water due to lining = 25% of spout source

Ans: Benefit: Cost = 2.71:1

9.8 Calculate the seepage index (SI) and percent to total SI from the segments of a water channel to be lined for saving a 75% of seepage losses.

Outlet No.	Length of channel section, m	Time for each section runs, h	Seepage loss, $m^3/m^2/h$
A	40	15.2	0.15
B	75	14.5	0.14
C	65	12.5	0.12
D	90	10.5	0.14
E	70	9	0.15
F	80	3.5	0.16
G	125	2.5	0.18

Ans: Seepage index (SI):2. 28,2.03,1.5,1.47,1.37,0.56,0.45

Percent to total SI = 8,23.65,44.71,60.27,75.52,89.52,95.33,100

Length of channel to be lined = 266.93m of sections A, B, C&D

9.9 The channel lengths to be lined at 70% saving of seepage loss and cumulative seepage index at different outlet of a canal irrigation system were as below. Determine the priority values and priority ranking of the outlets.

Outlet No.	Name of the channel	Length to be lined (L_0) Index	Seepage
A	A_1	522	25.1
	A_2	1000	110.5
	A_3	2100	150
B	B_1	325	35
	B_2	235	74.1
	B_3	376	314.8
	B_4	1400	223.09
C	C_1	750	43.2
	C_2	225	125
	C_3	478	118.3

Ans: Priority: 0.25, 0.197 & 0.079 of B, C, A respectively.

9.10 Study of seepage loss was conducted by ponding water in a precast concrete channel for a period of 20 hours. Determine the seepage rate with the following data:

Width of channel = 25cm

Side slope = 0.5:1

Length of channel = 50m

Depth at u/s at the beginning of study = 15m

Depth at u/s at the end of study = 10cm

Depth at d/s at the beginning of study = 25cm

Depth at d/s at the end of study = 20cm

Pan evaporation = 3.5mm/day

Ans: $3.08x10^{-2}$ $m^3/m^2/day$

9.11 The upstream and downstream depths of water recorded 20cm, 15cm and 22cm, 17cm respectively in a half-round 45cm diameter earthen tiled channel at the beginning and end of study of seepage by ponding of water in a 60m section. The ponding of water continued for 25 hours when the average rate of evaporation was 3.0mm/day. Determine the seepage rate of the channel and the possible discharge through it at 18cm depth of flow. Assume Manning's n = 0.02

Ans: Seepage rate: $2.34x10^{-2}$ $m^3/m^2/day$

9.12 Select the appropriate answer from the followings:

1. Wetted perimeter of 30cm half round section is

a) 27cm b) 35cm

c) 47cm d) 57cm

2. Water flows through a half round channel of length 50m for a period of 11 hours. The average wetted perimeter of the channel is 45cm.If the seepage rate is 10cm/day, the loss of water during the period of flow through the channel is about

i) $1m^3$ b) $2m^3$

c) $3m^3$ d) $4m^3$

3. In ponding method of measuring the seepage loss through channel of length 50m in 0.2% field slope records 15cm water depth at upstream section. The depth of water at the downstream section is

a) 15cm b) 20cm

c) 25cm d) 30cm

4. A half round channel section of 30cm diameter contains 10cm the depth of water. The wetted area of the section is

   a) $122cm^2$ b) $156cm^2$

   c) $185cm^2$ d) $206cm^2$

5. A brick made water filled reservoir of size 3mx3mx3m lost 6.5cm water in a day. If evaporation rate is 3.5mm/day, the seepage through the reservoir is,

   a) $0.012m^3/m^2/day$ b) $0.055m^3/m^2/day$

   c) $0.10m^3/m^2/day$ d) $0.15m^3/m^2/day$

Ans.

1. c) 2. a) 3. c) 4. d) 5. a)

9.13 Write True or False of the following statements

1. Lining increases the velocity of flow and decreases the silting in the channel
2. Repair works is a headache to lining the channels
3. Bentonite is a good quality factory made lining material
4. Salvage value is the resale value at the end of the service life
5. In straight line method depreciation of second year is more than the first year

Ans.

1. True 2. True 3. False 4. True 5. False

References

Biswas, R.K. and S.Mallick (1997). Performance of locally made burnt clay tiles in controlling seepage losses. J. of Agril. Engg. 34(4):47-54.

Biswas, R.K. (1999). Application of low cost non-conventional lining materials for seepage control in some irrigation channels. J. Agril. Engg.23(3-4):1-14.

Biswas, R.K. and S.Mallick (2000). Performance of burnt clay tiles in lining small irrigation channels of different geometric shape. Karnataka J.Agril.Sci.13(3):682-686.

Chawla,A.S. (1992). Canal lining and maintenance. Seminar on Irrigation Water Management (WMF), New Delhi.

Hegade,Channappa, T.C. & B.K. Anand Ran (1981). Water harvesting recycling in the red soils of Karnataka. Ind. J. of Soil Conservation, 9(2-3):107-111.

INCID (2000). Manual of Canal Lining. Indian National Committee on Irrigation and Drainage. New Delhi.

Karma, S.K. & Dhrubanarayan, B. (1987). Chemical sealant for seepage controlling farm pond. The role of Agricultural Engineering in dry land agriculture. Proceedings of 23rd convention of ISAE, Jabalpur, India,9-11 March.

Khair,A., Ahmed,M. & Datta, S.C. (1984). Development of low cost indigenous technology to minimize water losses due to seepage in irrigation canal. AMA, XV(1):77-81.

Kool,Y.M., Kandwal,B.K. and Gupta, R.K. (1993). Performance evaluation of different canal lining system in heavy clay soils. Indian J. of Agril. Engg., 3(1-2):72-76.

Lenka, D. (1985). Lining of irrigation channels. Kalyani Publication, NewDelhi.

Luthra, S.D.L.(1986). Measure to minimize conveyance losses in irrigation canal net works. Proceedings of International Seminar on Water Management in Arid and Semi-arid zones. Nov 27-29, held at HAU, Hissar.

Misra, G.N. & Panda, R.K. (1990). Economics of channel lining-a case study.

Murthy, V.V.N. (1988). Land and Water Management Engg. Kalyani Publication, Ludhiana.

Ranade,D.H.,Gupta,R.K.;Jain.L.K.and Cool,D.K.M.(1993). Comparative performance of different sealant materials for seepage control. Ind.J.of Soil Conservation. 21(2):1-5.

Sur, H.S. and Rachpal Singh (1993). A procedure for working out lining priorities of field water courses. Ind.J.of Agril. Engg. 3(1-2):47-50.

Shally,H.L.(1985). Lining of earthen channel. Asia Publishing House, Bombay.

Talley, S.N. & Kohale, S.K. (1986). Performance of the irrigation lining materials. PVK J., 10(2):138-140.

10

Salt Problem and Irrigation Water Quality

Poor quality of water is one of the important factors causes gradual accumulation of salts on the ground or in the root zone of crops leads to saline and alkaline condition. This results in loss of permeability, toxicity and anaerobic conditions. In salt affected soils plants face constraint to derive their nutrients at the required rate due to high soil moisture tension, which affects adversely the crop growth and productivity.

10.1 Causes of Salt Problems

The following are the reasons for causing the soils become saline or alkaline.

1. Rise in water table: A rise in water table and high evaporation and transpiration rates with poor drainage lead to build up dissolved salts.
2. Quality of irrigation water: Irrigation water contains the dissolved salts exceeding certain limit cause accumulation of salts in soil profile.
3. Climatic condition: In high rainfall area accumulated salts can be leached out through high rate of infiltration. In arid and semi-arid regions rainfall is less and leaching is limited leads to soil remain saline or alkaline.
4. Nature of soil: Heavy soils are of poor in permeability and leaching. It is therefore more susceptible to salinity and alkalinity.
5. Under irrigation: In sprinkler and drip irrigation there are no allowance for leaching of water may cause of accumulation of salts in upper layer of soils.
6. Seawater intrusion: In coastal area seawater remains below the fresh water in the ground. When this seawater is extracted and applied to field causes saline or alkaline problem of the soils.

Contents in irrigation water

Whatever may be the source of water there remains some dissolved salts in irrigation water. The main soluble contents are calcium, magnesium, sodium and potassium and the anions of sulphate, chloride, bicarbonate, carbonate and silicate. Other than these lithium, silicon, bromine, iodine, copper, nickel, cobalt, fluorine, boron, zircomium, titanium, vanadium, barium, rubidium, arsenic, antimony, and organic matter may be available in minor quantities. Among the soluble constituents, calcium, magnesium, sodium, chloride, sulphate, bicarbonate and boron are of prime importance in determining quality of irrigation water. Excess amount of boron, selenium, molybdenum and fluorine available in irrigation water taken by the plants is harmful to animals that are feed by such forage or drink such water.

10.2 Terminology

Parts per million (ppm): One part of solute dissolved in million parts of water (1 ppm) (1 gm/1,000,000 gm of water). One ppm is numerically equivalent to milligram per liter.

Equivalent weight: It is the weight in grams of an ion or compound that combines with or replace one gram of hydrogen.

$$\text{Equivalent weight} = \frac{Atomic\ weight}{Valency}$$

For example, equivalent weight of $Na^+ = \frac{23}{1} = 2$

$$Ca = \frac{40}{2} = 20, \text{ etc.}$$

Milliequivalent per liter (meq/l): Milliequivalent of an ion or compound in one liter of water.

$$\text{Milliequivalent weight} = \frac{\text{Equivalent weight}}{1000}$$

$$\text{Milliequivalent weight of } Na^+ = \frac{23}{1000} = 0.023$$

$$Ca^+ = \frac{20}{1000} = 0.02, \text{etc.}$$

Cation exchange: Interchange of cation in solution with another cation on surface-active material.

Exchangeable cation: A cation that is adsorbed on the exchange complex and which is capable of being exchanged with other cations.

Exchangeable complex: The surface-active constituents of soil (both organic and inorganic) that is capable of cation exchange.

Alkali or sodic soil: A soil that contains sufficient exchangeable sodium which can affect the plant growth with or without the appreciable quantities of soluble salts. Usually, exchangeable sodium more than 15% is called as alkali soil.

Alkaline soil: A soil that shows alkaline reaction, i.e., a soil of whose saturated paste gives the p^H value more than 7.

Alkalisation: It is the process by which the exchangeable sodium content in soil is increased.

Dispersed soil: Soil in which the clay particles form the colloidal solution.

Cation exchange capacity (CEC): The total quantity of cations that a soil can absorbs by cation exchange. It is usually expressed as milliequivalent per 100 gms.

Exchangeable sodium percentage (ESP): It indicates the extent of saturation of the soil exchange complex with sodium. It may be calculated as

$$\text{ESP} = \frac{\textit{Exchangeable sodium}\left(\textit{milliequivalent per}\,100\textit{gm}\right)}{\textit{Cation exchange capacity}\left(\textit{milliequivalent per}\,100\textit{gm soil}\right)}$$

Molar solution = A solution of salt concentration equal to one gram molecular weight per litre. Millimole=Molar solution/1000. For NaCl molecular weight=58.44. So, 1 mole of NaCl=58.44g/liter.

Non-soluble alkali soil: A soil that has exchange sodium percentage (ESP) above 15 and EC of the solution extract less than 4 millimhos/cm. The p^H value of such soil is more than 8.5 and ranges 8.5 to 10.

Saline-sodic or saline-alkali soil: A soil of whose saturation extract has EC value more than 4 millimhos/cm at 25^{0}C, ESP value less than 15% or SAR at least 13 and the p^H value usually less than 8.5 and preponderance of chlorides and sulfates at sodium, calcium and magnesium.

Saline soil: A soil of whose saturation extract has EC value more than 4 millimhos/cm at 25^{0}C, ESP value greater than 15% and the p^H value usually less than 8.5.

Saturation extract: The solution extracted from a soil at its saturation percentage. The saturation percentage is the maximum percentage of a saturation paste expressed on dry weight basis.

Soil extract: The soil at any available moisture content.

Sodium adsorption ratio (SAR): A ratio to express in soil extracts and in irrigation water, the relative activity of sodium ions in exchange reactions with soil.

$$\text{SAR} = \frac{\text{Na}^{++}}{\sqrt{\left(\text{Ca}^{++} + \text{Mg}^{++}\right)/2}}$$

Ionic concentration is expressed in milliequivalent per litre. SAR is also known as SAR practical (SAR_p). Adjusted SAR is known as true SAR (SAR_t).

SAR_t=0.08+1.115(SAR_p).

p^H = This is the logarithm to the base 10 of the reciprocal of hydrogen ion concentration in water. The p^H value is 7 in distilled water. p^H value of 7 indicates a neutral solution, neither alkaline nor acid. A value of 7.5 to 8.0 usually indicates the presence of carbonates of calcium and magnesium, and a p^H value of 8.5 or above usually indicates appreciable exchangeable sodium.

Osmotic pressure: It is the equivalent negative pressure that influences the rate of diffusion of water through a semi impermeable membrane.

Specific surface: This is the surface area per unit weight of the soil (m^2/gm).

Salt concentration: High degree of correlation exists between EC, the total cation and osmotic pressure of soil-water extract. The relations are expressed as the following:

i) Salt concentration, mg/l or ppm

   = 640 x EC, millimhos/cm or mmhos/cm

ii) Total cation concentration, meq/l

   = 10 x EC millimhos/cm

iii) Osmotic pressure, atmospheres

   = 0.36 x EC millimhos/cm

iv) ppm/equivalent weight = meq/l

v) Milliequivalent per litre to ppm

   = sum of the product of the milliequivalents of each ion multiplied by equivalent weight.

vi) deci-Siemen (dS/m): According to SI units the EC of 1 millimhos/cm at 25°C is equal to 1 dS/m. Therefore, 1 millimhos/cm and 1 mmhos/cm is equal to 1dS/cm and mS/cm respectively.

Gypsum requirement (GR): The gypsum requirement is estimated in removing the soluble salts as below (Michael, 1985):

GR (meq/100 gm soil) = [Ca concentration of added gypsum solution (meq/l)-(Ca + Mg) concentration of filtrate (meq/l] x 2

10.3 Classification of Salt Affected Soils

The US Soil Salinity Laboratory Staff (1954) classified the salt affected soils based on soluble salt concentration (expressed in electrical conductivity, EC_e) and exchangeable sodium percentage (ESP) of soil saturation extract (Table 10.1). The saline and non-saline soils were demarcated by the electrical conductivity as 4dS/m. However, some workers found that some sensitive plants get salt affected for concentration in the range of 2-4dS/m. Accordingly, the Soil Science Society of America proposed the salt classification taking the lower limit of saline soil as 2dS/m at 25°C (Table 10.1). The salt affected soils were classified into three groups: saline soil, sodic soil and saline-sodic soil. Szaboles (1980) divided the salt affected soils into five major groups: saline soils, alkali soils, gypsiferous salt affected soil, acid sulfate soil (rich in $Al_2(SO_4)$ and $Fe_2(SO_4)_3$ salts) and other type of salt affected soils. Bhumba and Abrol (1979) proposed to classify the salt affected soils only to alkali and saline soils in consideration to management practices.

Table 10.1 Classification of salt affected soils

Classification	Normal soil	Saline soil	Sodic soil	Saline-sodic soil
US Soil Salinity Laboratory Staff classification	$EC_e < 4dS/m$ at 25°C, ESP <15	$EC_e > 4dS/m$ at 25°C, ESP <15	ESP >15	$EC_e > 4ds/m$ at 25°C, ESP >15
Soil Sci. Soc. of America	$EC_e < 2dS/m$ at 25°C, SAR < 13	$EC_e > 2dS/m$ at 25°C, SAR < 13	SAR > 13	$EC_e > 2dS/m$ at 25°C, SAR > 13

Source: Bandyopadhyay *et al.* (2001)

Example 10.1 Express 4800 ppm of salt concentration in to micromhos/cm, millimhos/cm, mhos/cm, dS/m and osmotic pressure in atmospheres.

Solution:

Salt concentration, ppm = 640 x EC, mmhos/cm

$$\therefore\ EC = \frac{ppm}{640} = \frac{4800}{640} = 7.5 \text{ mmhos/cm} = 7.5 \text{ dS/m}$$

$$= 7.5x1000 = 7500\mu mhos / cm$$

$$= \frac{7.5}{1000} = 0.0075 mhos / cm.$$

Osmotic pressure, atmospheres = 0.36 x EC millimhos/cm

= 0.36 x7.5

= 2.7

Example 10.2 Express the salt concentration of solution in ppm showing their EC values as 1.5 dS/m, 0.2 dS/cm, 1000 μmhos/cm and 10meq/l.

Solution:

Salt concentartion, ppm = $640xdS / m$

$\therefore\ 1.5dS/m = 640x1.5 = 960\, ppm$

$0.2dS / cm = 640x0.0x100 = 12800\, ppm$

$$1000\ \mu mhos / cm = 640x\frac{1000}{1000} = 640\, ppm$$

Example 10.3 If an irrigation water source has the concentration of Na^{++}, Ca^{++} and Mg^{-} as 28. 10 and 5 milliequvalents per liter, respectively, the Sodium adsorption ratio is... (GATE, 2015)

Solution:

$$SAR = \frac{Na^{++}}{\sqrt{\frac{(Ca^{++} + Mg^{++})}{2}}}$$

$$= \frac{28}{\sqrt{\frac{(10+5)}{2}}} = \frac{28}{\sqrt{7.5}} = 10.22$$

Example 10.4 Analysis of 10gm soil sample showed exchangeable sodium, calcium, magnesium and potassium as 1.46, 1.241, 0.657 and 0.292 meq, respectively. Determine the cation exchange capacity of each cation.

Solution:

Total cation in 10gm soil = 1.241 + 0.657 + 0.292 + 1.46 = 3.65meq

= 36.5meq/100gm soil

$\therefore$ *CEC = 36.5meq/* 100 *gm soil*

$$Na\% = \frac{1.46}{3.65} x100 = 40$$

$$Ca\% = \frac{1.24}{3.65} = 34$$

$$Mg\% = \frac{0.657}{3.65} = 18$$

$$K\% = \frac{0.292}{3.65} = 8$$

Example 10.5 A 5gm soil sample on shaking with 100 cc gypsum solution of 31.2 meq/l of calcium concentration showed 26.5 meq/l as Ca + Mg in the filtrate. Determine the gypsum requirement per 100gm soil and in tones/hectare for 15 and 30 cm depth of soil. Please note that1meq Ca/100gm requirement is equivalent to gypsum of 1.72t/ha for 15cm depth of soil.

Solution:

Weight of soil = 5gm

Volume of gypsum solution = 100ml

Concentration of Ca in gypsum solution = 31.2meq/l

Concentration of (Ca+Mg) after shaking with soil = 26.5meq/l

GR (meq/100 gm soil) = [Ca concentration of added gypsum solution (meq/l)- (Ca + Mg) concentration of filtrate (meq/l] x 2

Gypsum requirement (meq/100gm soil) = 2(31.2-26.5) = 9.4meq/100gm soil

= 9.4meq Ca/100gm soil

1 meq Ca/100gm soil = 1.72 t/ha gypsum for 15cm depth of soil

Gypsum requirement for 15cm depth of soil = 9.4x1.72= 16.17 t/ha

" " for 30cm , " " " = 16.17x2 = 32.34 t/ha

Example 10.6 On analyzing 5gm soil sample the total concentration of exchangeable cations and exchangeable sodium was found 1.5meq/l and 1.0meq/l respectively. Express the cation exchange capacity and exchangeable sodium percentage (ESP).

Solution:

The total exchangeable cation in 5gm soil = 1.5meq/l

$\therefore$ Total exchangeable cation = $\frac{1.5x100}{5}$

Therefore, cation exchange capacity = $30meq/100\ gm\ soil$

Exchangeable sodium percentage (ESP) = $\frac{Exchangeable\ Na}{Total\ exchangeable\ cation} x100$

$$\frac{1.0}{1.5} x100 = 66.67$$

10.4 Irrigation Water Quality and Effects on Soil and Crop

The physical and chemical properties of soils largely depend on quality of irrigation water. The factors like climatic condition, irrigation practices, soil-water retentivity characteristics, etc., as well are equally important in characterizing the composition of quality of soils. The following are the important characteristics for determining water quality.

Salinity hazard

It is measure of total quantity of soluble salts in water and expressed in electrical conductivity (EC). The EC of the water sample is determined by measuring the electrical resistance between two parallel electrodes in the solution. The electrical conductivity for unit cross-section and across unit distance is obtained by the following relationship:

$$EC = \frac{K}{R}$$

where, K = cell constant,

R = resistance.

The EC is conventionally expressed as millimhos per centimeter (mmhos/cm) or micromhos per centimeter (mmhos/cm) in the past. Now a day, it is expressed according to SI Unit as deci Siemen per meter (dS/m). The electrical conductivity 1 mmhos/cm at 25^{0}C is equal to 1 dS/m.

Most of the time irrigation water is good to excellent quality and causes no serius salanity constraints. But, salts get added to the soils when it is irrigated with water of high salinity causing the reduction of crop yield. The primary objective of irrigation is to provide a crop with adequate water at the time of need. When salts accumulates in soils plants require to exert additional force to extract water than it would requires in a soil not affected by excess of salts in root zone depth. This additional force is referred as osmotic potential. The plants in such a situation try to adjust internally to exert the required force to extract the water. But, if fails to extract sufficient water against their higher potential the plants suffer in water stress resulting poor yield.

Sodicity hazard

Other than the total concentration of soluble salts, irrigation water is also judged in consideration to predominance of sodium. Any water may be suitable for irrigation in respect to salt concentration but may be unsuitable if sodium content is high. The cations present in irrigation water take part in cation exchange reaction with the soil. The main cation exchange takes place between sodium ion in irrigation water and calcium ions in the exchange complex of soil. The sodic soils which are having exchangeable sodium percentage (ESP) greater than 15 and p^H values ranged between 8.5 and 10. This indicates that the irrigation water should be such that it does not bring the soil to ESP value of 15 and p^H value of 8.5. However, now a day, the sodic soil is defined on the basis of SAR value. The SAR greater than 15 is a sodic soil.

Sodium calcium activity ratio (SCAR): High salinity water having high Mg/Ca ratio and low in SAR if applied to soil, usually shows low SAR but much higher value of ESP in comparison to ESP as would be due to irrigation water of such SAR. This situation indicates that high salinity water promotes to increase in exchangeable sodium. However, if the SAR value is calculated simply as $Na/\sqrt{Ca}$, the value approximately corresponds to observed values of ESP. this suggests that SAR of saline irrigation water containing high Mg/Ca ratio should be calculated simply as $\sqrt{Ca}$ called as Sodium to Calcium Activity Ratio.

True SAR: SAR equation does not takes into account the ion pair formation. However, under some condition this affects the ratio of monovalent to divalent ions in the soil solution. The ion pair formation becomes very important to increase in soil solution. Therefore, the SAR or the practical SAR (SAR_p), which is incorrect, requires correction or ion pair complexes as 'true' SAR (SAR_t). Sposito and Mattigod (1977) gave the following equation for SAR_t (Gupta, 1990).

$SAR_t = 0.08 + 1.115 SAR_p$,

where, $SAR_t > SAR_p$.

Alkalinity hazard: Eaton (1950) stated the iomporatnce of residual sodium carbonate (RSC) for development of alkali soils (saline or non-saaline) by the irrigation water containing CO_3+HCO_3 higher than Ca+Mg and used in such quantity that little leaching occurs. The calcium and to some extent the magnesium has the tendency to precipitate as carbonate in the presence of high concentration of bicarbonate ion. The reaction follows as below:

$Ca^{++}+2HCO_3 = CaCO_3+H_2O+CO_2\uparrow$

The above equation continues to reduce the concentration of calcium and magnesium in comparison to sodium proportion. Thus, reaction process causes to decrease in salinity and increase in ESP. the residual bicarbonate reacts with sodium to form sodium bicarbonate. The sodium bicarbonate converts into sodium carbonate upon evaporation. The reaction follows as below:

$Na^{+}+HCO_3 \Leftrightarrow NaHCO_3$

$2NaHCO_3 \Leftrightarrow Na_2CO_3+H_2+CO_2\uparrow$

10.5 Quality and Classification of Irrigation Water

Quality of water is of equal importance to quantity. Using the water to a greater extent, combined with reuse of water, there is possibility of quality suffer unless recommendation is given to protect it. The quality required of a water simply depends upon its purpose; thus, the needs for drinking water, industrial water, and irrigation water vary widely. For the determination of quality criteria, physical, chemical and biological constituents of water must be specified as well as standard methods for reporting results for water quality (Todd, 1959).

The quality of irrigation water depends primarily upon its salt constituents. Among the most important salt factors in water quality are: (i) total concentration, (ii) proportion of sodium to other cations, and (iii) the presence of special toxic anions such as borate or for some crops, possibly chloride, sodium or bi-carbonate.

In classification of irrigation waters, it is assumed that the water will be used under average conditions with respect to soil texture, infiltration rate, drainage, quantity of water used, climate and salt tolerance of crops. Large deviations from the average for one or more of these variables may make it unsafe to use what, under average conditions, would be good water.

The Salinity Laboratory of the United States Department of Agriculture (Richards, 1954) put forth a rational approach to determine the quality of ground water. It recommends the sodium-adsorption ratio (SAR) because of its direct relation to the absorption of sodium by soil. It is defined by

$$SAR = \frac{Na^{+}}{\sqrt{\frac{\left(Ca^{++}+Mg^{++}\right)}{2}}}$$

Where, the concentrations of the constituents are expressed in milliequvalents per litre. Recommended water classification for SAR areas follows:

SAR	Water class
<10	Excellent
10-18	Good
18-28	Fair
>28	Poor

Actually water with SAR more than 10 need not be used on land.

Eaton (1953) proposed the conception of residual sodium carbonate (RSC) to evaluate the effect and existence of high carbonate in water. He defined the residual sodium carbonate as,

$$RSC = (Co^{--}_3 + HCo_3) - (Co^{++}_3 + Mg^{++})$$

in which the concentrations are expressed as milliequivalents per litre. The limits of RSC in classifying the water quality are:

RSC	Water class
<1.25	Good
1.25-2.50	Marginal
>2.5	Unsuitable

Sodium concentration is important in classifying irrigation water because sodium reacts with soil to reduce its permeability. Soils containing a large proportion of sodium with carbonate as the predominant anion are termed alkali soils; those with chloride or sulfate as the predominant anion, saline soils. Ordinarily, either type of sodium- saturated soil will support little or no plant growth. Sodium content is usually expressed in terms of percent sodium also known as soluble sodium percent (SSP), defined by

$$SSP = \frac{Na}{Ca + Mg + Na + K}$$

Where, all ionic concentrations are expressed in milliequivalent per liter. In place of rigid limits of salinity for irrigation water, classes of relative suitability commonly express quality. Table10.2 outlines a classification prepared by Wilcox (1955). Under this scheme the suitability is judged on measurement of electrical conductivity, sodium content reported as percent sodium and boron concentrations. The lowest rating of any of the three factors determines the water class.

Table 10.2 Quality classification of water for irrigation

Water class	Percent sodium	EC x 10^6 At 25°C	Boron, ppm		
			Sensitive crops	**Semi-tolerant crops**	**Tolerant crops**
Excellent	<20	<250	<0.33	<0.67	<1.00
Good	20-40	250-750	0.33-0.67	0.67-1.33	1.00-2.00
Permissible	40-60	750-2000	0.67-1.00	1.33-2.00	2.00-3.00
Doubtful	60-80	2000-3000	1.00-1.25	2.00-2.50	3.00-3.75
Unsuitable	>80	>3000	>1.25	>2.50	>3.75

Source: wilcox, l.v. (1955)

Quality diagram

The diagram proposed by U.S.Salinity Laboratory (Richards, 1954) for classifying irrigation waters based on electrical conductivity and sodium absorption ratios, but modified for higher salt classes as shown in Fig 10.1.The lines dividing sodium absorption ratio classes are empirical and are based on results of green house and field tests. The electrical conductivity scale is logarithmic (with base 10). The chart (Fig 10.1) is binomial in that C_1, C_2, C_3 & C_4 represent water classes with increasing hazards from total salt concentrations and that S_1, S_2, S_3 & S_4 represent water classes representing increasing hazards of exchangeable sodium accumulation in irrigated soils.

Electrical conductivity in micro-mhos/cm at 25°C and concentrations of sodium, calcium and magnesium in equivalents per million are required for use of the chart (Fig 10.1).

Salinity classification

Water can be divided in to the following six classes with respect to conductivity, the dividing points between classes being at 250, 750, 2250, 4000, and 6000 micro-mhos/cm as indicated in Fig.10.1.

C_1 **Low salinity water** (0 to 250 micro-mhos/cm)

Low salinity water can be used for irrigation with most crops on most soils with likelihood that soil salinity will develop. Some leaching is required, but this will occur under normal irrigation practices except in cases where the soil has extremely low permeability.

C_2 **Moderate salinity water** (250 to 750 micro-mhos/cm)

Moderate salinity water can be used for irrigation with all but extremely salt sensitive plants when grown on soils of high to medium permeability. With soils of low sensitivity, some leaching precautions, and at times the selection of plants of moderate salt tolerance, may be necessary. Usually ordinary irrigation practices will provide ample leaching.

C_3 **Medium to high salinity water** (750 to 2250 micro-mhos/cm)

Medium to high salinity water should be used only on soils of moderate to good permeability. Regular leaching is often needed to prevent serious salinity. Special arrangement for salinity control will often be needed, and plants with moderate to good salt tolerance should be selected.

C_4 **High salinity water** (2250 to micro-mhos/cm)

High salinity water can be used for irrigation only on soils of good quality water can be used for irrigation only on soils of good permeability and where special leaching is provided to excess salt. Only salt tolerant crops should be grown.

C_5 **Very high salinity water** (4000 to 6000 micro-mhos/cm)

Very high salinity water is generally undesirable for irrigation and should be used only on highly permeable soils, with frequent leaching and with plants of high salt tolerance.

C_6 **Excessive salinity water** (above 6000 micro-mhos/cm)

Excessive salinity water should not be used for irrigation.

The establishment of water quality classes from the standpoint of the sodium hazard is more complicated than for the salinity hazard. The classification of irrigation water with respect to SAR is based primarily on the effect of exchangeable on the physical condition of the soil (Richard, 1954). Classification of irrigation water with respect to sodium hazard can be made as:

S_1 **Low sodium water:** Low sodium water can be used on almost all soils with little danger of accumulation of harmful amounts of exchangeable sodium.

S_2 **Medium sodium water:** Medium sodium water will present appreciable sodium hazard in soils of high clay content and low organic matter, especially under low leaching conditions, unless gypsum is present in the soil. This water can be used readily on coarse textured soils with good permeability.

S_3 **High sodium water:** High sodium water may produce harmful levels of exchangeable sodium in most soils and will require special soil management, good drainage, high leaching and organic matter additions. Chemical amendments may be required for replacement of exchangeable sodium, except that amendments may not be feasible with waters of very high salinity.

S_4 **Very high sodium water:** Very high sodium water is generally unsatisfactory for irrigation purposes except at low and perhaps medium salinity, where the solution of calcium from the soil or use of gypsum or other amendments may make the use of these waters feasible.

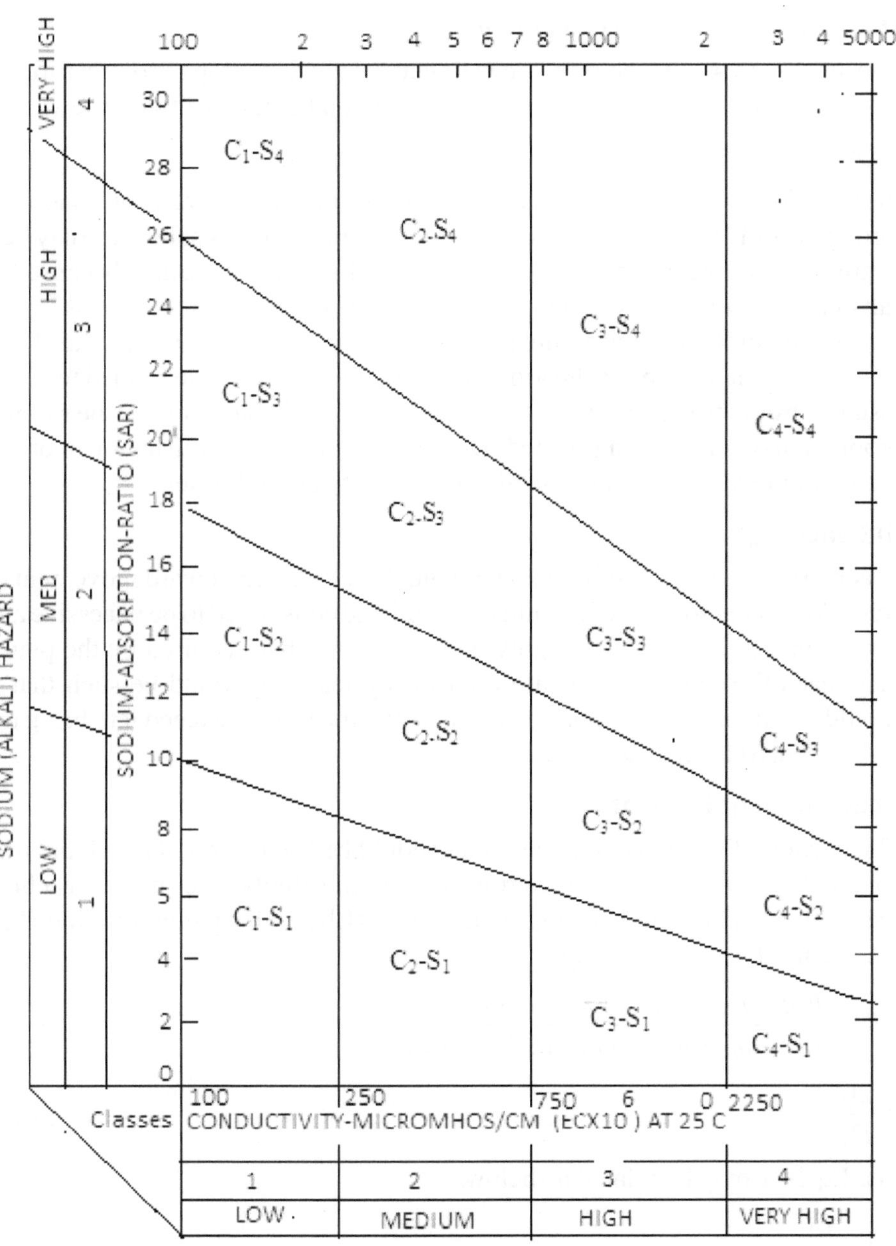

SALANITY HAZARD

Fig. 10.1 Diagram for the classification of irrigation waters

Source: Biswas (1980)

10.6 Control of Salinity Problem

The purpose of salinity control is to maintain the yields of crop to control salinity, which can be exercised each separately or in combination of two or few.

i) Drainage

In a shallow water table condition water rises to the active root zone area through capillary action. If the shallow water table contains salts, this may be a continuous source for causing deposition of salts to root zone area because the salts carried in capillary water is left to soil when evaporates or used by the plants. The problem of accumulation of salts in root zone is more in arid and semi arid climates. Unless the ground water table could be maintained to a safe distance from the soil surface (at least 2m) by any suitable drainage the salinity problem may remain complicated. Drainage can be done by using open or tile drain or drainage wells and safe disposal of salty ground water.

ii) Leaching

Leaching is the process of removal of soluble salts by downward movement of water. When the soluble salts build up in soil and considered to be excessive and detrimental to plant growth, applying more water than required by the plants may leach the salt out. The safe removal by leaching would be such that it should equalize the accumulation of salts over the time or exceed it to bring the soil to desired level of salts in it.

Leaching requirement

The portion of the irrigation water that should be leached out through the root zone to keep the soil at a predetermined level of salinity is termed as leaching requirement. The term leaching requirement (LR) and leaching fraction (LF) may be used interchangeably.

$$LR = \frac{depth\ of\ water\ leached\ below\ root\ zone}{depth\ of\ water\ applied\ at\ the\ surface}$$

$$= \frac{D_d}{D_w} \tag{10.1}$$

The Eq.10.1 may be related as below,

$$LR = \frac{EC_w}{EC_d} \tag{10.2}$$

$$\therefore D_w = \frac{D_d EC_d}{EC_w} \tag{10.3}$$

Where,

D_d = the depth of water leached out below the root zone

D_w = the depth of water applied on the soil surface

EC_w = electrical conductivity of the applied water

EC_d = electrical conductivity of the drainage water.

In estimating the leaching requirement the crop tolerance to soil salinity (EC_e) and salinity level to irrigation water (EC_w) are to be known. The useful crop tolerance data can be used from Table 10.3. The minimum leaching requirement to control the salts within the tolerance (EC_e) of the crop (usually within10 percent or less yield loss) can be calculated by using the following equation (Rhodes, 1974; Rhodes and Merril, 1976) (FAO, 1985).

$$LR = \frac{EC_w}{5(EC_e) - EC_w} \qquad (10.4)$$

The total depth of water over the season that needs to be applied to satisfy the requirement of crop demand and the leaching requirements may be calculated as below,

$$D_w \frac{ET}{1 - LR} \qquad (10.5)$$

Where, ET = evapotranspiration or crop water demand (mm/season)

LR = leaching requirement, fraction.

The depth of water applied (D_w) is equal to the consumptive use (D_c) and the depth of drainage water (D_d).

Therefore,

$$D_w = D_c + D_d \qquad (10.6)$$

$$\therefore \; D_w = LRD_w + D_c$$

$$or,\; D_w = (1\text{-LR}) = D_c$$

$$or,\; D_w = \frac{D_c}{1\text{-}LR} \qquad (10.7)$$

From **Eq.10.3**, $D_w = \dfrac{D_d EC_d}{EC_w}$

$$= (D_w - D_c)\frac{EC_d}{EC_w}$$

$$or, D_w\left(\frac{ECd}{EC_w}\right) = -D_c \frac{EC_d}{EC_w}$$

$$\therefore D_w = \frac{D_c \dfrac{EC_d}{EC_w}}{\dfrac{EC_d}{EC_w} - 1}$$

$$= \left(\frac{EC_d}{EC_d - EC_w} \right) D_c \qquad (10.8)$$

The Eq.10.8 relates the depth of water applied to depth of water lost by ET and the salt concentration of irrigation and drainage water. The EC_w can be measured by chemical analysis of irrigation water and EC_d can be estimated on the basis of salt tolerance limit of crop. The soil-water salinity (EC_{sw}) is the average root zone salinity to which the plant is exposed. Thus, the EC_{sw} should be equal to EC_d. Salinity measurement is normally done on the basis of saturation extract of the soil and called the soil salinity (EC_e) and is approximately equal to one-half of the soil-water salinity ((EC_{sw} or EC_d). In a situation of 15-20% leaching requirement, the salinity of the applied water (EC_w) may be used to predict the soil salinity (EC_w) by using the following equations (FAO, 1985),

$$EC_{sw} = 3\ EC_w \qquad (10.9)$$

$$EC_e = 1.5 EC_w \qquad (10.10)$$

$$EC_{sw} = 2\ EC_e \qquad (10.11)$$

The depth of applied water per unit depth of soil which required to produce any specific increase in soil salinity or any given conductivity of applied water can be calculated by the following equations (Michael, 1984):

$$\frac{D_w}{D_s} = \frac{d_s}{d_w} \frac{SP}{100} \frac{\Delta EC_e}{EC_w} \qquad (10.12)$$

Where,

D_w = depth of irrigation water

D_s = depth of soil

d_s = density of soil (bulk density)

d_w = density of irrigation water

SP = saturation percentage of soil

ΔEC_e = increase in electrical conductivity of saturation extract of soil

EC_w = electrical conductivity of applied irrigation water (usually assumed 1 dS/m).

When there is high water table and soil salinity increases due to evaporation of ground water, the Eq. 10.12 can be used as below,

$$\frac{D_g}{D_s} = \frac{d_s}{d_w}\frac{SP}{100}\frac{\Delta EC_e}{EC_g} \quad (10.13)$$

$$or, \Delta EC_e = \frac{D_g}{D_s}\frac{d_w}{d_s}\frac{EC_g}{SP}100 \quad (10.14)$$

Where,

D_g = depth of ground water evaporated

EC_g = electrical conductivity of ground water

Table 10.3 Crop tolerance and yield potential of selected crops as influenced by irrigation water salinity (EC_w) or soil salinity (EC_e

	100%		**90%**		**75%**		**50%**		**0%, "maximum"**	
Field Crops	**EC_e**	**EC_w**	**EC_e**	**EC_w**	**EC_e**	**EC_w**	**EC_e**	**EC_w**	**EC_e**	**EC_w**
Barley	8.0	5.3	10	6.7	13	8.7	18	12	28	19
Cotton	7.7	5.1	9.6	6.4	13	8.4	17	12	27	18
Sugarbeet	7.0	4.7	8.7	5.8	11	7.5	15	10	24	16
Sorghum	6.8	4.5	7.4	5.0	8.4	5.6	9.9	6.7	13	8.7
Wheat	6.0	4.0	7.4	4.9	9.5	6.3	13	8.7	20	13
Wheat, drum	5.7	3.8	7.6	5.0	10	6.9	15	10	24	16
Soybean	5.0	3.3	5.5	3.7	6.3	4.2	7.5	5.0	10	6.7
Cowpea	4.9	3.3	5.7	3.8	7.0	4.7	9.1	6.0	13	8.8
Groundnut (peanut)	3.2	2.1	3.5	2.4	4.1	2.7	4.9	3.3	6.6	4.4
Rice (paddy)	3.0	2.0	3.8	2.6	5.1	3.4	7.2	4.8	11	7.6
Sugarcane	1.7	1.1	3.4	2.3	5.9	4	10	6.8	19	12
Corn (maize)	1.7	1.1	2.5	1.7	3.8	2.5	5.9	3.9	10	6.7
Flax	1.7	1.1	2.5	1.7	3.8	2.5	5.9	3.9	10	6.7
Broadbean	1.5	1.1	2.6	1.8	4.2	2.0	6.8	4.5	12	8.0
Bean	1.0	0.7	1.5	1.0	2.3	1.5	3.6	2.4	6.3	4.2
Vegetable Crops										
Squash, zucchini	4.7	3.1	5.8	3.8	7.4	4.9	10	6.7	15	10
Beet, red	4.0	2.7	5.1	3.4	6.8	4.5	9.6	6.4	15	10
Quash, scallop	3.2	2.1	3.8	2.6	4.8	3.2	6.3	4.2	9.4	6.3
Broccoli	2.8	1.9	3.9	2.6	5.5	3.7	8.2	5.5	14	9.1
Tomato	2.5	1.7	3.5	2.3	5.0	3.4	7.6	5.0	13	8.4
Cucumber	2.5	1.7	3.3	2.2	4.4	2.9	6.3	4.2	10	6.8
Spinach	2.0	1.3	3.3	2.2	5.3	3.5	8.6	5.7	15	10
Celery	1.8	1.2	3.4	2.3	5.8	3.9	9.9	6.6	18	12
Cabbage	1.8	1.2	2.8	1.9	4.4	2.9	7.0	4.6	12	8.1
Potato	1.7	1.1	2.5	1.7	3.8	2.5	5.9	3.9	10	6.7
Corn, sweet (maize)	1.7	1.1	2.5	1.7	3.8	2.5	5.9	3.9	10	6.7
Sweet potato	1.5	1.0	2.4	1.6	3.8	2.5	6.0	4.0	11	7.1
Pepper	1.5	1.0	2.2	1.5	3.3	2.2	5.1	3.4	8.6	5.8
Lettuce	1.3	0.9	2.1	1.4	3.2	2.1	5.1	3.4	9.0	6.0
Radish	1.2	0.8	2.0	1.3	3.1	2.1	5.0	3.4	8.9	5.9

Onion	1.2	0.8	1.8	1.2	2.8	1.8	4.3	2.9	7.4	5.0
Carrot	1.0	0.7	1.7	1.1	2.8	1.9	4.6	3.0	8.1	5.4
Bean	1.0	0.7	1.5	1.0	2.3	1.5	3.6	2.4	6.3	4.2
Turnip	0.9	0.6	2.0	1.3	3.7	2.5	6.5	4.3	12	8.0
Forage Crops										
Wheatgrass, tall	7.5	5.0	9.9	6.6	13	9.0	19	13	31	21
Wheatgrass, fairy crested	7.5	5.0	9.0	6.0	11	7.4	15	9.8	22	13
Bermudagrass	6.9	4.6	8.5	5.6	11	7.2	15	9.8	23	15
Barley (forage)	6.0	4	7.4	4.9	9.5	6.4	13	8.7	20	13
Ryegrass, perrenial	5.6	3.7	6.9	4.6	8.9	5.9	12	8.1	19	13
Trefoil, narrowleaf birdsfoot	5.0	3.3	6.0	4.0	7.5	5.0	10	6.7	15	10
Harding grass	4.6	3.1	5.9	3.9	7.9	5.3	11	7.4	18	12
Fescue, tall	3.9	2.6	5.5	3.6	7.8	5.2	12	7.8	20	13
Wheatgrass, standard crested	3.9	2.3	6.0	4.0	9.8	6.5	16	11	28	19
Vetch, common	3.0	2.0	3.9	2.6	5.3	3.5	7.6	5	12	8.1
Sudan grass	2.8	1.9	5.1	3.4	8.6	5.7	14	9.6	26	17
Wildrye, beardless	2.7	1.8	4.4	2.9	6.9	4.6	11	7.4	19	13
Cowpea (forage)	2.5	1.7	3.4	2.3	4.8	3.2	7.1	4.8	12	7.8
Trefoil, big	2.3	1.5	2.8	1.9	3.6	2.4	4.9	3.3	7.6	5.0
Sesbania	2.3	1.5	3.7	2.5	5.9	3.9	9.4	6.3	17	11
Sphaerophysa	2.2	1.5	3.6	2.4	5.8	3.8	9.3	6.2	16	11
Alfalfa	2.0	1.3	3.4	2.2	5.4	3.6	8.8	5.9	16	10
Lovegrass	2.0	1.3	3.2	2.1	5.0	3.3	8	5.3	14	9.3
Corn (forage) (maize)	1.8	1.2	3.2	2.1	5.2	3.5	8.6	5.7	15	10
Clover, berseem	1.5	1.0	3.2	2.2	5.9	3.9	10	6.8	19	13
Orchard grass	1.5	1.0	3.1	2.1	5.5	3.7	9.6	6.4	18	12
Foxtail, meadow	1.5	1.0	2.5	1.7	4.1	2.7	6.7	4.5	12	7.9
Clover, red	1.5	1.0	2.3	1.6	3.6	2.4	5.7	3.8	9.8	6.6
Clover, alsike	1.5	1.0	2.3	1.6	3.6	2.4	5.7	3.8	9.8	6.6
Clover, ladino	1.5	1.0	2.3	1.6	3.6	2.4	5.7	3.8	9.8	6.6
Clover, strawberry	1.5	1.0	2.3	1.6	3.6	2.4	5.7	3.8	9.8	6.0
Fruit Crops										
Date palm	4.0	2.7	6.8	4.5	11	7.3	18	12	32	21
Grapefruit	1.8	1.2	2.4	1.6	3.4	2.2	4.9	3.3	8.0	5.4
Orange	1.7	1.1	2.3	1.6	3.3	2.2	4.8	3.2	8.0	5.3
Peach	1.7	1.1	2.2	1.5	2.9	1.9	4.1	2.7	6.5	4.3
Apricot	1.6	1.1	2.0	1.3	2.6	1.8	3.7	2.5	5.8	3.8
Grape	1.5	1.0	2.5	1.7	4.1	2.7	6.7	6.5	12	7.9
Almond	1.5	1.0	2.0	1.4	2.8	1.9	4.1	2.8	6.8	4.5
Plum, prun	1.5	1.0	2.1	1.4	2.9	1.9	4.3	2.9	7.1	4.7
Blackberry	1.5	1.0	2.0	1.3	2.6	1.8	3.8	2.5	6.0	4.0
Boysenberry	1.5	1.0	2.0	1.3	2.6	1.8	3.8	2.5	6.0	4.0
Strawberry	1	0.7	1.3	0.9	1.8	1.2	2.5	1.7	4.0	2.7

Table 10.4 Relative salt tolerance to agricultural crops

Tolerant	Moderately tolerant	Moderately sensitive	Sensitive
Fibre, Seed and Sugar	Fibre, Seed and Sugar	Fibre, Seed and Sugar	Fibre, Seed and
Crops	Crops	Crops	Sugar Crops
Barley	Cowpea	Broadbean, Castorbean	Guayule
Cotton	Oats	Maize, Flax,	Sesame
Jojoba	Rye	Millet, foxtail	Vegetable crops
Sugarbeet	Safflower	Groundnut/Peanut	Bean
Grasses and Forage	Sorghum	Rice/paddy	Carrot
Crops	Soybean	Sugarcan,Sunflower	Okra
Alkali grass, Nuttal	Triticale	Grasses and Forage	Onion
Alkali sacation	Wheat	Crops	Parsnip
Bermuda grasss	Wheat, Drum	Alfalfa,	Fruit and nut
Kallat grass	Grasses and Forage Crops	Bentgrass, Bluestem,	crops
Saltgrass, desert	Barley (footage)	Angleton Brome, smooth	Almond
Wheat grassfairy	Brome, mountain	Bufflegrass, Bernet,	Apple
crested wheat grass	Canary grass, reed	Clover, alsike	Apricot
tall wildrye Altai	Clover, Hubam	Clover, Berseem	Avocado
Wildrye, Russian	Clover, sweet	Clover, ladino	Blackberry
	Fescue, meadow	Clover, red	Boysenberry
	Fescue, tallHarding grass	Clover, strawberry	Cherimoya
	Panic grass, blue	Clover, white Dutch	Cherry, sand
	Rape,Rescue grass,	Corn (forage) (maize)	Currant
	Rhodes grass	Cowpea (forage)	Gooseberry
	Ryegrass, Italian	Dallis grass	Grapefruit
	Ryegrass, perennial	Foxtail, meadow	Lemon
	Sudan grass	Gramma, blue	Lime
	Trefoil, narrowleaf birds	Lovegrass	Loquat
	foot	Milvetch, Ciecer	Mango
	Trefoil, broadleaf	Oatgrass, tall	Orange
	birdsfoot	Oats (forage)	Passion fruit
	Wheat (forage)	Orchard grass	Peach
	Wheatgrass, standard	Rye (forage)	Peach
	crested	Sesbania,Siratro,	Pear
	Wheatgrass, intermediate	Sphaerophysa	Persimmon
	Wheatgrass, slender	Timothy	Plums:Prume
	Wheatgrass, western	Trefoil, big	Pummelo
	Wildrye, beardless	Vetch, common	Raspeberry
	Wildrye, Canadian	Vegetable Crops	Rose
	Vegetable crops	Broccoli,Brussels	apple'Sapote,
	ArtichokeBeet, red	sprouts	white
	Squash, zucchini	Cabbage,Cauliflower,	Strawberry
	Fruit and Nut Crops	Celery,Corn, sweet	Tangerine
	FigJujube	Cucumber, Eggplant,	
	Olive	Kale, Kohlrabi, Lettuce	
	Papaya	Muskmelon,Pepper,	

Pineapple Pomegranate	Potato, Pumpkin,Radish, Spinach, Squash, scallop Sweet potato,Tomato,Turnip Watermelon Fruit and Nut Crops Grape

Source: FAO (1995)

Example 10.7 About 1000mm irrigation water (EC=1.0dS/m) was applied during a season, which caused the soil in to saline condition. The bulk density of the soil is 1.3g/cm^3; the density of water is 1.0g/cm^3 and the saturation percentage of the soil is 38. Determine the depth of soil profile, which get saline condition due to irrigation. Assume the crop cultivated is of semi-tolerant type whose yield loss begins at salinity 3-6 dS/m.

Solution:

Depth of irrigation water, D_w = 1000mm

Bulk density of the soil, d_s=1.3g/cm^3

Density of water, d_w = 1.0g/cm^3

Saturation percentage of soil = 38

Electrical conductivity of the irrigation water, EC_w = 1dS/m

Electrical conductivity of the saline soil, $EC_e = \frac{3+6}{2} = 4.5$ dS/m

Depth of soil, D_s = ?

Using Eq. 10.13, $D_s = \frac{d_w}{d_s}\frac{D_w}{SP}\frac{EC}{\Delta EC_e}100$

$= \frac{1.0}{1.3} x \frac{1000}{38} x \frac{1}{45} x\, 100$

$= 449.8\text{mm} \cong 45\text{cm}$

Example 10.8 A soil is subject to evaporation of 10cm groundwater whose electrical conductivity is 20 millimhos/cm. Root zone depth of soil is 50cm, bulk density of soil is 1.45g/cm^3, saturation percent of soil is 65 and density of groundwater is 1.01g/cm^3. Determine the change of salinity level of root zone depth soil.

Solution:

Depth of ground water evaporates, $D_g = 10cm$

EC of ground water, $EC_g = 20 millimhos / cm$

Bulk density of soil, $d_s = 1.45g/cm^3$

Density of ground water, $d_w = 1.01g/cm^3$

Saturation percent of soil, $SP = 65$

Root zone depth of soil, $D_s = 50cm$

Change of soil salinity, $\Delta ECe = ?$

Using Eq. 10.14 , $\Delta EC_e \frac{D_g}{D_s} \frac{d_w}{d_s} \frac{EC_g}{SP} 100$

$$= \frac{10cm x 1.01g/cm^3 x 20 millin hos/cm}{50cm x 1.45g/cm^3 x 65} x100$$

$= 4.29$ *millim hos / cm*

Example 10.9 The irrigation water ($EC_w = 1ds/m$) applied to a field to meet up the requirement of 850mm crop water demand (ET) over the season. The leaching requirement is 15%. The crop water use pattern is 40-30-20-10. Determine the salinity of soil-water draining from each zone of root zone quarter and average soil-water salinity of the root zone.

Solution:

Leaching factor = 15%, $LF = \frac{Water\ leached}{Water\ applied}$

$$EC_{sw} = \frac{EC_w}{LF}$$

EC_{sw} = EC of soil-water (percolated water from any quarter)

EC_{sw} = EC of applied water to any quarter

Water to be applied to satisfy ET (850mm) = $\frac{850}{1-LF} = \frac{850}{1-0.15} = 1000mm$

For the bottom of the first quarter:

$$LF_1 = \frac{Water\ leached}{Water\ applied} = \frac{1000-850x0.4}{1000} = 0.66,$$

$$\therefore EC_{sw_1} = \frac{EC_w}{LF_1} = \frac{1}{0.66} = 1.515dS/m$$

For the bottom of the second quarter:

$$LF_2 = \frac{(1000-850x0.4)-850x0.3}{1000} = \frac{405}{1000} = 0.405,$$

$$\therefore EC_{sw_2} = \frac{EC_w}{LF_2} = \frac{1}{0.405} = 2.47dS/m$$

For the bottom of the third quarter:

$$LF_3 = \frac{(1000 - 850x0.4) - 850x0.3\text{-}850x0.2}{1000} = \frac{235}{1000} = 0.235$$

$$\therefore\ EC_{sw_3} = \frac{EC_{sw}}{LF_3} = \frac{1}{0.235} = 4.25dS/m$$

For the bottom of the fourth quarter:

$$LF_4 = \frac{(1000 - 850x0.4) - 850x0.3 - 850x0.2 - 850x0.1}{1000} = \frac{150}{1000} = 0.15,$$

$$EC_{sw_4} = \frac{EC_w}{LF_4} = \frac{1}{0.15} = 6.67dS/m$$

The average salinity of the root zone soil-water may be calculated by taking the salinities of the four root zone quarters and below the fourth quarter (drained water).

Average soil-water salinity, $\frac{1.0 + 1.515 + 2.47 + 4.25 + 6.67}{5}$

$$= \frac{15.905}{3} = 3.18dS/m$$

Example 10.10 A wheat crop grown in uniform loam soil is irrigated by border irrigation by using river water, which has an EC_w = 1.25dS/m. The crop evapotranspiration (ET) is 900mm/season and the irrigation application efficiency is 70 percent. Determine the requirement of additional water for leaching.

Solution:

$EC_w = 1.25dS/m$

From **Eq. 10.10**, $EC_e = 1.5EC_w = 1.25x1.5 = 1.875dS/m$

From Eq. 10.4 Leaching fraction,

$$LF = \frac{EC_w}{5(EC_e) - EC_w} = \frac{1.25}{5x1.875 - 1.25} = \frac{1.25}{8.1.25} = 0.15$$

Leaching requirement, LR = 900 x 0.15 = 135mm

So, the actual amount of water needed to satisfy ET and leaching requirement = 900mm + 135mm = 1035mm

Since the application efficiency is 70%, to satisfy the requirement of ET, the depth of water needs to be applied = 900/0.7 = 1286mm. Out of this 1286mm of irrigation water, it can be presumed that 251mm (1286-1035 = 251mm) which is much higher than leaching requirement gets lost through percolation. Owing to deep percolation it satisfies the purpose of leaching. Therefore, no additional water is required to control the soil salinity in the present case.

iii) Crop tolerance

Crops respond to salinity in different manner. Some crops are sensitive to salinity and produce increasing low yield with the increase of salinity and some crops less sensitive or tolerant to salinity and capable of drawing water from soil against the additional force due to osmotic pressure. For controlling the salinity the usual practice may be adoption of appropriate leaching. However, when leaching goes above 0.25-0.30 may not be considered practical because of excessive amount of water required. In such a situation, by changing a comparatively more tolerant crop, the leaching requirement may be brought to desirable limit.

The rate of growth of plants decrease linearly as salinity increases above an initial threshold salinity at which the growth rate first begins to decrease. The following equation expresses straight-line salinity effect on yield (Mass & Hoffman, 1977) (FAO, 1985),

$$Y = 100 - b(EC_e - a) \tag{10.15}$$

Where,

Y = relative crop yield, percent

EC_e = salinity of the soil saturation, dS/m

a = salinity threshold value

b = yield loss per unit increase in salinity.

The value of 'a' is the EC_e value for 100% yield potential i=n Table 10.3. The value of 'b' can be determined by using Table 10.3 as follows:

$$b = \frac{100}{EC_e at 0\% yield - EC_e at 100\% yield}$$

For an example, for calculation the value of b for barely crop the EC_e values may be taken as 8 and 28 for 0% and 100% yield respectively from Table 10.3. Thus,

$$b = \frac{100}{28-8} = 5.$$

The soil salinity tolerance of crops as stated in Table 10.3 applies primarily from late seedling stage to maturity. Tolerance during the germination of seed and early seedling stage may have different impact to salinity. Climate also effects crop tolerance to salinity and drought. In general, cooler climate have higher salt tolerance than from during wormer and dried period. Fertilization has little effect on salt tolerance.

iv) Cultural practices

The yield of drops often decreases due to poor crop standing during the germination of seeds or early seedling stage on saline condition. The following management practices may help in avoiding the impact of salinity on temporal basis.

a) Land smoothing

In an undulated field the irrigation water avoids the high spots and accumulates to low spots. This causes the high spots shortage of water for germination of seeds and high salinity due to low penetration of water and the low spots water logging. This problem is mostly with the flooding irrigation (borders & checks); the sprinkler and drip irrigation has the less effect.

The process of leveling the land by the redistribution or spreading of top soils is land smoothing. The land smoothing may favours in uniform distribution of water. Undulation and unevenness usually develops for cultivation practices. The land smoothing is suggested at the beginning when a crop is changed in a field. However, land smoothing some time may not be adequate to bring the field in desired condition unless a major work is done with the soil to provide a regular shape i.e., land grading. A land smoothing is also required after land grading. In fact, smoothing is a seasonal and grading is one-time extensive work towards leveling of the land.

b) Timing of irrigation

The selection of suitable time for irrigation may reduce the salinity and water stress between the irrigations. Increasing the frequency of irrigation, the irrigation interval may get reduced and provides better and continuous soil-water availability. When high saline water is used for irrigation or saline ground water is at the close proximity to plant root zone, during the non-crop period there is great scope of accumulating salts in root zone area by the evaporation process of capillary water. In such a situation, a pre-plant irrigation is better for lowering of salinity from topsoil layers.

c) Placement of seed

In saline soil or in moderately saline irrigation water the germination of seeds and satisfactory stand of furrow-irrigated crops is difficult. Growers may increase the seed rate to obtain the increased germinated seeds. But this will cost for thinning and even the germination may not be uniform and good yields cannot be assumed.

In furrow irrigation methods, crops are planted on raised bed where water moves from the furrows. Salts carried by the movement of water get accumulated in the upper center of the raised bed, usually exactly where the seeds in a single row are placed (Fig.10.2). In a double-row planting the seeds are placed in two rows near the shoulder and thereby avoid direct contact of high salt accumulation in the center of the bed (Fig 10.3). In alternate row irrigation, the salts accumulate to the shoulder away from the watered furrow and the center of the bed remains saved (Fig. 10.4). Better salt management can also be achieved by planting the seeds on sloping bed just above the water line (Fig. 10.5).

Fig 10.2 Single bed row **Fig 10.3** Double row bed

Fig. 10.4 Alternate irrigation

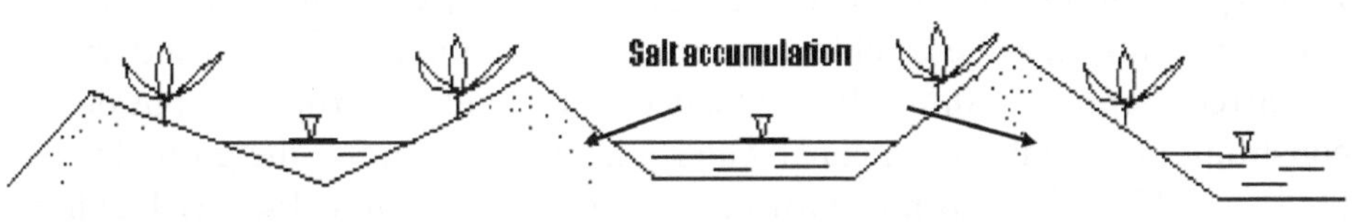

Fig. 10.5 Salinity control with sloping bed

d) Fertilization

Fertilizer, manure and soil amendments include many soluble salts in high concentration. Application of fertilizer very close to the young germinated seedlings or plants at growing stage may cause increase of salinity and toxicity problem to a large extent. Therefore enough care should be taken in timing and placement of fertilizers.

v) Changing method of irrigation

The flood or sprinkler irrigation use to apply uniform depth of water to the entire field and thereby increases salt problem uniformly over the area. The furrow irrigation builds up the salts in soil similar to flood irrigation but salts also accumulates in the areas covered by water. In flood irrigation the salinity in soil increases with the depth. In furrow irrigation the salinity is more in soils in ridges. In drip irrigation the salts accumulate at the edges of the wetted area is usually spherical in shape and the salts accumulate at the outer periphery. Each irrigation method has its own advantages and disadvantages. In consideration to all perspective one can attempt to change the method suitable to irrigation for better controlling the salinity problem.

vi) Changing or blending water supplies

The poor quality water that causes the salinity problem may be changed if any better quality water is available. Practically, sufficient quantity good quality water may not be available where such problem arises. Therefore, it is better to use the poor quality water blending with the good quality water. On blending the quality of water may be determined by the following equation:

Concentration of the blended water = [Concentration of water (a) x Proportion of water (a) used] + [Concentration of water (b) used x Proportion of water (b) used] (10.16)

Using the blended water is not a common practice. The poor and good quality water may be advantageously used in alternately. Usually, the good quality water being used at the early days of planting when the plants are more sensitive to salts and the later the poor quality water when the crops have developed some crop tolerance.

Example 10.11 Maize crop is cultivated by using canal water ($EC_w = 0.25 dS / m$) where achievable leaching fraction (LF) is 0.15 and annual water requirement 750mm. It is intended to expand the crop area but due to constraint in availability of canal water it is proposed to use the ground water ($EC_w = 3.25 dS / m$) along with canal water. Determine the (a) leaching requirements and depths of water to be applied for canal and ground water, and (b) proportion of blending of these two sources of water for expansion of crop area.

Solution

Canal water EC_w = 0.25 dS/m

Ground water EC_w = 3.25dS/m

Water requirement (ET) of crop = 750mm

Leaching fraction (LF) = 0.15

By using **Table 10.3**, for 90% yield potential of maize the EC_e and EC_w should be 2.5dS/m and 1.7dS/m respectively.

Leaching requirement is calculated by using Eq. 10.4:

$$LR = \frac{EC_w}{5(EC_e) - EC_w}$$

$$\therefore \text{ LR of canal water} = \frac{0.25}{5(2.5)-0.25} = \frac{0.25}{12.25} = 0.02$$

$$\text{LR of ground water} = \frac{3.25}{5(2.5)-3.25} = \frac{3.25}{9.25} = 0.35$$

From the leaching requirements it appears that a leaching fraction of 0.15 is excess to requirement. However, the LR of less than 0.15 is difficult to achieve. If ground water is used alone, it will increase the requirement of water to a large extent. The depths of water required may be calculated by using Eq. 10.5.

$$D_w = \frac{ET}{1-LR}$$

$$\text{Depth of canal water, } D_w = \frac{750}{1-0.15} = 882.35 mm/yr$$

$$\text{Depth of ground water, } D_w = \frac{750}{1-0.35} = 1154 mm/yr$$

The maximum EC_w of the blended water should not be more than 1.7dS/m at 0.15 leaching fraction. Thus, using the Eq. 10.16:

Concentration of the blended water = [Concentration of water (a) x Proportion of water (a) used] + [Concentration of water (b) used x Proportion of water (b) used]

Since a + b = 1, a = b-1, then the above equation is

1.7 = 0.25 x (b-1) + 3.25b

or, 0.25 + 3b = 1.7

or, b = 1.45/3 = 0.48

$\therefore$ a = 1-0.48 = 0.52

From the above it appears that the depth of canal water 882.35 mm/ha/year presently being used if continued the area will remain same. It may be blended with up to 48 percent of ground water leading to an increase the additional 48 percent crop area.

Example 10.12 In a canal command maize crop is grown in an area of 30ha. The crop evapotranspiration (ET_c) of maize is 840mm per season and the effective rainfall during growing season is 20mm. It is irrigated with water having salinity of 1.1dS/m by a surface irrigation method. If the leaching efficiency of the field soil is 0.8 and the average soil salinity tolerated by the maize crop for 100% yield is 1.7dS/m, the depth of irrigation water in mm per season required to meet the seasonal ET_c and leaching requirement will be - (GATE, 2015)

Solution:

Salinity of irrigation water, $EC_w = 1.1$ dS/m

Soil salinity tolerated by the crop, EC_e=1.7dS/m

Eq.10.4 for leaching requirement,

$$LR = \frac{EC_w}{5(EC_e) - EC_w} = \frac{1.1}{5(1.7) - 1.1} = \frac{1.1}{8.5 - 1.1} = 0.149$$

Crop evapotranspiration, EC_c= 840mm

Effective rainfall, R=20mm

Irrigation water required by the crop for evapotranspiration(ET)

= EC_w-R=840-20= 820 mm

Using leaching requirement the depth of irrigation water required (Eq.10.5),

$$D_w = \frac{ET}{1 - LR} = \frac{820}{1 - 0.149} = \frac{820}{0.851} = 963.57mm$$

Using leaching efficiency (LE) irrigation water required,

$$D_w = \frac{ET}{LE} = \frac{820}{0.8} = 1025mm$$

Therefore, seasonal irrigation water requirement = 1025cm

Example 10.13 A crop grown over an area of 50 ha can tolerate 5 decisiemens/m of electrical conductivity in the drainage water. The consumptive use of crop is 1.0m, 30% of which is obtained from the rainfall and the remainder is met from the irrigation water having electrical conductivity of 2 decisiemens/m. For efficient crop production, the leaching requirement and the quantity of water that must be drained from the area are:

a) 40% and 0.23 Mm^3 b) 60% and 0.23 Mm^3

c) 40% and 0.21 Mm^3 d) 60% 0.21 Mm^3

(GATE, 2011)

Solution:

Area of the field=50 ha

Electric conductivity of the drainage water, EC_d=5dS/m

Electric conductivity of the irrigation water, EC_w=2dS/m

Consumptive use of crop, ET=1.0m

Rain water used in ET=1.0x0.3=0.3m

Irrigation water used in ET=1.0-0.3=07m=70 cm

Leaching requirement, $LR = \frac{EC_w}{EC_d}$

$= \frac{2}{5} = 0.4 \text{ or } 40\%$

Depth of irrigation water,

$D_w = \frac{ET}{1-LR}$

$= \frac{70cm}{1\text{-}0.4}$

$= \frac{70}{0.6} = 116.67cm$

Therefore, depth of drainage water, D_d=116.17-70 cm=46.67 cm

Volume of water drained = 46.67 cm x 50 ha

= 46.67/100 x 50 x 10000m^3

= 0.23 Mm^3

10.7 Infiltration Problem

Infiltration is the entry of water in to the soil. The rate at which water enters in to soil is referred as rate of infiltration and often expressed in terms of mm/hr. An infiltration problem in soil reduces the water put in to soil body for future use by the crops. An infiltration rate in 3mm/hr is considered low while a rate above 12mm/hr is relatively high. Infiltration rate can be affected by other than the water quality, the physical characteristics of the soil such as soil texture, type of clay minerals, etc. the rate of infiltration generally increases with the increasing salinity and decreases with the decreasing salinity or increasing sodium content relative to sodium and magnesium (FAO, 1985). The sodium proportion in respect to calcium and magnesium is expressed by sodium adsorption ratio (SAR). Therefore, infiltration characteristics need to be evaluated in consideration to SAR and salinity.

Excessive presence of sodium in irrigation water promotes dispersion and plugging and sealing of the surface pores, breakdown the soil aggregate and soil structure. However, the severe impact of sodium is experienced only when the ratio of sodium and calcium exceeds 3. Poor infiltration rate of soil makes the soil difficult in getting enough water to satisfy the crop demand. Other than this, soil crusting, poor germination of seeds, poor aeration, susceptible to plant and root disease, difficulties in weed control and mosquito problem are the result of poor infiltration rate.

Similar to high sodium content, if irrigation water contains very low sodium (<0.5dS/m and especially <0.2dS/m) it is equally detrimental on reducing the infiltration rate. Low saline water leach out the soluble minerals and salts, calcium in particular, which destabilize the formation of soil aggregates and structure. The soil gets dispersed and small pores are filled up by finer soil. Thus, high and low sodium in irrigation water is equally bad in respect to infiltration rate in soil.

Management of infiltration problem

The practices of overcoming the infiltration problem may be both physical and chemical. The physical practices include the cultural practices that can improve or maintain the infiltration rate at the expected level. The chemical practice generally is the application of chemical amendments, such as, gypsum, to the applied water or to the soil, in some cases blending of water of two or more sources, which can reduce the salinity hazard.

i) Soil and water amendments

The chemical amendments are added to soil or water for improving the low infiltration due to the low salinity or high SAR in applied water. Therefore, the amendment should increase the salinity in case of low salinity and incrase the soluble calcium content to reduce the SAR in high salinity water. The most common soil and water amendments supply calcium directly (gypsum) or indirectly through acid or acid forming substance (sulphuric acid and sulphur). The amendments, which indirectly release the calcium, react with soil lime ($CaCo_3$) to the soil solution. Thus, indirect release of calcium by application of acid or acid forming substance is possible only when enough lime is present in soil.

Gypsum

The gypsum may be applied to irrigation water or to soil. Water application is more efficient if the water salinity is low (EC $\succ 0.5 dS/m$). It is difficult to counter the sodium present in high salinity water by application of gypsum as because usually 1 to 4 ml/l Ca is required to be released in fast moving irrigation stream. Fine-ground pure grade gypsum ($\prec 0.25 mmdia$) dissolves rapidly and suitable for water application.

ii) Organic residues

The crop residues or any other organic matter is very much helpful in improving the infiltration in soil. The organic matters get decomposed in soil and forming voids and channel to permit easy entrance of water. The organic matter which less easily decompose and in character less fibrous provide better performance in penetration of water viz., rice, wheat, barely, sorghum, jute, etc. Opposite to this vegetable residues have less impact in improving infiltration characteristics of soils. The organic matter is preferred to be supplied on topsoils because infiltration is a surface phenomenon. Deeper application helps in forming soil structure and deeper percolation of applied water. Organic matter may be applied at the rate of 40 to 400 t/ha and in the range of 10-30 percent in the upper 15cm soil layer (FAO, 1985).

iii) Irrigation Management

i) Irrigation more frequently: Frequent application of water may avoid to a good extent the infiltration problem due to low salinity or high SAR. It also avoids the problem of water logging and poor aeration. Frequent application of water avoids the water stress, which may cause due to long interval irrigation.

ii) Pre-plant irrigation: A pre-plant irrigation may provide the opportunity to wet the deeper soil in difficult soil (poor infiltration rate) and availability of water for the plants during the time of need after planting.

iii) Changing irrigation system: In a more difficult soil the surface irrigation system may be changed to sprinkler or drip system. Sprinkler or drip system may adjust the rate of application equal or less than infiltration rate. Sprinkler system may not be that much suitable in clayey soil but suitable in sandy and sandy loam soils. Drip performs better in loamy and heavy or clayey soil.

Example 10.14 Use of low salinity water (EC_w = 0.15*dS/m*) has caused infiltration problem attributed to ponding of water. It is proposed to use the gypsum with irrigation water for increasing the infiltration and reducing the water logging for a cultivated area of 10 ha. The depth of irrigation water is to be applied is 100 mm and it is desired that the gypsum will increase the calcium content in water by 3 ml/l. Calculate the gypsum requirement if it is 75% pure.

Solution:

$EC_w = 0.15\ dS/m$

$Area = 10\ ha$

$Gypsum = 75\%\ pure$

$Volume\ of\ water\ required = 10ha x 100mm$

$$= 10x10000/m^2x\frac{100}{1000}m$$

$$= 10000m^3$$

Equivalent weight of pure gypsum ($CaSO_4,2H_2O$) = 86

$$\text{Milliequivalent/l of gypsum} = \frac{86}{1000} = 0.086g/l$$

Equivalent weight of calcium = 20

$$\therefore \text{ 1 milliequivalent per liter of calcium} = \frac{20g}{1000}/l = 0.02g/l$$

$\therefore\ 3 = 0.02x3 = 0.06g / l$

For 0.06g Ca/l water, the content of gypsum = 0.086 x 3g/l = 0.258g/l

Gypsum (100% pure) requirement for 10000m^3 water = 0.258g x 1000 x 10000

= 258g x 10000

= 2580kg

Gypsum requirement (75% pure) = $\frac{2580}{0.75} = 3440kg$

10.8 Toxicity

The toxicity problem arises out of the accumulation of certain ions in the leaves carried through the soil-water during the process of transpiration to the extent that cause damage to the plants. The degree of damage depends on the time, concentration, crop sensitivity and crop water use. The toxic ions that can be present are chloride, sodium and boron. In other than these ions, there are many trace elements, viz., Al, As, Cr, F, Fe, Li, etc.; which are in very low concentration may cause damage to the plants.

Chloride: Chloride is the most common toxic element in irrigation water. It is not absorbed by the soils, and therefore, easily taken by the plants to the leaves by the transpiration process. Leaf burn or drying of leaf tissues are the symptom of chloride toxicity. Quality of irrigation water alone is not responsible for chloride uptake, but also depends on the presence of soil chloride and the crop ability to exclude chloride.

Sodium: Sodium requirement in plant is so small that frequently it is called as non-essential. Sodium toxicity is the cause of high sodium concentration (Na or SAR) in water. The typical symptoms of sodium toxicity are leaf burn, scorch and dead tissues along the outside edges of the leaves. The symptoms first appear in older leaves and advance to leaf center with the accumulation of more toxic concentration. Crop sensitivity to sodium varies much. The deciduous fruits, nuts, citrus, avocados and beans, mung, cowpeas, etc., are the sensitive crops. The carrot, lettuce, rice, rye, tomato, wheat, etc., are semi-tolerant and alfalfa, barely, beet, cotton, sugar beet, etc., are sodium tolerant crops. For the plants, sodium in the leaf tissue in excess of 0.25 to 0.5 percent on dry weight basis use to show toxic effect.

Boron: Boron is an essential element to plant growth. It is required in small amount. However, a little more amount available in water than the required amount causes toxicity. Boron content 1 to 2mg/l in water may be toxic but 0.2 mg/l is essential. Surface water is usually free from presence of excess boron. But the ground water or spring water originated from near the geothermal areas and earthquake faults contains toxic amount of boron. Boron toxicity symptoms start from older leaves as yellowing, spotting and drying of leaf tissue at the tips and edges. There is a wide range of tolerance for boron toxicity among the plants. Most of the crop shows toxicity symptoms if boron concentration exceeds 250-300mg/kg (dry weight).

Management of Toxicity Problem

The most perfect corrective measure for occurrence of toxicity problem is to select the irrigation water, which does not have the potential to cause toxicity. However, this may not be possible in most of the cases. Therefore, one has to adopt the management options to reduce the toxicity. Leaching may be one of the suitable options. The leaching requirement depends on the type and concentration of toxic ions. But if the leaching requirement is found excessive and uneconomic the farming practices may be changed or the poor irrigation water may be applied blending with comparatively safe good quality water.

Questions and Problems

10.1 Why salt in soils is a problem? What are the causes of salt problem in soil?

10.2 Define alkaline soil, saline soil, exchangeable sodium percentage (ESP), sodium adsorption ratio (SAR), specific surface, cation exchange capacity (CEC).

10.3 Discuss the salinity, sodicity and alkalinity hazard in water.

10.4 Discuss the principle of blending poor quality water in irrigation.

10.5 Discuss with necessary sketches the practice of placement of seeds to reduce the effect of salts in irrigation water.

10.6 Describe the management practice to protect the crop from the effect of salinity.

10.7 Classify the irrigation water based on the electrical conductivity and discuss the possible measures to use it in agriculture.

10.8 Describe the irrigation water classified with respect to sodium hazard.

10.9 Covert 3500ppm salt concentration in to dS/m and 500¼ S/cm in to ppm.

Ans: 546.8dS/m, 320ppm

10.10 Irrigation water caused saline condition to soil of depth 50cm. the bulk density of soil is 1.4g/cm^3, the density of water is 1.01g/cm^3, and the saturation percentage of soil is 33. Determine the depth of water applied if the crop starts to suffer at salinity 5dS/m.

Ans: 114.36 cm

10.11 Evaporation of ground water causes to change the soil salinity 7.5 millimhos/cm. the root zzone depth of soil is 1.35g/cm^3 and saturation percentage of soil is 55. The electrical conductivity and density of ground water is 25 millimhos/cm and 1.01g/cm^3 respectively. Determine the amount of water evaporates.

Ans: 11.03 cm

10.12 Determine the leaching requirements of a reasonably good (EC_w = 1.1dS/m and a poor quality (EC_w = 8.0dS/m irrigation water for wheat crop at 75% yield potential and blending proportion at 15% leaching requirement.

Ans: LR of good quality water = 2.4%

LR of poor „ „ = 20%

Proportion of good quality water for blending = 7%

Proportion of poor „ „ „ „ = 91%

10.13 The irrigation water (EC_w=1dS/m) of 1000mm is applied in a crop season. If the crop water uses pattern follows 40-30-20-19 and leaching requirement 15%, what is the salinity of drain out water?

Ans: 6.67dS/m

10.14 It is proposed to to use the gypsum to improve the infiltration and reducing the water logging for an area of 10Ha. The depth of irrigation water is to be applied 80mm and it is desired to increase the calcium content by 2.5milliequivalent/l. how much gypsum (80% pure) is required?

Ans: 2150kg

10.15 Select the appropriate answer from the following:

1. Irrigation water contains EC value of 3000 μmhos/cm can be said as

a) Low salinity water b) Medium salinity water

c) High salinity water d) Very high salinity water

2. The leaching requirement is 20% and the soil salinity 2.5dS/m, the salinity of irrigation water is

a) 0.75dS/m b) 1.0dS/m

c) 1.25dS/m d) 1.5dS/m

3. Irrigation water is applied whose electrical conductivity is 1.3dS/m. If the crop evapotranspiration during the season is 950mm, the leaching requirement is

a) 11% b) 15%

c) 18 d) 25%

4. On application of saline irrigation water, maximum soil-water salinity develops in root zone depth's

a) 1st quarter b) 2nd quarter

c) 3rd quarter d) 4th quarter

5. The soil-water salinity to soil salinity ratio is approximately

   a) 0.5 b) 1

   c) 1.5 d) 2

6. Application of gypsum in irrigation water

   a) Increases the infiltration b) Decreases the infiltration

   c) Increases the salinity hazard d) Decreases the soil fertility

7. The USDA classification of irrigation water with regard to alkali amd salanity hazards is based on

   a) Exchangeable sodium percentage and p^H

   b) Electrical conductivity and sodium adsorption ratio

   c) Electrical conductivity and p^H

   d) Sodium percentage and p^H (GATE, 2015)

8. The EC value of 0.1dS/cm is equal to

   a) 6000ppm b) 6200ppm

   c) 6300ppm d) 6400ppm

9. The salt concentration of 9600 ppm when expressed in osmotic pressure in atmosphere is

   a) 2.7 b) 3.6

   c) 5.4 d) 6.3

10. A crop field using irrigation water of salt concentration 1.2dS/m and crop tolerance to soil salinity is 1.75dS/m. The leaching fraction is

    a) 0.16 b) 0.17

    c) 0.18 d) 0.20

Ans.

1. c) 2 c) 3. a) 4. d) 5. d) 6. a) 7. b) 8. b)
9. c) 10. a)

10.16 Write True or False of the following statements:

1. Boron content >3.75 ppm is not suitable to crop
2. Soil water salinity is usually much more to salinity of irrigation water
3. According to FAO, EC of soil water should be equal to thrice EC of crop tolerance
4. Saturation extract of a saline soil has EC value less than 4 millimhos/cm
5. Milliequivalent weight of Ca is 0.01

Ans.

1. True 2. True 3. False 4. False 5. False

References

Biswas, R.K.(1980). A \Study on Quality of Irrigation Water on Physico-Chemical Properties of Soil in Kashimpur Area. A B. Tech. dissertation submitted to the Faculty of Agricultural Engineering, Bangladesh Agricultural University, Mymensingh, Bangladesh.

Bandyopadhyay, B.K. et al (2001). Saline & Alkali Soils and Their Management. ISCAR Manograph Series 1. Indian Society of Coastal Agricultural Research, West Bengal, India. Eaton, F.M.(1950). Significance of carbonate in irrigation water. Soil Sci., 69:123-33.

FAO (1985). Irrigation and Drainage Paper-29

Gupta, I.C. (1990). Use of Saline Water in Agriculture (Revised Edition), Oxford & IBH Publishing Co. Pvt. Ltd., New Delhi.

Michael, A.M. (1978). Irrigation Theory and Practices. Vikas Publishing House Pvt. Ltd., New Delhi.

Richards, L.A. (Ed.) (1954). Diagnosis and Improvement of Saline and Alkali Soils. USDA Hbk 60.

Todd, D.K. (1959). Ground Water Hydrology, Wiley International Edition, John Wiley & Sons, Inc.

Groundwater potential available for domestic & industries	:	71 BCM
Per capita water availability	:	1730.6 m^3
Storage available due to completed projects	:	213.03 km^3
Major &		
Medium (Live capacity more than 10Mm^3)	:	184.0 km^3
Storage of projects under construction/ consideration Land		
Land Resources		
Total cultivable land (2000-02) (P)	:	183 Mha
Ultimate irrigation potential	:	140 Mha
Cumulative irrigation potential created up to (2004-05)	:	99.30 Mha
Gross sown area (2001-02) (P)	:	190.3 Mha
Net sown area (2001-02) (P)	:	141.3 Mha
Gross irrigated area (2001-02) (P)	:	76.4 Mha
Net irrigated area (2001-02) (P)	:	55.9%
Cropping intensity	:	135.3%
Food grain production during 1950-01	:	50.8 M ton
Food grain production during 2003-04	:	213.5 M ton
Hydro Power		
Utilisable hydro-power potential (estimated) (01.01.2006)	:	84044 MW at 60% LF
Installed capacity as on (31.03.2005)	:	29507 MW
Potential developed as on (p1.01.2005)	:	16032 at 60% LF
Potential under development (01.01.2005)	:	4577 MW

Source: INCID News, Nov-Dec, 2006

The water resource potential of the country has been assessed by different agencies in different time (Table 11.1). It is observed that the latest estimate of 1869bcm (1993) are within the little variation of the estimates since 1954. The utilizable water resources of the country are assessed 1123bcm by CWC (1993), of which 690bcm from surface water and 433bcm from ground water sources. This 690bcm can be harnessed if matching storage structures are built. The surface water utilizable potential may increase by 220bcm if inter-basin transfer of water is taken up at full extent as proposed. The irrigation potential of the country has been estimated 140Mham. It may reach to 175Mham if inter-basin transfer of water is executed.

Table 11.2 Estimates of water resources in India

S.No.	Agency	Estimates, bcm	Deviation from 1869 bcm
1	First Irrigation Commission (1902-03)	1443	-23%
2	Dr.A.N.Khosla (1949)	1673	-10%
3	Central Water and Power Commission (1954-66)	1881	+0.6%
4	National Commission on Agriculture (1988)	1850	-1%
5	Central Water Commission (1988)	1880	+0.6%
6	Central Water Commission (1993)	1869	-

Source: 11th Five Year Plan, Govt. of India

11

Water Resources and Irrigation Development in India

10.1 Land and Water Resource

India has got the world's 2.45% area and 16.87% of population with only the 4% of available fresh water (Table 11.1). This indicates the necessity of development of water resources and optimal use of it. India at macro level does not have short of water. But the shortage of water is due if not properly managed by conservation and sector wise utilization.

Table 11.1. India-land & water resources at a glance

General		
Geographical area	:	329 Mha
Area as % of world area	:	2.45
Location	:	Latitude 8^0-4' & 37^0-6' North
Forest Cover	:	21.11%
Population (As on 01.03.2005)	:	1097.1M
Population as percentage of world population	:	16.87%
Average annual rainfall	:	1170mm
Rainfall variation	:	100mm in Western most regions to 11000 mm in Eastern most regions
Temperature range	:	50^0C in Western Desert in summer to Sub-zero in Northern most areas in winter
Major river basins (Catchments area more than20, 000 sq.km)	:	12 Nos. having catchments area 253 Mha
Medium river basins (Catchments area less than20,000 sq. km)	:	46 Nos. having catchments area 25 Mha
Total navigable length of important rivers	:	15604 km
Coastline of the country	:	7516.16 km (including A&N Islands & Lakhshdeep islands)
Boundary with adjacent countries	:	15200 km
Water Resources		
Average annual precipitation	:	4000 BCM
Average annual precipitation during monsoon (June-Sept.)	:	3000 BCM
Natural runoff	:	1869.35 BCM
Estimated utilizable surface water sources	:	690 BCM
Total replenishable groundwater resources	:	433 BCM
Total annual utilisable water resources	:	1123 BCM
Available groundwater resources of irrigation	:	361 BCM

India has got some inherent hydrological problems which cause uneven distribution of water. The rivers of the Central and Southern India depend only on the rainfall. The rainfall occurs very unevenly with respect to time and space. Therefore, these rivers remain almost dry in most of the time. The State of Rajasthan, Gujarat, Karnataka, Andhra Pradesh etc. are feed with these rivers and identified as drought prone area. On the other hand the rivers of North India are originated from the Himalaya and carry plenty of water throughout the year due to high rainfall as well as melting of ice within their respective catchments (Roy, 2008).

Table. 11.3 Major river basins of the country

S. No.	Name of the River	Origin	Length (Km.) (Sq. Km.)	Catchment Area (Sq. Km.)
1.	Indus	Mansarovar (Tibet)	1114 +	321289 +
2.	a) Ganga	Gangotri (Uttar Kashi)	2525 +	861452 +
	b) Brahmaputra	Kailash Range (Tibet)	916 +	194413 +
	c) Barak & other rivers flowing into Meghna, like Gomti, Muhari, Fenny etc.			41723 +
3.	Sabarmati	Aravalli Hills (Rajasthan)	371	21674
4.	Mahi	Dhar (Madhya Pradesh)	583	34842
5.	Narmada	Amarkantak (Madhya Pradesh)	1312	98796
6.	Tapi	Betul (Madhya Pradesh)	724	65145
7.	Brahmani	Ranchi (Bihar)	799	39033
8.	Mahanadi Pradesh)	Nazri Town (Madhya	851	141589
9.	Godavari	Nasik (Maharashtra)	1465	312812
10.	Krishna	Mahabaleshwar (Maharashtra)	1401	258948
11.	Pennar	Kolar (Karnataka)	597	55213
12.	Cauvery	Coorg (Karnataka)	800	81155
		Total		2528084

Table 11.4 Medium river basins

S. No.	Name of the River	Village/Distt. (Origin)	State	Length (Km.)	Catchment Area (Sq. Km)
		West Flowing Rivers			
1	Ozat	Kathiawar	Gujarat	128	3189
2	Shetrunji	Dalkania	Gujarat	182	5514
3	Bhadar	Rajkot	Gujarat	198	7094
4	Aji	Rajkot	Gujarat	106	2139
5	Dhadhar	Panchmahal	Gujarat	135	2770
6	Purna	Dhosa	Maharashtra	142	2431
7	Ambika	Dangs	Maharashtra	142	2715
8	Vaitarna	Nasik	Maharashtra	171	3637
9	Dammanganga	Nasik	Maharashtra	143	2357

10	Ulhas	Raigarh	Maharashtra	145	3864
11	Savitri	Pune	Maharashtra	99	2899
12	Sastri	Ratnagiri	Maharashtra	64	2174
13	Washishthi	Ratnagiri	Maharashtra	48	2239
14	Mandvi	Belgaum	Karnataka	87	2032
15	Kalinadi	Belgaum	Karnataka	153	5179
16	Gangavati or Bedti (in upper reaches)	Dharwar	Karnataka	152	3902
17	Sharavati	Shimoga	Karnataka	122	2209
18	Netravati	Dakshina Kannada	Karnataka	103	3657
19	Chaliar or Baypore	Elamtalvi Hills	Kerala	169	2788
20	Bharathapuzha (known as Ponnani)	Annamalai Hills	Tamil Nadu	209	6186
21	Periyar	Sivajini Hills	Kerala	244	5398
22	Pamba	Devarmalai	Kerala	176	2235
		East Flowing Rivers			
23	Burhabalang	Mayurbahanj	Orissa	164	4837
24	Baitarni	Keonjhar	Orissa	365	12789
25	Rushikulya	Phulbani	Orissa	146	7753
26	Bahuda	Ramgirivillage	Orissa	73	1248
27	Vamsadhara	Kalahandi	Orissa	221	10830
28	Nagavali	Kalahandi	Orissa	217	9410
29	Sarda	Vishakhapatnam	Andhra Pradesh	104	2725
30	Eleru	Vishakhapatnam	Andhra Pradesh	125	3809
31	Vogarivagu	Guntur	Andhra Pradesh	102	1348
32	Gundlakamma	Kurnool	Andhra Pradesh	220	8494
33	Musi	Nellore	Andhra Pradesh	112	2219
34	Paleru	Nellore	Andhra Pradesh	104	2483
35	Muneru	Nellore	Andhra Pradesh	122	3734
36	Swarnamukhi	Koraput	Orissa	130	3225
37	Kandleru	Vinukonda	Andhra Pradesh	73	3534
38	Kortalaiyar	Chinglepet	Tamil Nadu	131	3521
39	Palar (including tributary Cheyyar)	Kolar	Karnataka	348	17871
40	Varahandi	North Arcot	Tamil Nadu	94	3044
41	Ponnaiyar	Kolar	Karnataka	396	14130
42	Vellar	Chithri Hills	Tamil Nadu	193	8558
43	Vaigai	Madurai	Tamil Nadu	258	7031
44	Pambar	Madurai	Tamil Nadu	125	3104
45	Gundar	Madurai	Tamil Nadu	146	5647
46	Vaippar	Tirunolvolli	Tamil Nadu	130	5288
47	Tambraparni	Tirunolvolli	Tamil Nadu	130	5969
48	Subarnarekha	Nagri/Ranchi	Bihar	395	19296

Water requirement and availability

The population of the country was 102.7 crores according to census report of 2001. This population will gradually increase to around 158 crores by the year 2050. Naturally, it will have great pressure on the availability of water and other natural resources. The annual flow through the rivers of the country is 1953km^3. Out of this surface flow, 690km^3 and 396km^3 from total 433 km^3 dynamic ground water potential will be available for use if proper storages are made. It has been estimated that an additional amount of 169km^3 may made available by reverse flow from irrigation. Thus, the availability of usable water in the year 2050 will be in the order 1255km^3 as below.

a) Ground water = 396 km^3

b) Surface water = 7690 km^3

c) Return flow from irrigation = 169 km^3

Total = 1255 km^3

The National Commission on Integrated Water Resources Development (NCIWRD) has assessed the water requirements for various sectors in the year 2000. The basis of this assessment was on the assumption of increase of irrigation efficiency to 60% from the present 35-40%. The Standing Committee of Ministry of Water Resources also assesses it periodically. These are shown in Table 11.5.

Table 11.5 Projected water requirements for various sectors

Sector	Water demand in km^3 or bcm					
	Standing Sub-Committee of MoWR			NCIWRD		
	2010	2025	2050	2010	2025	2050
Irrigation	688	910	1072	557	611	807
Domestic	56	73	102	43	62	111
Industry	12	23	63	37	67	81
Power	5	15	130	19	33	70
Others	52	72	80	54	70	111
	813	1093	1447	710	843	1180

According to international norms a country is identified as water stressed if per capita water availability is less than 1700m^3 and water scared if less than 1000m^3. The availability of water will be 794m^3 in 2050 nearing the water scared situation. At present even after construction of 4525 large and small dams the present per capita water storage is 213m^3 as against 6103m^3, 4733m^3, 1964m3 and 1111m^3 respectively

11.2 History of Irrigation in India

Agriculture practice is as old as civilization. It has been established beyond doubt that agriculture technique for secured food supply was invented around

7,000 B.C. in the river valleys of Tigris, Euphrates and parts of Baluchistan. The irrigation as a dependable and efficient means of increasing food production was invented sometime around 6, 500 B.C. Over this period of more than 8, 000 years, lot of modification, up gradation and indigenization have taken place in irrigation techniques.

The evidences of ruined bunds (small dams) and canals in Indus valley civilization (Harappa-Mohenjodaro) and places like Lothal (Gujrat), Inamgaon (Maharashtra) and other places in northern and western India proofs the developed irrigation system in that time. Kautilya's *Arthashastra* contains the idea of principles and management of irrigation system. About 2,000 years ago, before the rule of Satvahana king the surface water was mainly used in irrigation by erecting the embankment to store rain water and later conveyed to the fields through the channels. They were the first to introduce the brick and ring wells for the purpose of using surface and ground water simultaneously. The villagers of southern states had the local governance over the local water sources much long before. The local management institution in the form of *nattamaikar* or *manyams* where the villagers were empowered by the kings to repair and management of local water bodies since centuries. It has been observed that the rulers of the southern India always patronized the decentralization of irrigation structures and embankments but the north Indian rulers showed the great trust on king-owned management. Thus, social management institution were lacking in north India.

At the initiation of Muslim rule in India, Muhammad Bin Tughlaq (1315-1351) encouraged irrigation by dug wells. Feroze Shah Tuglaq (1351-1388) extended irrigation in vast dry land in Northern India. Taimur Long (1414-1451) substantially destroyed the agricultural base. In the Southern India, Vizianagara Kingdom (1336-1546) had the great contribution to waste land reclamation. Anatarajasagar or Poruma milla tank at Cuddapah district (AP) was constructed in his tenure. King Shri Krishnadevaraya is famous for his remarkable work Korrangal dam and its distributory channels. Sultan Zain Uddin (1420-1470), the ruler of Kashmir introduced irrigation networks of Upalpur, Nadashail, Bijbihara etc. In Mugal period, the Charasa and Persian wheels were extensively used for lift irrigation. Shahjahan (1628-1658) restored the defunct Western Yamuna canal with a new canal up to Red fort and constructed the Doab or Hasli canal. The Eastern Yamuna canal was constructed at the instance of Rangila Muhammad Sha's rule (1719-1748) (Bagchi, 1995).

During the beginning of eighteenth century the Britishers started to grab the total control over Indian sub-continent. In the early part of nineteenth century extensive canal irrigation was promoted by them which certainly gave partial coverage against famine and drought. But the motto of the Britishers was to earn more revenue out of this rather than security of the farmers. The application of British idea and practice in irrigation, the local management institutions who used to look after the irrigation structures was become defunct. The age-old techniques which the villagers inherited over the generations and with which they had the great emotional bondage- were able to withstand against this adverse administrative change and tried their best to save those resources. The situation was more grave in directly British govern provinces than the princely states. The Britishers always tried for strong centralized management. The same idea was taught to the so called educated and enlighten youth during the 19th century. During the period of Lord Curzon (1899-1903 A.D.), British government made the first Irrigation Commission in India with Sir Collin Scott-Moncrieff as chairman to know the Indian irrigation scenario and future plan. The committee recommended for (i) proper maintenance and extension of the existing irrigation projects, and (ii) initiation of a large number of new irrigation schemes. Accordingly, the Upper Chenab canal, Upper Jhelum and Bari Doab canals were sanctioned and completed for irrigating 8.9 lakh ha land area in Punjab province (1905 A.D.). The idea of this Triple Canal Project was conceived and initiated by Sir Thomas Benton, a great engineer and canal expert. As a result of follow up action of recommendations of First Irrigation Commission the total irrigated area under public and private works reached to 16.0mha in 1920-21 (Michael, 1995).

In the 1st quarter of 20th century the major irrigation projects in the princely states are Krishnarajasagar dam in the Mysore state with irrigation capacity more than 1.30 lakh ha, Nizamsagar dam in the Hyderabad state of Nizam's completed in 1924 with irrigation capacity 1.11 lakh ha, and the Ganga canal initiated by the Maharaja Ganga Singh of Bikaner (Rajasthan) which was the first irrigation project in his state in 1922 and irrigation capacity was made 2.25 lakh ha. The Sardar Canal Project (UP) was initiated in 1915 and completed by 1926 with the irrigation capacity of 4.4 lakh ha.

Sir Ganga Ram, a renowned engineer, agriculturist and philanthropist of Punjab who first came up with noble idea of lift irrigation and initiated the Renala Irrigation Schemes. In this project a 2 ft fall in a canal was converted to 6 ft and a huge gas turbine suitable to such small fall was fitted to it to generate hydro-electric power. He could have converted 32, 374 ha wasteland in to productive agricultural land by using the generated power in operating the pumps. This project was proved very remunerative.

The idea of using deep tube well for irrigation was first conceived by Sir William Stampe in the earlier part of 20's. He found that it would have been a good alternative to canal irrigation for upland areas where extension of canal was not possible for irrigating dry tracts of the north-western province. However, this project was not successful as because during that time unrest developed for independence movement and the British government was also reluctant to spend in any public works in that situation. Overall, during 1923-1947 little expansion and new invention took place in the irrigation sector. It is very heartening for India that during the partition of India in 1947, Pakistan got the 31% of fertile irrigated crop land while the India had to bear the 80% pre-partition population. This had the severe impact to production of irrigation dependant crops like oilseeds, cereals and fibres. India had to face 4.0 million tones of food grain deficiency during 1947. The problem further aggravated due to trans border movement and huge migration of refugees from Pakistan made lakhs of hectares land fellow for years together.

Table 11.6 Net sown and irrigated area (mha) in India and Pakistan during the time of partition (1947)

Cropped area	India	Pakistan	Undivided India
Net sown area	98.5	18.3	116.8
Net irrigated area	19.4	8.8	28.2(100)
Percentage	19.7	48.1	24.1
Govt. canals	6.3(32.5)	6.8(77.2)	13.1(46.5)
Pvt. canals	1.9(9.8)	0.2(2.3)	2.1(7.4)
Total	8.2(42.3)	7.0(79.5)	15.2(53.9)
Wells	5.3(27.3)	1.3(14.8)	6.6(23.4)
Tanks	3.3(17.0)	-	3.3(11.7)
Others	2.6(13.4)	0.5(5.7)	3.1(11.0)

Source: Bagchi, K.S. (1995) (p.48)

The Second Irrigation Commission of India (1972) gave the progressive increase in irrigated area in Indian Union during 1910-1950 (Table 11.7). The growth rate of irrigation during this period was estimated about 2% per annum to irrigated area under Govt. canals, 0.54% in well irrigation and 0.54% in respect of all sources of irrigation.

Table 11.7 Net area irrigated in Indian Union from 1910 to 1950

Period	Area irrigated from		Total (all sources), Mha	Irrigated area as % of sown area
	Govt. canals, Mha	Wells, Mha		
1910-11 to 1914-15	4.4	4	14.5	17.9
1915-16 to 1919-20	4.1	4.7	15.7	19
1920-21 to 1924-25	4.4	4.7	16	17.4
1925-26 to 1929-30	4.6	4.8	16.2	17.1
1930-31 to 1934-35	5	4.8	17.1	17.6
1935-36 to 1939-40	5.6	5.2	18	18.6
1940-41 to 1944-45	6	5.4	19	19.2
1945-46 to 1949-50	6.4	5.3	19.4	19.1

Source: Michael, A.M. (1995) (p.18)

11.3 Irrigation Development Under the Five Year Plan

The planning commission in India was established in 1950. Subsequently, the Ministry of Irrigation and Power along with the apex bodies like Central Board of Irrigation and Power and Central Water and Power Commission to assist the Planning Commission in planning and formulating projects in irrigation and power sectors. The irrigation projects were categorized as major (over Rs. 50million), medium (2 to 50 million) and minor (less than Rs.1.5 million). The above stated apex bodies were given the responsibility for execution of major and medium projects and the Ministry of Agriculture to the minor ones.

Due to severe shortage of food grain the irrigation projects got the top priority during the first five year plan (1951-56). The important projects started during this time are shown in Table 11.8. Some of the projects started during this time were completed only after third five year plan.

Table 11.8 Important irrigation projects initiated in first five year plan

States	Name of the project
Andhra Pradesh	Nagarjunasagar Project
West Bengal	Mayurakkshi Project (1946-1956)
Orissa	Hirakud Project (1948-56)
Bihar	Damodar Valley Project (1950-1958) & Koshi Project
Uttar Pradesh	Rihand Project (1952-56), Matatila Project
Punjab	Harike
Karnataka	Tungabhadra Project (1945-1954)
Maharashtra	Konya Project (1954-1955)
Gujarat	Kakrapara Project (1949-1955)
M.P.& Rajasthan	Chambal Project
Tamil Nadu	Lower Bhavani Project

The potential created were 1.25mha also much less than the expected potential 12.5mha area due to different constraints (Bagchi, 1995). In the second five year plan (1956-61) 195 new irrigation projects were undertaken aiming 6.1mha additional area to be irrigated. Some of the important irrigation projects are the Rajasthan Project; Inter-State Gandak Project (Bihar & U.P.); Ramganga (U.P.); Parambikulam Aiyar (T.Nadu); Kabini(Mysore); Kansabati(W.Bengal); Kadana, Ukai & Broach(Narmada) in Gujarat and the Purna, Girna & Mula in Maharashtra. In the third five year plan only 9 additional major irrigation projects were included. Among these projects, Beas in Punjab, Malaprabha & Upper Krishna in Mysore are well known. Eighty numbers medium irrigation projects were also taken up. The irrigation potential created 2.1mha against the estimated 5.2mha. During the Annual Plans 1966-69 no significant addition in irrigation potential could be achieved in national level. Only the Maharashtra initiated 6 new projects. They are Bhima, Jayakwadi, Krishna, Warna, Upper Godavari and Kukadi. During the fourth five year plan emphasis was shifted to the completion of ongoing projects. In the fifth plan (1974-78) attention was drawn to the widening gap between the potential creation and utilization. Accordingly, Command Area Development (CAD) project was initiated. Sixth plan had a new start and emphasized on completion of irrigation project. At the end of the eighth plan (1996-97) central assistance has been provided under AIBP to the State Governments for early completion of the projects.

Table 11.9 Plan wise cumulative irrigation potential created and utilized (mha)

Plan	Potential created				Total	Potential utilized				Total
	Major & Medium	SW	Minor GW	Total	-	Major & Medium	SW	Minor GW	Total	-
Up to 1951 (pre plan)	9.7	6.4	6.5	12.9	22.6	9.7	6.4	6.5	12.9	22.6
First 1951-56 Plan	12.2	6.43	7.63	14.06	26.26	10.98	6.43	7.63	14.06	25.04
Second Plan 1956-61	14.33	6.45	8.3	14.75	29.08	13.05	6.45	8.3	14.75	27.8
Third Plan 1961-66	16.57	6.48	10.52	17	33.57	15.17	6.48	10.52	17	32.17
Annual Plan 1966-69	18.1	6.5	12.5	19	37.1	16.75	6.5	12.5	19	35.75
Fourth Plan 1969-74	20.7	7	16.5	23.5	44.2	18.39	7	16.5	23.5	41.89
Fifth Plan 1974-78	24.72	7.5	19.8	27.3	52.02	21.16	7.5	19.8	27.3	48.46
Annual Plan 1978-80	26.61	8	22	30	56.61	22.64	8	22	30	52.64
Sixth Plan 1980-85	27.7	9.7	27.82	37.52	65.22	23.57	9.01	26.24	35.25	58.82
Seventh Plan 1985-90	29.92	10.9	35.62	46.52	76.44	25.47	9.97	33.15	43.12	68.59
Annual Plan 1990-91	30.74	11.46	38.89	50.35	81.09	26.31	10.29	36.25	46.54	72.85

Eighth Plan 1992-97	32.95	12.51	40.8	53.31	86.26	28.44	11.07	37.7	48.77	77.21
Ninth Plan 1997-2002	37.05	13.6	43.3	56.9	93.95	31.01	11.44	38.55	49.99	81
Tenth Plan 2002-2007	42.35	14.31	46.11	60.42	102.8	34.42	12	40.81	52.81	87.23

Source: 11th Five Year Plan, GoI

Ultimate irrigation potential

The anticipated irrigation potential created up to March, 2007 is 102.77Mha. This is 73.46% of ultimate irrigation potential (Table 11.9 & 11.10). The irrigation potential created 42.35Mha under MMI projects against 140Mha UIP. MI have the UIP of 81.43 Mha against which the potential created is 60.42 Mha.

The ultimate irrigation potential (UIP) created and utilized till the end of tenth plan was assessed by a Committee constituted by MoWR in 1977 (Table 11.10).

Table 11.10 Ultimate irrigation potential, potential created and potential utilized (Mha)

Sector	Ultimate	Potential Created		Potential Utilized Till end of	
	Irrigation Potential	Tenth Plan	Anticipated in Tenth Plan	Till end of Ninth Plan	Anticipated in Tenth Plan
MMI	58.47	37.05	5.3	31.01	3.41
MI					
Surface water	17.38	13.6	0.71	11.44	0.56
Ground water	64.05	43.3	2.81	38.55	2.26
Subtotal	81.43	56.9	3.52	49.99	2.82
Total	139.9	93.95	8.82	81	6.23

Growth of Irrigation Potential (M ha)

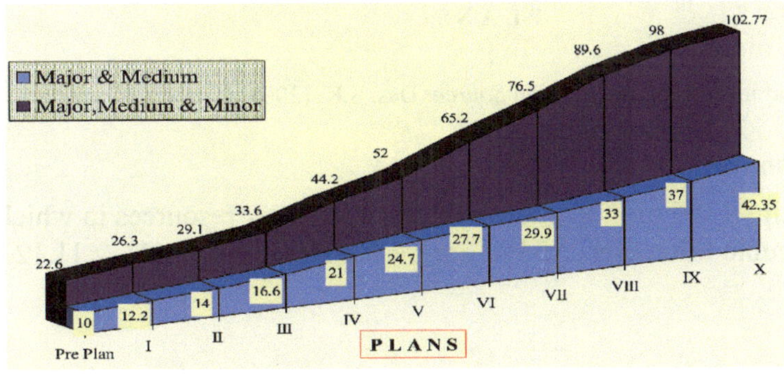

Cumulative Investment in Irrigation (Rs. Billion)

Investment in Irrigation Sector

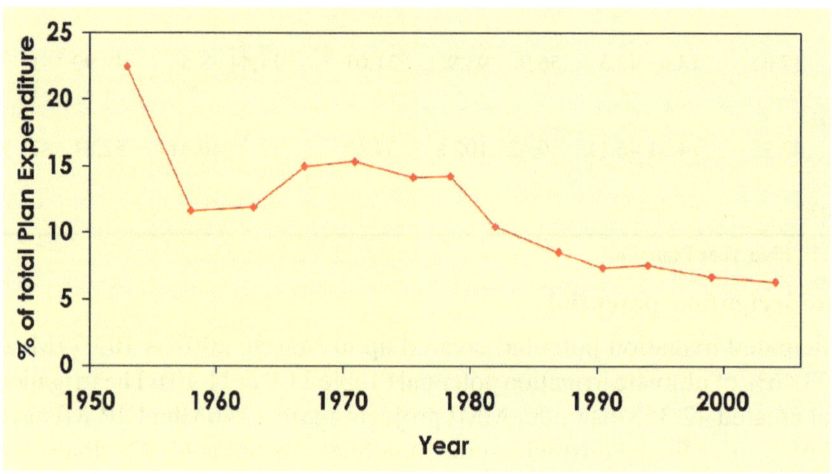

Cumulative Investment in Irrigations (Rs. Billion)

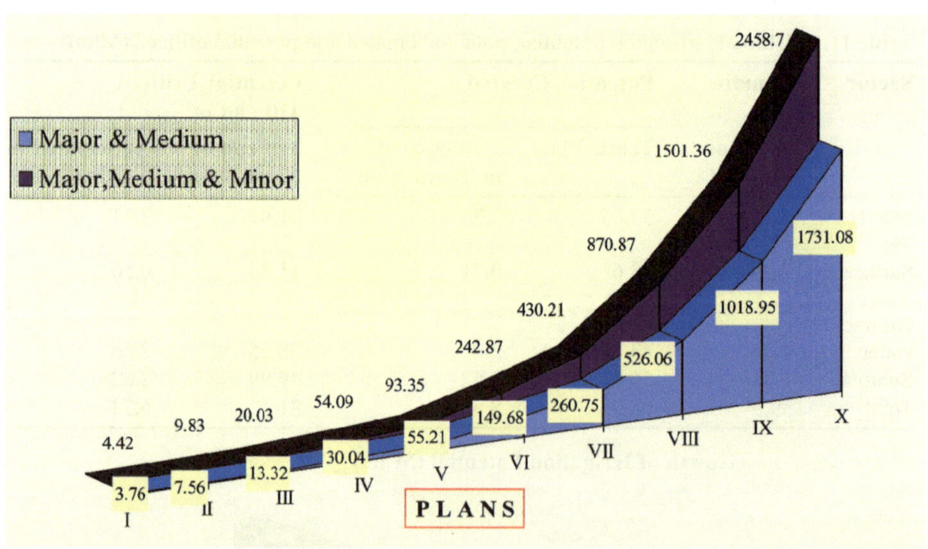

Fig. 11.1 Growth and investment in irrigation *Source:* Das, S.K. (2008) Courtesy: Das (2008)

11.4 Water Resources of the World

The world contains 1386 million cubic-kilometer of water resources in which only 35 million cubic-kilometer (2.53%) are fresh water (Table 11.11 & 11.12.)

Table 11.11 Water resources of the world

S. No.	Types of water	Volume (Km^3x10^6)	(%)
1	Sea water	1338	96.5
2	a) Fresh water	10.53	0.76
	b) Saline ground water	12.57	0.94
3	Soil moisture	0.0175	0.001
4	Glacier and permanent snow pack	24.064	1.74
5	Underground ice in permafrost zone	0.3	0.02
6	Fresh water lakes	0.091	0.007
7	Salt water lakes	0.085	0.006
8	Swamps	0.0115	0.0008
9	Rivers	0.0021	0.0002
10	Water in biosphere	0.0011	0.00008
11	Water in atmosphere	0.0129	0.0009

Table 11.12 Sources of fresh water

S.No.	Source	Volume (Km^3x10^6)	Percent of total storage
1	Fresh ground water	10.53	0.76
2	Glacier and permanent snow pack	24.064	1.74
3	Underground ice in permafrost zone	0.3	0.022
4	Fresh water lakes	0.091	0.007
5	Swamps	0.0115	0.0008
6	Rivers	0.0021	0.0002
7	Biosphere	0.0011	0.00008
8	Atmosphere	0.0129	0.0009
9	Soil moisture	0.0175	0.001
	Total	35.0301	2.53198

All the sources of fresh water are not usable. The usable sources of fresh water are shown in Table 11.13.

Table 11.13 Useable sources of fresh water

S.No.	Source	Volume (Km^3x10^6)	%
1	Sweet water aquifer	10.53	0.76
2	Fresh water lakes	0.091	0.007
3	Rivers	0.0021	0.002
	Total	10.6231	0.7672

11.5 Quality of Ground Water Resources of India

Ground water plays the vital role in our life support system. Ground water is utilized for domestic, irrigation and industrial purpose. The rapid urbanization, industrialization as well as agricultural activities have necessitated more exploitation to ground water resources leads to deterioration of its quality Excessive exploitation of ground water accounts for lowering of ground water level and sea water intrusion. Urbanization without adequate attention to sewage and waste disposal causes ground water pollution. Industrialization without the provision of treatment disposal wastes and affluent is one of the major sources of water pollution. A suitable portion of the applied inorganic fertilizers and pesticides to agricultural fields join to ground water through percolated water. Indiscriminate use of these agricultural inputs along with application of excess water due to inefficient irrigation practice causes quick deterioration to ground water particularly in intensive cultivation zone.

Factors affecting ground water quality

The quality of water is determined by the presence of dissolved solids and micro organism. The following parameters may be considered to judge the ground water quality (Mehta & Srivastava, 2008).

i) **Dissolved gases:** The gases like CO_2, CH_4, H_2S are often exist in ground water because these are the gases which are the products of the bio-geological processes in non-aerated subsurface zone. Excess pressure of these gases makes the ground water unusable.

ii) **Reaction of carbonic acid:** Water containing of high concentration of CO_2 while passes through soil zones through percolation can dissolve the minerals under the influence of CO_2 or H_2CO_3 (or H^++HCO_3) results in dissolution of calcite or dolomite and raises the p^H content and bicarbonate content of water.

iii) **Oxidation and reduction process:** Rain water contains oxygen equilibrium with the atmosphere. Hydro-chemical and bio-chemical process reduces the oxygen in ground. The reaction process still continues to organic matter when entire dissolve oxygen when entire dissolved oxygen is consumed but the oxidizing agents (NO_3^-,MnO_2,Fe(OH),SO_4^{-1}) get reduced and leads poor ground water environment.

iv) **Ion exchange capacity:** The ion exchange process is almost exclusively limited to colloidal particles as because these particles are electrically charged.

v) **Silicate mineral reaction:** Crystalline rocks of igneous and metamorphic origin have appreciable amount of quartz and alumina silicate minerals which are thermodynamically unstable and tends to dissolve in contact with water. This water acquires dissolve ions by this dissolution process.

vi) **Saline water intrusion:** Sea water and over that the fresh ground water remains in a hydraulically stable state in the aquifer. When over pumping reduces the ground water level the sea water level rises and the usual discharge of ground water to the sea may be reversed. This is the intrusion of sea saline water in to the aquifer.

Ground water quality in India

The ground water of shallow aquifers is usually of good quality and suitable for all purpose. However, sometime it is also rich with calcium and sodium chloride. The quality of water of deeper aquifers depends on the places of occurrence. Still exploration is going on to find out the quantity and quality of water in different aquifers. The contamination in ground water is mainly geogenic origin.

i) Inland salinity

It has been estimated that about 1.93 lakh square kilometer area in the arid and semi arid regions in Rajasthan, Haryana, Punjab, UP (partly), Delhi, MP, Maharashtra, Karnataka to be affected by saline water of electrical conductivity (EC) more than 4000μS/cm (Table 11.14). In some places of Haryana and Rajasthan the electrical conductivity of ground water goes above 10000 and the water becomes non-potable. In some pocket of Rajasthan the ground water are so saline that it is being used directly for manufacturing salts by solar drying.

Table 11.14 Saline status of different States

S. No.	State	Total area of the State (km^2)	Area underlain by Saline ground water (km^2)	Percent of affected area in the State
1	Haryana	44212	11438	25.87
2	Punjab	50353	3058	6.07
3	Delhi	1485	140	9.43
4	Rajasthan	342239	141036	4.12
5	Gujarat	196024	24300	12.4
6	Uttar Pradesh	294411	1362	0.46
7	Karnataka	191791	8804	4.59
8	Tamil Nadu	130058	3300	2.54
Total		1250573	193438	15.47

Source: (Mehta & Srivastava, 2008)

ii) Coastal salinity

India has a dynamic coast line of about 7500 km stretched from Rann of Kutch to Konkan, Malabar coast to Kanyakumari in the south to northwards along the Coromondal coast to Sunderbans in West Bengal. In the coasts the sea and the land meet and the aquifers of these zones have their end boundaries to sea water. The withdrawal of excess fresh water from the aquifer causes disequilibrium in static balance to fresh and sea water and the situation permits the intrusion of saline water to inland. In coastal zones situation may arise like, (i) fresh water aquifer overlying saline water,(ii) saline water overlying fresh water, and (iii) alternate sequence of fresh water and saline water aquifers.

Salinity problems have been observed in different places of coastal zones of India. Minjur of Tamil Nadu and Mangrol-Chorwad-Porbander belt along the

Saurashtra have the problem of salinity ingress. The upper aquifers contain the saline horizons which decrease landwards about an 8-10 km wide belt of Subarnarekha, Salandi, Brahamani out fall regions in the proximity of the coast.

iii) Fluoride

Fluorite is (CaF_2) is a common fluoride mineral occurs in both igneous and sedimentary rocks. Apatite Ca_5 (CL, F, OH)$(PO_4)_3$ commonly contains fluoride. The fluorides are sparingly soluble and present in natural waters in small amounts. (Mehta & Srivastava, 2008). High concentration of fluoride in ground water beyond the limit of 1.5mg/l is a major health problem in India. About 90% of the population of the country uses ground water for drinking and other domestic purposes. Excess fluoride in ground water has been identified in 199 districts of 19 States (Table 11.15).

Table. 11.15 Fluoride in ground water of India

S. No.	State	No. of districts affected	Districts
1	Andhra Pradesh	16	All districts except Adilabad, Nizamabad, West Godavari, East Godavari, Vizag, Srikakulam, Vizianagaram
2	Assam	2	Karbi Anglong, Nagaon
3	Bihar	5	Daltongang, Gaya, Rhotas, Gopalganj, Paschim Champaran
4	Chattishgar	2	Durg, Dhantenwalan
5	Delhi	7	Central, South, West, East, South West, North West, North East Zones
6	Gujarat	18	All districts except Dang
7	Haryana	12	Rewari, Faridabad, Karnal, Sonipat, Jind, Gurgaon, Mohindgar, Rohtak, Kurukshetra, Kaithal, Bhiwani
8	J&K	1	Doda
9	Jharkhand	4	Gridhi, Palamau, Pakur, Sajabganj
10	Karnataka	16	Dharwad, Gadga, Bellary, Belgam, Raichur, Bijapur, Gulbaarga, Chikmaglur, Mandya, Bangalore rural, Mysore, Manglore, Kolar, Bhimoga
11	Kerala	3	Palghat, Allepy, Vamanapuram
12	Maharashtra	10	Chandrapur, Bhandara, Nagpur, jalgaon, Buldhana, Amravati, Akola, Yavatmal
13	MP	14	Shivp[uri, Jhabua, Mandla, Dindori, Chhindwara, Dhar, Vidhisha, Sehore, Raisen, Mandsour, Neemuch, Ujjain, Seoni
14	Orissas	18	Phulbani, Koraput, Dhenkenal, Angur, Boudh, Nayagarh, Puri, Balasore, Bhadrak, Bolangir, Ganjam, jagatpur, jaipur, Kalahandi, Keonjhar, Kurda, Mayurbhanj, Rayagada

15	Punjab	17	Mansa, Faridkot, Bhatinda, Muktsar, Moga, Sangrur, Ferozpur, Ludhiana, Amritsar
16	Rajasthan	32	All districts
17	Tamil Nadu	8	Salem, Pariyar Ferode, Dhrampuri, Coimbtore, Tiruchirapalli, Vellore, Madurai, Virudinagar
18	UP	7	Unnao, Agra,Meerut, Mathura, Aligarh, Raiberali, Allahabad
19	West Bemgal	7	Birbhum, Bardhaman, Bankura, Purulia, Malda, Uttar Dinajpur & Dakshin Dinajpur

(Mehta & Srivastava, 2008)

iv) Arsenic

Arsenic, a mettaloid released from natural and anthropogenic sources, ranks as the number one toxic environment contaminant (http:www.atsdr.cdc.gov/).The common valencies of arsenic in unpolluted ground water of geogenic origin are +III & +V as hydrolysis species.

The arsenic problem in India was first reported in 1980 in West Bengal. It is one of the most affected areas in the world and has been a global concern over the last twenty years. The affected areas confined to Ganga-Brahmaputra delta, expose nearly 10 million people of West Bengal and 30-35 million people in adjoining Bangladesh.

The safe and permissible limit of arsenic for human consumption is 0.01 and 0.05 mg/l respectively as per WHO standard. It has been reported that 85 Blocks in 8 districts have arsenic in ground water beyond safe limit of 0.01mg/l and affecting 42.7 million people. Soil acts as the arsenic sink and thereby reduces the availability of toxicant to the crop. It has been reported by some researchers that they found arsenic content in rice grain of this area to the extent of 0.2-1.2mg/kg. The arsenic problem is become more acute because the plants are known to convert the less toxic arsenate (As+V) into more toxic arsenite (As+III).

The most affected districts in West Bengal are on the eastern side of the Bhagirathi river in the districts of Malda, Murshidabad,Nadia, North 24 Parganas and South 24 Parganas and western side of the river districts are Howrah, Hugli and Bardhaman. Arsenic in ground water mainly occurs in the intermediate aquifer between 20-100m, the deeper aquifers are free of arsenic problem (Mehta & Srivastava, 2008). Apart from West Bengal, arsenic contamination in ground water has been found in the State of Bihar, Chhatisgarh, Uttar Pradesh and Assam. The places of occurrence of arsenic in different States are shown in Table 11.16.

Table. 11.16 Arsenic of ground water in some States

S.No.	District	Blocks
Assam	Dhemaji	Bordolani,Dhemaji, Simlborgaon, Murkongselek
Bihar	Bhojpur	Barhar, Shahpur, Koilwar, Ara, Bihiya, and Udawantnagar
	Patna	Maner,Danapur, Bhaktiarpur and Bath
	Begusarai	Matihani, Begusarai, Baruani, Balia, Sabehpur Kamal, Bachwara
	Khagaria	Khagaria, Mansi,Gorgi, Parbatta
	Samastipur	Mahinuddin Nagar, Mohanpur, Patroi, Vidyapati Nagar
	Bhagalpur	Jagidishpur, Sultamganj, Nathnagar
	Saran	Sonepur, Dighwara, Chapra Sadar, Ravelganj
	Munger	Jamalpur, Dharhara, Bariarpur, Munger

Questions and Problems

11.1 Give the brief statement of land and water resources of India.

11.2 What are the major river basins in India.

11.3 Discuss the pre-independence history of irrigation in India.

11.4 What are the important irrigation projects initiated in 1st 5-year plan of India? What is the ultimate irrigation potential in India?

11.5 Describe in brief the water quality in India.

11.6 Describe the factors which affect ground water quality.

11.7 Select the appropriate answer from the following multiple-choice questions.

1. India has got world's percent of area

a) 1.25 b) 2.45

c) 3.65 d) 4.35

2. India's percent of world's population is approximately

a) 5 b) 7

c) 17 d)

3. India's percent of world's fresh water

a) 1 b) 2

c) 3 d) 4

4. India's average precipitation is

a) 1000 BCM b) 2000 BCM

c) 3000 BCM d) 4000 BCM

5. The origin of river Brahmaputra is

a) Tibet b) Uttar Pradesh

c) Nepal d) China

6. The origin of river Narmada is
   a) Bihar b) Uttar Pradesh
   c) Madhya Pradesh d) Jharkhand
7. The percent of water use in agricultural sector is about
   a) 40 b) 60
   c) 70 d) 80
8. The Cauvery river basin said to be
   a) Small b) Medium
   c) Major d) Large
9. Muhammad Bin Tughlaq encouraged irrigation by
   a) Water pond b) Canal water
   c) Dug well d) Tube well
10. The important irrigation project initiated in Orissa in 1st 5 year plan is
   a) Damodar b) Mayurakshi
   c) Hirakud d) Nagarjunasagar

Ans.

1. a) 2. c) 3. d) 4. d) 5. a) 6. c) 7. c) 8. d)
9. c) 10. c)

11.8 Write True or False of the following statements:

1. Geographical area of India is 329Mha
2. The estimated irrigation potential of India is 130 Mham
3. Tungabhadra project in Karnataka was initiated in second five year plan
4. Average annual rainfall in India is 1270mm.
5. Origin of the Indus river is Mansarovar (Tibet)
6. Longest river in India is the Brahmaputra.
7. Command Area Development (CAD) project was initiated in fourth five year plan
8. Water in biosphere is more than atmosphere
9. India has a dynamic coast line of about 7500 Km
10. The river Bhagirathi is in the State of Bihar

Ans.

1. True 2. False 3. False 4. False 5. True 6. False 7. True
8. True 9. True 10. False

References

Bagchi, K.S. (1995). Irrigation in India. History and Potential of Social Management. Upalabdhi Trust for Development Initiatives, New Delhi.

Das,S.K.(2006). Speech as a session Chairman of Symposium on Environment and Water (NSEM-2008) organized by Indian Association of Hydrologist at Jadavpur University, West Bengal on 18th April, 2008.

GoI . 11th Five Year Plan. Vol. III, Chapter 2: Water Management and Irrigation. INCID (2007). (INCID) News, A Quarterly Bulletin of Indian National Committee on Irrigation and Drainage, New Delhi.Vol.7, Issue-2, April-May-June, 2007.

Meheta & Srivastava (2008). Speech as invited Speaker in National seminar on Ecorestoration of Soil and Water Resources Towards Efficient Crop Production, held on June 6-7, 2007 at Bidhan Chandra Krishi Viswavidyalaya, West Bengal.

Michael, A.M. (1995). Irrigation Theory and Practice. Vikas Publishing House Pvt. Ltd. Co., New Delhi.

Roy, A. (2008). Water. National Symposium on Environment and Water. Organized by Indian Association Hydrologists, West Bengal Regional Center, 18th April, 2008.

12

Surface Drainage

12.1 Introduction

Agricultural drainage is the removal of excess water from the farmland for the purpose of providing favorable soil air environment for plant growth. Removal of water may take place form the surface or below the surface of the land. The practice of lowering ground water table below the root zone depth to improve the plant growth or to reduce the accumulation of salts in soil is also called as drainage.

Drainage system is broadly classified as surface & subsurface drainage. Removal of excess water over the land surface is known as surface drainage & that of removing water below the surface is known as subsurface drainage. The flow of water in the form of evaporation or evapotranspiration also play important role in improving drainage situation. It gives the idea of the time required to get the wetland workable though it is not called a type of drainage as because this is very slow process & cannot be controlled by the human activities. However, using of certain tree plants e.g., Eucalyptus, Poplar, Casurina or similar species in wet fields, which transpires at high rate and improves the drainage problem by lowering the shallow water table is the new idea known as bio-drainage.

Sources of excess water

To know the source of water causing water logging to farmland is important before going for planning to any drainage project. Otherwise, it may be a wasteful effort and loss of money. There are many factors responsible to water logging. Major factors are listed below:

i) Excess rainfall or irrigation.

ii) Seepage water from reservoir, ditches and unlined canals.

iii) Accumulation of seepage water in valleys due to excess irrigation in the high land.

iv) Inundation of low-lying areas from floodwater.

v) High groundwater table due to poor drainage scope.

vi) Low capacity disposal channels which causes ponding on the land for damaging period.

vii) Uneven land surface with pockets and ridges causing retardation of natural runoff. Problem may intensify in slow permeable soil.

viii) High level of water in drainage outlet causing hindrance to designed drain out water.

Effect of water logging

Water logging condition arises when water table is very close (less than 3m) to the surface and there is ample scope of rapid capillary rise of groundwater to the surface. The ill effects of water logging are:

i) Water leads to poor soil aeration. Water filled soil pores not only displace the air but also obstructs gases, which are given off by the roots.

ii) Anaerobic decomposition of organic matter takes place after the completion of dissolved oxygen in waterlogged soils. It leads to build up of reduced organic compounds such as carbon dioxides, methane, ethane, and profane in the soil, which are harmful to the plant root system. Toxic concentration of ferrous and sulfide ions may develop within a few days of submergence. Anaerobic condition slow downs the decomposition and release of nitrogen. It affects much to the growth of the plants.

iii) Poor root development or decay of roots and restricted uptake of nutrients. The systems, which develop under these circumstances, are yellowing, reddening, or a scorched or stippled appearance of the leaves.

iv) Soil temperature decreases; soil structure is destroyed and reduced microbiological activities.

v) Plant nutrients get lost through leaching of water.

vi) Significant reduction or even total damage of crop.

vii) Provide unhygienic condition to the human habitation.

Benefits of drainage

The ultimate benefit of expenditure towards drainage system is the increase in crop production for maintaining a favourable soil environment. Different benefits of drainage are listed below:

i) Drainage facilitates early working condition of the soil. Field operations are possible without destroying the soil structure. Ensure early planting and good crop yield. Increases the cropping intensity.

ii) Drainage receives aeration in root zone soil which is very much essential for microorganism activities leads to converting soil organic matter and fertilizer to favourable plant growth.

iii) Removal of excess water from soil provides favourable soil temperature, which accelerates microorganism activities.

iv) Leaching out of harmful salt from the root zone provides favourable salt-water balance.

v) Root zone depth of plant increases. Higher root zone depth provides more nutrient and better anchorage of the plants.

vi) Well-drained lands are saved from breeding of agents that cause disease to men and livestock such as malaria, dengue, foot rot and liver fluke. A healthy and pleasant environment persists.

vii) Higher infiltration capacity of the soil reduces erosion problem.

viii) Intercultural operation is possible even in the rainy days and damage of crops in harvesting time is minimized.

12.2 Surface Drainage System

The method of removal of excess water depends on the type and magnitude of drainage problem. Primarily the slope of a free water surface is developed over the land surface to outlet. Surface drainage not only includes the removal of excess water from the affected area by the methods of land grading, field drains, etc., but also the diversion of water by constructing suitable structures so that the land under consideration is saved from drainage problem. Whatever may be the case, the system is divided into three functional parts:

I. **Collection system:** It is the part of the system which picks up the water from the land by the surface field ditches, diversion ditches, bedding, etc.

II. **Disposal system:** This part of the system conveys the collected water to the outlet usually through the open channel.

III. **Outlet:** It is the last part, which control the exit of drain water.

The factors, which influence the choice of a particular drainage system, depend on soil, topography, crops and farmers preference.

The following are the various type of surface drainage system.

i) Land smoothing

The soils, which require surface drainage usually, have many depressions of different shape and size. For the purpose of uniform movement of water over the land surface it is required to fill up the depressions and level the high points in the field. Larger depressions are not smoothened but connected to a ditch. It is generally required once in a year after completion of all tillage operation for seedbed preparation.

ii) Land grading

It is the practice of cutting, filling and smoothing of land surface to provide a designed continuous level slope to aid in surface drainage. Land grading involves considerable cost. A topographic survey is required to fix up the cut and fill at different locations for a minimum earthwork and designed grade. The grade should be minimum of 0.05%, preferably 0.1% to a maximum of 0.5%. The grade and length of the sloping field depends on the type of soil. Land smoothing is essential operation after land grading or any other land farming practices to minimize surface irregularities (Fig 12.1).

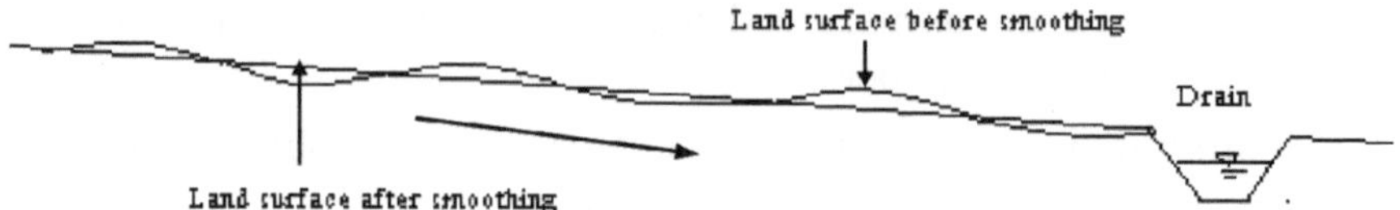

Fig. 12.1. Land smoothing and grading

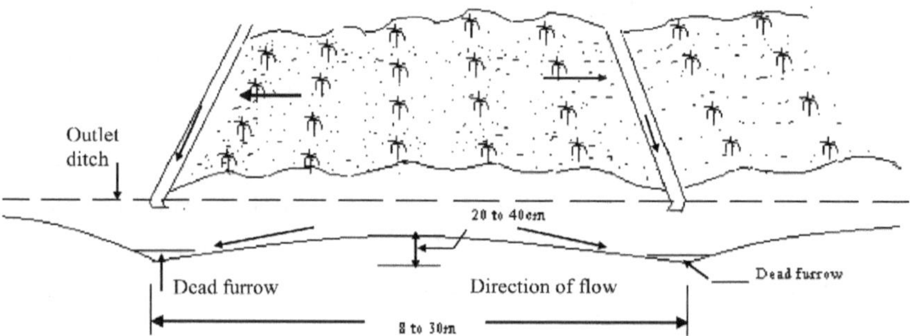

Fig. 12.2. Bedding system of surface drainage

iii) Bedding system

This system uses the dead furrows in the direction of prevailing slope in poorly drained soil of slope up to 1%. Subsurface drainage is not suitable to this area. Ploughing is done parallel to these furrows but other farming operations can be done in either direction. In general, the crops, which mature early, cultivated across the furrow and the row crops, which mature late should be planted parallel (Fig 12.2). The field drain is made at the lower end of the field to receive water from the furrows. The field drains deliver the water to the diversion ditch. The area between the dead furrows called 'bed' is sloped in such a way that the water is collected in the furrow end whose length depend upon the land use, slope of the field, soil permeability, farming operation and width of farm machinery. The recommended bed widths for different permeability of soil are presented in Table 12.1.

Table 12.1. Recommended bed width

Drainage condition	Permeability, cm/day	Bed width, m
Very slow internal drainage	0.5	8 to 12
Slow internal drainage	5 to 10	15 to 17
Fair internal drainage	10 to 20	20 to 30

Source: Mal (1995)

The length of the bed is made 100 to 300m, the maximum bed height 40 cm for pasture land and 20 cm for arable land, the field drains are shallow with average width of 25 cm, side slopes are 6:1 to 10:1 and the longitudinal slope at least 0.1% (Mal, 1995).

iv) Random ditch

The land which have depressions too deep or too large for land smoothing are connected by ditches from one to another for collecting and carrying the water in to a lateral outlet ditch (Fig. 12.3).

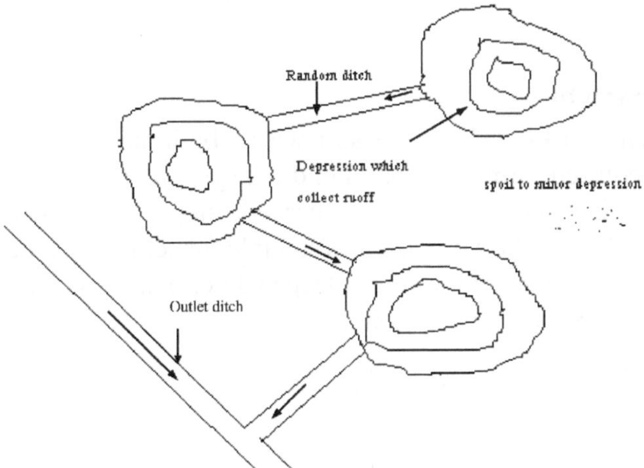

Fig. 12.3. Random ditch system

The ditches constructed for drainage should have the sufficient capacity to remove the water rapidly and completely. The spoil of the ditches may be used to fill up the minor depressions in the field. Side slope of the ditches should not be less than 4:1 where farming operations are parallel and 8:1 where farming operations cross the ditches. In a poorly permeable soil the random ditch can be adopted in combination to bedding system to get more effective drainage.

v) Parallel ditch system

In parallel ditch system the ditches are made parallel to each other usually at spacing 100 to 30 m depending upon hydraulic conductivity of the soil, crop grown, topography and lateral slope (Fig. 12.4). This system is effective in poorly drained flat land with numerous depressions. The land between the two ditches is required to proper smoothing by eliminating the minor depressions and to provide a gentle slope.

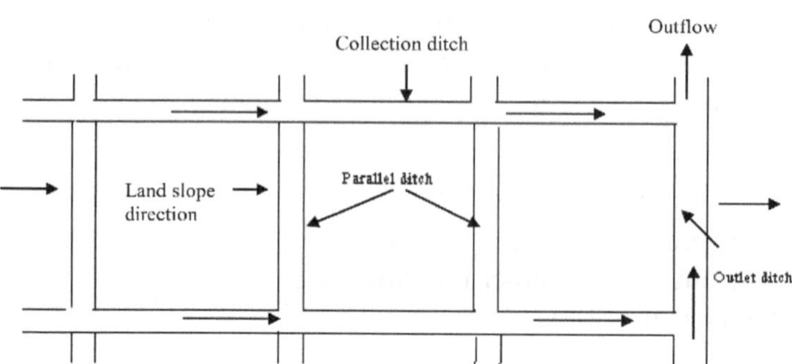

Fig. 12.4. Parallel ditch system

vi) Parallel deep ditch system

Parallel deep ditch system works well in the soil where both surface and subsurface drainage are required. Similar to parallel ditch system, the ditches are made parallel to each other but deeper and steeper side slope where farm implements cannot cross it (Fig. 12.5). The system is suitable for controlling the water table and permits sub irrigation if the soil is deep and permeable with an impermeable bed below to check deep percolation.

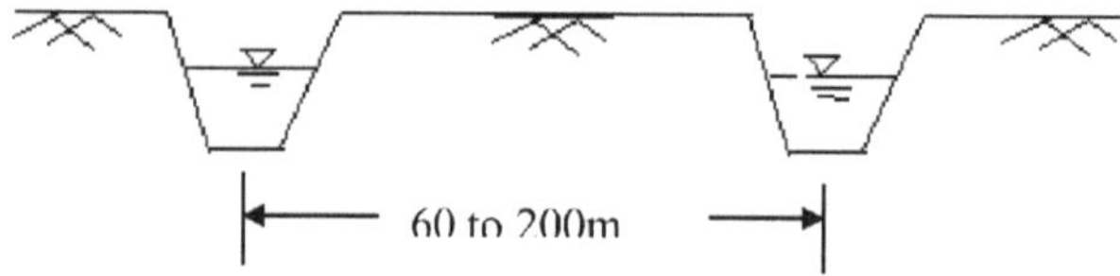

Fig. 12.5. Parallel deep ditch system

vii) Cross-slope/ interception ditch system

In this system the ditches are made across the slope resembles terracing that's why sometime it is referred as drainage type terrace. The system is suitable to soils with poor internal drainage, numerous depressions to collect rainwater and land slope is too great for bedding and subsurface drainage is not feasible. The area between the ditches are smoothed or graded by eliminating the depressions by the excavated soil. The side slope of the ditch is usually not less than 10:1. The ditches should be constructed as straight as possible to avoid the difficulties in farming (Fig. 12.6).

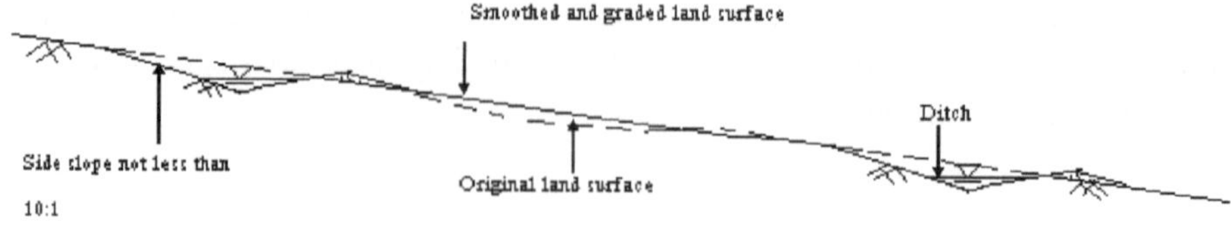

Fig. 12.6. Cross-slope ditch

viii) Pipe outlet

Sometime the pipe may be used to divert the ponded water to the disposal channel (Fig. 12.7).

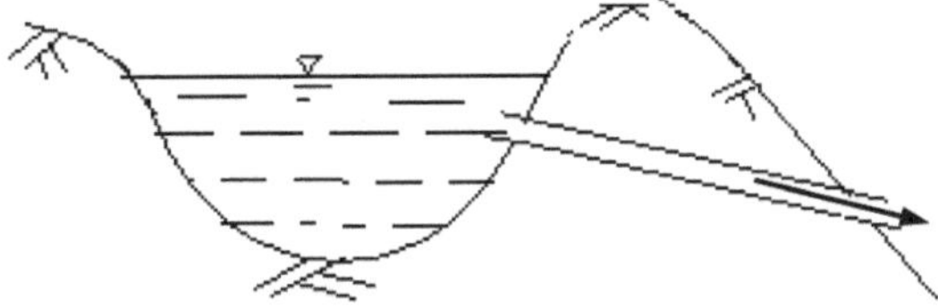

Fig. 12.7. Pipe outlet

ix) Surface inlet

The surface inlet is a device or structure, which carry the pit water to the sub-surface drain. At the inlet of the pipe mesh or beehive grate is provided to check the entry of the debris inside the drain (Fig. 12.8).

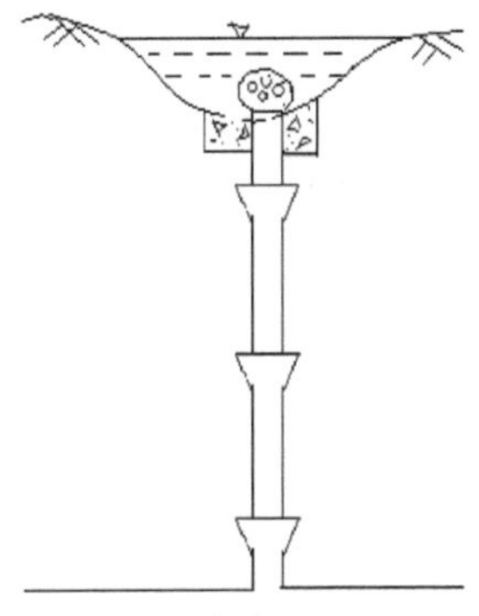

Fig. 12.8. Surface inlet

Open ditch

Open ditch is the conduit with flows of free water surface. It is widely used for irrigation as well as drainage purpose because of its simplicity in construction and low cost. However, there is some difference in respect of construction of drainage and irrigation ditches. Irrigation ditches follow the higher contour whereas the drainage ditches follow the lowest contour. Similarly, opposite to irrigation ditch the capacity of the drainage ditch should increase downward. The designed water level in drain channel usually lies below the ground surface level but in irrigation channel designed water level lies above the ground surface level. Irrigation channels carry water to deliver it over the land surface whereas the drain channels collect it from the land surface. This difference in practices leads to some differences in constructional features (Fig. 12.9). An open ditch drainage system consists of a network of ditches drain out water from a given watershed or large area. The contour map of the area may suggest the location of the field ditches and the outlet. The outlet of the drainage system is very important and whose selection should be appropriate so that an economic drainage system can be designed. The natural water way is the best selection for an outlet.

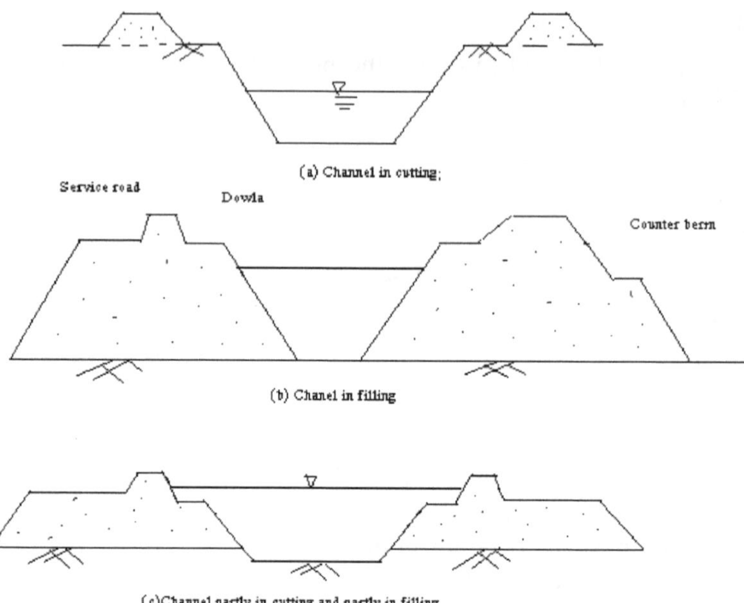

Fig. 12.9. Typical irrigation channel cross-sections

The cross-section of an open ditch may be triangular, rectangular, trapezoidal or parabolic. The trapezoidal ditches are more preferred because in course of time it takes the shape of a parabolic channel, which is akin to the natural shape of the channel. The ditches are constructed with ploughs, V-ditchers, bulldozers, etc. (Fig.12.9).

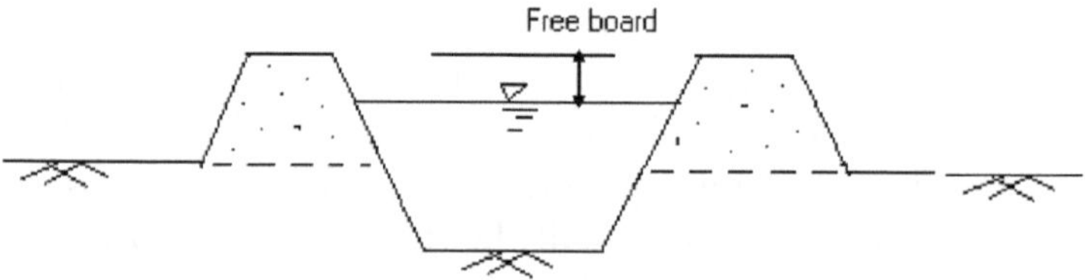

a) Typical irrigation channel section

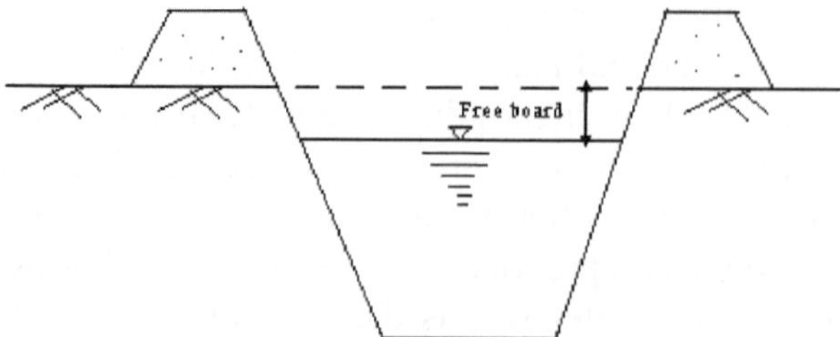

b) Typical drainage channel section

Fig. 12.10. Typical irrigation channel cross-sections

12.3 Design of Surface Drainage System

The primary purpose of surface drainage is the removal of excess water over the land surface is cropped area. The removal of excess water takes place through a network of channels. The channels of the network should be properly designed and correctly laid so that the excess water can be intercepted and disposed of before it can sufficiently damage the crops. Secondly, well functioning surface drainage network also controls the water table rise and salinity hazard in soils.

In designing surface drainage the drainage coefficient is to be known first. It is the rate at which the excess water needs to be removed to avoid the ill effect of water logging. Secondly, the design involves the determination of the geometry and lay out of the channel. In doing so the outlet condition, disposal system and other essential structure of the system should have due consideration. Thus, the surface drainage system has two parts, hydrologic design for establishing the drainage coefficient and hydraulic design for facilitating the drainage. Hydrologic design is more difficult to hydraulic design because it deals with the variable like rainfall and runoff, which are very much uncertain of its occurrence in respect of time, space and magnitude. The crop response to excess water varies much crop-to-crop and even place-to-place. Therefore, it is very difficult to decide upon all agreed removal rate of excess water.

Drainage coefficient

Drainage coefficient is defined as the amount of water need to be removed from a cropland in a day of 24hrs so that crop does not suffer much. It is usually expressed in depth. The preferable unit is mm/day.

In practice, the drainage coefficient represents a flow rate lower than the peak flow of the hydrograph. The flow rate is adjusted so judiciously that the duration it takes to remove the volume of water represented by the direct runoff chemical hydrograph does not cause significant harm of the plants. Drainage coefficient is the key factor for hydraulic design of the drainage system. Fixing the channel dimension is one important item of hydraulic design. As it is stated above that the flow rate following the drainage coefficient is lower than the peak rate flow the drainage channel may overflow its bank for certain period. However, the overflow cannot be allowed. Therefore, the flow of water should not be underestimated. It should be restricted to different unit of the drained area so that water cannot reach to peak value and flow rate remains within the designed rate. In determining the drainage coefficient so many parameters are involved which are characteristically uncertain. These are:

i) Runoff estimation is based on several assumption and seldom gives correct result.

ii) Crop tolerance to excess water is not well known. It is greatly related to environment of the root zone of the soil. Excess water and excess salt in root zone is more harmful. Opposite to this standing water remove the salt beyond the root zone if there is downward flow gradient. One has to settle down the suitable drainage coefficient in consideration to these contradictory statements.

iii) Crop suffers from excess water depends on its stage of growth. It is more harmful at the seedling and harvesting stage.

iv) Due to enormous variation of topography in a watershed the depth of accumulation is not same. Upland gets free of excess water sooner than the lowland.

v) Varieties of crops are grown in a watershed having different level of tolerance to submergence. It is difficult to select a suitable drainage coefficient.

The drainage designer has to consider all the complexities and uncertainities to find a reasonable drainage coefficient to develop the hydraulic design of the drainage network.

Estimation of runoff

Rainfall is the primary causative factor of runoff. The occurrence of rainfall with certain magnitude is a matter of chance, similarly uncertain the runoff. However, the surface drainage system is designed on the basis of estimated runoff based on rainfall data analyzed statistically. Long-term data gives better result to estimation. Runoff production of a watershed influenced by other many factors along with the rainfall. There are many empirical relationships between runoff and the causative factors developed based on observed or field experimentally generated data. The simplest of these is the linear regression between rainfall and runoff. The complex relationships are the mathematical models that include the physical process like space-time distribution of rainfall, infiltration, evapotranspiration, overland flow, etc.

Empirical relations

The very commonly known empirical equation for estimating the peak rate of runoff is the rational formula, which was proposed by C.E. Ramser of USA and expressed as

$$Q = CIA \tag{12.1}$$

Where,

Q = Peak rate of drainage at the outlet, m^3/s.

I = Rainfall intensity of the watershed for a desired recurrence interval for a period of time of concentration of the watershed, m/s.

C = A dimensionless constant whose value theoretically ranges between zero and 1.

A = Area in m^2

The rational formula more conveniently expressed as, $Q=\frac{1}{36}CIA$ when rainfall intensity is in cm/h and area of watershed in hectare. This formula is said rational in the sense that it is dimensionally homogeneous. It is widely used for estimating the peak rate of flow. However, there is great uncertainty to get the uniform I. Infact, uniform I is not possible for a large watershed. Rational formula is not recommended for watershed larger than 15 Km^2. The value of C depends on the rainfall and the characteristics of watershed. It is very difficult to assess the value from the knowledge of watershed characteristics unless experimentally determined. Even in same watershed the value of C is not constant with respect to time.

Ramser originally suggested the value of C by studying the rainfall-runoff of small watershed. Later on, the other research workers also established some more number of values of C to represent more specifically the slope, type of soil, and vegetative cover. These values of C are tabulated in Table 12.2.

Table 12.2 Values of C in the rational formula =

Vegetation cover & slope	Soil texture		
	Sandy loam	**Clay and silt loam**	**Stiff clay**
Wood land			
0-5% slope	0.1	0.30	0.40
5-10% „	0.25	0.35	0.60
10-15% „	0.35	0.50	0.60
Pasture land			
0-5% slope	0.1	0.30	0.40
5-10% „	0.16	0.36	0.55
10-15% „	0.22	0.42	0.60
Cultivated land			
0-5% slope	0.30	0.50	0.60
5-10% „	0.40	0.60	0.70
10-15% „	0.52	0.72	0.82

Source: Michael & Ojha (1999)

A watershed may have different sub areas of distinct difference to vegetative cover, slope and soil type. The weighted average runoff coefficient of such watershed is calculated by dividing the n-sub areas by

$$C = \frac{C_1A_1 + C_2A_2 + C_3A_3 + \ldots\ldots\ldots\ldots C_nA_n}{A_1 + A_2 + \ldots\ldots\ldots\ldots An} \tag{12.2}$$

In which A_1, A_2, A_3, ...A_n are the sub-areas in watersheds and C_1, C_2, C_3 ...C_n are the respective runoff coefficients of the sub-areas.

The rational formula needs the time of concentration to determine I. Time of concentration (T_c) is the time required by the runoff water to travel from most remote point of the watershed to its outlet. There are number of empirical equations available for the estimation of time of concentration. The most widely used equation given by Kirpich (1940), which is,

$$T_c = 0.0195 L^{0.77} S^{-0.385} \quad (12.3)$$

$$or, \quad T_c = 1.525\left(\frac{A^2}{S}\right)^{0.2}\left(\frac{L}{D}\right)$$

where,

T_c = Time of concentration, min.

L = Maximum length of travel of runoff water, m.

S = Slope of the watershed, H/L.

H = Difference in elevation between the most remote point of the watershed to its outlet, m.

A = Watershed area, ha.

D = Diameter of the circle whose area is equal to that of watershed, m. The time of concentration can also be determined as:

$$T_c = 0.1947(K)^{0.77} \quad (12.4)$$

where,

$K = \sqrt{L^3/H}$, Tc, L and H are same as above.

The time of concentration is also determined by dividing the length of run to the average velocity of flow based on the slope of the channel (Table 12.3).

Table 12.3 Average velocity based on channel slope

Channel slope, %	Velocity, m/s
1-2	0.6
2-4	0.9
4-6	1.2
6-10	1.5

Source: Das (2000)

For using the rational formula it is required to know the rainfall intensity (cm/h) for the period of time of concentration. This can be obtained from Fig. 12.11 or by using the equations of the curves of this figure for different duration's rainfall intensity as stated below.

i) 5 min, $Q = 0.0112 R^2 + 0.1266R - 0.055$

ii) 15 min, $Q = 0.01R^2 + 0.34R - 0.025$

iii) 30 min, $Q = 0.0087R^2 + 0.5637R - 0.0225$

iv) 60 min, $Q = 0.0033R^2 + 1.0224R - 0.0221$

v) 120 min, $Q = -0.078R^2 + 2.0761R - 0.2901$

vi) 240 min, $Q = -0.1528R^2 + 3.224R + 0.0452$

vii) 480 min, $Q = 0.2R^2 + 3.35R + 4E - 15$

viii) 960 min, $Q = 0.6R^2 + 5.7R + 5E - 15$

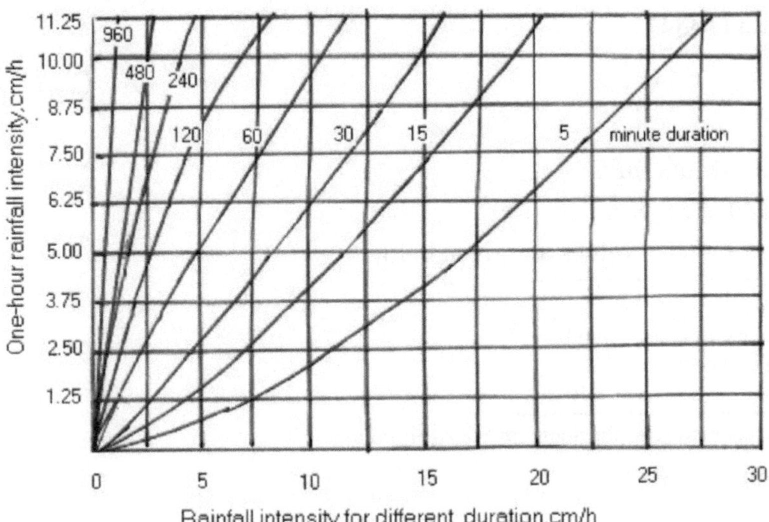

Fig. 12.11. Relation between one-hour rainfall intensities and rainfall intensities of different duration (Redrawn from Das (2000)

Ram Babu *et al.* (1979) developed the following relationships to estimate the rainfall intensity of any duration for different zones of India as below:

$$Zone\,1(Northern): I = 5.9143(T)^{0.1623} / (t + 0.5)^{1.0127} \tag{12.5}$$

$$Zone\,2(Central): I = 7.4645(T)^{0.1712} / (t + 0.75)^{0.9599} \tag{12.6}$$

$$Zone\,3(Western): I = 3.974(T)^{0.1647} / (t + 0.5)^{0.7327} \tag{12.7}$$

$$Zone\,4(Eastern): I = 6.933(T)^{0.1353} / (t + 0.5)^{0.8891} \tag{12.8}$$

$$Zone\,5(Southern): I = 6.311(T)^{0.1523} / (t + 0.5)^{0.9465} \tag{12.9}$$

In the above equations, I is the rainfall intensity in cm/h, T is the recurrence interval in years and t is the duration of rainfall in hours.

Some other empirical equations for estimating peak runoff

The following relationships have been developed in India to apply in the regions for which they are developed.

Dickens formula (1865)

$$Q_p = C_d A^{3/4} \tag{12.10}$$

Where,

Q_p = Peak runoff rate, m^3/s

C_d = A constant ranging from 6 to 30

A = Watershed area, km^2.

Ryves formula (1884)

$$Q_p = C_r A^{2/3} \tag{12.11}$$

Where,

Q_p = Peak runoff rate, m^3/s

A = Watershed area, Km^2.

C_r = A constant whose values are given below.

Region	C_r
Within 80 Km from east coast	6.8
80-160 Km from east coast	8.5
Hills	10.2

Ryves formula is recommended for Southern States of India.

Khosla's formula (1960)

$$Q_m = P_m - L_m \tag{12.12}$$

and $L_m = 0.48T_m$ for $T_m > 4.5°C$

Where,

Q_m = Monthly runoff, cm

P_m = „ Rainfall, cm

L_m = „ Runoff lost, cm

T_m = Mean monthly temperature of the catchment, ^0C

For $T_m \leq 4.5$ °C, the L_m may provisionally be assumed as:

T^0C	4.5	-1	-6.6
L_m	2.17	1.78	1.52

Annual runoff = ΣR_m

Inglis and Desouza formula (1929)

1. For Ghat region of Western India

$$R = 0.85P - 30.5 \tag{12.13}$$

2. For Deccan plateau

$$R = \frac{1}{254} P(P - 17.80) \tag{12.14}$$

Where,

R = Annual runoff, cm

P = „ Rainfall, cm.

Barlow's Tables: Barlow (1915) by studying the small catchments (area~130 Km^2) in Uttar Pradesh expressed runoff R as

$$R = K_b P \tag{12.15}$$

Where K_b = runoff coefficient which depends upon the type of watershed and nature of monsoon rainfall values of K_b are given in Table 12.4.

Table 12.4. Barlow's coefficient K_b in percentage (developed for use in UP)

Class	Description of catchment	Values of K_b (percentage)		
		Section I	Section II	Section III
A	Flat, cultivated and absorbent soils	7	10	15
B	Flat, partly cultivated, stiff soils	12	15	18
C	Average catchment	16	20	32
D	Hills and plain with little cultivation	28	35	60
E	Very hilly, steep and hardly any cultivation	36	45	81

Section I: Light rain, no heavy downpour

Section II: Average or varying rainfall, no continuous downpour

Section III: continuous downpour

Source: Subramanya (1994)

Strange's Table's (1928)

Strange studied the rainfall-runoff data of border areas of Maharashtra and Karnataka and estimated the runoff coefficient

$$K_s = R/P \tag{12.16}$$

as a function of catchment character. He also categorized the catchment as 'zero', 'average' and 'bad' and gave a table for K_s values of these under different monsoon rainfall.

Table 12.5 Strange's tables of runoff coefficient K_s in percent. (For use in the area of Maharashtra and Karnataka).

Total monsoon rainfall (cm)	Runoff coefficient, K_s, %		
	Good catchment	Average catchment	Bad catchment
25	4.3	3.2	2.1
50	15	11.3	7.5
75	26.3	19.7	13.1
100	37.5	28	18.7
125	47.6	35.7	23.8
150	58.9	44.1+	29.4

Source: Subramanya (1994)

Example 12.1 Calculate the most efficient bottom width, velocity and discharge capacity of the open ditch if the flow occurs 2m deep in clay soil. Assume the gradient of the channel as 0.03%.

Solution:

Depth of flow = 2m

Ditch gradient, S=0.03%

Side slope of the ditch, s = 1:1, since the ditch is made in clay soil (Table 5.1)

$\therefore \theta = 45^\circ$ & $\tan \theta/2 = \tan 22.5^\circ = 0.4142$

The maximum permissible velocity in clay soil = 0.7m/s **(Table 5.2)**

Bottom width of ditch, $b = 2d \tan \theta/2 = 2 \times 2 \times 0.4142 = 1.66$m

Wetted x-section of the ditch, $A = bd + zd^2$

$= 1.66 \times 2 + 1 \times 2^2 = 7.314 m^2$

Wetted perimeter, $P = b + 2d\sqrt{z^2 + 1}$

$= 1.66 + 2 \times 2\sqrt{1+1} = 7.316$m

Hydraulic radius, $R = \dfrac{A}{P} = \dfrac{7.314}{7.316} = 1$

Manning's equation, $V = \dfrac{1}{n} R^{2/3} S^{1/2}$, roughness coefficient, n = 0.035 (Table 5.3)

$$= \frac{1}{0.035} \times 1 \times (0.003)^{1/2}$$

$$= \frac{0.01732}{0.035} = 0.49 m/s$$

Discharge capacity, $Q = A \times V = 7.314 \times 0.59 = 4.32 m^3/s$

Example 12.2 Design a drainage channel for 500ha land of drainage coefficient 2cm/day. The area has a natural land slope 0.1% and the soil is silty loam.

Solution:

Required discharge capacity of the channel = 500ha x 2cm / day

$= 500 \times 10000 \times 2/100\ m^3/day$

$= 100000\ m^3/day = 1.157\ m^3/s$

The maximum permissible velocity of the channel in silty loam soil = 0.65m/s.

The permissible side slope of the channel in silty loam soil = 1.5:1.

$\therefore$ z = 1.5 and θ = 33.69°

Manning's roughness coefficient, n = 0.04

Let us try to design the channel for most economic section. Let, d = 1.2m, which is about the minimum depth of drainage channel.

$\therefore$ b = 2d tan $\theta/2$ = 2 x 1.2 x 0.302= 0.725 m

A = bd + zd^2 = 0.725 x 1.2 + 1.5 x 1.22^2 = 3.03m^2

$$P = b + 2d\sqrt{z^2+1} = 0.725 + 2\times1.2\sqrt{1.5\times1.5+1} = 5.05m$$

$$R = \frac{A}{P} = \frac{3.03}{5.05} = 0.6m$$

Manning's equation, $V = \frac{1}{n}R^{2/3}S^{1/2} = \frac{1}{0.04}(0.6)^{2/3}(0.004)^{1/2} = 0.56m/s$

Q = A x V = 3.03 x 0.56 = 1.69m^3 / s

The designed discharge capacity 1.69m^3/s is much greater than the required discharge capacity (1.157m^3/s). Therefore, the channel should be designed on trial and error method keeping the depth 1.2m fixed because this is about the minimum drainage channel depth.

Trial 1: Let, b = 0.4m

$$\therefore A = bd + zd^2 = 0.4x1.2 + 1.5x1.2^2 = 2.64m^2$$

$$P = b + 2d\sqrt{z^2+1} = 0.4 + 2\times1.2\sqrt{1.5\times1.5+1} = 4.726m$$

$$R = \frac{A}{P} = \frac{2.64}{4.726} = 0.55m$$

$$V = \frac{1}{n}R^{2/3}S^{1/2} = \frac{1}{0.04}(0.55)^{2/3}(0.001)^{1/2} = 0.53m/s$$

$Q = A\ x\ V$ = 2.64 x 0.53 = 1.4m^3 / s

$$1.44m^3/s > 1.157m^3/s$$

Trial 2: Let b = 0.25m

$\therefore$ A = bd + zd^2 = 0.25 x 1.2 + 1.5 x 1.2^2 = 2.46m^2

$$P = b + 2d\sqrt{z^2+1} = 0.25 + 2\times1.2\sqrt{1.5\times1.5+1} = 4.576m$$

$$R = \frac{A}{P} = \frac{2.64}{4.576} = 0.53m$$

$$V = \frac{1}{n}R^{2/3}S^{1/2} = \frac{1}{0.04}(0.53)^{2/3}(0.001)^{1/2} = 0.51m/s$$

$Q = A \times V = 2.46 \times 0.51 = 1.25 m^3/s$

This discharge is little higher to 1.157m³/s and b is almost nominal. Therefore, b = 0.25m is accepted.

∴ Designed dimension : b = 0.25m, d = 1.2m & z = 1.5

Example 12.3 Design a most efficient trapezoidal drainage channel for an area of 10Km² assuming Manning's roughness coefficient 0.04, maximum permissible velocity 1.4m/s and side slope 1:1. The rate of water removal is calculated by $Q = 2.5A^{0.6}$ (Q in cumec and A in Km²). Suggest the suitable dimension of the spoil bank made by excavated soil on one side of the channel and used as pathway. Take 2:1 side slope for spoil bank and channel bed slope 0.3%.

Solution:

$Q = 2.5A^{0.6} = 2.5\,(10)^{0.6} = 9.95 m^3/s$

Side slope of the channel, z = 1, ∴ tan θ = 45°

For most efficient cross-section, $b = 2d\,tan\,\theta/2 = 2d \times tan\left(45/2\right)^{o} = 0.8284d$

$A = bd + zd^2 = 0.8234d \times d + 1 \times d^2 = 1.8234d^2$

$$P = b + 2d\sqrt{z^2+1} = 0.8234d + 2d\sqrt{1+1} = 3.65d$$

$$R = \frac{A}{P} = \frac{1.8284d^2}{3.65d} = \frac{d}{2}$$

$$Q = A \times V = 1.8284d^2 \times \frac{1}{n} R^{2/3} S^{1/2}$$

$$or,\ 9.95 = 1.8284dx\frac{1}{0.04}(0.5d)^{2/3}(0.003)^{1/2} = 1.576d^{8/3}$$

∴ d = 2.0m ∴ b = 0.8284 x 2 = 1.657m ∴ R = 0.5 x 2 = 1.0m

$V = \frac{1}{n}R^{2/3}S^{1/2} = \frac{1}{0.04}(1.0)^{2/3}(0.003)^{1/2} = 1.37 m/s$. The velocity of flow is within the permissible limit of 1.4m/s.

Let, the free board to be used 20% of the channel depth. So, ultimate channel depth, D = 2 x 1.2 = 2.4m.

Cross-sectional area, $A = bD + zD^2 = 1.657 \times 2.4 + 1 \times 2.4^2 = 9.737 m^2$

Therefore, the volume of the spoil per meter length of the channel is 9.737m². The same volume of soil is used for construction of 1m length of the spoil bank. Let, there is a top width of 3m of the bank (pathway) and depth d.

The cross-sectional area of the pathway, $A = bd + zd^2 = 3 \times d + 2 \times d^2 = 9.727$

Let, d = 1.5m, ∴ 4.5 + 2 x 2.25 = 9 $\neq RHS$

d = 1.6m, ∴ 4.8 + 5.12 = 9.92 $\neq RHS$

d = 1.55, ∴ 4.65 + 4.805 = 9.455 $\neq RHS$

Since, 5cm is assumed to be the incremental change in dimension the depth of the pathway may be fixed as 1.55m.

Example 12.4 Determine the diameter of the tile drain on a grade of 0.25% draining from an area of 10ha. The drainage coefficient is 2cm.

Solution: The required discharge through the drain, Q = 10ha x 2cm/day

$$= \frac{10 \times 10000 \times 2cm/day}{24 \times 3600} = 0.023 m^3/s$$

Let, d is the diameter of the tile drain.

$$Q = A \times V = \pi \frac{d^2}{4} x \frac{1}{n} R^{2/3} S^{1/2} = 0.023$$

$$or,\ \pi \frac{D^2}{4} x \frac{1}{0.01} \left(D/4\right)^{2/3} (0.0025)^{1/2} = 0.023$$, (assuming Manning's roughnes coefficient, n = 0.01)

$$or, 1.55 D^{8/3} = 0.023$$

$$or, D = (0.0148)^{3/8}$$

$$\therefore D = 0.205m \cong 20cm$$

Example 12.5 The intensity of 1-hour rainfall expected during the recurrence interval of 10 years is 7.5 cm/h of a 150ha watershed. The watershed is of cultivated land of sandy loam soil and 4.5% slope. The maximum length of travel of water particle is 1500m. Estimate the peak rate of runoff.

Solution: For cultivated sandy loam soil and 5% slope the value of C from Table 12.2 is 0.3.

The time of concentration, $T_c = 0.0195\ L^{0.77}\ S^{-0.385}$

$= 0.0195\ (1500)^{0.77}\ (4.5/100)^{-0.385}$

= 0.0195 x 278.99 x 33 = 17.95 min

From **Fig. 12.11** the rainfall intensity is 14.25cm/h for duration of 17.95min.

Rational formula,

$$Q = \frac{1}{36} CIA$$

$$= \frac{1}{36} \times 0.3 \times 14.25 \times 150 = 17.81 m^3/s$$

Example 12.6 A watershed of 100ha of which 20ha is pasture on 6% slope and sandy loam soil, 15ha woodland of 8% slope and silt loam soil and 65ha cultivated land of 3% slope in silt loam soil. The expected rainfall intensity is 6.25cm/h for a recurrence interval of 10 years. The maximum length of run of water is 2200m. Calculate the peak runoff rate.

Solution

20ha of pasture land on 6% slope and sandy loam soil, C = 0.16

15ha of woodland on 8% slope and silt loam soil, C = 0.35

65ha of cultivated land on 3% slope and silt loam soil, C = 0.50

Weighted average of $C = \dfrac{C_1A_1 + C_2A_2 + C_3A_3}{A_1 + A_2 + A_3}$

$= \dfrac{0.16 \times 20 + 0.35 \times 15 + 0.50 \times 65}{100} = 0.4095$

Weighted average of slope, $S = \dfrac{0.06 \times 20 + 0.08 \times 15 + 0.03 \times 65}{100}$

$= \dfrac{1.2 + 1.2 + 1.95}{100} = 0.0435$

Time of concentration, $T_c = 0.0195\ L^{0.77}\ S^{-0.385}$

$= 0.0195\ (2200)^{0.77}\ (0.0435)^{-0.385}$

$= 0.0195 \times 374.68 \times 3.343 = 24.43$ min

From **Fig. 12.11** the rainfall intensity is 11.95 cm/h for duration of 24.43min.

$Q = \dfrac{1}{36} CIA$

$= \dfrac{1}{36} \times 0.4095 \times 11.95 \times 100 = 13.59 m^3/s$

Example 12.7 A 50ha watershed of general slope 0.8%, maximum run of water 2025m. The maximum depth of rainfall recorded for different duration for a recurrence interval of 15 year of this area is given below.

Duration of rain (min)	5	10	15	20	25	30	40	50
Rainfall depth (mm)	10	17	22	30	35	35	40	44

The watershed area is divided to equal two parts for favoring the cultivation due to the soils are distinct to silt and sandy loam. Calculate the peak runoff rate. Assume runoff coefficient C= 0.04.

Solution:

Time of concentration, $T_c = 0.0195\ L^{0.77}\ S^{-0.385}$

$T_c = 0.0195\ (2025)^{0.77}\ (0.08)^{-0.385}$

$= 0.0195 \times 351.5 \times 6.42 = 43.98\text{min}$

The time of concentration 43.48min falls between the duration 40 and 50 minutes. It is 3.98min after the 40min period. At 40min the rainfall depth is 40mm.

Rainfall depth for 3.98min after the 40min

$$= \frac{44-40}{10} \times 3.98 = 1.592mm$$

Therefore, rainfall for the period of 43.98min

$= 40 + 1.592 = 41.592\text{mm}$

The average rainfall intensity, I

$$= \frac{\text{Rainfall depth}}{T_c}$$

$$= \frac{41.592}{43.98} \times 60 = 56.74mm/h$$

Runoff coefficient, $Q = \frac{1}{36} CIA$

$$Q = \frac{1}{36} \times 0.04 \times 5.674cm/h \times 50ha = 0.315m^3/s$$

Example 12.8 A watershed of 100ha of which 20ha is pasture on stiff clay and 12% slope, 25ha is cultivated land on 3% slope in silt loam, 30ha is cultivated land on 7.5% slope in sandy loam and 25ha is wood land on 15% slope in sandy loam soil. If the expected 1-hour rainfall intensity is 7.5cm/h for the watershed and the length of run is 2500m, calculate the peak runoff rate.

Solution:

The weighted value of $C = \frac{C_1A_1 + C_2A_2 + C_3A_3 + ...C_nA_n}{A_1 + A_2 + ...A_n}$

$$= \frac{0.6 \times 20 + 0.5 \times 25 + 0.4 \times 30 + 0.35 \times 25}{20 + 25 + 30 + 35}$$

$$= \frac{12 + 12.5 + 12.0 + 8.75}{100} = \frac{45.25}{100} = 0.45$$

The weighted value of slope, $S = \frac{20 \times 0.12 + 25 \times 0.03 + 30 \times 0.075 + 25 \times 0.15}{100}$

$$= \frac{2.3 + 0.75 + 2.25 + 3.75}{100} = \frac{9.15}{100} = 0.0915$$

The velocity of flow for 0.0915-weighted value of slope from **Table 12.3** is 1.5m/s.

The time of concentration, $T_c = \dfrac{\text{Length of run}}{\text{velocity}} = \dfrac{2500m}{1.50/s} = 27.78\,min.$

From **Fig. 12.11** the rainfall intensity is 11.7cm/h for duration of 27.78min.

$$\text{Peak runoff rate, } Q = \frac{1}{36} CIA$$

$$= \frac{1}{36} \times 0.45 \times 11.7 cm/h \times 100 ha = 14.62 \text{m}^3/\text{s}$$

Example 12.9 A watershed of 75ha and maximum length 1600m in eastern India having average land slope 0.9% and silt loam soil. The land is mainly used for cultivation purpose. Calculate the peak runoff rate for 10 and 15-year recurrence interval.

Solution:

The time of concentration, $T_c = 0.0195\ L^{0.77}\ S^{-0.385}$

$= 0.0195 \times 293.21 \times 6.13 = 35.06$ min.

Using Eq.12.8, the rainfall intensities for 10 and 15-year recurrence interval are calculated as:

$$I = \frac{6.933(T)^{0.1353}}{(t+0.5)^{0.8801}}$$

$$I_{10} = \frac{6.933(10)^{0.1353}}{\left(\frac{35.06}{60} + 0.5\right)^{0.8801}} = \frac{9.467}{1.0739} = 8.82 cm/h$$

$$I_{15} = \frac{6.933(15)^{0.1353}}{\left(\frac{35.06}{60} + 0.5\right)^{0.8801}} = \frac{10.1}{1.0739} = 9.64 cm/h$$

From Table 12.3 for 0.9% slope of land in silt loam soil, the runoff coefficient, C = 0.5.

$$\therefore Q_{10} = \frac{1}{36} CIA$$

$$= \frac{1}{36} \times 0.5 \times 8.82 \times 50 = 6.13 m^3/s$$

$$Q_{15} = \frac{1}{36} \times 0.5 \times 9.642 \times 50 = 6.69 m^3/s$$

Example 12.10 A drainage channel is excavated for bottom width 2m, depth of water 2.4m, side slope 1.5:1 with 20% free board. The spoil bank made by the excavated soil is used for the pathway. Using 2:1 side slope of the pathway suggest a suitable dimension of it.

Solution:

Area of x-section of the channel = bd + zd^2 = 1.5 x 2.4 + 1.5(2.4)^2 = 12.24m^2

Let top width of the pathway, b = 2m

X-sectional area of the pathway= bd + zd^2 = 12.24m^2

or, 2d+2d^2=12.24

or, d^2+d-6.12=0

$$\therefore d = \frac{-1 \pm \sqrt{1+4\times1\times6.12}}{2} = 2.02m$$

Example 12.11 A watershed has 1.8km^2 of cultivated area, 2.2km^2 of forest land and 1.4km^2 of grassed area. The runoff coefficient of cultivated area, forest land and grassed area are 0.25, 0.15 and 0.30, respectively. The main drainage channel has a fall of 25m in the total length of 2.5km. The Intensity-Duration-Frequency relationship for the watershed is expressed as,

$$I = \frac{70T^{0.25}}{(t_c+15)^{0.4}}$$

Where, I-intensity in cm/h, T-recurrence interval in years and t_c-time of concentration in minutes.

For a recurrence interval of 20 years, the peak rate of runoff for the watershed will be – (GATE, 2014)

Solution:

Cultivated area, A_1= 1.8km^2, Runoff coefficient, C_1=0.25

Forest area, A_2=2.2km^2, Runoff coefficient, C_2=0.15

Grassed area, A_3=1.4km^2, Runoff coefficient, C_3=0.30

Drainage channel length, L=2.5km

Drainage channel fall, H=25m

Slope of the watershed, S=H/L

Recurrence interval, T=20 years

The runoff coefficient of the watershed, $C = \frac{C_1A_1 + C_2A_2 + C_3A_3}{A_1 + A_2 + A_3}$

$$= \frac{0.25\times1.8 + 0.15\times2.2 + 0.3\times1.4}{1.8 + 2.2 + 1.4} = \frac{1.2}{5.4} = 0.22$$

Kirpich (1940) equation for time of concentration, $Tc = 0.0195L^{0.77}\,S^{-0.385}$

$$= 0.0195 \text{ x } 2500^{0.77} \times \left(\frac{25}{2500}\right)^{-0.385}$$

$$= 0.0195 \text{ x } 413.44 \text{ x } 5.89 = 47.48\text{min}$$

$$I = \frac{70T^{0.25}}{(t_c+15)^{0.4}}$$

$$= \frac{70(20)^{0.25}}{(47.48+15)^{0.4}}$$

$$= \frac{70x2.11}{5.23} = 28.24cm/h$$

Total area, A = 5.4 x 100 = 540ha

Rational equation, $Q = \frac{1}{36}CIA$

$$= \frac{1}{36} \times 0.22 \times 28.24 \times 540 = 93.19m^3/s$$

Example 12.12 The peak runoff volume from the catchment of a 2.5% slope is 972.5m³. The contour bunds have top width of 500mm, height 600mm, side slope 2:1, vertical interval 1.0m and length 250m. If the crest of the overflow weir is at a height of 300mm from the ground and time available for the excess water to flow through the weir is 20 minutes, discharge in m³ per minute will be

a) 15.0 b) 20.0

c) 25.0 d) 48.6

(GATE, 2009)

Solution:

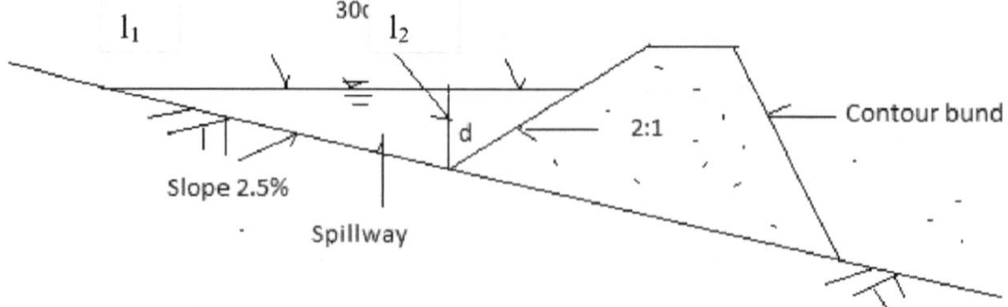

As shown in Fig. here, $S = d/l_1$, $l_2 = 2d$

Area of ponded water at the bund, $A = \frac{1}{2}dl_1 + \frac{1}{2}dl_2$

$$= \frac{1}{2} d.\frac{d}{S} + \frac{1}{2} d.2d$$

$$= \frac{1}{2} x \frac{302}{\frac{2.5}{100}} + \frac{1}{2} \times 30 \times 2 \times 30$$

= 18000 + 900 = 18900cm^2 = 18900/100 x 100 = 1.89m^2

Volume of water ponded at contour bund, V_1=AXL

L = length of the contour bund = 250m

V = 1.89m^2 x 250m = 472.5m^3

Total runoff volume = 972.5m^3

The excess runoff water = 972.5-472.5 = 500m^3

Time of flow of excess runoff water = 20min

Rate of flow = 500m^3/20min = 25m^3/min

Example 12.13 A hydraulically efficient trapezoidal drainage channel with a side slope of 2:1 has been designed in a sandy loam soil for a catchment of 600ha. Taking a drainage coefficient of 16mm, the flow velocity in mm/s in the drainage channel with a flow depth of 1m is

a) 427 b) 450

c) 497 d) 527

(GATE, 2009)

Solution:

For a side slope 2:1, z = 2

Catchment area, A= 600ha

Drainage coefficient, d_c=16mm

Depth of flow, d =1m

Discharge from the channel, Q = $\frac{A \times d_c}{Time} = \frac{600 \times 1000 m^2 \times \frac{16}{1000} m}{24 \times 3600 s} = 1.11 m^3/s$

For hydraulically efficient trapezoidal channel,

Half of the top width = Length of the side

$$or, \frac{b}{2} + zd = d\sqrt{z^2+1}$$

$$or, \frac{b}{2} = d\sqrt{z^2+1} - zd$$

$$or, \frac{b}{2} = 1\sqrt{4+1} - 2\times1$$

$$or, \frac{b}{2} = 2.236 - 2 = 0.236$$

or, b = 0.472m

Area of the trapezoidal channel, A = bd + zd^2

= 0.472 x 1 + 2 x 1 = 2.472 m^2

Velocity of the chennel

$$V = \frac{Q}{A} = \frac{1.11}{2.472} = 0.449m/s = 449mm/s \cong 450mm/s$$

12.4 Grassed Waterways

Grassed waterways are natural or manmade channels for safe disposal of excess runoff water at safe velocities from the crop land to some outlet, such as rivers, reservoir, stream etc. Constructed waterways are usually used for the terraces, graded bunds, diversions channels, spillways, furrows etc. The channel surfaces use adequate erosion resistant vegetation. The channels are constructed along the direction of slope.

The design of grassed waterways is similar to design the irrigation channel. Off course, the size of the grassed waterways depends on the expected runoff. Invariably, the rational formula and a 10 year recurrence interval is used to calculate the maximum runoff. The runoff increases gradually towards the outlet. To accommodate the runoff suitably and to economize the construction cost of channel the grassed waterways are given smaller cross-sectional area at the upstream and greater at the downstream near the outlet.

The common grassed waterways are triangular, trapezoidal and parabolic. The shape of the waterways depends on the land and the equipment used for the construction. In course of time the waterways converts to almost parabolic shape due to erosion of water flow and sediment deposition. It is suggested to construct the waterways before one season for establishment of grasses in the waterways before the terraces, bunds etc. A grade approximately 5% is recommended for vegetated waterways and grade 10% is not recommended.

The permissible velocity in the grassed waterways depends on the type and condition of vegetation. The approximate values of permissible velocity in different grass cover are given and the design dimension of parabolic channel is given in Table 12.6 & 12.7.

Table 12.6. Recommended velocities of flow in vegetated channels

Type of vegetation cover	Flow velocities (m/s)	
	Type	Magnitudes
Sparse grass cover	Low velocity	1.1-1.5
Good quality cover	Medium velocity	1.5-1.8
Excellent quality cover	High velocity	1.8-2.5

Table 12.7 Design dimension for parabolic cross-section

Cross-sectional area, a	Wetted perimeter, P	Hydraulic radius, $R = A/P$	Top width
$\frac{2}{3}tD$	$t + \frac{8d^2}{3t}$	$\frac{t^2 d}{1.5t^2 + 4d^2}$	$t = \frac{A}{0.67d}$
		or $\frac{2}{3}d\ approx$	$T = t\left(\frac{D}{d}\right)^{\frac{1}{2}}$

Example 12.14 Design a parabolic shape grassed waterways for a flow of 3.0m³/s in a land of 2.5% slope for maximum permissible velocity 1.8m/s. Assume the values of Manning's n as 0.04.

Solution:

Discharge = 3.0m³/s

Land slope, S = 2.5%

Velocity of flow, V = 1.8m/s

Manning's n = 0.04

Area of cross-section, of cross-section, A = $\frac{2}{3}td$

Let,

Width of channel, t = 5m

Depth of water in channel, d = 0.5m

$$\therefore A = \frac{2}{3}td = \frac{2}{3} \times 5 \times 0.5 = 1.67m^2$$

Perimeter. $P = t + \frac{8d^2}{3t} = 5 + \frac{8x(0.5)^2}{3x5} = 5.13m$

$$\therefore R = A/P = 1.67/5.13 = 0.325m$$

Using Manning's equation, $V = \frac{1}{n}R^{2/3}S^{1/2}$

$$= \frac{1}{0.04}(0.325)^{2/3}(0.025)^{1/2} = 1.87m/s$$

The velocity exceeds the permissible limit.

Let, t = 6m, d = 0.45m

$$\therefore A == \frac{2}{3} \times 6 \times 0.45 = 1.8m^2$$

$$P == 6 + \frac{8 \times (0.45)^2}{3 \times 6} = 6.2m$$

$$\therefore R = {}^{1.8}\!/_{6.2} = 0.29m$$

$$V = \frac{1}{0.04}(0.29)^{2/3}(0.025)^{1/2} = 1.73m/s$$

The velocity is within the permissible limit.

$Q = AV = 1.8x1.73 = 3.11m^3/s$

The discharge is satisfactory. The assumptions for t & d are O.K.

Example 12.15 Design a waterways with good grass cover for a discharge of $3.5m^3/s$ in a land slope of 4%. Assume Manning's n as 0.04 and maximum depth of flow 0.4m.

Solution

Discharge, $Q = 3.5m^3/s$

Land slope, S = 4%

Manning's n = 0.04

Maximum depth of flow, d = 0.4m

Maximum average permissible velocity for good cover grass **(Table 12.6)**, V = (1.5+1.8)/2 = 1.65m/s (approx.)

Using Manning's equation, $V = \frac{1}{n} R^{2/3} S^{1/2}$

$$1.65 = \frac{1}{0.04} R^{2/3} (0.04)^{1/2}$$

$$\text{or, } R^{2/3} \frac{1.65 \times 0.04}{(0.04)^{1/2}} = 0.33$$

$$\therefore R = (0.33)^{3/2} = 0.19$$

$$R = \frac{t^2 d}{1.5t^2 + 4d^2}$$

Let, t = 10m, d = 0.4

$$\therefore RHS = \frac{0.4 \times 100}{1.5 \times 100 + 0.64} = 0.26m \neq LHS$$

Let, t = 10m, d = 0.3m

$$\therefore RHS = \frac{0.3 \times 100}{1.5 \times 100 + 0.36} = 0.199m = LHS \text{ (approx)}$$

Using Manning's equation, $V = \frac{1}{n} R^{2/3} S^{1/2}$

$$= \frac{1}{0.04}(0.199)^{2/3}(0.04)^{1/2} = 1.7\text{m/s}$$

Velocity 1.7m/s is within the permissible velocity.

$$Q = AV = \frac{2}{3} td \times V = \frac{2}{3} \times 10 \times 0.3 \times 1.7 = 3.4 m^3/s$$

The discharge is less than required.

Let, t = 10m,d = 0.35m

$$\therefore RHS = \frac{0.35 \times 100}{1.5 \times 100 + 0.49} = 0.23m = LHS \text{ (approx)}$$

$$V = \frac{1}{n} R^{2/3} S^{1/2}$$

$$= \frac{1}{0.04}(0.23)^{2/3}(0.04)^{1/2}$$

= 1.87m/s

The velocity exceeds the permissible limit.

Let, t=11m, d=0.3

$$\therefore RHS = \frac{0.30 \times 121}{1.5 \times 121 + 0.36} = 0.199m = LHS \text{ (approx)}$$

$$V = \frac{1}{n} R^{2/3} S^{1/2}$$

$$= \frac{1}{0.04}(0.199)^{2/3}(0.04)^{1/2}$$

= 1.7m/s

$$Q = AV = \frac{2}{3} tdxV = \frac{2}{3} \times 11 \times 0.3 \times 1.7 = 3.74 m^3/s$$

The discharge is little more than the desired discharge of $3.5 m^3/s$. This is accepted

Example 12.16 A triangular shaped grassed waterway with a longitudinal slope of 2.5% is to carry a discharge of 1.5m³/s with a permissible velocity of 1.2m/s. The side slope of the channel is 1.5:1 (Horizontal: Vertical). Without considering free board, the top width of the channel in m is – (GATE, 2014)

Solution:

Discharge, Q = 1.8m³/s

Permissible velocity, V = 1.2m/s

Side slope of the channel, s = 1.5:1

Let, top width = T

Let, depth of water = D

T/2:D = 1.5:1

or, T/2=1.5D

or,T=1.5x2D=3D

Area of the triangular channel,

$$A = \frac{1}{2} TD = \frac{1}{2} \times 3D \times D = 1.5\ D^2$$

Discharge,

Q= Area of channel (A) x Velocity (V)

So, A=Q/V=1.5m²/s/1.2m/s=1.25m²

Therefore, 1.5D²=1.25

or, D²=1.25/1.5=0.83m²

∴D=0.91m

T = 3D = 3 x 0.91 = 2.73m

Example 12.17 A trapezoidal grassed waterway is constructed along a longitudinal gradient of 4%. If the cross-sectional area of flow is 152m², wetted perimeter is 12.5m and Manning's η for the waterway is $0.04m^{-1/3}s$, the flow through the waterway in m³/s is –

a) 1.9 b) 2.1

c) 2.3 d) 2.5

(GATE, 2012)

Solution:

Longitudinal slope,	S=4%
Cross-sectional area of floOw of waterway,	A=1.52m²
Wetted perimeter,	P=12.5m
Manning's	n=0.04

Velocity of flow,

$$V = \frac{1}{n} R^{2/3} S^{1/2}$$

$$= \frac{1}{n} \text{x} \left(A/P\right)^{2/3} S^{1/2}$$

$$= \frac{1}{0.04} \left(1.52/12.5\right)^{2/3} (0.04)1/2$$

= 25 x 0.245 x 0.2 = 1.225m/s

Discharge, Q = AV

= 1.52 x 1.255 = 1.86 m^3/s

Example 12.18 The parabolic grassed water channel 8m wide at the top and 60cm deep is laid on a slope of 3%. Assuming the value of 'n' in Manning's formula as $0.04m^{-1/3}$ s, the discharge capacity (in m^3/s) of the channel *(rounded off to two decimal places)* is .... (GATE, 2019)

Solution:

Top width,	t=8m
Depth of water in channel,	d=60cm=0.6m
Channel slope,	S=3%
Manning's	n=0.04

$$Area,\ A = \frac{2}{3} td = \frac{2}{3} \times 8 \times 0.6 = 3.2m^2$$

Hydraulic radius, $R = \dfrac{t^2 d}{1.5t^2 + 4d^2}$

$$= \frac{8^2 \times 0.6}{1.5 \times 8^2 + 4 \times 0.6^2} = \frac{38.4}{97.44} = 0.39$$

Velocity, $V = \dfrac{1}{n} R^{2/3} S^{1/2}$

Using Manning's equation, $V = \dfrac{1}{n} R^{2/3} S^{1/2}$

$$= \frac{1}{0.04} (0.39)^{2/3} (0.04)^{1/2} = 2.31\text{m/s}$$

Discharge, Q=AV=3.2x2.31=7.39m/s

12.5 Curve Number Method

The hydrologic soil cover complex method or more popularly Curve Number (CN) method was developed by the Soil Conservation Services of the United States Department of Agriculture for estimating runoff volume for small agricultural watersheds. The method indirectly considers the soil type, soil wetness, vegetation coverage, level of growth of vegetation and nature of cultivation practices. It is practically impossible to quantify the impact of individual effect of the above parameters on runoff. The combined effect of the parameters is reflected on CN method.

Each watershed would have a curve of its own depending on its features and different from the others. However a same watershed may have different curve number at different time depending on the change of features at the time of consideration. The curve number varies from 0 and 100. When curve number is 0, it is assumed that rainfall has occurred on a flat gravel bed of deep water table and there is no runoff, all rainwater has been absorbed by the soil profile. In other case, when CN is 100 it is assumed that rainfall has occurred on an impermeable layer from where all the rainwater has come out as runoff. In fact these two extreme cases are rare occurrence. In extreme case, if the rainfall (P) and runoff (Q) are in graph paper it will produce a straight line whose slope is 1:1 and pass through the origin. In real behavior of a watershed the P & Q relation curve will not be straight. The rainwater has to satisfy the initial abstraction (I_a), which consists of interception losses, depression storage and infiltration before the runoff occurs. Therefore, initially there is no runoff and the runoff gradually increases as the soil profile fulfils all of its voids and the rainfall continues. At the later part of the rainfall, the rainfall may produce equal amount of runoff and the P-Q relation curve may be asymptotic to 45^0 (Fig.12.12). The method is simple because it uses a small number of watershed features parameters, which are usually available. The limitation of the method lies in the fact that it does take in to consideration the simplest factors like slope and rainfall intensity.

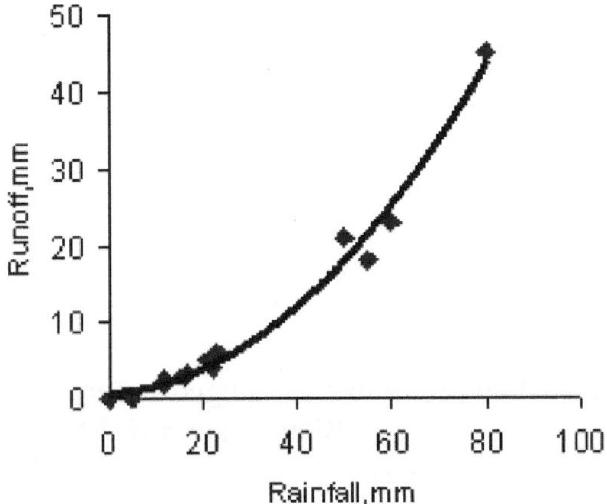

Fig. 12.12. Asymptotic nature of P-Q relation

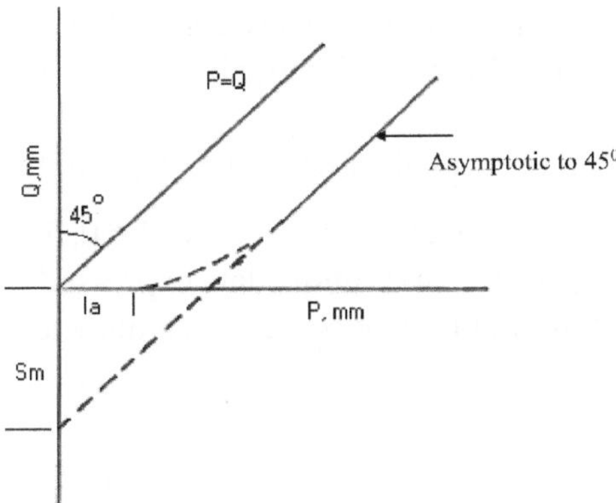

Fig. 12.13. Relationship among the rainfall (P), runoff (Q), initial abstraction (I_a) and maximum abstraction (S_m) or reservoir capacity

Development of curve number

The development of the CN method starts with the assumption that the actual to maximum possible runoff is equal to the ratio of actual to maximum abstraction. Assuming,

The actual moisture abstraction = $P-I_a-Q$

The possible maximum abstraction = S_m (where, $S_m \geq (P-I_a-Q)$)

The actual runoff = Q

The possible maximum runoff = $P-I_a$ (where $(P-I_a) \geq Q$)

The ratios between the actual and maximum possible abstraction and the actual to possible maximum runoff is equal and expressed as

$$\frac{P-I_a-Q}{S_m}=\frac{Q}{P-I_a} \tag{12.17}$$

Solving Eq. 12.17, Q x Sm = $(P - I_a)^2 - Q (P - I_a)$

or, $Q \times S_m + Q (P - I_a) = (P-I_a)^2$

or, $Q (S_m + P - I_a) = (P - I_a)^2$

$$\therefore Q=\frac{(P-I_a)^2}{(S_m+P-I_a)} \tag{12.18}$$

The US Soil conservation Service used large number of gauged watershed to develop relation between I_a and S_m as $I_a = 0.2S_m$ (Bhattacharya & Michael, 2003). Therefore, equation (ii) can be rewritten as

$$Q=\frac{(P-0.2S_m)^2}{P+0.8S_m} \tag{12.19}$$

As stated earlier, the quantitative value of curve number of a watershed ranges as 0 < CN < 100. The curve number (CN) and the maximum abstraction (S_m) may be related as $CN=\frac{1000}{10+S_m}$. When there is no abstraction i.e., $S_m = 0$, the CN value is 100; and when no surface runoff occurs, $S_m = \alpha$, and CN = 0. The original CN method was developed using rainfall and runoff data expressed in inches. This multiplied by 25.4 (1 inch = 25.4mm) to get the S_m value in millimeters. Therefore, the equation CN can be rewritten as

$$CN=\frac{25400}{254+S_m} \tag{12.20}$$

Runoff relationship

In estimating the runoff by using Eqs. 12.19 & 12.20 at first CN to be determined based on watershed characteristics. The value of S_m is determined from CN versus S_m relationship. By selecting the appropriate relation between I_a and S_m, the runoff equations can be used for estimating runoff for a given rainfall. The CN depends on watershed features and soil moisture condition at the time of occurrence of rainfall. It can be evaluated from tables as a function of hydrologic soil group (Table 12.8), antecedent moisture condition (Table 12.9), land use pattern, density of plant cover and cultivation practices (Table 12.10 & 12.11). It may be stated that the CN value of Table 12.9 is valid for medium wet condition of watershed (AMC-II). If the watershed condition is too dry (AMC-I) or too wet (AMC-III) the CN value of Table 12.11 to be converted to CN value AMC-I or AMC-III as the case may be following the Table 12.10.

Table 12.8. Hydrologic soil group (USSCS, 1964)

Soil Group	Description
A	Lowest runoff potential. Includes deep sand with very little clay and silt, and also deep soils.
B	Moderately low runoff potential. Mostly sandy soil less deeper than A, and soils less deeper or less aggregated than A, but the group as a whole has above average infiltration after thorough wetting.
C	Moderately high runoff potential. Comprises shallow soil and soil containing considerable clay and colloids, though less than those of group D. the group has below average infiltration after pre-saturation.
D	Highest runoff potential. Includes, mostly clay of high swelling percent, but the group also includes some shallow soil with nearly impermeable sub-horizon near the surface.

Source: Das (2000)

Table 12.9. Seasonal rainfall limits for antecedent moisture condition classes (USSCS, 1964)

Antecedent moisture condition (AMC) class	5day total antecedent rainfall (cm)	
	Dormant season	Growing season
I. Optimum soil condition from about lower plastic limit to wilting point	Less than 1.25	Less than 3.5
II. Average value for annual floods	1.25 to 2.75	3.5 to 5.25
III. Heavy rainfall or light rainfall and low temperature during five days preceding the given storm	Over 2.75	Over 5.75

Source: Das (2000)

Table 12.10. Conversion of CN from AMC II to AMC I and AMC III

Curve number at AMC II	Factor to convert the Curve Number at AMC II to	
	AMC I	AMC III
10	0.4	2.22
20	0.45	1.85
30	0.50	1.67
40	0.55	1.50
50	0.62	1.40
60	0.67	1.30
70	0.73	1.21
80	0.79	1.14
90	0.87	1.07
100	100	1.00

Source: Das (2000)

Table 12.11. Curve number for hydrologic cover complexes for water shed condition II, and (USSCS, 1964)

Land use cover	Treatment	Hydrologic condition	Hydrologic soil group			
			A	B	C	D
Fallow	Straight row	Poor	77	86	91	94
	Straight row	Poor	72	81	88	91
	Straight row	Good	67	78	85	89
	Contoured	Poor	70	79	81	86
	Contoured	Good	65	75	82	86
	Contoured and terraced	Poor	66	74	80	82
	Contoured and terraced	Good	62	71	78	81
Small grain	Straight row	Poor	65	76	84	88
	Straight row	Good	63	75	83	87
	Contoured	Poor	63	74	82	85
	Contoured	Good	61	73	81	84
	Contoured and terraced	Poor	61	72	79	82
	Contoured and terraced	Good	59	70	78	81
Close seeded	Straight row	Poor	66	77	85	89
legumes or	Straight row	Good	58	72	81	85
rotational meadow	Contoured	Poor	84	75	83	85
	Contoured	Good	55	69	78	83
	Contoured and terraced	Poor	63	73	80	83
	Contoured and terraced	Good	51	67	76	80
Pasture range		Poor	68	79	86	89
		Fair	48	69	79	84
		Good	39	61	74	80
	Contoured	Poor	47	67	81	88
	Contoured	Fair	25	59	75	83
	Contoured	Good	6	35	70	79
Meadow	-	Good	30	58	71	78
(Permanent)						
Woodland	-	Poor	45	66	73	83
		Fair	36	60	73	79
		Good	25	55	70	77
Farmsteads	-	-	59	74	82	86
Forest	-	Dense	26	40	58	61
		Open	28	44	60	64
Orchard	-	Average	40	54	68	72
Paddy	-	-	95	95	95	95
Roads, dirt	-	-	72	82	87	89
Road, hard surfaces	-	-	74	84	90	92

Source: Das (2000) & Bhattacharya & Michael (2003)

Runoff estimation using CN graphs

Runoff may be estimated by using the graphs (Fig. 12.14, 12.15 & 12.16). The figures are used by selecting the appropriate CN based on the features of the watershed as described earlier and the rainfall. It is not required to calculate $S_{m.}$ The natures of the curves show that a lower CN meet the rainfall axis at points away from those of higher CN. This explains that initial abstraction is higher for low runoff producing watershed than high runoff watershed. The CN graphs of Fig 5.19a & 5.19b for the relationship $I_a = 0.1S_m$ or 0.3 S_m have been adopted by the Central Unit of Soil Conservation (Hydrology & Sedimentation), Soil Conservation Division, Ministry of Agriculture, Govt. of India (Bhattacharya & Michael, 2003). The Fig. 12.16 is the original CN graphs developed in USA for $I_a = 0.2S_m$.

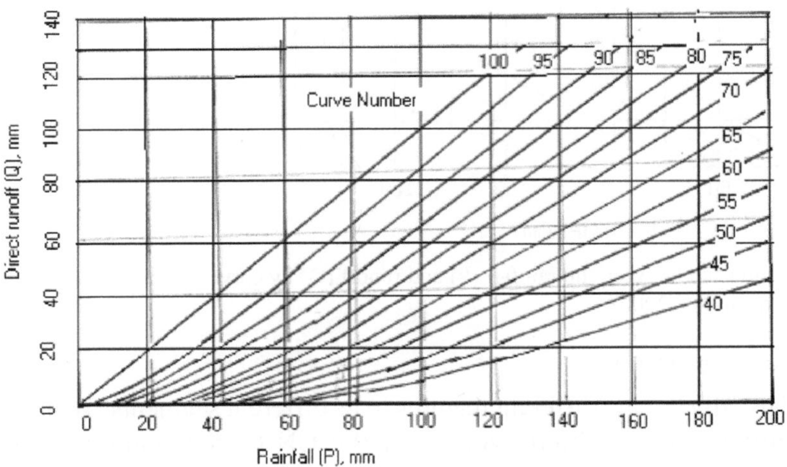

Fig. 12.14. Relationship between the rainfall (P) and runoff (Q) by curve numbers, when, $I_a = 0.1S_m$ (Redrawn from Bhattacharya & Michael (2003))

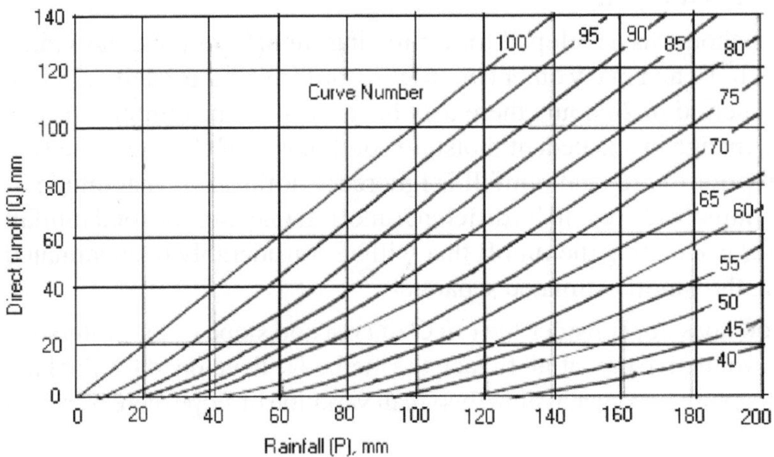

Fig. 12.15. Relationship between the rainfall (P) and runoff (Q) by curve numbers, when, $I_a = 0.3S_m$ (Redrawn from Bhattacharya & Michael (2003))

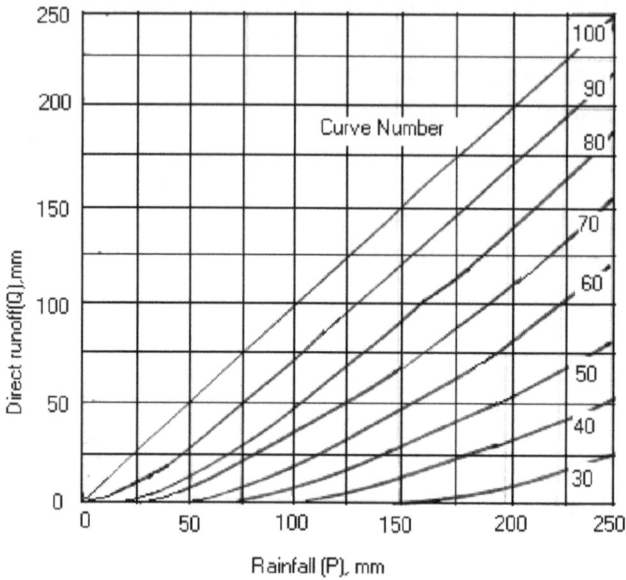

Fig. 12.16. Relationship between the rainfall (P) and runoff (Q) by curve numbers, when, $I_a = 0.2S_m$(Redrawn from Das (2000))

Weighted curve number

A watershed usually having different features to its different segments. To have a representative CN, weighted average of the CN of the segments are considered. For a paddy field a standard CN of 95 is assumed because it requires standing water and bunded condition during its growth period.

Limitation of CN method

1. The CN method was developed for estimating runoff from one individual rainfall. In India and many other countries records rainfall for daily (24 hrs) basis. In a period of 24 hours there may be more than one rainfall. In such cases, the change in antecedent moisture condition (AMC) before the first rainfall to prior the second rainfall is ignored. For the same magnitude of these two rainfalls there will be different runoff. If one uses the total rainfall of a day for calculating the runoff that will be considerably overestimated comparing the runoff estimated separately.
2. CN method does not consider the effect of area and slope of the watershed as runoff productive potential. In fact, the method is developed for flat land (0.0-0.5% slope); however, within this small variation of slope there is impact on runoff.
3. It dose not take in to account the impact of rainfall intensity and duration of rainfall. Both these are important factors in causing runoff.

Determination of peak runoff rate and time of concentration by CN method

Ogrosky and Mockus (1957) suggested the method to determine the peak runoff rate by using the curve number method. The method determines peak runoff rate by the following formula using the 6-hour rainfall as the design frequency of small watersheds (Das, 2000).

$$Q_p = \frac{0.0208 \times A \times Q}{T_p} \tag{12.21}$$

Where,

Q_p = Peak rate of runoff, m^3/s

A = Area, ha

Q = Runoff, cm

T_p = Time of peak, h

$= 0.6T_c + \sqrt{T_c} = \frac{1}{2}$ (duration of excess rainfall) + $0.6T_c$

T_c = Time of concentration, h

The time of concentration can be determined by the following formula (Schwab *et al.*, 1993) (Das, 2000).

$$T_c = \frac{L^{0.8}\left(\frac{1000}{CN} - 9\right)^{0.7}}{4407(S_g)^{0.5}} \tag{12.22}$$

Where,

L = Longest flow length, m

S_g = Average slope of the watershed, m/m

Example 12.19 A 40ha watershed of Group B soil has 5ha under fair pasture, 5ha dense forest, 10ha average orchard and rest 20ha good contoured fallow. It received a rainfall of 30 mm in one day after a monsoon break of 10 days. Select the appropriate curve number and estimate the runoff for $I_a = 0.3S_m$.

Solution:

By using the **Table 12.11** the curve numbers for the constituent areas are: pasture:69; forest;40; orchard;54; contour cultivation:75.

The weighted average of CN = $69 \times \frac{5}{40} + 40 \times \frac{5}{40} + 54 \times \frac{50}{40} + 75 \times \frac{20}{40}$

$$= \frac{2585}{40} = 64.63$$

Since there were 10 days monsoon break before the rain, the preceding 5 days rainfall was zero, hence AMC-I was prevailing. By using the Table 12.10 through interpolation the equivalent CN is 45 of AMC-I to 64.63 of AMC-II. For a CN value of 45 and 30mm rainfall, there is no runoff ((Fig.12.15)). Alternatively, we may calculate S_m as,

$$CN = \frac{25400}{254 + S_m}$$

$$\therefore S_m = \frac{25400 - 45 \times 254}{45} = 310.44$$

Putting the S_m value in **Eq.12.18** i.e., $Q = \frac{(P - I_a)^2}{(S_m + P - I_a)}$ and $I_a = 0.3S_m$,

$$Q = \frac{(30 - 0.3 \times 310.44)^2}{310.44 + 30 - 310.44 \times 0.3}$$

$$= \frac{(30 - 93.132)^2}{310.44 + 30 - 310.44 \times 0.3}$$

We see that P-I_a is negative. Therefore, there is no runoff.

Example 12.20 The following data are available for an 80ha of group C soil watershed.

Rainfall depth	=	80mm
Antecedent moisture condition	=	AMC I
Small grain of straight row good condition	=	50ha
Good pasture range	=	30ha.

Determine the runoff of the watershed. Also determine the surface runoff if another rain of 50mm occurs after the rainfall of 80mm.

Solution:

By using the Table 12.11 the CN for constituent areas:

Small grain = 83, and pasture range = 74

The weighted value of the CN

$$= \frac{83 \times 50 + 74 \times 30}{80} = \frac{6370}{80} = 79.625$$

Since the prevailing condition is AMC I, converting the CN to AMC I (Table 12.10) = 62.72

Using the equation, $CN = \frac{25400}{254 + S_m}$

$$S_m = \frac{25400 - 254 \times CN}{CN}$$

$$= \frac{25400}{CN} - 254 = 404.97 - 254 = 150.974\text{mm}$$

From **Eq.12.19**, $Q = \frac{(\text{P - 0.25m})^2}{\text{P} + 0.85m}$

$$= \frac{(80\text{-}0.2 \times 150.974)^2}{(80 + 0.8 \times 150.974)} = \frac{(49.81)^2}{201.66} = 12.36\text{mm}$$

For the second rainfall of 50mm after the previous day rainfall of 80mm, the antecedent condition AMC III exists. Therefore, CN of 62.55 to be converted to AMC III by using **Table 12.10**. The converted CN is 90.775

$$\therefore S_m = \frac{25400}{90.775} - 254 = 25.81\text{mm}$$

$$Q = \frac{(50\text{-}0.2 \times 25.81)^2}{(50 + 0.8 \times 25.81)} = \frac{2010.45}{70.65} = 28.45mm$$

It is important to note that for the second rainfall the runoff is 28.45mm much more than the runoff 12.36mm of first rainfall though the first rainfall depth (80mm) was much higher than the second (50mm). This explains the importance of antecedent moisture condition in soil.

Example 12.21 Compare the runoff of a watershed under AMC II, rainfall 80mm and CN 80 for three assumed condition of $I_a = 0.1S_m$, $I_a = 0.2S_m$ and $I_a = 0.3S_m$.

Solution: $S_m = \frac{25400 - 254 \times CN}{CN} = \frac{25400}{CN} - 254 = 63.5$

Runoff, $Q = \frac{(P - I_a)^2}{(S_m + P - I_a)}$

$$\therefore Q_1 = \frac{(80\text{-}0.1 \times 63.5)^2}{(80 + 0.9 \times 63.5)} = \frac{5424.32}{137.15} = 39.55mm$$

$$Q_2 = \frac{(80\text{-}0.2 \times 63.5)^2}{(80 + 0.8 \times 63.5)} = \frac{4529.29}{130.8} = 34.627mm$$

$$Q_3 = \frac{(80\text{-}0.2 \times 63.5)^2}{(80S_m + 0.7 \times 63.5)} = \frac{3714.9}{124.45} = 29.85mm$$

Example 12.22 A watershed of general land slope 0.3%, area 40ha, longest flow length 1300m and soil cover condition gives the CN value 65. Determine the time of concentration and peak runoff rate for 7.5cm of runoff from the watershed.

Solution:

Using **Eq.12.22,**

$$\text{Time of concentration, } T_c = \frac{L^{0.8}\left(\frac{1000}{CN}-9\right)^{0.7}}{4407(S_g)^{0.5}}$$

$$= \frac{1300^{0.8}\left(\frac{1000}{65}-9\right)^{0.7}}{4407(0.003)^{0.5}} = \frac{309.85\times 3.66}{241.38} = 4.7h$$

$$\text{Peak runoff rate } \textbf{(Eq.12.21)}, \ Q_p = \frac{0.0208\times A\times Q}{T_p}$$

$$= \frac{0.0208\times A\times Q}{0.6T_c+\sqrt{T_c}} = \frac{0.0208\times 40\times 7.5}{0.6\times 4.7+\sqrt{4.7}}$$

$$= \frac{6.24}{2.82+2.17} = 1.25m^3/s$$

Questions and Problems

12.1 Define agricultural drainage. What are the sources and ill effects of drainage?

12.2 What are the functional parts of surface drainage? Describe with neat sketches the type of surface drainage systems.

12.3 How land smoothing and land grading is differentiated? What are the characteristic differences in irrigation and drainage channels? Explain with suitable sketches.

12.4 What is drainage coefficient? Discuss the components of surface drainage with relative importance.

12.5 Discuss the rational formulae for estimating peak runoff rate? What are the drawbacks of rational formula?

12.6 Derive the relation for estimating runoff in small watershed. What are the limitations of CN method?

12.7 The intensity of 1-h rainfall is expected to be 7.5cm/h in a recurrence interval of 10 years. What is the intensity of rainfall for duration of 15 minutes?

Ans: 16.00cm/h

12.8 The expected rainfall intensity of a 50ha watershed is 6.25cm/h for recurrence interval of 10 years and the maximum length of run is 1800m. The watershed have 4 equal sub units of slopes 3.5, 7, 10 & 15% and soils clay, silt loam, stiff clay & sandy loam which are used for cultivation, pasture, cultivation and forest respectively. Calculate the peak runoff rate.

Ans: 7.9m³/s

12.9 Calculate the peak runoff rate of 10-year recurrence interval for a watershed of 100ha in Southern India if the maximum length of which 2500m, the average land slope 0.5% and the soil is cultivated sandy loam.

Ans: 4.98m³/s

12.10 In a drainage area of 15ha, the slope and drainage coefficient are 0.4% and 11mm/day, respectively. The value of Manning's roughness coefficient is 0.016. The inside diameter (in mm) of the corrugated plastic tubing used for drainage is

a) 200.51 b) 205.52

c) 209.51 d) 215.23

Ans: (c) (GATE, 2019)

12.11 An earthen roadway of top width 3.5m, depth 1.0m with side slope 2:1 is constructed by the excavated soil of a drainage channel. What is the depth of the drainage channel if its bottom width is 0.5 m and side slope 1.5:1.

Ans: 1.76m

12.12 Design a waterways with sparse grass cover in a discharge of 3.0m³/s in a land slope of 3%. Assume Manning's n as 0.03 and maximum depth of flow 0.3m.

Ans: t= 15m, d=0.235m

12.13 Estimate the runoff of a watershed under the condition AMC II, rainfall 50mm, CN 75 and $I_a = 0.2S_m$. What will be the CN if same runoff occurs under $I_a = 0.1S_m$?

Ans: Q = 9.29mm, CN = 62.18.

12.14 Determine the time of peak and peak runoff rate from the following characteristics of a watershed.

Area	=	50ha
Slope	=	1 in 250
CN	=	70
Longest flow length	=	1400m
Runoff depth	=	5.0cm

Ans: Q_p = 1.23m³/s, T_c = 3.79h

12.15 Select the appropriate answer of the followings:

1. CN method for estimating runoff was originally developed by considering initial abstraction to theoretical maximum abstraction as
   a) 0.1 b) 0.2
   c) 0.3 d) 0.4
2. CN method for estimating runoff does not consider the effect of
   a) Rainfall intensity b) Antecedent moisture content
   c) Infiltration characteristic of soil d) Land use.
3. CN method uses
   a) AMC I b) AMC II
   c) AMC III d) AMC IV.
4. The rational formula was developed by
   a) Dickens b) Kirpich
   c) Ramser d) Weisbach
5. The value of runoff coefficient in rational formula is
   a) One b) Less than one
   c) More than one d) None of these
6. CN method was developed to estimate
   a) Peak rate of runoff b) Runoff volume
   c) Volume of precipitation d) The evapotranspiration
7. CN method is more applicable to
   a) Small watershed b) Medium watershed
   c) Medium to large watershed d) Large watershed
8. CN method uses the rainfall duration equal to that of a
   a) Storm b) Day
   c) Week d) Month
9. Runoff coefficient is more when watershed has
   a) Mild slope b) Nearly flat slope
   c) Steep slope d) Dry and vegetated field
10. A watershed of area 80ha has a runoff coefficient of 0.3. A storm of intensity 5cm/h occurs for duration more than the time of concentration of

the watershed. The peak discharge in m^3/s (rounded off in two decimal places) is

a) 2.55 b) 3.33

c) 3.75 d) 3.85

(GATE, 2019)

11. The incorrect statement of the following is
    a) The peak runoff from an agricultural watershed is generally less than that from an urban watershed of the same area
    b) The horizontal hydraulic conductivity of soil is less than its vertical hydraulic conductivity
    c) The magnitude of a 75-year flood is less than that of a 100- year flood
    d) The rating curve due to an unsteady flood event from a loop

(GATE, 2015)

12. Highest runoff potential is available in soil group

    a) A b) B

    c) C d) D

Ans.

1. b) 2. a) 3. b) 4. c) 5. b) 6. b) 7. a) 8. a)
9. c) 10. b) 11. b) 12. a)

12.16 Write True or False of the following statements

1. The time of peak runoff in a watershed is more than time of concentration.
2. Drainage coefficient can be determined only by knowing the topography of the watershed.
3. Good vegetation in watershed is suitable for higher runoff rate and volume.
4. In some cases abstraction and precipitation may be same.
5. CN method produces high runoff if two or more rainfall occurs in a day 24 hrs and considered as one rainfall for calculating runoff instead of separately.
6. Rainfall intensity and rainfall duration are considered in CN method.
7. Dicken's formula was developed in 1865.
8. Drainage coefficient does not fepends on stage of crop.
9. The surface inlet is structure to carry the pit water to the sub-surface drain.
10. CN method is used in determining peak runoff rate suggested by Ogrosky and Mockus (1957).

Ans.

1. True 2. False 3. False 4. True 5. True 6. False 7. True
8. False 9. True 10. True

References

Bhattacharya, A.K. and Michael, A.M. (2003). Land Drainage Principles Methods and Applications. Konark Publishers Pvt. Ltd., New Delhi.

Das, G. (2000). Hydrology and Soil Conservation Engineering. Prentice-Hall of India Pvt. Ltd.,New Delhi.

Ogrosky, H.O. and Mockus, V. (1957). The Hydrology Guide. National Engineering handbook. Section 4, SCS,USDA.

Michael, A.M. and Ojha. T.P. (1999). Principles of Agricultural Engineering. Vol. II. Jain Brothers, New Delhi.

Ram Babu, Tejwani, K.G., Agarwal, M.C., and Bhushan, L.S. (1979). Rainfall intensity- duration-return period equations and nomographs of India. Bulletin No. 3. Central Soil and Water Conservation Research and Training Institute (ICAR), Dehradun.

Subramanya, K. (1994). Engineering Hydrology. Tata McGraw-Hill Publishing Co. Ltd., New Delhi.

13

Sub-Surface Drainage

The purpose of a sub-surface drainage system is to lower the ground water table and removal of salts beyond the crop root zone. The ground water beyond table may rise due to commonly practiced flood irrigation method, seepage from unlined channels, percolation of rain and runoff water and sub-surface flow from higher areas. Sub-surface drainage consists of drains laid at a specified depth, spacing and grade below the ground surface. The size, spacing and depth of drains depend largely on the soil-water properties viz., infiltration, saturated hydraulic conductivity, drainable porosity; and the requirement of removal of water through drains in certain time period fixed on the basis of rainfall and crop cultivated. The drains are usually placed in a pervious stratum. Suitable filter of coarse sand and gravel may be used when it is placed in a low permeable stratum. Sub-surface drainage serves the important job of removal of concentrated salt solution from the crop root zone particularly in arid and semi-arid region.

Advantage of sub-surface drainage system

1. Since sub-surface drainage are underground, no loss of land for cultivation. Some part of the land goes out of use for making drains in surface drainage.
2. No hindrance to movement of farm machinery or other farming operations.
3. Lowering of water table provides greater root zone depth vis-à-vis plants get more volume of soil for available nutrient.
4. Once constructed requires less maintenance attention compared to surface drains. Continuous supervision and maintenance is required to keep the surface drains functional.
5. Reduces soil erosion due to reduction of overland flow.

13.1 Design Consideration

The channels in surface drainage are designed on the basis of steady uniform flow after knowing the drainage coefficient. In sub-surface drainage, the design of channels may be done following steady state or non-steady flow condition. In steady state condition it is assumed that steady constant flow occurs through the soil in to the drains, quantity of water drained is equal to amount recharge by rainfall and hydraulic head causing flow remains constant. In non-steady state or falling head condition both these parameter vary with respect to time. In surface drainage, drainage coefficient is the required rate of removal of excess water, whereas, in sub-surface drainage it is not the desired rate of water removal but the desired rate of removal of water table. Steady-state drainage procedure was evolved in European countries where rainfall distribution is fairly uniform and rainfall intensity is low. Under such condition one may expect equilibrium between recharge and water table. But in our country where rainfall distribution is non-uniform and intensity is high for short duration an equilibrium condition between rainfall recharge and water table cannot exists. The sub-surface flow is much slower to rainfall received in a given time and obviously water table will rise after a high rainfall or irrigation. In such cases, in steady-state design the infiltration component of the incoming water is distributed over a longer period and obtains the design rate of removal of water. This is simplified approach and under several assumptions gives satisfactory result.

Hydraulic design parts select the combination of spacing and depth of the drain network inconsideration to diameter of the drain. The drainage through drains is estimated by using Manning's formula. The length of the drain is generally determined by the field geometry. The filters around the drains are selected based on the particle size distribution around the drains.

Drainage coefficient

In steady state sub-surface drainage design the hydrologic consideration requires to fix up an appropriate value of drainage coefficient, which also subsequently uses a relevant input to hydraulic design. Unlike the surface drainage system the drainage coefficient cannot be modified in sub-surface drainage system. The least of the following factors determine the drainage coefficients.

1. If rainfall intensity is less than the infiltration rate, then the rainfall intensity is the drainage coefficient.
2. If the rainfall intensity is higher than the infiltration rate then the infiltration rate controls the intake of water and it becomes the drainage coefficient. For a short duration rainfall, when basic infiltration rate has been established, an average infiltration rate is considered with respect to rainfall rate.
3. If the saturated hydraulic conductivity of the soil is the limiting factor then this will be the drainage coefficient.
4. If saturated hydraulic conductivity of the soil around the drain and below the soil surface differs then lowest of these is the drainage coefficient.

5. If the soil has salt problem and the excess salt to be removed by leaching then the amount of water needed to leach out the excess salt in consideration to quality of applied water and quality of water in soil profile desired and the time within which the leaching is to be completed gives the drainage coefficient.
6. If the field remain inundated such as paddy field the percolation rate is relevant in selecting drainage coefficient.

   Suppose following are the test data of a field:

   - Rainfall: 0.05 m/day
   - Hydraulic conductivity: 0.1 m/day
   - Leaching requirement: 0.004 m/day
   - Percolation rate: 0.01 m/day

The least of the above is the leaching requirement 0.004m/day, which will be drainage coefficient. If the area is not relevant to leaching requirement then the next lowest 0.01 m/day the percolation rate to be considered as drainage coefficient.

Empirical formula

USDA (1973) proposed the following empirical relation for estimating drainage coefficient for irrigated lands (Bhattacharya & Michael, 2003).

$$D_c = (D_s + L_r + C_s)\frac{D_i}{I_i}$$

Where,

D_c = Drainage coefficient, mm/day,

D_s = Deep seepage, expressed as a fraction of irrigation water

L_r = Leaching requirement expressed as a fraction of irrigation water, mm

C_s = Canal seepage expressed as a fraction of irrigation water

D_i = depth of irrigation water, mm

I_i = irrigation interval, days

Steady State Sub-surface Drainage Design

Dupuit-Forchheimer (D-F) assumptions

The earliest theory related to sub-surface drainage based on the assumptions of parallel flow lines through the saturated soil profile towards vertical sided drain of negligible width compared to spacing of the drains. This is explained in **Fig 13.1**, which is a case of steady state sub-surface drainage. The assumptions of this flow behavior are listed below and known as Dupuit-Forchheimer assumptions.

i) All streamlines of flow towards a drain are horizontal.

ii) The velocity of flow along the streamlines is proportional to the slope of the phreatic line and is also independent of the depth of the streamline

In first assumption inconsistency lies in the fact that if the streamlines are horizontal there won't be any hydraulic gradient and consequently no flow. Second assumption is in contradiction to first assumption as it claims slope of the phreatic line, which was assumed horizontal in first assumption. However Fig. 13.1 describes these assumptions with the streamlines parallel and if a vertical line is drawn intersecting the streamlines, the tangents drawn to the streamlines at the point of their intersection with the vertical line will give same slope of the tangents means same hydraulic gradient with the depth. When water level declines faster in drain in comparison to water table in the soil profile, the hydraulic gradient become very high violates the D-F assumptions and even the Darcy's law. In a fully penetrating drain of deeper depth the flow lines are more or less parallel. In a shallow depth drain compared to depth of saturated soil profile the flow lines are neither horizontal nor parallel but quite curved. This is feature of small diameter tile drain. However, the strength of the D-F assumptions lies in the fact that the resulting equation gives an almost accurate value (within 10% of the true value) (Luthin, 1970).

Ellipse equation

This is a theoretical consideration of steady flow towards the parallel drains in homogeneous soil made vertical sided with depth up to the impermeable layer (Fig 13.1). The equation relates the coordinates of the phreatic lines, the depth of water in the drain, the spacing between the drains, the constant recharge rate and the saturated hydraulic conductivity. Referring the Fig 13.1, the origin of coordinates is taken at the mid point of the left hand side drain and above the impermeable layer. The drains are evenly spaced and S apart. Width of the drain is negligible compared to spacing of the drains. Depth of water in drain is h and the maximum depth of water table at the middle of adjacent drain i.e.,halfway of spacing (S/2) is h_m from the impermeable layer. The constant recharge rate is R. In case of drain outflow is less or more than the R water will rise or fall leading to unsteady state condition.

By examining the Fig. 13.1, it appears that if a vertical plane is drawn between the centers of the drains, the water from left side of the plane will contribute to the left drain and similarly, the water on right side of the plane will goes to right drain. Let us consider a vertical plane A-A at a distance x from the origin. The vertical plane intersects the phreatic line at point P and whose height y. All the water entering to the soil to the right of this plane must pass through it on the way to drain. Therefore, the flow passing per unit length of the plane perpendicular to the paper towards the drain on left hand side may be written as,

q_x =-(v)(y)(1) (13.1)

where, q_x flow passing through the cross- sectional area yi.e., unit thickness of soil in unit time,

v = velocity of flow (assuming uniform throughout the depth according to D-F assumption)

According to Dupuit-Forchheimer it is stated that the velocity of flow is proportional to gradient of phreatic line and applicable to all depth of flow.

Therefore, $v \propto \frac{dy}{dx}$.

According to Darcy's law the proportionality constant is the hydraulic conductivity, k and $v = -k\frac{dy}{dx}$. The negative sign implies that y decreases along the direction of flow. Hence, for a homogeneous soil, the Eq.13.1 becomes

$$q_x = K\frac{dy}{dx}.\text{y}.1 \qquad (13.2)$$

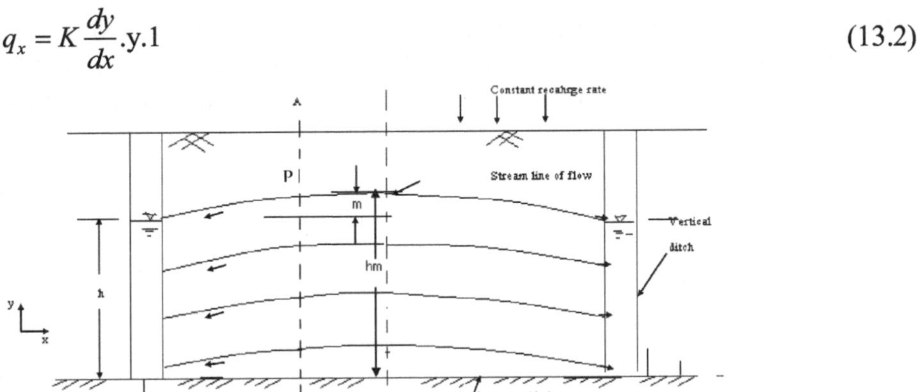

Fig. 13.1 Sub-surface flow towards vertical parallel ditches

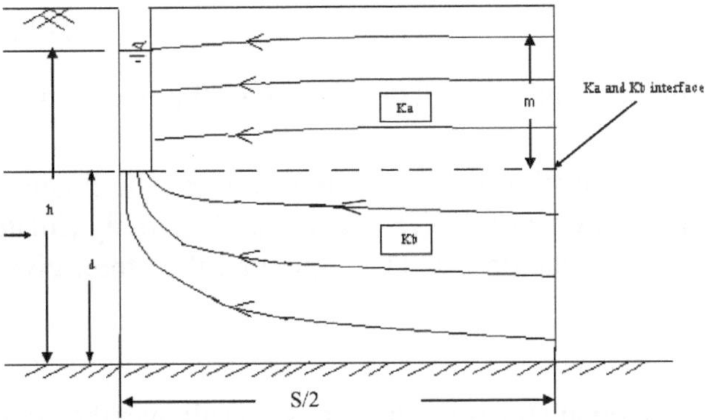

Fig 13.2 Streamline elongation in a partially penetrating drain

It may be stated that the recharge rate R is uniformly distributed over the entire drain area. Therefore, the quantity of water passing through the A-A plane will be equal to R multiplied by the surface area from the plane to the midpoint between the drains. This area is equal to , $\left(\frac{S}{2}-x\right)$ where 1 stands for 1 unit length of drain. The quantity of water passing the plane per unit time is given by

$$q_x=\left(\frac{S}{2}-x\right).R.1 \tag{13.3}$$

Equating Eqs.13.2 & 13.3

$$\text{or,}\quad Kydy=\left(\frac{S}{2}-x\right)dxR$$

$$\text{or,}\quad \frac{R}{K}\left(\frac{S}{2}-x\right)dx=ydy \tag{13.4}$$

This is a first order linear differential equation and can be integrated with the lower limit x = 0; y = h and upper limit of x & y is related to equation of the phreatic line is a relation between the distance from the drain and the height of the water table from the impermeable layer. Thus,

$$\frac{R}{K}\int_{x=0}^{x=x}\left(\frac{S}{2}-x\right)dx=\int_{y=h}^{y=y}ydy$$

$$\text{or,}\quad \frac{R}{S}\left|\frac{S}{2}x-\frac{x^2}{2}\right|_0^x=\left|\frac{y^2}{2}\right|_h^y$$

$$\text{or,}\quad \frac{R}{K}\left[\frac{S}{2}x-\frac{x^2}{2}\right]=\frac{\left(y^2-h^2\right)}{2}$$

$$\text{or,}\quad y^2-h^2=\frac{R}{K}\left(Sx-x^2\right) \tag{13.5}$$

Eq.13.5 is referred as the Ellipse equation and it describes the elliptical shape of the phreatic line

The shape of the water surface at any time in between the drains can be obtained by using Eq.13.5 with the known values of R, K, S and h. Eq.13.5 grossly violate the parallel flow assumption near the drains. This equation, therefore, gives higher value than the actual.

Hooghoudt's equation

Ellipse equation enables one to obtain the water table position all over the drain area, which may not be much concerned compared to the water table at the mid-way between the drains. The water table at mid-way should be below the root zone depth. It is obvious that water table anywhere is deeper than the water table at the mid-way. For integrating Eq.13.4 between the fixed lower and upper limit of x and y, i.e., x = 0; y = h and x = S/2, y = h_m, we get:

$$\frac{R}{S}\int_{x=0}^{S/2}\left(\frac{S}{2}-x\right)dx=\int_{y=h}^{h_m} ydy$$

$$\text{or, } \frac{R}{K}\left|\frac{3}{2}x-\frac{x^2}{2}\right|_0^{S/2}=\left|\frac{y^2}{2}\right|_h^{h_m}$$

$$\text{or, } \frac{R}{S}\left|\frac{S^2}{4}-\frac{S^2}{8}\right|=\frac{h_m^{\ 2}-h_2}{2}$$

$$\text{or, } \frac{R}{K}S^2=4\left(h_m^{\ 2}-h^2\right)$$

$$\text{or, } S^2=\frac{4K}{R}\left(h_m^{\ 2}-h^2\right) \tag{13.6}$$

Eq.13.6 is known as Hooghoudt's equation and derived on the reasoning of low, constant and uniform rate of infiltration of rain or irrigation water. The equation is widely used for calculating the sloping of sub-surface parallel drains and gives good result if the drains are of negligible width compared to the drain spacing, shallow and fully penetrating drains with small difference between h and h_m such that the nearly parallel flow condition exists and the soil is homogeneous (Bhattacharya & Michael, 2004).

Hooghoudt's equation (Eq.13.6) was derived for constant water table with constant rainfall or irrigation water with all the assumption of ellipse equation (Eq.13.5) and consideration to the situation as stated above. However, summary of the Hooghoudt's assumptions are listed below following Luthin (1970).

1. The soil is homogeneous and of hydraulic conductivity k.
2. The drains are evenly spaced a distance S apart.
3. The hydraulic gradient at any point is equal to the slope of the water table above the point, $\frac{dy}{dx}$
4. Darcy's law is valid for the water through soils.
5. An impermeable layer underlines the drain at a depth d.
6. Rain is falling or irrigation water is applied at a rate R.
7. The origin of co-ordinate is taken on the impermeable layer below the centre of one of the drain.

Luthin (1970) stated that it was the Eq.13.6, which had been used for design purpose in Holland, Australia and USA either completely or partially penetrating open ditch or sub-surface drain such as tile drain. It was assumed that the drain is empty of water for practical purpose. Thus h and h_m of Eq.13.6 are become d and d+m respectively (Fig. 13.1 &13.2). Putting this value in Eq.13.6 we get,

$$S^2 = \frac{4k}{R}\left[\left(d+m^2\right)-d^2\right]$$

$$= \frac{4K}{R}\left[2dm+m^2\right]$$

$$= \frac{4km}{R}(2d+m) \qquad (13.7)$$

Eq.13.7 is largely used for quick estimation of drain spacing.

Modification for layered soil

Hooghoudt modified the Eq.13.6 for layered soil i.e., soils of dissimilar hydraulic conductivity within the saturated soil profile contributing to flow towards the drain. One layer is considered from above the drain water level and the other below the drain water level by factorizing the term (h_m^2 - h^2) of Eq.13.6. Thus,

$$R = \frac{4k}{S^2}(h_m+h)(h_m-h)$$

$$= \frac{4k}{S^2}(2h+m)(m)$$

$$= \frac{8khm}{S^2}+\frac{4km^2}{S^2} \qquad (13.8)$$

When h = 0 in Eq.13.8, $R = \frac{4km^2}{S}$ implies that there is no flow below the drain i.e., the drain on the impermeable layer. Therefore, it may be said that the first term of Eq.13.8 is the flow from below the drain water level and the second term for the flow above the drain water level. If the saturated hydraulic conductivity of the zones above and below the drain water level are represented by k_a and k_b respectively,

$$R = \frac{8k_b hm}{S^2}+\frac{4k_a m^2}{S^2} \qquad (13.9)$$

When the flow is strictly parallel, m = 0, hence R = 0. If h = m, $k_a = k_b$; the flow contributes from below the drain water level is double than the above. Such a situation may exists if the drain depth is 1.5m and the root zone depth 0.5m. Drain depth usually do not exceed 1.5m in agricultural field and in most of sub-surface drainage, the flow from below the drain water level is comparatively larger than the flow from above the drain water level.

Modification for partially penetrating drain

The flow occurs nearly parallel in a fully penetrating small hydraulic head vertical sided drain and streamline lengths are practically equal to S/2. In a partially penetrating drain the flow also occurs below the drain and the length of the streamlines depend on the depth of the impermeable layer (Fig. 13.2). More the depth more will be the length because streamline has to cover a long path before it enters the drain. If the depth of impermeable layer below the drain is d, d/h proportion of streamlines will follow the extended paths of different lengths. The shortest streamline has an approximate length $\frac{S}{2}$ and the longest streamline has the approximate length of $\left(\frac{S}{2}+d\right)$. Thus, the minimum and maximum elongations of streamline due to partial penetration are 0 and d with the average of d/2. The streamlines of d/h proportion are subjected to this elongation. Thus, the average length of streamlines is $\left(\frac{S}{2}+\frac{d}{h}\cdot\frac{d}{2}\right)$. Rearranging Eq.13.6,

$$\frac{S}{2}=\frac{\mathrm{K}\left(h_m{}^2\right)}{\left(S/2\right)R}.$$

The left hand side of the equation is the half of the drain spacing or the length of the streamlines in a fully penetrating drain. The terms $\left(S/2\right)R$ in the RHS is the half of the total discharge intercepted by the per unit length of the drain. This may be denoted by $Q_{1/2}$. Therefore, the above equation becomes,

$\frac{S}{2}=\frac{k\left(h_m{}^2-h^2\right)}{Q_{1/2}}$. In case of elongation of the streamlines are as shown in Fig 13.2, the streamline length S/2 may be substituted by

$$\frac{S}{2}+\frac{d}{d}\cdot\frac{d}{2}=\frac{k\left(h_m{}^2-h^2\right)}{Q_{1/2}}$$

$$\text{or. } S=\frac{2k\left(h_m{}^2-h^2\right)}{Q_{1/2}}-\frac{d^2}{h} \qquad (13.10)$$

Eq.13.9 denotes that for maintaining same h_m, h and $Q_{1/2}$ the drain spacing will get reduced. This is due to more loss of head and thereby reduced flow capacity through the soil profile. The concepts of elongation may be applied to two-layered soil having different K values above and below the drain bottom. Referring Eq.13.9,

$$R = \frac{2k_b hm}{\left(S/2\right)^2} + \frac{k_a m^2}{\left(S/2\right)^2}$$

$$\text{or, } \frac{S}{2} = \frac{2k_b hm}{R\left(S/2\right)} + \frac{k_a m^2}{\left(S/2\right)R}$$

$$= \frac{2k_b hm}{Q_{1/2}} + \frac{k_a m^2}{Q_{1/2}}$$

Substituting $\left(\frac{S}{2} + \frac{d^2}{2h}\right)$ against the $\left(S/2\right)$ as before,

$$\frac{S}{2} + \frac{d^2}{h} = \frac{2k_b hm}{Q_{1/2}} + \frac{k_a m^2}{Q_{1/2}} - \frac{d^2}{2h}$$

$$\text{or, } S = \frac{4k_a hm}{Q_{1/2}} + \frac{2k_a m^2}{Q_{1/2}} - \frac{d^2}{h} \tag{13.11}$$

13.2 Replacement of the Ditch by Tile Drain and Concept of Equivalent Depth

Large-scale use of vertical sided sub-surface drains are not feasible because it requires frequent maintenance, land area is lost for cultivation and create hindrance to movement of farm equipments. These problems are overcome by using buried conduit usually made of baked clay tiles and perforated flexible PVC materials. In using such drainage conduit the flow area near the drain reduces drastically and the streamlines are more concentrated near the drain compared to the situation in fully or partially penetrating open ditches (Bhattacharya & Michael, 2003). The flow patterns are more radial and parallel near the drain violating the Dupuit's assumptions. The streamlines are shown in Fig. 13.3. More loss of energy occurs due to this elongation and convergence of streamlines near the drain. Loss of energy may be less if the drains are close to impermeable layer and the drain is at shallow depth (i.e., h_m is small).

To overcome the energy loss due to elongation and convergence of streamlines near the drain Hooghoudt's proposed the concept of equivalent depth assuming that as if the flow occurs only through a reduced depth of the profile (smaller than h). Thus, h is replaced by d_e, the equivalent depth (Fig. 13.3).

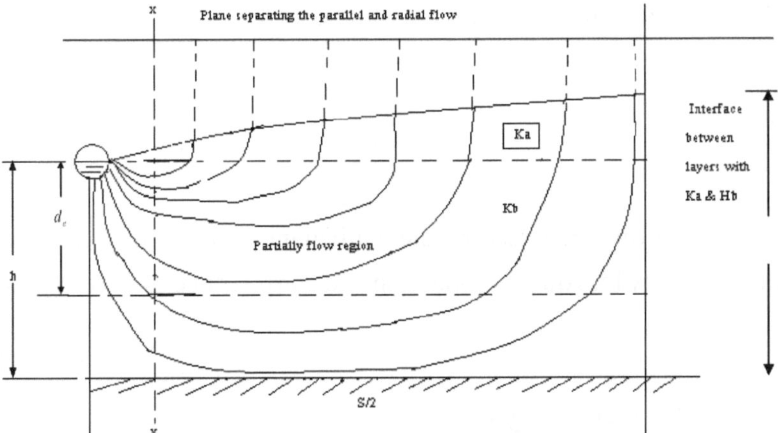

Fig. 13.3 Elongation and convergence of streamlines in a tile drain

In a parallel flow assumption, there is no potential difference between two streamlines and the equipotential lines are vertical. In case of flow towards the tile drain the streamlines are parallel to certain distance and after that gradually radial near the drain (Fig. 13.3). Hooghoudt separated this zone of different flow by a vertical plain x-x as shown in Fig. 13.3. Hooghoudt found the location of this plane x-x as 0.7h from the drain centre. Hooghoudt further assumed that the recharge water R is entering into soil profile vertically and move vertically before becoming parallel to the impermeable layer and moves towards the drain. This leads to assume that very little horizontal flow occurs from above the drain and neglecting this flow Hooghoudt obtained the relation:

$$m = \frac{RS}{K}.F_H \tag{13.12}$$

Where m, R, S, K are as defined earlier and F_H is a complex and unwieldy logarithmic and trigonometric function involving the depth of impermeable layer from the tile drain centre (h), drain spacing (S) and the radius of tile drain r_0 (Bhattacharya & Michael, 2003).

$$F_H = A + \frac{1}{\pi}\ln\left[\left\{B\left(C - D^{1/2}\right)\right\}/Z\right] + F \qquad (13.13)$$

and, $A = \left(\frac{1}{\pi}\right)\ln\left\{\frac{h}{r_0\sqrt{z}}\right\}; B = \sin\{(\pi/2)\}\left(\sqrt{2}\right)(h/S);$

$$C = \cosh\{(4)(\pi)(h/S)\}; D = \cos\left\{(\pi)\left(\sqrt{2}\right)(h/S)\right\}$$

$$E = (\pi)(h/S)\left[\sinh\{(2\pi)(h/S)\}\right]; F = \left\{1-(h/S)\left(\sqrt{\sqrt{2}}\right)\right\}^2 / \{(8)(h/S)\}$$

Neglecting the flow above the drain the Eq.13.8 becomes

$$R = 8k_b \frac{hm}{S^2},$$

$$or, m = \frac{RS^2}{8k_b h}$$

If h in the above equation is replaced by the equivalent depth d_e ($d_e \prec h$) to compensate the energy loss due to elongation and convergence of streamlines,

$$m = \frac{RS^2}{8k_b d_e} \quad (13.14)$$

Eliminating m from Eq.13.12, we get,

$$\frac{RS}{k}.F_H = \frac{RS^2}{8k_b d_e} or, d_e = \frac{S}{8k_H} \quad (13.15)$$

On realizing the impracticality of use of Eq.13.15 for calculating the F_H value, Hooghoudt proposed table or nomograph of equivalent depth relating h and S for some specific values of r_0. Bhattacharya & Michael (2003) worked out the equivalent depth tables of all probable diameter of the tile drain used in the drainage purpose and presented in Table 13.1a-13.1h.

Table 13.1a Equivalent depths (d_e) for drain radius (r_0) of 4cm

				S, meter				
h, meter	**10**	**20**	**30**	**40**	**50**	**60**	**80**	**100**
				Equivalent depth (d_e), meter				
0.4	0.365	0.382	0.388	0.391	0.393	0.394	0.395	0.396
0.6	0.467	0.545	0.562	0.571	0.577	0.581	0.585	0.59
0.8	0.599	0.686	0.721	0.793	0.751	0.759	0.769	0.775
1	0.676	0.809	0.865	0.895	0.915	0.928	0.945	0.956
1.2	0.735	0.915	0.994	1.039	1.068	1.088	1.114	1.131
1.4	0.779	1.005	1.111	1.172	1.212	1.24	1.277	1.30
1.6	0.813	1.083	1.216	1.294	1.346	1.383	1.432	1.463
1.8	0.839	1.15	1.343	1.407	1.472	1.518	1.581	1.62
2.0	0.858	1.208	1.395	1.510	1.589	1.646	1.723	1.772
3.0	0.904	1.398	1.706	1.915	2.066	2.18	2.342	2.450
4.0	↓	1.491	1.892	2.193	2.405	2.578	2.832	3.009
6.0	↓	1.559	2.077	2.489	2.824	3.10	3.532	3.851
8.0	↓	1.569	2.148	2.635	3.049	3.405	3.983	4.433
10.0	↓	2.173	2.707	3.175	3.588	4.482	4.842	4.842
∝	0.904	1.601	2.173	2.746	3.315	3.853	4.837	5.711

Table 13.1b Equivalent depths (d_e) for drain radius (r_0) of 5cm

				S, meter				
h, meter	**10**	**20**	**30**	**40**	**50**	**60**	**80**	**100**
				Equivalent depth (d_e), meter				
0.4	0.373	0.386	0.39	0.393	0.395	395	0.397	0.397
0.6	0.512	0.553	0.568	0.576	0.581	0.584	0.589	0.59
0.8	0.62	0.7	0.731	0.747	0.757	0.764	0.773	0.778
1.0	0.703	0.828	0.879	0.907	0.924	0.936	0.951	0.961
1.2	0.767	0.939	1.013	1.055	1.081	1.1	1.123	1.138
1.4	0.815	1.035	1.135	1.192	1.229	1.255	1.288	1.309
1.6	0.852	1.118	1.245	1.319	1.367	1.402	1.447	1.475
1.8	0.881	1.189	1.343	1.436	1.497	1.541	1.599	1.635
2.0	902	1.251	1.433	1.543	1.618	1.672	1.744	1.79
3.0	0.953	1.456	1.762	1.969	2.116	2.226	2.381	2.485
4.0	0. 960	1.557	1.962	2.253	2.472	2.642	2.89	3.061
6.0	↓	1.631	2.162	2.58	2.917	3.194	3.622	3.938
8.0	↓	1.642	2.239	2.738	3.159	3.518	4.099	2.547
10.0	↓	2.266	2.815	3.294	3.715	4.417	4.979	4.979
∝	0.96	1.677	2.266	2.857	3.445	3.999	5.009	5.902

Table 13.1c Equivalent depths (d_e) for drain radius (r_0) of 6cm

				S, meter				
h, meter	**10**	**20**	**30**	**40**	**50**	**60**	**80**	**100**
				Equivalent depth (d_e), meter				
0.4	0.38	0.39	0.939	0.395	0.396	0.397	0.397	0.398
0.6	0.524	0.56	0.573	0.58	0.584	0.586	0.59	0.592
0.8	0.638	0.712	0.739	0.754	0.763	0.769	0.776	0.781
1.0	0.727	0.844	0.891	0.916	0.932	0.943	0.957	0.965
1.2	0.795	0.96	1.03	1.068	1.092	1.109	1.131	1.144
1.4	0.847	1.06	1.155	1.209	1.243	1.267	1.298	1.317
1.6	0.8987	1.148	1.269	1.339	1.385	1.417	1.459	1.485
1.8	0.918	1.223	1.372	1.46	1.518	1.559	1.614	1.648
2.0	0.942	1.288	1.465	1.572	1.643	1.694	1.762	1.805
3.0	0.997	1.507	1.812	2.015	2.158	2.265	2.415	2.514
4.0	1.005	1.615	2.024	2.314	2.53	2.697	2.939	3.105
6.0	↓	1.695	2.237	2.66	2.998	3.275	3.7.0	4.011
8.0	↓	1.707	2.32	2.828	3.254	3.617	4.199	4.645
10.0	↓	↓	2.348	2.91	3.398	3.825	4.533	5.097
∝	1.005	1.745	2.348	2.955	3.558	4.126	5.159	6.069

Table 13.1d Equivalent depths (d_e) for drain radius (r_0) of 7.5cm

h, meter	S, meter 10	20	30	40	50	60	80	100
				Equivalent depth (d_e), meter				
0.4	0.388	0.394	0.396	0.397	0.398	0.398	0.399	0.399
0.6	0.54	0.57	0.58	0.585	0.588	0.59	0.592	0.594
0.8	0.662	0.726	0.75	0.762	0.769	0.774	0.781	0.784
1	0.758	0.865	0.907	0.929	0.942	0.951	0.963	0.97
1.2	0.833	0.987	1.05	1.084	1.106	1.121	1.14	1.151
1.4	0.89	1.093	1.181	1.23	1.261	1.282	1.31	1.327
1.6	0.935	1.186	1.3	1.365	1.407	1.436	1.474	1.498
1.8	0.969	1.267	1.409	1.491	1.545	1.582	1.632	1.663
2.0	0.995	1.337	1.507	1.607	1.674	1.721	1.784	1.824
3.0	1.056	1.574	1.876	2.074	2.212	2.315	2.457	2.55
4.0	1.065	1.693	2.104	2.392	2.605	2.768	3.002	3.161
6.0	↓	1.781	2.336	2.764	3.104	3.38	3.8	4.104
8.0	↓	1.794	2.426	2.946	3.379	3.745	4.328	4.771
10.0	↓	↓	2.457	3.036	3.535	3.968	4.684	5.249
∝	1.065	1.836	2.457	3.085	3.708	4.294	5.355	6.285

Table 13.1e Equivalent depths (d_e) for drain radius (r_0) of 8cm

h, meter	S, meter 10	20	30	40	50	60	80	100
				Equivalent depth (d_e), meter				
0.4	0.391	0.396	0.397	0.398	0.398	0.399	-	0.399
0.6	0.545	0.572	0.581	0.586	0.589	0.591	0.593	0.594
0.8	0.67	0.731	0.753	0.764	0.771	0.776	0.593	0.785
1.0	0.768	0.871	0.911	0.932	945	0.954	0.782	0.922
1.2	0.844	0.995	1.056	1.089	1.11	1.124	0.965	1.154
1.4	0.903	1.103	1.189	1.216	1.266	1.287	1.142	1.33
1.6	0.949	1.198	1.31	1.373	1.413	1.442	1.314	1.502
1.8	0.985	1.28	1.419	1.5	1.552	1.589	1.479	1.668
2.0	1.012	1.352	1.519	1.618	1.683	1.729	1.638	1.829
3.0	1.075	1.595	1.896	2.092	2.229	2.33	1.791	2.561
4.0	1.084	1.717	2.219	2.416	2.627	2.789	2.469	3.178
6.0	↓	1.807	2.366	2.196	3.136	3.411	3.021	4.132
8.0	↓	1.82	2.459	2.982	3.417	3.784	3.83	4.809
10.0	↓	↓	2.491	3.074	3.576	4.012	4.367	5.294
∝	1.084	1.864	2.492	3.124	3.754	4.345	4.729	6.351

Table 13.1.f Equivalent depths (d_e) for drain radius (r_0) of 10cm

h, meter	10	20	30	S, meter 40	50	60	80	100
				Equivalent depth (d_e), meter				
0.4	0.399	0.4	→	→	→	→	→	0.4
0.6	0.563	0.582	0.588	0.591	0.593	0.594	0.596	0.596
0.8	0.696	0.746	0.764	0.773	0.778	0.782	0.786	0.789
1	0.803	0.893	0.927	0.945	0.955	0.963	0.972	0.977
1.2	0.887	1.024	1.078	1.106	1.124	1.136	1.152	1.161
1.4	0.952	1.139	1.216	1.258	1.285	1.303	1.326	1.34
1.6	1.003	1.24	1.343	1.4.0	1.437	1.462	1.494	1.515
1.8	1.043	1.329	1.459	1.533	1.58	1.614	1.657	1.684
2.0	1.073	1.406	1.564	1.656	1.716	1.758	1.814	1.848
3.0	1.145	1.67	1.967	2.156	2.287	2.382	2.513	2.598
4.0	1.156	1.805	2.218	2.502	2.708	2.865	3.087	3.236
6.0	↓	1.905	2.477	2.911	3.252	3.525	3.937	4.232
8.0	↓	1.92	2.579	3.114	3.555	3.924	4.507	4.944
10.0	↓	↓	2.614	3.215	3.728	4.17	4.893	5.459
∝	1.156	1.968	2.614	3.269	3.921	4.532	5.631	6.589

Table 13.1.g Equivalent depths (d_e) for drain radius (r_0) of12cm

h, meter	10	20	30	S, meter 40	50	60	80	100
				Equivalent depth (d_e), meter				
0.4	0.4	→	→	→	→	→	→	0.4
0.6	0.578	0.59	0.593	0.595	0.596	0.597	598	0.598
0.8	0.719	0.759	0.773	0.78	0.784	786	0.79	0.792
1.0	0.834	0.912	0.943	0.955	0.964	0.97	0.977	0.982
1.2	0.925	1.049	1.096	1.121	1.136	1.146	1.16	1.168
1.4	0.996	1.17	1.239	1.277	1.30	1.316	1.336	1.349
1.6	1.052	1.277	1.371	1.423	1.456	1.479	1.508	1.525
1.8	1.096	1.371	1.492	1.56	1.604	1.634	1.673	1.697
2.0	1.129	1.454	1.603	1.689	1.744	1.782	1.833	1.864
3.0	1.209	1.737	2.028	2.211	2.336	2.427	2.55	2.63
4.0	1.221	1.884	2.297	2.577	2.778	2.93	3.143	3.285
6.0	↓	1.993	2.575	3.013	3.353	3.624	4.029	4.316
8.0	↓	2.009	2.686	3.231	3.676	4.047	4.628	5.06
10.0	↓	↓	2.724	3.339	3.861	4.309	5.036	5.6
∝	1.221	2.062	2.724	3.398	4.069	4.696	5.821	6.796

Table 13.1.h Equivalent depths (d_e) for drain radius (r_0) of 15cm

h, meter	10	20	30	40	50	60	80	100
				Equivalent depth (d_e), meter				
0.4	0.4	→	→	→	→	→	→	0.4
0.6	0.597	0.6	0.6	0.6	0.6	0.6	0.6	0.6
0.8	0.75	0.776	0.784	0.788	0.791	0.792	0.794	0.796
1.0	0.875	0.937	0.958	0.968	0.975	979	0.984	0.987
1.2	0.976	1.081	1.119	1.139	1.151	1.159	1.169	1.175
1.4	1.056	1.21	1.269	1.30	1.32	1.333	1.349	1.359
1.6	1.119	1.325	1.408	1.453	1.481	1.5	1.524	1.539
1.8	1.169	1.426	1.536	1.596	1.634	1.66	1.693	1.714
2.0	1.207	1.516	1.653	1.73	1.779	1.813	1.857	1.884
3.0	1.298	1.828	2.109	2.283	2.40	2.484	2.597	2.67
4.0	1.312	1.99	2.401	2.675	2.869	3.013	3.215	3.348
6.0	↓	2.113	2.707	3.148	3.48/6	3.753	4.148	4.425
8.0	↓	2.131	2.83	3.386	3.837	4.208	4.785	5.21
10.0	↓	↓	2.873	3.505	4.038	4.492	5.223	5.784
∝	1.312	2.191	2.873	3.57	4.267	4.915	6.072	7.069

Kirkham's equation

When depth of impermeable layer is much more compared to depth of water table above the drain, m, the flow from above the drain may be neglected and the Eq.13.14 becomes,

$R=8k_b d_e m/S^2$

From Eq.13.15, putting, $d_e = \dfrac{S}{8F_H}$

$$R = 8k_b m \frac{S}{8F_H} / S^2$$

$$= \frac{k_b m}{SF_H}$$

$$or, m = \left(\frac{RS}{k_b}\right) F_H, and\ S = \frac{k_b}{RF_H} \tag{13.16}$$

Kirkham (1958) gave a similar expression for hydraulic head excepting he replaced F_H by F_k and F_k is defined as,

$$F_k = \frac{1}{\pi}\left[A + \sum_{n=1}^{\infty}\left(\frac{1}{n} BC\right)\right] \tag{13.17}$$

where, $A = \ln\left(\dfrac{S}{\pi r_0}\right)$; $B = \cos(2n\pi r_0 / S) - \cos(n\pi)$

$C = \coth(2n\pi h/S)-1$

Kirkham *et al.* (1974) gave the following equation when the flow above the drain is considered (Bhattacharya & Michael, 2003),

$$m = \frac{RS}{K_b}\left(\frac{1}{1-R/k}\right)/F \qquad (13.18)$$

Bhattacharya & Michael (2003) worked out F_k values for a few selected values of h, S and r_0 considering the sum of the first three terms in the summation series, which are given in Table 13.3a-13.3c.

Dagan's equation

In similar way to Hooghoudt and Kirkham, Dagan (1964) considered combination of radial and horizontal sub-surface flow towards the tile drains in a homogeneous soil and neglected the flow above the drain and gave the expression for hydraulic head as,

$$m = \frac{RS}{K_b}F_D \qquad (13.19)$$

where, $$F_D = \frac{1}{4}\left(\frac{S}{2h} - \beta\right) \qquad (13.20)$$

and $$\beta = \frac{2}{\pi}\left[\ln\left\{2\cosh\left(\pi r_0 / h\right) - 2\right\}\right] \qquad (13.21)$$

Bhattacharya & Michael (2003) worked out F_D values for a range of h and S and for four values of drain radius and given in Table 13.3a-13.3d.

Table 13.2 Convergence of the summation series at the right hand side of Eq.13.17

h m	**S** m	r_0 m	**Sum of the series in Eq.13.17 of the terms for the following n:**				
			$n = 1$	1,2	1,2,3	1,2,3,4	1,2,3,4,5
1	20	0.04	4.5716	4.5715	4.80918.	4.8098	4.8459
1	60	0.04	17.1606	17.1606	18.6844	18.6844	19.1166
1	100	0.04	29.86	29.86	32.7704	32.7704	33.6847
4	20	0.05	0.3522	0.3522	0.3529	0.3529	0. 3529
4	60	0.05	3.0489	3.0488	3.1662	3.1662	3.1785
4	100	0.05	6.1215	6.1214	6.5002	6.5002	6.5706
8	20	0.1	0.0264	0.0264	0.0264	0.0264	0.0264
8	60	0.1	0.9205	0.9205	0.9293	0.9293	0.9295
8	100	0.1	2.307	2.307	2.3756	2.356	2.3808
10	20	0.04	0.0075	0.0075	0.0075	0.0075	0.0075
10	60	0.05	0.5612	0.5612	0.5637	0.5637	0.5637
10	100	0.1	1.5902	1.5902	1.6216	1.6216	1.6231

Courtesy: Bhattacharya & Michael (2003)

Table 13.3a Kirkham's F_k values when r_0 (based on sum of the first three terms of Eq.13.17

h, meter	S, meter 20	40	60	80	100
		Kirkham's F_k values (vide Eq.13.17)			
1	3.115	5.617	8.045	10.451	12.843
2	2.219	3.376	4.612	5.838	7.055
4	1.725	2.35	2.974	3.596	4.216
6	1.643	2.063	2.479	2.895	3.311
8	1.621	1.946	2.258	2.571	2.883
10	1.615	1.891	2.142	2.392	2.642

Table 13.3b Kirkham's F_k values when r_0 (based on the sum of the first three terms of Eq.13.17

h, meter	S, meter 20	40	60	80	100
		Kirkham's F_k values (vide Eq.6.17)			
1	3.115	5.617	8.045	10.451	12.843
1	3.084	5.546	7.974	10.38	12.772
2	2.058	3.305	4.541	5.767	6.984
4	1.654	2.279	2.903	3.525	4.145
6	1.572	1.992	2.408	2.824	3.24
8	1.55	1.875	2.187	2.5	2.812
10	1.554	1.82	2.071	2.321	2.571

Table 13.3c Kirkham's values when (based on the sum of the first three terms of Eq.13.17

h, meter	S, meter 20	40	60	80	100
		Kirkham's values (vide Eq.13.17)			
1	2.863	5.325	7.753	10.159	12.551
2	1.838	3.084	4.32	5.546	6.764
4	1.434	2.058	2.682	3.305	3.924
6	1.351	1.771	2.188	2.604	3.019
8	1.33	1.654	1.967	2.279	2.591
10	1.324	1.6	1.85	2.1	1.622

Table 13.4a Dagan's F_D values in Eq.13.20 for a drain radius of 4cm

h, meter	S, meter 20	30	40	50	60	80	100
			Dagan's F_D values (vide Eq.13.20)				
0.5	5.438	7.938	10.438	12.938	15.438	20.438	25.348
1	3.16	4.41	5.66	6.91	8.16	10.66	13.16
2	2.13	2.755	3.38	4.005	4.63	5.88	7.13
3	1.843	2.259	2.676	3.093	3.509	4.343	5.176
4	1.726	2.038	2.351	2.663	2.976	3.601	4.226
5	1.672	1.922	2.172	2.422	2.672	3.172	3.672
6	1.647	1.855	2.063	2.272	2.48	2.897	3.313
8	1.634	1.79	1.947	2.103	2.259	2.572	2.884
10	1.634	1.768	1.893	1.018	2.143	2.393	2.643

Table 13.4b Dagan's F_D values in Eq.13.20 for a drain radius of 5cm

h, meter	S, meter 20	30	40	50	60	80	100
		Dagan's F_D values (vide Eq.13.20)					
0.5	5.367	7.867	10.367	12.867	15.367	20.367	25.367
1	3.089	4.339	5.589	6.839	8.089	10.589	13.089
2	2.059	2.684	3.309	3.934	4.559	5.809	7.059
3	1.772	2.188	2.605	3.022	3.438	4.272	5.105
4	1.655	1.967	2.28	2.592	2.905	3.53	4.155
5	1.601	1.851	2.101	2.351	2.601	3.101	3601
6	1.576	1.784	1.992	2.201	2.409	2.826	3.242
8	1.563	1.719	1.876	2.032	2.188	2.501	2.813
10	1.563	1.697	1.822	1.946	2.072	2.322	2.572

Table 13.4c Dagan's F_D values in Eq.13.20 for a drain radius of 7.5cm

h, meter	S, meter 20	30	40	50	60	80	100
		Dagan's F_D values (vide Eq6.21a)					
0.5	5.236	7.736	10.236	12.736	15.236	20.236	25.236
1	2.959	4.209	5.459	6.709	7.959	10.459	12.959
2	1.93	2.555	3.18	3.805	4.43	5.68	6.93
3	1.643	2.059	2.476	2.893	3.0309	4.143	4.976
4	1.526	1.838	2.151	2.463	2.776	3.401	4.026
5	1.472	1.722	1.972	2.222	2.472	2.972	3.472
6	1.447	1.655	1.863	2.072	2.28	2.697	3.113
8	1.443	1.59	1.747	1.903	2.059	2.371	2.684
10	1.443	1.578	1.693	1.818	1.943	2.193	2.443

Table 13.4d Dagan's F_D values in Eq.13.20 for a drain radius of 10cm

h, meter	S, meter 20	30	40	50	60	80	100
			Dagan's values (vide Eq.13.20)				
0.5	5.143	7.643	10.143	12.643	15.143	20.143	25.143
1	2.867	4.117	5.367	6.617	7.867	10.367	12.867
2	1.839	2.464	3.089	3.714	4.339	5.589	6.839
3	1.551	1.968	2.384	2.801	3.218	4.051	4.884
4	1.434	1.747	2.059	2.372	2.684	3.309	3.934
5	1.38	1.63	1.88	2.13	2.38	2.88	3.38
6	1.355	1.563	1.772	1.98	2.188	2.605	3.022
8	1.342	1.499	1.655	1.811	1.967	2.28	2.592
10	1.342	1.476	1.601	1.726	1.851	2.101	2.351

Example 13.1 Determine the diameter of the tile drain on a grade of 0.25% draining from an area of 10ha. The drainage coefficient is 2cm. Assume Manning's n = 0.01.

Solution: The required discharge through the drain,

$Q = 10ha x 2cm / day$

$$= \frac{10x10000x2/100}{24x3600}$$

$= 0.023 m^3 / \sec$

Let, d is the diameter of the tile drain.

$$Q = AxV = \pi\frac{d^2}{4}x\frac{1}{n}R^{2/3}S^{1/2} = 0.023$$

$$or, 1.55D^{8/3} = 0.023$$

$$or, D = (0.0148)^{2/8}$$

$$\therefore\ D = 0.205m \cong 20cm$$

Example 13.2 Determine drain spacing from an area where impermeable layer exists 4.5m below ground surface. The water surface should be 0.5m below ground level. The rate of discharge per unit area of land surface is 0.72cm/day and coefficient of permeability 1.25cm/h. Tiles are placed at a depth of 1.5m from the ground surface.

Solution:

By using the Eq.13.7 with appropriate notation and as shown the diagram given here,

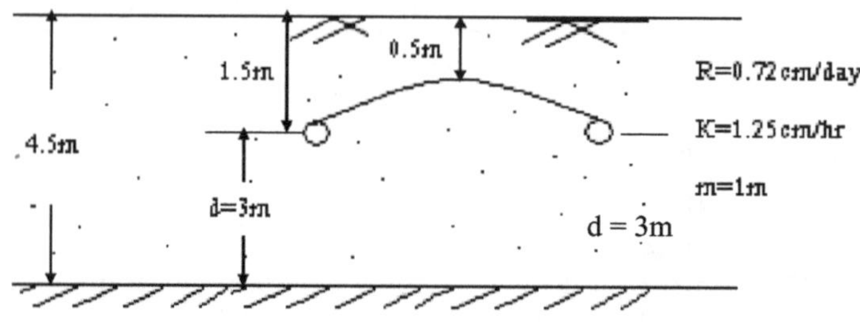

$$S^2 = \frac{4Km}{R}(m + 2d)$$

$$= \frac{4x1.25cm / hx(1m + 2x3m)}{\frac{0.72}{24}cm / h}$$

$$= \frac{4x1.25x24x7}{0.72}$$

$= 1166.66$

$$\therefore S = \sqrt{1166.66} = 34.15m$$

Example 13.3 How many hectares will a 25cm tile drain with a slope of 0.3% control if tiles are placed in an area where drainage coefficient is 2cm?

Solution:

We have, $S = 0.3\%$, $D = 25cm$ drainage requirement $= 2cm / day$

Let, Manning's roughness coefficient, $n = 0.01$

Using Manning's equation, $V = \frac{1}{n} R^{2/3} S^{1/2}$

$= \frac{1}{0.01}(D/4)^{2/3}(0.003)^{1/2}$

$= \frac{1}{0.01}(0.25/4)^{2/3}(0.003)^{1/2}$

$= 0.86\ m/s$

Rate of discharge, $Q = AxV = \pi\frac{D^2}{4}x0.86 = \pi\frac{(0.25)^2}{4}x0.86 = 0.042m^3/s$

Considering 1ha of land, the flow requirement $= 10{,}000m^2 x\ 2cm / day$

$= \frac{10{,}000x0.02}{3600x24} m^3/s$

$= 0.0555/24\ m^3/s$

$= 2.3148x10^{-3}$

Flow of $0.042m^3/$ will control the area $\frac{0.042}{2.3148x10^{-3}}$

$= 18.144\ H_a$

Example 13.4 Determine the outflow from 250m length of tiles spaced 15m apart laid at a depth of 2m above the impermeable layer if the water table is maintained at a height of 5m from the impervious layer. Assume soil hydraulic conductivity as 20cm/h.

Solution:

We have,

Lenght of drain, $L = 250m$,

Spacing of drain, $S = 15m$,

Hydraulic conductivity, $K = 20cm / hr$

Hydraulic head, $m = 3m$,

Depth of drain, $d = 2m$

Using Hooghoudt's equation,

$$S^2 = \frac{4Km}{R}(m+2d)$$

$$\therefore R = \frac{4Km}{S^2}(m+2d)$$

$$= \frac{4x20cm/hx3m(3m+2x2m)}{15mx15m}$$

$$= \frac{4x0.2m / hx3mx7m}{225m^2}$$

$= 0.07464$m/h

Drain area, A $= 250mx15m = 375m^2$

Outflow, $Q = AxR$

$= 0.07464m / hx3750m^2$

$= 280m^3 / h$

$= 6720m^3 / day$

Example 13.5 (a) In a flat area where the rainfall rate is expected to be 4cm/h and the hydraulic conductivity of the soil is 3m/day. Sub surface drains are to be installed to control the water table level. If the impermeable layer is located at a depth of 4m and the water table should not be closer than 1m from the soil surface, suggest a suitable depth and spacing of a tile drain system. (b) Calculate the outflow from the system per hectare of area under steady state condition. (c) After an irrigation without rainfall if the water table takes rises within 60cm of the soil surface, how long will it take to drop from the soil surface? Assume the needed. (GATE, 1999).

Solution:

a) The problem can be solved by trial and error method.

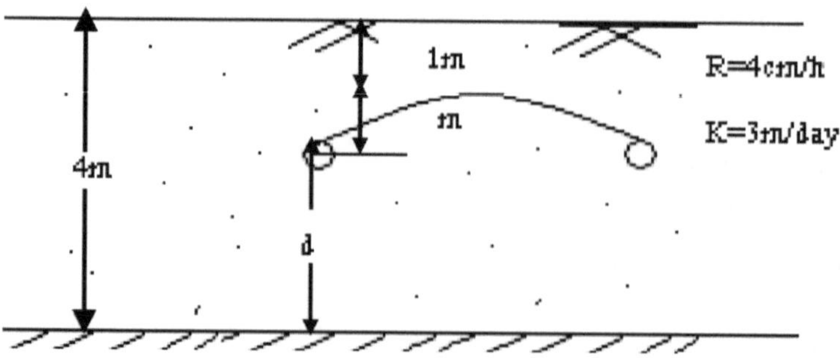

Trial 1: We have, $d + m = 4 - 1 = 3m$

Let, $d = 2m$ & $m = 1m$

Hooghoudt's equation, $S^2 = \frac{4km}{R}(m+2d)$

$$= \frac{4x3m/dayx1m(1m+2mx2m)}{\frac{4x24m/day}{100}}$$

$$= \frac{12x5}{96/100} = \frac{6000}{96} = 62.5$$

$\therefore$ $S = 7.9$m

Trial 2: Let, d = $1m$, $\therefore$ $m = 2m$

$$\therefore S^2 = \frac{4x3x2(2+2)}{96/100} = \frac{2400x4}{96} = 100$$

$S = 10m$

The spacing of 7.9m in Trial 1 may be selected because the tiles are placed at 2m depths from the ground surface, which appears apparently economical than tiles at 3m depth of Trial 2. However, an overall study is required to estimate the cost and feasibility at different depth consideration.

b) Outflow from the system,

$Q = 4cm/hx10000\ m^2$

$= 400m^3/day$

$= 0.11m^3/\text{s}$

c) Water table has risen to 60cm from the ground surface, i.e., (100-60) = 40cm extra rise of water table has taken place. Let, the porosity of soil = 40%. So, 40cm rise of water table = 40 x 0.4 = 16cm net depth of water rise.

Volume of irrigation water for 1ha area = 10000m^2 x 16cm=1600m^3

Discharge capacity = 9600m^3/day

So, the required time = 1600m^3/9600m^3/day = 0.166day = 4h

Example 13.6 A flat area is to be drained by deep open drains. The water table is to be lowered by 30cm uniformly assuming a drainable porosity 40%. Calculate spacing of the open drains considering that the capacity of the open drain may not exceed 1m^3/s.

Solution:

$Q = 1m^3/s$

30cm water table to be lowered i.e., 30x0.4=12cm net depth of water

$\therefore$ R= 12cm/day

Let, bottom width of open drain, b = 1.0m,

Side slope = 1:1, i.e., z = 1 or cotθ = 1

Longitudinal slope, S = 1%,

Manning's roughness coefficient, n = 0.035

For economic cross-section of a trapezoidal channel,

$$b = 2d \tan \theta/2$$

$$\therefore d = \frac{b}{2 \tan \theta/2} = \frac{1.0}{2x0.414} = 1.21m$$

Wetted cross-sectional area of drain, $A = bd + zd^2 = 1.0x1.21 + 1.(1.21)^2 = 2.67m^2$

$$p = b + 2d\sqrt{z^2 + 1}$$

$$= 1 + 2x1.21\sqrt{1+1} = 4.42m$$

$$R = \frac{A}{P} = \frac{2.67}{4.42} = 0.605m$$

$$V = \frac{1}{n} R^{2/3} S^{1/2} = \frac{1}{0.035}(0.605)^{2/3}(0.001)^{1/2} = 0.65m/s$$

Velocity is within acceptable limit.

$Q = AxV = 2.67mx0.65 = 1.74m^{3/}s$

The discharge $1.74m^3/s$ is more than the desired discharge ($1.0m^3/s$). The drainage channel depth is taken minimal of 1.2m. So, the bottom width of the channel needs to be reduced. But, in such condition the most economic section is not possible as because at a reduced value of b, the d value will get reduced to less than 1.2m. Therefore, the channel section to be designed on trial and error method.

Trial 1.Let, d = 1.2m, b = 0.4m

$A = bd + zd^2 = 0.4x1.2 + 1.2^2 = 1.92m^2$

$$P = b + 2d\sqrt{z^2 + 1} = 0.4 + 2x1.2\sqrt{1+1} = 3.79m$$

$$R = \frac{A}{P} = \frac{1.92}{3.79} = 0.506m$$

$$V = \frac{1}{n} R^{2/3} S^{1/2} = \frac{1}{0.035}(0.506)^{2/3}(0.001)^{1/2} = 0.57m/s$$

$Q = AxV = 1.92m^2x\ 0.57m = 1.10m^3/s \succ 1.0m^3/s$

Trial 2. Let, d = 1.2m, b = 0.3m

$A = bd + zd^2 = 03x1.2 + 1.2^2 = 1.8m^2$

$P = b + 2d\sqrt{z^2+1} = 0.3 + 2x1.2\sqrt{1+1} = 3.69m$

$$R = \frac{A}{P} = \frac{1.82}{3.69} = 0.49m$$

$$V = \frac{1}{n}R^{2/3}S^{1/2} = \frac{1}{0.035}(0.49)^{2/3}(0.001)^{1/2} = 0.56m/s$$

$Q = AxV = 1.8m^2x0.56m = 1.008m^3/s \approx 1.0m^3/s$

Again, let the depth of water table remains equilibrium at 0.5m from the ground level. Let the impermeable layer be at 4m from the ground level and hydraulic head 0.5m.

Using Eq.13.6, $S^2 = \frac{4K}{R}\left(h_m^{\ 2} - h^2\right)$

$$= \frac{4x1.5m/day}{0.12m/day}\{3.5 - 3^2\}, assuming\ K = 1.5m/day$$

$= 50x3.84$

$= 192$

$\therefore S = 13.86\ m$

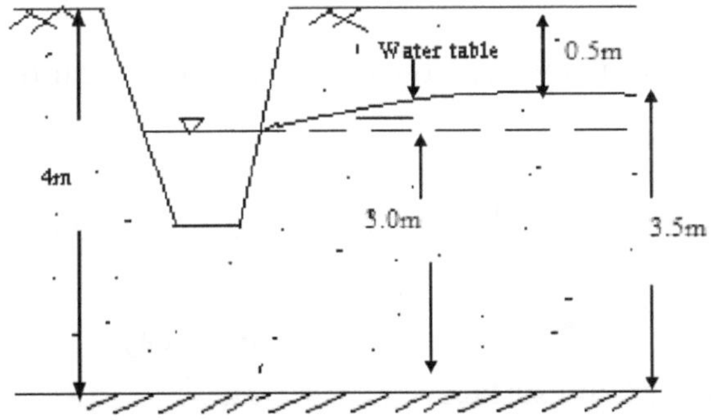

Example 13.7 A tile drainage system having 200mm diameter lateral of 400m length is used to drain an area with a drainage coefficient of 40mm. Manning's roughness coefficient for the drain pipe is 0.01. Drain pipes are laid at 0.3% slope. The spacing of the tile drain in m is -

(GATE, 2012)

Solution:

Velocity through the lateral, $V = \frac{1}{n}R^{\frac{2}{3}}S^{\frac{1}{2}}$

$$=\frac{1}{0.01}\left(\frac{200}{4x1000}\right)^{\frac{2}{3}}(0.003)^{\frac{1}{2}}$$

=100x0.135x0.0548

=0.74m/s

Discharge, Q=AxV

$$=\pi\frac{D^2}{4}xV$$

$$=\pi\frac{\left(\frac{200}{1000}\right)^2}{4}x0.74$$

=0.023m^3/s

Spacing(S)x Length of lateral(L)x Drainage coefficient (D)= Q

$$\therefore\ Spacing, S=\frac{Q}{LxD}$$

$$=\frac{0.023m^3/s}{400mx\frac{40}{1000x3600x24s}m}=124.2m$$

Example 13.8 The following information was obtained from an irrigation command:

Irrigation interval	=	12 days
Gross water application	=	7.0cm/irrigation
Application efficiency	=	65%
Hydraulic conductivity of soil	=	0.5m/day
Depth of impermeable layer from ground surface	=	5m
Diameter of the drain	=	10cm
Thickness of the filter around the drain	=	2.5cm
Depth of tile drain from the ground surface	=	1.1m
Root zone depth	=	0.6m.

Calculate the drain spacing for flow both above and below the drain and flow only below the drain.

Solution:

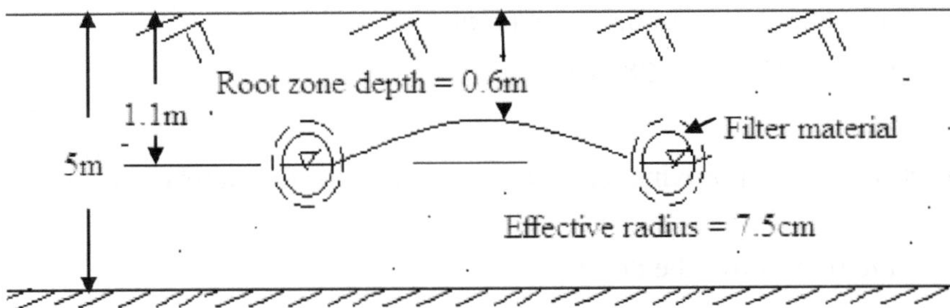

Net water application = $7.0x\frac{65}{100} = 4.55cm$

Percolation loss = 7.0 - 4.55 = 2.45 *cm*

Rate of water removal or drainage coefficient for constant water table

$= \frac{2.45cm}{12days} = 0.002m/day$

Depth of impermeable layer from the tile drain = 5 - 1.1 = 3.9*m*

Effective radius of drain, $r_0 = 5cm + 2.5cm = 7.5cm$

By using the Eq.13.8 in terms of standard notations and replacing h by equivalent depth d_e

$$S^2 = \frac{8kd_e m}{R} + \frac{4km^2}{R}$$

$$= \frac{8x0.5xd_e(1.1-0.6)}{0.002} + \frac{4x0.5x(1.1\text{-}0.6)^2}{0.002}$$

$= 1000\ d_e + 250$

Trial 1: Let S=80m

From **Table 13.1d** for h = 3.9m, d_e = (3.002-2.457)*x* 0.1

= 3.002 - 0.0545 = 2.9475*m*

$\therefore\ S^2 = 1000x2.9475 + 250$

= 03197.5

$\therefore\ S = 56.54m\ (\prec\ assumed)$

Trial 2: Let S = 60m

d_e = 2.768 - (2.768-2.315) *x* 0.1 = 2.7227

$\therefore\ S^2\ 1000\ x\ 2.7227 + 250 = 2972.7$

$\therefore\ S = 54.52m$

Trial 3: Let S = 55m

$d_e = 2.6865\text{-}(2.6865\text{-}2.2625)\, x\, 0.1 = 2.6441m$

$\therefore\ S^2 = 1000\, x\, 2.6441{+}250 = 2894.1$

$\therefore\ S = 53.79\text{m}$

The correct spacing is approximately 54m considering the flow above and below the drain.

Neglecting the flow above the drain,

$$S^2 = \frac{8\text{kd}_\text{e}\text{m}}{\text{R}} = \frac{8\times 0.5d_e \times 0.5}{0.002} = 1000d_e$$

Trial 1: Let S = 50m

$d_e = 2.605 - (2.605 - 2.212)\, x\, 0.1{=}2.5657\text{m}$

$S^2 = 1000\, x\, 2.5657 = 2565.7$

$\therefore$ S = 50.65m

The correct spacing will be approximately 50.5m when the flow above the drain is neglected.

Example 13.9 Tile drains have to be installed in an agricultural land having soil permeability of $2.3\text{x}10^{-3}$mm/s. An impermeable stratum exists at 3.2m below the land surface, and it is desired to keep the water level at least 1.0m below the surface. The average discharge of the drainage system is 2.0mm/day. If the drains are planned to be placed at 1.5m below the land surface, the drain spacing in m, assuming the equivalent depth to be the same as the tile depth, is

a) 10.6 b) 12.4

c) 13.9 d) 19.7

Solution

Soil permeability, K=$2.3\text{x}10^{-3}$mm/s	=	198.72mm/day
Depth of impermeable layer, d	=	3.2m
Drainage depth	=	2.0mm/day
Depth of the drain from the ground	=	1.5m
Depth of water over the drain, m=1.5-1.0	=	0.5m

Following Eq.13.14, $m = \dfrac{\text{RS}^2}{8\text{k}_\text{b}\text{d}_\text{e}}$

$$S^2 = \frac{m8\text{K}_\text{b}\text{d}_\text{e}}{R}$$

$$= \frac{0.5m \times 8 \times 198.72mm/day \times (3.2\text{m-}1.5\text{m})}{2.0mm/day} = 675.65\text{m}^2$$

$\therefore$ S = 25.99m

Example 13.10 Find out the spacing of vertical ditches and tile drains from the following:

Drainage coefficient = 2.5mm/day

K_a = 0.75m/day

K_b = 0.5m/day

Depth of impermeable layer from vertical ditches bottom and tile drain centre = 5.5m

Radius of tile drain = 7.5cm

Hydraulic head = 0.5m

Depth of water in vertical ditch is negligible.

Solution:

$R = 0.0025m/day, k_a = 0.75m/day, k_b = 0.5m/day, h = 5.5m, r_0 = 7.5cm, m = 0.5m$

Vertical ditches

$$S^2 = \frac{8k_b hm}{R} + \frac{4k_a m^2}{R}$$

$$= \frac{8x0.5x5.5x0.5}{0.0025} + \frac{4x0.75x(0.5)^2}{0.0025}$$

= 4400+300

= 4700

$\therefore$ $S = 68.55m$

Assuming elongation of the stream lines,

$$\frac{S}{2} = \frac{2k_b hm}{R\left(\frac{S}{2}\right)} + \frac{k_a m^2}{\left(\frac{S}{2}\right)} - \frac{d^2}{2h}$$

$$or, S = \frac{8k_b hm}{RS} + \frac{4k_a m^2}{RS} - \frac{d^2}{h}$$

$$or, S = \frac{8x0.5x5.5x0.5}{0.0025S} + \frac{4x0.75x(0.5)^2}{0.0025S} - \frac{(5.5)^2}{5.5}$$

or, $S^2 + 5.5S = 4400 + 300 = 4700$

$\therefore$ $S = 65.86m$

Tile drain

$$S^2 = \frac{8K_b d_e m}{R} + \frac{4K_a m^2}{R}$$

$$= \frac{8x0.5xd_e x0.5}{0.0025} + 300$$

$= 800\ d_e + 300$

Trial 1: Let S = 45m and h = 5.5m

From Table 13.1b, d_e = 2.934 - (2.934-2.4985) $x\frac{0.5}{2}$

$= 2.8251m$

$\therefore\ S^2 = 800x2.8251+300$

$= 2560.1$

$\therefore\ S = 50.59m\ (\succ assumed)$

Trial 2: Let S = 50m

$\therefore\ d_e$ = 3.104 - (3.104 - 2.605) $x\frac{0.5}{2}$

$= 2.979$

$\therefore S^2 = 800x2.9792 + 300 = 2683.4$

$\therefore S = 51.8m.$

The correct spacing will be approximately 51m.

Example 13.11 Calculate the spacing of the tile drains by using Hooghoudt, Kirkham, and Dagan's equation form the following data.

Constant recharge rate, R = 2mm/day

Hydraulic head = 0.5m

Hydraulic conductivity, $K = K_a = K_b\ 1m\ /\ day$

Depth of impermeable layer from the tile drain centre, h = 5m

Radius of tile drain, $r_0 = 5cm$

Solution

Hooghoudt's method

$$S^2 = \frac{8k_b hm}{R} + \frac{4k_a m^2}{R}$$

$$= \frac{8x1xd_e x0.5}{0.002} + \frac{8x1.0x(0.5)^2}{0.002}$$

$= 2000\ d_e + 500$

Trial 1: Let S = 60m

From **Table 13.1b** for h = 5m

$$d_e = 3.194 - \left(3.194 - 2.642x\frac{1}{2}\right)$$

$= 2.918m$

$S^2 = 2000x2.918 + 1000 = 6838.92$

$\therefore S = 82.69m(\succ assumed)$

Trial 2: Let S = 80m

$$\therefore d_e = 3.622\text{-}(3.622\text{-}2.8902)\,x\frac{1}{2}$$

$= 3.256m$

$S^2 = 2000x3.3.256 + 1000 = 7512$

$\therefore S = 86.67m(\succ assumed)$

Trial 3: Let S = 85m

$$\therefore d_e = \frac{3.622+3.938}{2} - \left(\frac{3.622+3.938}{2} - \frac{2.890+3.061}{2}\right)x\frac{1}{2}$$

$$= 3.78 - (3.78 - 2.976)\ x\frac{1}{2} = 3.378m$$

$S^2 = 2000x3.378 + 1000 = 7756$

$\therefore S = 88.07m$

From the above few trials it may be concluded that the tile spacing would be close to 88m.

Kirkham's method

By using Eq.13.18, where,

$$m = \frac{RS}{Kb}\left[\frac{1}{\{1-(R/Ka)\}}\right]F_k$$

$$or, S = \frac{mK_b}{RF_k} / \left[\frac{1}{1-(0.002/1)}\right]$$

$$= \frac{250}{F_K}/1.002 = \frac{249.5}{F_K}$$

Trial 1 Let S= 80m

From Table 13.3b for $r_0 = 5cm\ and\ h = 5m$

$F_K = 2.824 + 0.305 = 3.1745$

$$\therefore S = \frac{249.5}{3.1745} = 78.59m$$

The assumed value of S (80m) is close to 78.59m. The correct spacing would be approximately 79m.

Dagan's method

By using Eq.13.19, where,

$$m = \left(\frac{RS}{K_b}\right)F_D$$

Trial 1: Let S = 80m

From **Table 13.4b**, F_D = 3.101

$$\therefore\ S = \frac{mK_b}{RF_D} = \frac{250}{3.101} = 80.619m$$

So, assumption is O.K. and the correct value would be approximately 80.5m.

Example 13.12 On analyzing the hydrologic data it is estimated that there is an expected rainfall of 650mm in a region for period of 3 months. Out of this rainfall on and average 35% and 30% gets lost as runoff and evapotranspiration respectively. The rest contributes to ground water table. The location of the impermeable layer is at 6m from the ground surface. The top 60cm of the soil profile to be maintained out of reach of the water table, hydraulic conductivity of the homogeneous soil is 0.6m/day; calculate the spacing of the tile drains by assuming the average drain depth 1.2m and diameter of the drain 10cm.

Solution

Rainfall contributes to ground water

$$= 650 - \frac{35+30}{100}x650 = 650 - 422.5 = 227.5mm$$

Average rate of percolation = $\frac{227.5mm}{3months} = \frac{227.5}{3x30} = 2.528mm/day$

Using Dagan's drain spacing formula,

$$m = \frac{RS}{K_b}F_D$$

where, $F_D = \frac{1}{4}\left(\frac{S}{2h} - \beta\right)$

and $\therefore \beta = \frac{2}{\pi}\left[\ln\left\{2\cosh\left(\frac{\pi r_0}{h}\right) - 2\right\}\right]$

In this problem,

$m = 1.2\text{-}0.6 = 0.6m$

$r_0 = 0.05m$

$K = 0.6m/day$

$R = 0.002528m/day$

$$\therefore \beta = \frac{2}{\pi}\left[\ln\left\{2\cosh\left(\frac{\pi x 0.05}{4.8}\right) - 2\right\}\right]$$

$$= \frac{2}{\pi}\left[\ln\left\{2\cosh(0.0327) - 2\right\}\right]$$

$$= \frac{2}{\pi}\left[\ln\left\{2x1.0005352 - 2\right\}\right]$$

$$= \frac{2}{\pi}\left[\ln\left\{2x1.00107\right\}\right]$$

$$= \frac{2}{\pi}(-6.84)$$

$$=\text{-}\ 4.355$$

$$F_D = \frac{1}{4}\left(\frac{S}{2x4.8} - (-4.355)\right)$$

$$= \frac{S}{38.4} + 1.08875$$

Hence the expression for hydraulic head (m),

$$m = \frac{RS}{K_b}F_D$$

$$\text{or, } 0.6 = \frac{0.002528S^2}{K_b}\left(\frac{S}{38.4} + 1.08875\right)$$

$$= \frac{0.002528S^2}{23.04} + 0.004587S$$

$$\therefore\ 0.002528S^2 + 0.1057S = 13.824$$

$$or,\ S^2 + 41.812S\text{-}5468.35 = 0$$

$$\therefore S = \frac{\text{-}\ 41.812 \pm \sqrt{(41.812)^2 - 4x1x(-5468.35)}}{2x1}$$

$$or, S = \frac{\text{-}41.812 \pm 153.69}{2} = \frac{111.878\, or \text{ - } 195.502}{2} = 55.94\text{or} - 97.75$$

Neglecting the negative root, S = 55.94 $\cong 56\ m$

13.3 Ernst Equation

In Art.13.1 Hooghoudts equation for layered soil has been derived assuming that interface between the layers coincides to the drain level. The Ernst equation describes flow either the drain above or below the interface of layers. Ernst divided the flow to the drain, in to vertical, horizontal and radial component. Therefore, the total available head (h) is the sum of the head losses caused by vertical flow (h_v), horizontal flow (h_h), and the radial flow (h_r).

$$h = h_v + h_h + h_r \tag{13.22}$$

Vertical flow

The vertical flow occurs between the water table and the drain level. According to Darcy's law

$$q = K_v \frac{h_v}{d_v}$$

$$or, h_v = q\frac{d_v}{K_v} \tag{13.23}$$

Where

h_v = thickness of the layer considered for vertical flow, m

K_v = vertical hydraulic conductivity, m/day

Due to difficulty in measuring hydraulic conductivity it is usually replaced by horizontal hydraulic conductivity i.e. K_v to K_h.

Horizontal flow

Horizontal flow is assumed to take place only below the drain (Fig.13.4). Therefore, Eq.13.9 can be rewritten as

$$R = q = \frac{8\sum(Kd)_h h_h}{S^2}$$

$$\therefore h_h = \frac{qS^2}{8\sum(Kd)h} \tag{13.24}$$

Where $(Kd)_h$ = transmissivity of horizontal soil layer of flow, m²/day

If the impervious layer is very deep the value of $\sum(kd)h$ in Eq.13.24 is infinity and the h_h decreases to zero. It is therefore maximum thickness of the soil below the drain is restricted to 1/4S.

Radial flow

It is assumed that the radial flow takes place only below the drain level (Fig.13.4) and expressed as

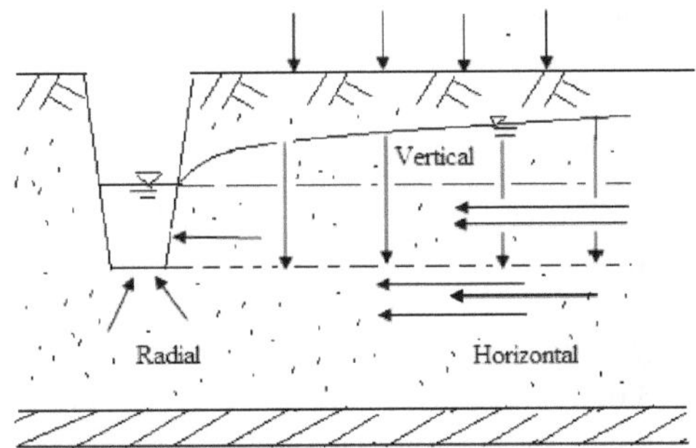

Fig.13.4 Two-dimensional flow towards drains following Ernst

$$q = \pi K_r h_r \frac{1}{S} / \ln \frac{ln\, ad_r}{u}$$

$$or, h_r = q \frac{S}{\pi K_r} \ln \frac{ad_r}{u} \tag{13.25}$$

Where

K_r = radial hydraulic conductivity, m/day

a = geometry factor of the radial resistance

d_r = thickness of the layer for radial flow, m

u = wet perimeter of the drain, m

Similar to the horizontal flow radial flow is restricted to maximum 1/4S. The geometry factor (a) is dependent on the position of the drain and soil profile. In homogeneous soil the value of (a) is on. If the drain is in bottom layer the flow is assumed restricted to thin layer and the value of (a) is one. When the drain is in top layer the value of (a) is the ratio of hydraulic conductivity to bottom (K_b) and top (K_t). Ernst (1962) stated following situation:

$0.1 \prec \frac{K_b}{K_t} \prec 50$,the bottom is considered impervious soil profile a =1.

1. $\frac{Kb}{K_t} \prec 0.1$, (a) depends on the ratios $\frac{K_b}{K_t}$ and $\frac{d_b}{d_t}$ and (Table 13.5).

$\frac{Kb}{Kt} \succ 50, a = 4$

By adding the expression for vertical, horizontal and radial flow

$$h = q \frac{d}{K_v} + q \frac{S^2}{8 \sum (Kd)_h} + q \frac{S}{\pi K_r} \ln \frac{ad_r}{u} \tag{13.26}$$

The Eq.13.26 is the Ernst equation.

Table 13.5 The geometry factor (a) obtained by relaxation method (after Van Beers 1979)

$\frac{K_b}{K_t}$	$\frac{d_b}{d_t}$					
	1	**2**	**4**	**8**	**16**	**32**
1	2.0	3.0	5.0	9.0	15	30
2	2.4	3.2	4.6	6.2	8.0	10
3	2.6	3.3	4.5	5.5	6.8	8.0
5	2.8	3.5	4.4	4.8	5.6	6.2
10	3.2	3.6	4.2	4.5	4.8	5.0
20	3.6	3.7	4.0	4.2	4.4	4.6
50	3.8	4.0	4.0	4.0	4.2	4.6

Source: Ritzema (1994)

In using Ernst equation it is to be noted that this is mainly used for two-layered soil profile when the top layer has a lower hydraulic conductivity than the bottom layer ($K_t \prec K_b$).

Referring to Fig.13.4 we may have the following simplifications.

i) Vertical resistance in the bottom layer can be neglected since hydraulic conductivity in the bottom layer is higher than top layer.

ii) Transmissivity of the top layer is neglected because ($K_t \prec K_b$) and usually ($d_t \prec d_b$). Thus the component $\sum(K_d)h$ in Eq.13.24 can be replaced by $K_b d_b$

iii) Radial flow is assumed to take places only in layer below the drain. Thus, a =1

iv) With consideration to above Eq.13.26 is converted to

$$h = q\left(\frac{d_v}{K_t} + \frac{S^2}{8K_b d_b} + \frac{S^2}{\pi K_b}\ln\frac{d_r}{u}\right) \tag{13.27}$$

v) When the drain is situated in the top layer there is no vertical flow in the bottom layer. Therefore, $d_v = h$.

vi) If the horizontal flow is taken in to consideration transmissivity of the top layer cannot be neglected. Then $\sum(Kd)h = K_b d_b + K_t d_t$ in which $d_t = d_r + 1/2h$.

vii) If the radial flow is restricted top soil layer zone, the geometry factor depends on the ratio of the hydraulic conductivity of the top and bottom layer and the Eq. 13.25 is reduced to

$$h = q\left(\frac{d_v}{Kt} + \frac{S^2}{8(K_b d_b)} + \frac{S}{\pi K_b}\ln\frac{ad_r}{u}\right) \quad 13.28$$

When the drain is in bottom layer

$$h = q\left(\frac{d_v}{K_t} + \frac{S^2}{8K_b d_b} + \frac{S}{\pi K_b}\ln\frac{d_r}{u}\right) \quad 13.29$$

Example 13.13 Tile drain of 10cm diameter is to be installed in a soil profile of two distinct layers. Determine the spacing of the drains and different heads of flow with the following data.

Root zone depth = 0.75m

Constant recharge rate = 2.0mm/day

Drain depth = 1.5m from the ground level and 1.0m above the interface between the two layers

Depth of bottom layer = 4.5m

Hydraulic conductivity of top layer = 0.6m/day

Hydraulic conductivity of bottom layer = 2.5m/day

Solution

We have,

$q = 2.0mm/day = 0.002m/day$

$d_v = h = 1.5\text{-}0.75 = 0.75m$

$K_t = 0.6m/day$

$K_b = 2.5m/day$

$d_b = 4.5m$

$d_r = 1.0m$

$r_0 = 0.05m$

$d_t = d_r + 1/2h = 1.0+1/2x0.75 = 1.375m$

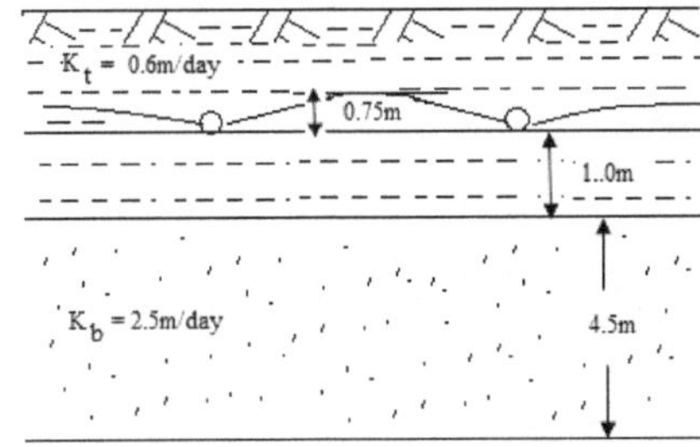

$$\frac{K_b}{K_t} = \frac{2.5}{0.6} = 4.17$$

$$\frac{d_b}{d_t} = \frac{4.5}{1.375} = 3.27$$

From **Table 13.5** through interpolation:

For $\frac{K_b}{K_t} = 4.17$ and $\frac{d_b}{d_d} = 2 \& 4$, the value of

$a = (3.5-3.3)\frac{1.17}{2}+3.3 = 3.417 and = 4.4+(4.5-4.4)\frac{0.83}{2} = 4.442$ a respectively.

Again for

$\frac{d_b}{d_t} = 3.27 \text{ and } \frac{K_b}{K_t} = 4.17,\ a = \frac{4.442-3.417}{4-2}x1.27+3.417 = 0.65+3.417 = 4.068$

$K_b d_b + K_t d_t = 2.5x4.5 + 0.6x1.375 = 12.075 m^{2/}day$

$u = \pi r_0 = \pi x 0.05 = 0.157m$

$$\therefore h = q\left(\frac{d_v}{K_t}+\frac{S^2}{8(K_b d_b + K_t d_t)}+\frac{L}{\pi K_b}\frac{ad_r}{u}\right)$$

$$= 0.002\left(\frac{0.75}{0.6}+\frac{S^2}{8x12.075}+\frac{S}{\pi x0.6}\ln\frac{4.068x1}{0.157}\right)$$

$$=0.002\left(1.25+\frac{S^2}{96.6}+1.727S\right)$$

$= 0.002\ (1.25 + 0.01S^2 + 1.727S)$

$or,\ 375 = 1.25 + 0.01S^2 + 1.727S$

$or,\ 0.01S^2 + 1.727S - 373.75 = 0$

$$\therefore S = \frac{-1.727 \pm \sqrt{1.727^3 + 4x0.01(-373.75)}}{2x0.01} = \frac{1.727+4.234}{0.02} = \frac{2.507}{0.02} = 125.35m$$

Different heads of flow:

Vertical head, $h_r = q\frac{d}{K_v} = \frac{0.002x0.075}{0.6} = 0.025$

Radial head, $h_r = q\frac{S}{\pi K_r}\ln\frac{ad_r}{u} = \frac{0.002x125.25}{\pi x06}\ln\frac{4.068xx1.0}{0.157} = 0.43m$

Horizontal head, $h_h = \frac{qS^2}{8\sum(Kd)h} = \frac{0.002x(125.25)^2}{8x12.075} = 0.325m$

13.4 Non-steady State Sub-surface Drainage

The non-steady state sub-surface drainage refers the situation where water table is not at a constant level with respect to time. It usually fluctuates much at the time of rain and irrigation or leaching. However, it is necessary to ensure that soil profile should not be waterlogged for a considerable time due to cumulative impact of these or rise of water table. Thus, the purpose of sub-surface drainage is to quick removal of excess water after a rain so that it cannot build up any cumulative effect on rise of groundwater. There are many researchers who have worked to relate drain depth, spacing, rate of water table fall and certain soil physical parameters for non-steady state sub-surface drainage. The equation of continuity is the basis of all these works.

Equation of continuity

Let 'z' be the height of water table above a reference plane on the impermeable layer and it is variable against time. In case of steady-state drainage the height of water table above the drain was not a variable. The spacing of drain and the depth of impermeable layer below the drain are represented by S and h similar to steady-state drainage. The Fig.13.5(a&b) shows the phreatic surface above a reference plane (X-Y plane) and the height of water table above the reference axis (X-axis) at any instant of time. Both these are subjected to change with respect to space and time. The X-Y plane is also considered as the upper boundary of an impermeable layer. As shown in Fig 13.5a the control volume is $\Delta x \Delta yz$ on the impermeable layer.

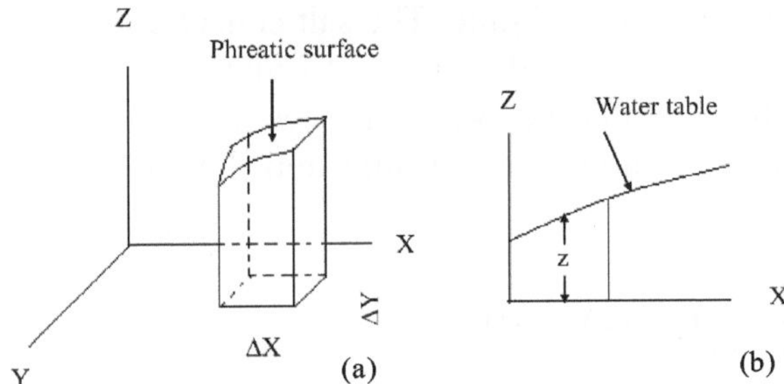

Fig. 13.5 Schematic diagram for the derivation of equation of continuity

At any instant V_x and V_y be the velocity components of flow in X and Y direction respectively and ρ the mass density. The velocity is changing at a rate of $\frac{\partial V}{\partial x}$ in X-direction and has changed to $V_z + \frac{\partial V_z}{\partial x}\Delta x$ at the exit face of the control volume. Therefore, the inflow and outflow in the X-direction:

Mass inflow rate: $(V_z)(z)(\Delta y)(\Delta \rho)$

Mass outflow rate: $(V_z)(z)(\Delta y)(\Delta \rho) + \frac{\partial}{\partial x}(V_z Z)(\Delta x \Delta y.\rho)$

Inflow-outflow = Storage rate = $-\Delta x.\ \Delta y.\ \rho.\ \frac{\partial}{\partial x}(V_y z)$

Similarly, the storage rate in the Y-direction = $-\Delta x.\ \Delta y.\ \rho.\ \frac{\partial}{\partial y}(V_y z)$

Thus, the net mass of water gained or lost within the control volume in X and Y direction

$$=(\Delta x\Delta y.\rho)\left[\frac{\partial}{\partial y}(V_x z)+\frac{\partial}{\partial y}(V_y z)\right]$$

Also, the net mass of water gained or lost within the control volume at any time

$$=\frac{\partial}{\partial t}(f\Delta x\Delta y.z.\rho)$$

where, f is the drainable porosity of the soil. In a saturated soil volume of $(z.\Delta x.\Delta y)$ there will be net volume of $(f.z\Delta x.\Delta y)$ gravitational water. Now, Δx and Δy are constants. The drainable porosity f is variable. However, it is considered as a constant for describing sub-surface flow in the drain. The mass density ρ is a variable and which changes with pressure, temperature and concentration of soluble salts. The water is incompressible fluid and sub-surface flow takes place under atmospheric pressure and which changes a little. Sub-surface water temperature fluctuation is negligible. The salt concentration of seawater is expressed in dS/m and its average density is 1.02g/cm^3. Soil water salinity is usually much less to this and that also vary little over a small range. Therefore, water density may be treated as a constant and the following relationship can be established:

$$(\Delta x\Delta y.\rho)\left[\frac{\partial}{\partial x}(V_x z)+\frac{\partial}{\partial y}(V_y z)\right]=\frac{\partial}{\partial t}(f.\Delta x\Delta y.z.\rho)$$

$$\text{or, } \frac{\partial}{\partial x}(V_x z)+\frac{\partial}{\partial y}(V_y z)=f\frac{\partial z}{\partial t} \tag{13.30}$$

As per Darcy's law $V_x=-K_x\frac{\partial z}{\partial x}$ and $V_y=-K_y\frac{\partial z}{\partial y}$, where, K_x and K_y are the saturated hydraulic conductivity in X and Y direction respectively. In a homogeneous and isotropic medium $K_x = K_y = K$. Therefore, the Eq.13.30 can be written as,

$$\frac{\partial}{\partial x}\left(z\frac{\partial z}{\partial x}\right)+\left(z\frac{\partial z}{\partial y}\right)=\frac{f}{K}\frac{\partial z}{\partial t} \tag{13.31}$$

This is the equation of continuity for non-steady state sub-surface flow in two dimensions in a homogeneous and isotropic soil. Eq.13.31 has been derived on the basis of depth of impermeable layer zero below the X-Y plane. If the impermeable layer is at the depth h below the X-Y plane, the z of Eq.13.31 will be replaced by (z+h). Thus,

$$\frac{\partial}{\partial x}\left\{(z+h)\frac{\partial}{\partial x}(z+h)\right\}+\frac{\partial}{\partial y}\left\{(z+h)\frac{\partial}{\partial y}(z+h)\right\}=\frac{f}{K}\frac{\partial}{\partial t}(z+h)$$

When, $h\rangle\rangle z$, the z may be neglected excepting where the z under the differential sign. Therefore, the above expression reduces to,

$$\frac{\partial}{\partial x}\left\{h\frac{\partial}{\partial x}(z+h)\right\}+\frac{\partial}{\partial y}\left\{h\frac{\partial}{\partial y}(z+h)\right\}=\frac{f}{K}\frac{\partial}{\partial t}(z+h)$$

or, $\frac{\partial}{\partial x}(z+h)\frac{\partial h}{\partial x}+\frac{\partial^2}{\partial x^2}(z+h)+\frac{\partial}{\partial y}(z+h)\frac{\partial h}{\partial y}+\frac{\partial^2}{\partial y^2}(z+h)=\frac{f}{K}\left(\frac{\partial z}{\partial t}+\frac{\partial h}{\partial t}\right)$

Since, h is a constant, differential of h = 0. Therefore,

$$\frac{\partial^2 z}{\partial x^2}+\frac{\partial^2 z}{\partial y^2}=\frac{f}{Kh}\frac{\partial^2 z}{\partial t} \qquad (13.32)$$

Eq.13.32 represents the non-steady state flow equation in two dimensions and the flow in two zone. One zone consists of a constant depth h and the other the variable zone z and $h\rangle\rangle z$. Such is the usual situation in a sub-surface ditch/tile drainage system. The impermeable layer lies at a large distance from the ditch/tile bottom. This constant distance is h and the variable height of water table above the drain is z. Flow takes place from these two zones. However, there should not be any flow when z is zero.

Again, in a system of parallel drain the flow takes place in one direction from mid point between the drains. Therefore, Eq.13.32 can be reduced to

$$\frac{\partial^2 z}{\partial x^2}=\frac{f}{Kh}\left(\frac{\partial z}{\partial t}\right) \qquad (13.33)$$

Glover solved the non-steady state continuity equation (Eq.13.33) with certain initial boundary conditions. He replaced h of Eq.13.33 by D, where, $D=\left[h+\frac{z_0}{2}\right]$ for the case when the depth to impermeable layer below the drain is large $(h\rangle\rangle z)$ compared to hydraulic head at any time in the sub-surface drainage process. The Z_0 and $z_{s/2}$ are the maximum and average depth of flow zone above the drain respectively. Based on these assumptions Glover deduced the following equation relating the water table depth at midway between the drains and spacing between the drains (Bhattacharya & Michael, 2003).

Drain above the impermeable layer:

$$S=\pi\left[\frac{KDt}{f}\ln\left(\frac{4z_0}{\pi z_{S/2}}\right)\right]^{1/2} \qquad (13.34)$$

Drain on the impermeable layer:

$$S = \left[\frac{9Kz_0 t}{\left\{2f\left(z_0 / ^{z} {s/2} - 1\right)\right\}} \right]^{1/2} \tag{13.35}$$

The following are some other important equation in non-steady state sub-surface design process:

Luthin (1959) formula:

$$S = BK\left(t_2 - t_1\right) / \left\{f \ln\left(z_{t1} - z_{t2}\right)\right\} \tag{13.36}$$

In the previous equations the term s Z_0 and $z_{s/2}$ were defined as the mid-point water table depth and the mid-point water table depth after some time t respectively. These terms have been replaced by z_{t1} and z_{t2} from Eq. 13.34 to Eq.13.35 and accordingly, the term t has been replaced by $(t_2 - t_1)$. In all the equations the water table depth, equivalent depth or depth to impermeable layer in meters, hydraulic conductivity in meters per day, time in days and drainable porosity in fraction:

Schilfgaarde (1963) formula:

$$S = 3A\left[\mathrm{K}\left(t_2 - t_1\right)\left(d_e + z_{t_2}\right)\left(d_e + z_{t_1}\right) / \left\{(2f)\left(z_{t_1} - z_{t_2}\right)\right\}\right]^{1/2} \tag{13.37}$$

where, $A^2 = z_{t1}\left(2d_e + z_{t_1}\right) / \left(d_e + z_{t_1}\right)^2$

Schilfgaarde (1964) formula:

$$S = 3\left[Kd_e\left(t_2 - t_1\right)/f \ln\left\{z_{t1}\left(2d_e + z_{t_2}\right)/\left(zt_2\right)\left(2d_e + z_{t_1}\right)\right\}\right]^{1/2} \tag{13.38}$$

Integrated Hooghoudt or Bouwer and Schilfgaarde (1963) formula:

$$\frac{Kt}{f} = \left(C_f S^2 / 8d_e\right) \ln\left[\left\{z_0\left(z_t + 2d_e\right)\right\}\right] / \left\{z_t\left(z_0 + 2d_e\right)\right\} \tag{13.39}$$

where, C_f is a correction factor for conversion a steady state drainage formula to a non-steady state formula and whose values generally varies between 0.8 to 1.0. the other terms are defined earlier.

Lal et al (1992) formula:

$$S = Kt / \left[fC_f\left\{\left(S - h\sqrt{2}\right)^2\right\} / (8hS) + fC_f\left(\frac{1}{\pi}\right) \ln\left\{h / \left(r_0\sqrt{2}\right)\right\}\right] \left\{\ln\left(z_0 / z_t\right)\right\} \tag{13.40}$$

Example 13.14 It is proposed to apply 5 irrigation of 10cm each in an area at 10 days interval. The rate of evapotranspiration is 5mm/day, drainable porosity of soil is 0.1, field capacity of the soil is 20% and first irrigation is applied at 50% depletion of it, the initial water table depth and the depth of drain is 1.5m below the ground surface, the depth of impermeable layer is 3m below the proposed drain level, soil is homogeneous with a K value of 0.5m/day. Estimate the drain spacing using Glover's approach to ensure the water table does not lie within 1m of the ground surface for more than 2 days.

Solution:

After one irrigation and before the second irrigation, the depth of water retained in the soil profile from irrigation water

= 10cm-5mm/day x 10day = 5cm

This 5cm of water will cause to bring the soil profile depth in field capacity

$$= \frac{5cm}{field\ capacity\ x\ Moisture\ depletion}$$

$$= \frac{5cmx100x100}{20x50} = 50cm$$

Therefore, to bring the soil profile of 1.5m into field capacity 3 irrigation is required ($150cm/50cm = 3$). After the fourth irrigation, the application of 10cm water causes to rise the water table = $10cm/0.1 = 100cm$. However, before the fifth irrigation it will decline by 50cm due to evapotranspiration. After the fifth irrigation the water table will rise from initial position = 50cm + 100cm = 150cm. This means that it will just touch the ground surface. This is the situation when drainage requirement should be highest. Therefore, designing the system considering this situation will be conservative but on safe side to admit unforeseen odd situation.

The formula to be used, $S = \pi\left[\frac{KDt}{f}\ln\left(\frac{4z_0}{\pi z_{s/2}}\right)\right]^{1/2}$

We get, $z_0 = 1.5m$ (after 5th irrigation)

$z_{s/2} = 1.5\text{-}1.0 = 0.5m$

$D = h + z_0/2 = 3 + 1.5/2 = 3.75m$

$t = 2\ days,\ K = 0.5m\ /\ day,\ f = 0.1$

$$\therefore S = \pi[(0.5x3.75x2)/(0.1)]\ln\left(\frac{4x1.5}{\pi x0.5}\right)^{1/2}$$

$$= [37.5\ln 3.81]^{1/2}$$

$$= \pi x(50.256)^{1/2}$$

$$= 22.27 \cong 23m$$

Example 13.15 Estimate the spacing of drains of drainage system capable of lowering of water table 1.0m to 0.5m in 10 days from the drain level by using Lal et al (1992) formula and Integrated Hooghoudt's formula. The depth of impermeable layer is 3.5m below the drain, soil is homogeneous and K value of 0.4m/day, drainable porosity is 0.1, tile drain used is of 10cm diameter. The correction factor C_f for converting a steady state drainage formula to a non-steady state may be taken as 0.9.

Solution

Lal et al (1992) formula:

$$S = Kt/\left[fC_f\left\{\left(s - h\sqrt{2}\right)^2/(8hs)\right\} + fC_f\frac{1}{\pi}\ln\left\{h/\left(r_0\sqrt{2}\right)\right\}\right]\left\{\ln\left(z_0/z_t\right)\right\}$$

The formula may be written as below for favouring calculation,

$$S = \frac{Kt}{\left[(A+B)\left\{\ln\left(z_0/z_t\right)\right\}\right]}$$

Where, $A = fC_f\left(s - h\sqrt{2}\right)^2/(8hS)$ and $B = fC_f\left(\frac{1}{\pi}\right)\ln\left\{h/r_0\sqrt{2}\right\}$

We have, $f = 0.1$, $h = 3m$, $r_0 = 0.05m$, $z_0 = 1.0m$, $z_t = 0.5m$

Let us calculate the A and B for some assumed S values:

$$S = 25m, A = (0.1x0.9)\left\{\left(25 - 3\sqrt{2}\right)/(8x3.5x25)\right\} = 0.052$$

$$B = (0.1x0.9)\left(\frac{1}{\pi}\right)\ln\left\{(3.5)/\left(0.05\sqrt{2}\right)\right\} = 0.11$$

$$\therefore S = \frac{0.4xt}{\left[(0.052 + 0.11)\left\{\ln(1/0.5)\right\}\right]} = 3.56t$$

$$\text{or}, t = \frac{25}{3.56} = 7.02 days \langle 10 days$$

$$S = 30m, A = (0.1x0.9)\left\{\left(30 - 3.5\sqrt{2}\right)^2/(8x3.5x30)\right\} = 0.067$$

$$B = (0.1x0.9)\left(\frac{1}{\pi}\right)\ln\left\{(3.5)/\left(0.05\sqrt{2}\right)\right\} = 0.11$$

$$\therefore S = \frac{0.4t}{\left[(0.067 + 0.11)\left\{\ln\left(\frac{1}{0.5}\right)\right\}\right]} = 3.26t$$

$\therefore t = \frac{30}{3.26} = 9.2 days$

From the above two calculations we may decide upon the spacing of drain as 30m. In fact, it will be little less to 30m. Decide upon to 30m will be on safe side and capable of attending some additional drainage requirement.

Integrated Hooghoudt or Bouwer and Schilfgaarde (1963) formula:

$$\frac{Kt}{f} = \left(C_f S^2 / 8de\right) \ln\left[\left\{z_0\left(z_t + 2d_e\right)\right\}\right] / \left\{z_t\left(z_0 + 2d_e\right)\right\}$$

From Table 13.1b, for h = 3.5m, r_0 = 0.05m and S = 25m, equivalent depth, d_e = 1.684m

Let, S = 25m, $\therefore LHS = \frac{0.4 \times 10}{0.1} = 40m$

RHS = $(0.9 \times 25^2)/(8 \times 1.684)\ln[\{1(0.5 + 2 \times 1.684)\}/\{0.5(1+2 \times 1.684)\}]$

= 41.75 ln (1.77) = 23.85m (<<LHS))

S = 35m, LHS = 40m

From Table 13.1b, d_e = 1.986m

RHS = $(0.9 \times 35^2)/(8 \times 1.986)\ln[\{1(0.5+2 \times 1.986)\}/\{0.5(1+2 \times 1.986)\}]$

$= 40.74m$ ($\approx$ LHS)

Therefore, it may be concluded that drain spacing will be 35m.

Example 13.16 Using Schilfgaarde (1963) formula find the time required to lower the water table from a z_0 of 0.8m to a z_t of 0.5m from the following data:

Depth of tile drain	=	1.3m
Spacing of drain	=	40m
Hydraulic conductivity of homogeneous soil, K	=	0.75m/day
Drainable porosity, f	=	0.12
Depth of impermeable layer from the drain	=	3m
Diameter of tile drain	=	10cm

Solution:

The formula to be used is

$S = 3A\,[K(t_2 - t_1)(d_e + z_{t2})(d_e + z_{t1})/\{(2f)(z_{t1} - z_{t2})\}]^{1/2}$

Where, $A^2 = z_{t1}(2d_e + z_{t1}) / (d_e + z_{t1})^2$

For, S = 40m, h = 3m and r_0 = 0.05m, $z_0 = z_{t1}$ = 0.8m and $z_t = z_{t2}$ = 0.5m

The equivalent depth from Table 13.1b, d_e = 1.969m

$\therefore A^2 = 0.8\,(2 \times 1.969 + 0.8) / (1.969 + 3 \times 0.8)^2$

$$=\frac{3.79}{19.09}=0.198 \quad \therefore \quad A=0.44$$

Again, $S = 3 x\ 0.44\ [0.75(t_2 - t_1)(1.969+0.5)(1.969+0.8)/\{(2\,x\,.12)(0.8\text{-}0.5)\}]^{1/2}$
$= 1.32\ [0.75\ x\ 6.84\ (t_2 - t_1)/\{0.072\}]^{1/2}$

$$=1.32\left[\frac{5.13}{0.072}.t\right]^{1/2}=1.32(71.25)^{1/2}t^{1/2}=11.14t^{1/2}$$

or, $40 = 11.14t^{1/2}$ $\therefore$ $t = (14/11.14)^{1/2} = 12.89 \approx 13$ days

13.5 Layout of the Tile Drains

The importance in properly laying the tile drain is to ensure the smooth disposal of excess water from the crop field to the point of outlet. The network of the drains should be such that it gives the best drain effect at minimum cost. The lay out usually developed on the contour map of the area, which includes the orientation, the desired slope of the channels and off course, the natural stream where the drainage water is disposed off. Usually, the flow through the drains takes place under the influence of gravity. In some cases where the bed level or water level of the natural stream goes above the level of the drain outlet, the drain water may require to pump out. The drainage system is identified on the basis of orientation of the drainage network. Some of the common types of drainage system are described below.

1. **Natural system:** This is the system in which the main and the lateral drains follow the natural depression of land and the network gives the shape of frame of a tree (Fig.13.6). The system is suitable for uniform texture and poorly drained soils at little slope.

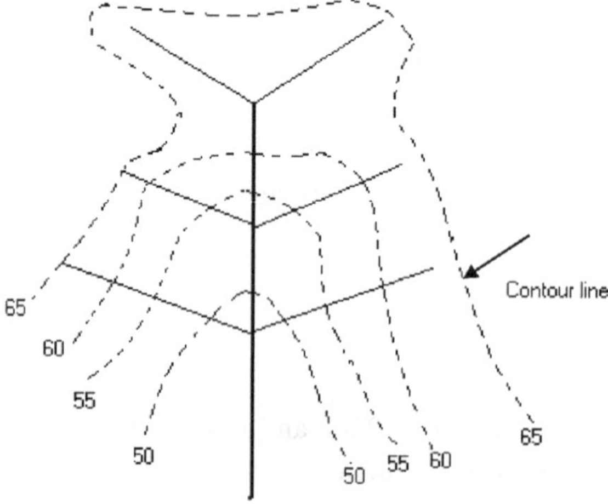

Fig. 13.6. Natural system

2. **Gridiron system:** The lands, which are uniformly wet and sloped towards a particular direction. The main runs through the bottom of the slope and the laterals join to it almost perpendicularly (Fig.13.7).

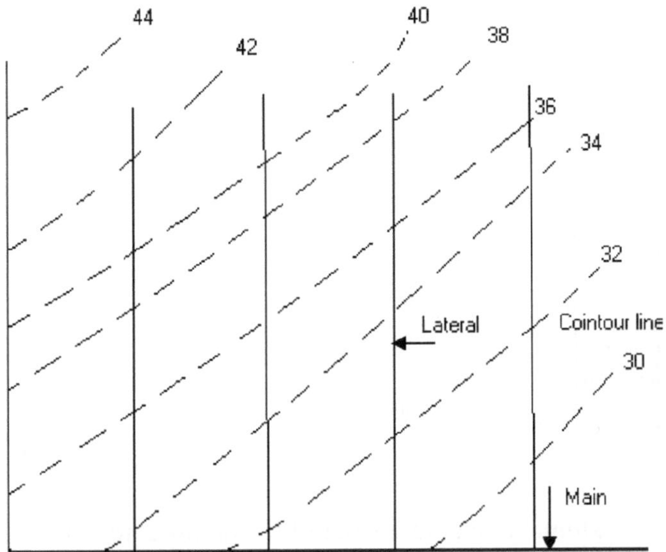

Fig. 13.7. Gridiron system

3. **Herringbone system:** In this system the main and sub-main line follow the natural depression and the laterals join to it laterally from both the sides to give the drain network a herringbone pattern. This system is not economical and adopted only where situation attracts (Fig.13.8).

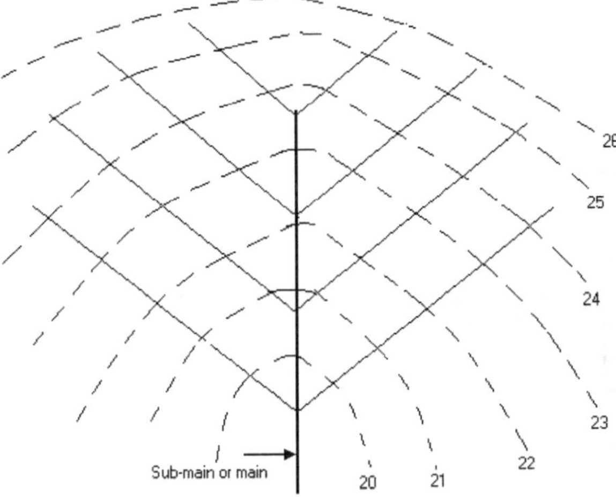

Fig.13.8. Herringbone system

4. **Double main system:** It is in essence adapting two gridiron systems side by side where the sub-mains run parallel to each other through a broad and flat depression. It is not commonly practiced because such situation rarely occurs (Fig.13.10).

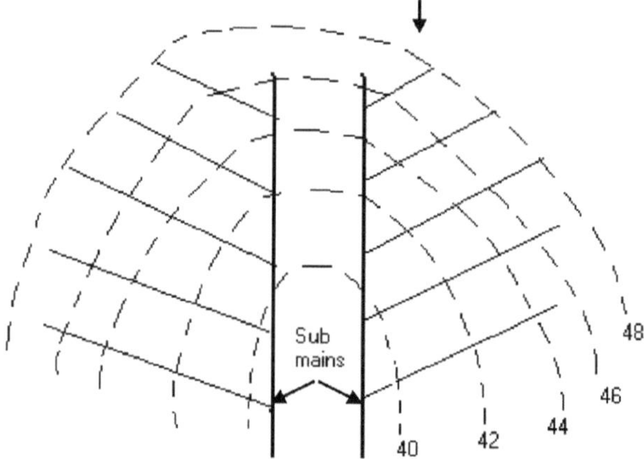

Fig. 13.9. Double main system

5. **Random or zigzag system:** A random lay out of the drains is required for small isolated water logged situation develops due to too much of undulation of land and where the land leveling is not feasible. The main line usually follows the natural drainage line and the laterals and sub-mains connect the individual water logged spots. Sometime it may require adopting one or more parallel system of sub-mains and laterals to provide the required drainage to the individual spots of large size (Fig.13.10).

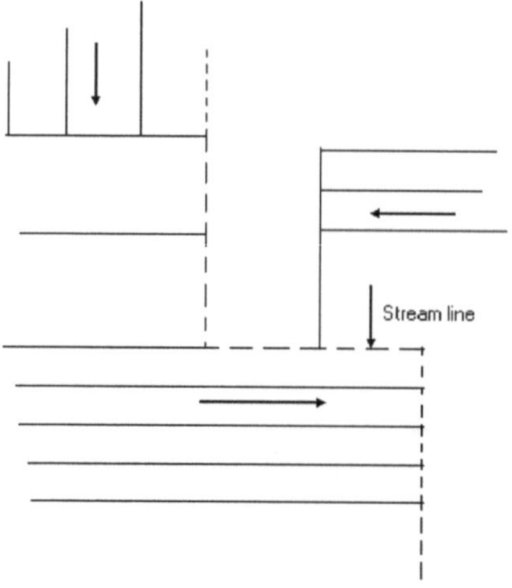

Fig. 13.10. Random or zigzag system

6. **Interceptor or sink drain:** The interceptor drain refers the drain that intercept the seepage water before it encroaches to the adjacent agricultural land. Interceptor drains are very common to earth embankments. Sub-surface interceptor drains are adopted in hilly areas. Usually, a sub-main is used to the bottom of the slope to intercept the seepage water (Fig.13.11).

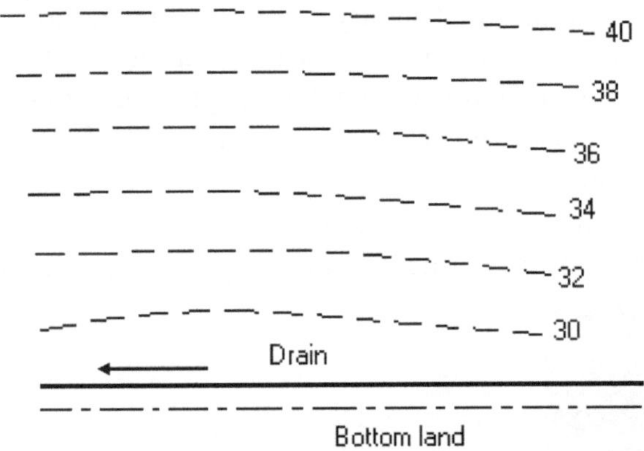

Fig. 13.11. Interceptor or sink drain

7. **Grouping system**: This is the system that forms when a few small systems are adopted to collect the drainage water from number of location. These locations have independent topographic and wetness characteristics (Fig.13.12).

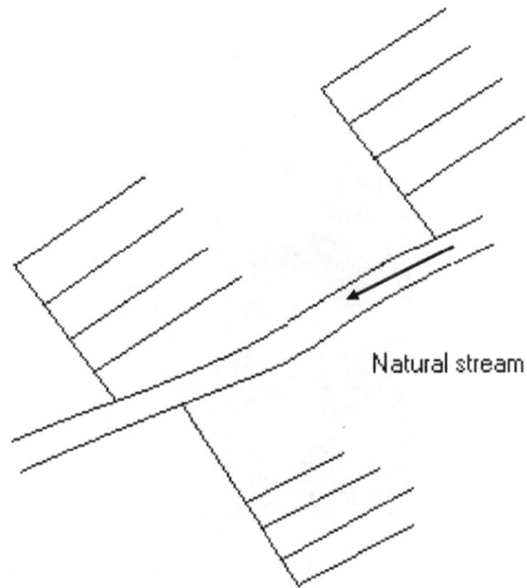

Fig. 13.12. Grouping system

13.6 Drainage Material

The drainpipe and the filter are the two most important materials in sub-surface drainage. The other materials are the fixtures on the drainpipe such as Tee joints, bends, outlet gates, end plug and impermeable plastic sheet. The plastic sheet is used to cover the filter in fine non-cohesive soil. The depth of drain usually varies from 1 to 2 meters. The performance of the drain largely depends on proper and accurate placement of the good quality drain materials. A shorter gap between the two pipes may cause inadequate flow or failure to the drain system and more than the required gap may cause choke the system due to soil flow in to the system. A mal functioning end gate may lead to allow the flowing trash inside the drain and choke it.

The function of the filter is to facilitate the easy and fast entry of gravitational water into the drain though screening and restricting the entry of soil particles.

Courtesy: AICRP on Agricultural Drainage (2002)

Fig.13.13 Clay tiles and coir fiber for sub-surface drainage

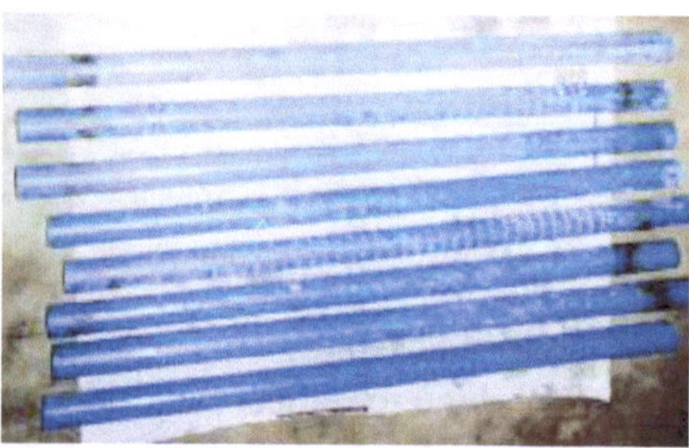

Courtesy: Roy & Rababi, 2003

Fig. 13.14 Perforated HDPE pipes for subsurface drainage

Fig. 13.15. Clay tiles installation

Fig. 13.16. Coir fiber and sand packing for sub-surface horizontal drainage

Courtesy: AICRP on Agricultural Drainage (2002), BCKV

Drain pipe: Drain pipes are made of asbestos cement, cement concrete, rigid or corrugated flexible PVC and baked clay (Fig. 13.13). The diameter of the pipes is usually 80 to 160mm. Higher diameter pipes are used for sub-main or main line of the system. Baked clay cement concrete or asbestos cement pipes are made of shorter lengths, usually, 30, 45 or 60cm. Asbestos pipes are not in use any more because of the asbestos fiber in it proved to be hazardous for health (Bhattacharya & Michael, 2003). Cement concrete or clay pipes can be made by wooden or metal form with appropriate dimensions. Clay pipes are also made in potter's wheel. Making good quality clay pipe deserves skill of the porter. A well baked and good burnt pipe in kiln shows red color and appropriate for use in drain system. Clay pipe should not absorb more than 15 percent of the mass of the pipe when soaked under water for the period of 24 hours in ambient temperature (Decks et al, 1972). Rigid PVC pipes are available in standard length. It is not usually used because it is become costly to have strong thick wall pipes. The flexible corrugation pipes have small wall thickness. The corrugation provides the strength and lesser wall thickness and leads to provide more length of pipe per unit mass of the material. The perforated PVC pipes are available in different diameter and length. The length and diameter depends on the demand of the demand of the user.

In case of short length drain such as cement concrete or baked clay pipes, which are placed one after another to have a continuous line, a certain gap usually less than 3mm provided between the two pipes are for the entry of gravitational water in to the pipes. In bell-mouth baked clay pipes apparently no gap is provided, however, there exists some gap to allow the water to enter into the pipe through the joints. For the purpose of better water entry clay pipes are made perforated on the pipe surface. The perforation may be narrow slit to any geometrical shape and evenly distributed throughout the surface either in clusters or individual manner.

The percentage of open area of a drain pipe usually varies between 0.15 to 2% (Cavelarrs, 1974). Ideally, a drain do not suffer from to water inflow if its perforation to the drainable porosity of the soil (Bhattacharya & Michael, 2003). However, this suggestion is not practicable to many soils. The drainable porosity of a sandy soil is 20%. It is simply not possible to open the drainpipe at this extent due to great risk in strength of the pipe.

Kumathe and Bhattacharya (1984) reported 70, 80 and 90 ml/min steady state discharge at 0.1, 0.2 and 0.4% open area in pipe respectively for a filter thickness of 5cm. For similar condition they found 115, 129 and 122.5 ml/min discharges respectively for 10cm thickness of filter and accordingly they suggested increasing of filter thickness particularly for rigid PVC or flexible pipes when percent open area cannot be increased beyond some limit. However, less increasing the filter thickness for clay or asbestos pipe may not yield much effect because these pipes usually have several times more open area than the rigid PVC or flexible pipes.

Mohammad & Skaggs (1982) studied the effect of varying open area from 0.05 to 2.4% and found increasing trend in discharge with the increase in the open area. However, rate of increase was tapered after certain range. Similar results were found by Roy & Rababi (2004) as a result of laboratory experimentation for discharges at different perforation level. They found no significant change in discharge after the 6% perforation level. They also reported the threshold head to flow 6.1, 5.8, 2.95, 2.2, 1.5 & 1.49cm for 0.5, 1.0, 3.0, 5.0, 6.0 & 7.0% open area respectively.

Diameter of drain pipe: The diameter of tile drain is calculated on the basis of discharge requirement at the outlet of the drain. T the discharge requirement is calculated by knowing the discharge coefficient, spacing and length of the drains and longitudinal slope of the drain line. The drain is assumed full flow without any hydraulic pressure. Flow occur uniform under the influence of gravity. Therefore, Manning's formula can be used for calculating velocity of flow. The value of Manning's roughness coefficient 'n' depends on the type of materials used and the manufacturing quality of the tiles. Bhattacharya & Michael (2003) mentioned Manning's n value 0.014 to 0.019 and 0.015 to 0.02 for PVC and clay tiles made by potters respectively. However, they suggested the safe n value of 0.0143 for design purpose for the pipes made of corrugated PVC, concrete and baked clay. The calculation of diameter of a drain following the procedure as stated above sometime may not be much useful; because, only a few diameter tiles are available in the market. However, by calculating the requirement of minimum diameter one can select either that exact diameter or next higher diameter pipe available in the market. In contrary to this, selecting a particular diameter one can calculate the capacity of flow and thereby the length of pipe.

Filter: It is the material more pervious than the soil and placed around the drain pipe. Filter serves the purpose of checking inflow of soil particles to the drain pipe providing zone of higher conductivity around the drain pipe and working as a bedding material to support the drainpipe and uniformly distribution of soil load. Filter material may be classified into three types, namely mineral (sand or gravel), organic (crop residues or coir fibres) and synthetic (sheet with micro perforations) (Fig.13.16). Mineral and synthetic fibres are not biodegradable. Therefore, they have long life. Organic fibres are biodegradable and naturally longevity is questionable. However, it is reported that organic and mineral filter can perform equally well over the year with practically no difference in their drainage characteristics and no change in their performance (Bhattacharya & Michael, 2003).

Filter design: Filter design refers the selection of appropriate grade of filter, which can effectively check soil inflow though it. The filter itself should not flow into the drainpipe through the perforation of drainpipe. Therefore, there is certain relation between the size of the filter and slot size of the perforated drainpipes. Design of organic filters is not based on soil particle size distribution. However, for mineral filters, there is a procedure for filter design (BIS, 1981). Examinations of many drainpipes with a deposit of sand shows that a great proportion of it entered at the time of installation or soon after or there after of a major storm. Soils with good spread of particle sizes do not offer serious siltation problem. The soils of particle sizes do not offer serious siltation problem. The soils of particle sizes lie between 50m to 100m causes the real problem. The uniformity of grading has a marked effect on the silting tendency. The uniformity of grading,

$U = d_{60} / d_{10}$

Where,

d_{60} = Particle size at which 60% passes sieve

d_{10} = Particle size at which 10% passes sieve

Regarding the influence of uniformity of grading,

U > 15: No tendency of siltation

U = 5 - 15: Limited tendency of siltation

U < 5 : High tendency of siltation

The clay-soil ratio also plays an important role to silting. Where the ratio exceeds 0.5, silting may be a problem. The silting tendency of different soils in drainpipes is shown in Table 13.6.

Table 13.6. Tendency mineral soils to cause siltation in drainage pipes

Soil type	Silting tendency	Particle size		
		Clay ((< 2μ))	Silt (2μ–20μ)	Sand (20μ–0.6mm)
Sand, Sandy loam, Loamy sand	Considerable	< 8%	< 25%	< 70%
Sand, Sandy loam, Loamy sand	Slight	< 8-10%	< 20%	> 70%
Loamy sand, Sandy loam, Loamy silt	Considerable	< 8%	25-55%	> 40-70%
Sandy loam, Loamy silt, Silt loam	Slight	< 8-12%	20-55%	> 40-75%

Source: Roy & Rababi, 2004

13.7 Mole Drain

Mole drains are the sub-surface unlined small diameter channels made in the cohesive soils for drainage of water. The moles are the small diameter cylindrical bullet nosed plug which are attached to the shank of the mole plough and drawn by the tractor to form similar diameter hole in the in soil at specified depth

(Fig.13.17). The shank of the plough is of small thickness and thus develops small fracture when drawn in soil. The sub-surface water joins to this fracture from the either directions to get delivered to mole drain. The mole drain then carries the water to discharge to the drainage ditch at the end of the field. The mole diameter varies between 7.5 to 10 cm, depth 50 to 75 cm and the spacing from 1.5 to 10m.

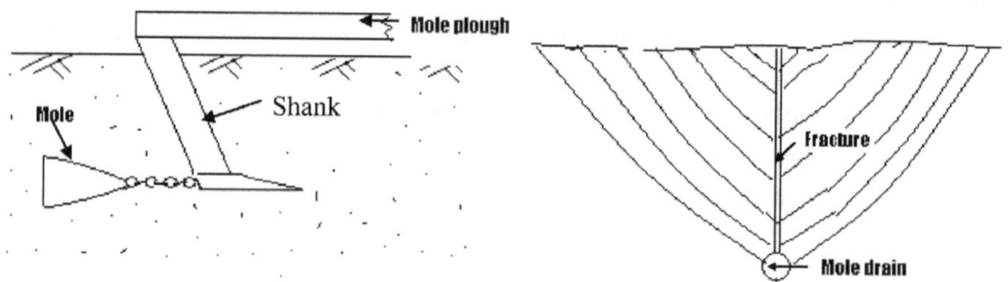

Fig. 13.17 Mole drain

Mole drains are not suitable to loose and heavy plastic soils. The mole drain will collapse in loose soil and in heavy soil the mole seals the soils that restrict the movement of water to the drain. Usually the durability of a mole drain is 3-5 years. However, the durability greatly influenced by the type of soil, length and slope, depth and spacing, adequacy of drain outlet and the conditions during the time of construction particularly the moisture in soil. The depth of the mole drain varies between 50 to 75cm and the spacing 1.5 to 10m. Mole drains are laid in similar way to tile drains where soil condition permits.

Questions and Problems

13.1 What is sub-surface drainage? What are the factors causes to increase the ground water level? What are the advantages of sub-surface drainage over the surface drainage?

13.2 What is the factors influence in determining drainage coefficient in sub-surface drainage system?

13.3 Derive the ellipse equation in sub-surface flow.

13.4 What are the assumptions in Hooghoudt's equation? Derive the Hooghoudt's equation for the water table in equilibrium with rainfall.

13.5 Derive the Hooghoudt's equation modified for partially penetrating drain in layered and unlayered soil.

13.6 Derive the expressions for flow in to tile drain accounting elongation and convergence of flow.

13.7 What are the important materials used for sub-surface drainage? Discuss the open area in drainpipe in relation to its stability and drain ability.

13.8 With a neat sketch discuss the use of filter materials around the drainpipe.

13.9 What are the suitable situations for a mole drain? How moles are constructed? What the factors influence the longevity of a mole drain?

13.10 Derive the expression $\frac{\partial^2 z}{\partial x^2} = \frac{\partial^2 z}{\partial y^2} = \frac{f}{Kh}\frac{\partial z}{\partial t}$ for non-steady state sub-surface flow.

13.11 From the following data draw the phreatic line of a sub-surface flow.

R = 2cm/day, K= 10cm/day, S = 100m, h = 0.5m.

Ans.	x(m)	0	5	10	20	30	40	50
	y(m)	0.5	9.76	13.43	17.89	20.49	21.91	22.36

13.12 What is the discharge capacity of a tile drain if it drains water from an area of 10ha with a discharge coefficient of 1.0cm?

Ans: $1.157 \times 10^{-3} m^3/s$

13.13 A tile drain is used for sub-surface drainage for an area of 5ha. The tile is laid to 0.15% gradient. The roughness coefficient of the tile material is 0.015. What is the size of the tile drain if drainage coefficient is 1.5cm?
Ans: D = 20cm

13.14 A25cm tile drain with a slope of 0.25% take care the sub-surface drainage of 20ha. What is drainage coefficient? Assume Manning's n 0.01. Ans: 1.67cm

13.15 The outflow of a tile drain is $500 m^3$/day, spaced 20m, water table is maintained at a height of 4m from the impervious layer and at 1.0m above the drain. What is length of the tile drain if the hydraulic conductivity of soil is 1.0m/day? Ans: 357m

13.16 The sub-surface tile drain is proposed to maintain the water table 1.0m below the ground level where the expected constant recharge rate is 5cm/h, hydraulic conductivity 1m/day and the impervious layer is at 6m. What is the depth of drain and outflow rate from a square area of 1ha?
Ans: S = 8.5m, Q = $42.5 m^3$/h, d=2.0m, m=3m

13.17 Design a deep open drain in a silty loam soil to drain out the ground water at a rate of $1.65 m^3/s$. The maximum permissible velocity of flow, Manning's n and drain laid at slope are 0.7m/s, 0.03 and 0.1% respectively.
Ans: b = 0.2m, d = 1.2.

13.18 Determine the spacing of vertical ditches and the tile drains for drainage coefficient 4mm/day, hydraulic conductivity K_a&K_b are 0.5&0.6m/day respectively, depth of impermeable layer at 5.0m from the bottom of the vertical ditch/tile drain, radius of tile drain 0.05m &hydraulic head 0.75m.

Ans: Vertical ditch, S = 64m, Tile drain, S = 50m

13.19 Calculate the spacing by using Hooghoudt's equation for the following data.

Hydraulic conductivity = 0.5m/day
Constant recharge rate = 2.5mm/day
Depth of impermeable layer from the tile drain center = 6m
Radius of tile drain = 6cm
Hydraulic head = 0.5m

Ans: S = 75m

13.20 Compare the spacing of tile drains designed following the Kirkham's & Dagan's equation by using the following data:

Hydraulic conductivity, $K_a = K_b = 0.5$m/day

Recharge rate = 3.00mm/day

Depth of impermeable layer from the tile drain center = 5cm

Radius of tile drain = 5cm

Hydraulic head = 1m

Ans: Kirkham's S = 61m,Dagan's S = 62m.

13.21 There are two distinct layers in a soil profile where pipe drain of 10cm diameter is to be installed in the top layer at 1.5m above the interface between the two layers. Determine the spacing of the drains and the radial component of flow with the following information.

q = 5mm/day
Depth of bottom layer = 3.5m
Hydraulic conductivity of the bottom layer = 2.0m/day
Hydraulic conductivity of the top layer = 0.4m/day
Available head of flow = 0.75m

Ans: S = 47m, h_r=0.65m.

13.22 Select the appropriate answer from the following.

1. Sub-surface drain
   a) Increases permeability of soil
   b) Increases hindrance to the movement of farm machinery
   c) Reduces the soil erosion
   d) Reduces possibility of root zone depth
2. In paddy field sub-surface drainage is relevant to
   a) Saturated hydraulic conductivity b) Rainfall intensity
   c) Percolation rate d) Infiltration rate

3. In D-F assumptions stream lines to a drain is
   a) Horizontal
   b) Vertical
   c) Downward
   d) Curvilinear
4. In Hooghoudt's equation rainfall rate is assumed
   a) Equal to rate of outflow
   b) More than the rate of outflow
   c) Less than the rate of outflow
   d) No relation between rainfall rate and outflow
5. Hooghoudt's equation is derived for
   a) Constant draining of water
   b) Constant rate of removal
   c) Constant water table
   d) Constant rainfall
6. A sub-surface tile drain installed for an area of 1ha. If the drainage coefficient is 5mm/day, the outflow rate is
   a) $20m^3$/day
   b) $30m^3$/day
   c) $40m^3$/day
   d) $50m^3$/day
7. The impermeable layer is at 5m from the bottom of the drain and depth of water in the drain is 0.5m. Average elongation of the stream line is
   a) 2.23m
   b) 2.25m
   c) 2.27m
   d) 2.29m
8. Excess gap between the sub-surface drainpipes may cause
   a) Inadequate flow
   b) Additional cost
   c) Chock the system
   d) Better filtration
9. Ernst equation assumes sub-surface flow towards pipe drain is only
   a) Horizontal
   b) Vertical
   c) Horizontal & vertical
   d) Horizontal, vertical & radial
10. Ernst equation estimates the spacing of drain when it is placed to
    a) On impervious layer
    b) On the interface of soil layer
    c) Above the interface of soil layers
11. Modified Hooghoudt's equation for the computation of drain spacing is applicable to
    a) Homogeneous soil
    b) Anisotropic soil
    c) Heavy clay soils only
    d) Layered soils

(GATE, 2017)

12. Mole drain is the most suitable drainage system for

a) Heavy clay soil b) Loam soil

c) Sandy soil d) Silty soil

(GATE, 2015)

13. Dupuit-Forchheimer assumptions are used for analyzing groundwater flow in

a) Confined aquifer

b) Leaky confined aquifers

c) Unconfined aquifers

d) Both confined and unconfined aquifers

(GATE, 2015)

14. The depth from the surface to subsurface tile drains, impermeable soil layer and the highest water table are measured as 2.8m, 5.0m and 0.8m, respectively. The effective hydraulic head for drainage in meter is

a) 0.8 b) 2.0

c) 2.2 d) 4.2

(GATE, 2018)

15. The thickness of the filter around a sub-surface drain of radius 7.5cm is 2.5cm. The effective diameter of the drain is

a) 15 cm b) 17.25 cm

c) 20.0 cm d) 22.5 cm

Ans.

1.	c)	2.	c)	3.	a)	4.	a)	5.	c)	6.	d)	7.	c)	8.	c)
9.	d)	10.	d)	11.	d)	12.	a)	13.	d)	14.	d)	15.	c)		

13.23 Write True or False of the following statements:

1. The radial component of sub- surface flow is more important than horizontal and vertical.
2. Ernst equation is an equation to estimate the spacing of drain on the impermeable layer.
3. Mole drains are the sub-surface unlined small diameter channels.
4. There is no relation between size of the filter pipe with its slot size.
5. In a fully penetrating drain of deeper depth the flow lines are more or less parallel.
6. In sub-surface drainage hydraulic design parts select the combination of spacing & depth of the drain network in consideration to diameter of the drain.
7. The drainage through sub-surface drains is estimated by using Manning's equation.

8. If rainfall intensity is higher than the infiltration rate then rainfall intensity is the drainage coefficient.
9. Drainage coefficient can be modified by using large diameter drains in sub-surface drainage system.
10. If saturated hydraulic conductivity of the soil around the drain and below the soil surface differs then highest of these is the drainage coefficient.

Ans.

1. True 2. False 3. True 4. False 5. True 6. True 7. True
8. False 9. False 10. False

References

Bhattacharya, A.K. and Michael, A.M. (2003). Land Drainage Principles Methods and Applications. Konark Publishers Pvt. Ltd., New Delhi.

BIS. (1981). Code for design and laying of mineral filters for tile drain system. IS 9979. Bureau of Indian Standards. New Delhi.

Bouwer, Herman and Schilfgaarde, Jan Van. (1963). Simplified method of predicting fall of water table in drained land. Trans. ASAE. 6(4).

Dagan, G. (1964). Spacing of drains by an approximate method. ASCE J. of Irr. And Drain. Div. Proc. Paper No. 3824. Pp 41-46.

Dumm, Lee D. (1954). New formula for determining depth and spacing of subsurface drains in irrigated lands. Agric. Engg.35.pp 726-30.

Kirkham, Don and Gaskell R.E. (1951). The falling water table in tile and ditch drainage. Soil Sc. Soc. of Amer. Proc. 15. Pp 37-42.

Kirkham, Don (1958). Seepage of steady rainfall through soil into drains. Treans. American Geophisical Union. 39(5). Pp 892-908.

Kumathe, S.S. and Bhattacharya, A.K. (1988). Model studies on the effect of drain tube inflow area and filter thickness on drain discharge. ISAE J. Agric. Engg. 25(1). Pp 36-45.

Lal, C., Chauhan, H.S. and Sewa Ram (1992). Developing a new drain spacing formula. Ind. J. Agril Engg. 2(2).pp 118-30.

Luthin, J.N. (1959). Falling water table in tile drainage-II. Proposed criteria for spacing the drains. Trans. ASAE. 2(1).

Luthin, J.N. (1970). Drainage Engineering. Wiley Eastern Pvt. Ltd., New Delhi.

Mohammad, F.S. and Skaggs, R.W. (1982). Effect of drain tube openings on transient drainage. Proc. 2nd International Drainage Workshop, Dec. 5-11. Corrugated Plastic Tube Association of the United States and Canada

Roy, D. & Rababi, S. (2004). Optimization of Performance of Sub-surface Drainage Pipes. A B.Tech. dissertation submitted to Bidhan Chandra Krishi Viswqavidyalaya, West Bengal.

Shilfgaarde, Jan Van. (1965). Design of tile drainage for falling water tables. J. Irri. & Drai. Div. Proc. Of ASCE. 89(IR 2) pp 1-13.

Shilfgaarde, Jan Van. (1965). Limitation of Dupuit-Forchheimer theory in drainage. Trans. ASAE. 8(4).pp 515-19.

Shilfgaarde, Jan Van, Kirkhan Don and Frevert, R.K. (1956). Physical and mathematical theories and ditch drainage and their usefulness in design. Agricultural Experiment Station. Iowa State College. Tes. Bul. No.436.

USDA (1973). Drainage of Agricultural Lands. Soil Conservation Service. United States Department of Agriculture. Water Information Center. Syosset, New York.

Index

A

Actual crop evapotranspiration 270, 334
Actual vapor pressure 281, 282, 283, 335
Advantage of sub-surface drainage system 547
Advantages of lining 395
Aeolian soils 193
Albedo 279, 284
Alkali or sodic soil 434
Alkaline soil 243, 465
Alkalinity hazard 465
Alkalisation 435
Alluvial soils 193
Apparent specific gravity of soil 22, 32, 48, 49, 131, 202
Aquiclude 339
Aquifer 129, 337, 338, 339, 340, 341, 342, 343, 344, 346, 347, 350, 351, 354, 355, 356, 357, 359, 360, 361, 362, 363, 364, 365, 366, 367, 368, 369, 370, 370, 371, 372, 373, 374, 375, 377, 481, 482, 483, 485
Aquifer Properties 337, 240, 265
Aquifuge 339
Aquitard 338
Asphalt concrete lining 399
Atmospheric pressure 207, 209, 269, 276, 279, 280, 283, 286, 338, 574

B

Bamboo-reinforced precast concrete 386, 410
Barlow's Tables 503
Base period 86, 87
Bazin's formula 145
Bedding system 492, 493
Benefits of drainage 490
Bentonite soil lining 401, 402
Black soils 193
Blaney-Criddle Method 333
Blending water supplies 458
Bottom contraction 74
Branch canal 135, 136, 187, 188
Brick lining 394, 395, 398
Buried membrane canal 402
Burnt clay tile lining 396

C

Can evaporimeter 118, 132
Capillarity 210, 217, 262
Capillary fringe 337, 338
Carrier canal 137, 188
Cation exchange 196, 434
Cement concrete lining 384, 385, 399, 400, 405
Cipoletti weir 74, 77,, 426, 427
Classification of Irrigation Water 441, 444, 467
Clay loam soils 194
Clay soils 194, 196, 200, 278, 431
Coastal salinity 483
Colluvial soils 193
Compound interest method 406, 407, 408, 427
Consumptive use Potential evapotranspiration 270
Control of Salinity Problem 446
Co-ordinate method 64
Critical stage of growth 114
Crop coefficient 117, 271, 272, 273, 295, 296, 297, 323
Crop evapotranspiration 270, 271, 282, 295, 305, 306, 308, 309, 310, 454459, 460
Crop factor 271, 292, 293
Crop period 86, 456
Crop tolerance 447, 449, 455, 458, 498
Cultural practices 456, 462
Current meter method 60
Curve fittings 222
Curve Number Method 520, 5227
Cutthroat flume 82, 83

D

Dagan's equation 541, 564
Daily consumptive use 98, 131
Darcy's law 239, 240, 256, 260337, 347, 359, 366, 375, 538539, 541, 568, 574
Darcy-Weisbach formula 145
Degree of saturation 202, 205, 262, 263, 344
Density index 202
Density of soil 105, 201, 263, , 448, 452, 453
Depreciation 406, 407, 408, 410, 411, 412, 413

Desert soils 194
Design of Surface Drainage System 497
Dethridge method 62, 90
Dew point temperature 280, 282, 302, 317, 335
Dickens formula 502
Diffusivity 242, 245, 248, 249, 250, 252, 257, 261, 263, 264
Dilution method 85
Disadvantages of lining 384
Dispersed soil 435
Double main system 582
Double ring infiltrometer 12, 221, 222, 263
Drainage coefficient 497, 498, 504, 507, 513, 547, 536, 537, 555, 559, 560, 561, 563
Drainage Material 584
Drainpipe 584, 585, 586, 587
Dry density of soil 201
Dry unit weight of soil 201
Dupuit-Forchheimer (D-F) assumptions 537
Duty of water 86

E

Earth lining 400
Economic (irrigation) efficiency 97
Economics of Canal Lining 405
Effect of water logging 490, 497
Effluent 339, 340
Effluent stream 339
Electrical resistance 113, 439
Ellipse equation 539, 540, 541
Empirical formula 36, 537
Empirical formulae 115, 270, 293
End contraction 74, 76
Energy Balance 275
Energy balance in ET 275
Equation of continuity 240, 355, 573, 574
Equation of Motion 355
Equivalent depth 544, 546, 547, 548, 549, 550, 561, 562, 576, 579
Equivalent weight 434, 436, 463
Ernst Equation 568, 569
Estimation of runoff 498
Evaporimeter 115, 117, 118, 293, 294
Evapotranspiration 95, 106, 115, 118, 269, 270, 271, 275, 279, 285, 289, 290, 292, 293, 294, 295, 297, 300, 301, 304, 305, 306, 308, 309, 310, 312, 315, 317, 323, 447, 454, 459, 460, 459, 498, 566, 576, 577
Exchangeable cation 434, 439
Exchangeable complex 434
Exchangeable sodium percentage 435, 439, 440
Exposed surface lining 384
Extraterrestrial radiation 283, 284, 285, 286, 309, 310, 311, 317, 324

F

Factors affecting ground water quality 482
Factors affecting infiltration 214
FAO-Monteith method 308
Feeder canal 137
Feel and appearance 111, 118
Fertilization 271, 456, 457
Filter 535, 560, 584, 585, 586, 587
Filter design 587
Float method 58, 59
Flow in confined aquifer 365, 375
Flow in Wells 365
Flumes 69, 8182, 84, 136, 425
Forest soils 194
Free board 46, 143, 394506, 511, 518

G

Garret's Diagram 176
Genesis of soil structure 197, 262
Ghosh's Model 220, 263
Glacial soils 193
Grassed Waterways 40, 514, 515
Gravimetric method 111, 133, 290
Gravitational potential 207, 208, 216
Gridiron system 581
Gross Irrigation Requirement 98, 99, 119, 131
Ground water quality in India 483
Gypsum requirement 436, 438, 439, 463, 464

H

Hargreaves method 309
Herringbone system 581
History of Irrigation in India 473, 486
Hooghoudt's equation 540, 541, 55, 557, 588, 590
Hose pipe 57
Hydraulic radius 142, 144, 146, 147, 148, 149, 156, 157, 164, 177, 187, 504, 515, 519
Hydraulic resistance 346
Hydraulic slope 17, 142, 144, 156, 157, 187
Hydro power 470
Hysteresis 229, 233

I

Increased population 118, 132
Indicator plant 114
Infiltration models 214

Infiltration Problem 461, 462, 463
Inflow-outflow method 292, 425
Influent stream 339, 340
Inglis and Desouza formula 502
Inland salinity 483
Interception ditch system 506
Interceptor or sink drain 583
Intermediate zone 337, 338
Intrinsic permeability 345, 346, 351
Inundation canal 135
Irrigation canal 135, 136, 163, 403
Irrigation Channels 6, 61, 80, 135
Irrigation development under the Five Year Plan 477
Irrigation efficiency 93, 99, 106, 101, 473
Irrigation Frequency or Irrigation Interval 119
Irrigation management 463
Irrigation Period 56, 98, 99, 120, 128, 129
Irrigation Water Quality 433, 441
IW/CPE ratio 115, 120, 129

K

Kennedy's formula 163, 165
Khosla's formula 502
Kirkham's equation 550
Kor period & Kor depth 86
Kutter,s formula 146

L

Lacey's formula 177
Lacey's shock theory 181
Lal et al (1992) formula 576
Land grading 456, 491
Land Resources 470
Land smoothing 456, 491, 492, 493, 530
Laplace's equation 240, 263
Latent heat of vaporization 279, 280
Lateritic soils 194
Leaching 29, 30m 118, 333, 441, 443, 444, 446, 447, 448, 453454455, 458, 459, 460, 461, 465, 491, 537, 472
Leaching requirement 446, 4447, 448, 453, 454, 455, 458, 459, 460, 461, 465, 537
Leaf temperature 114
Leaf water potential 114
Leakage factor 346
Leaky aquifer 339
Limitation of CN method 526
Lindley's formula 176
Lining irrigation channels 383, 396
Loam 9, 10, 12, 29, 34, 41, 45, 110, 111, 143, 156, 160, 177, 194, 196, 239, 244, 247, 260, 454, 463, 499, 504, 505, 507, 508, 509, 510, 513, 587
Loam soils 194
Luthin (1959) formula 576
Lysimeter 289, 290, 291, 293, 294, 323

M

Main canal 135, 136, 187
Major distributor 136, 187
Management of toxicity problem 465
Manning's formula 144, 146, 519, 536, 586
mass wetness 202, 262
Mean saturation vapor pressure 281, 282
Measurement of Evapotranspiration 289
Measurement of Seepage 416
Measuring Structures 55, 69
Meteorological parameter 115, 305
Meter gate 73
Milliequivalent per liter 434
Minor distributor 136, 187
Model of second parameter of Philip's equation 219
Modified Penman Method 305, 306
Molar solution 435
Mole Drain 587, 588, 589
Most economical channel section 147

N

Nappe 74
Natural system 580
Navigation canal 135
Net Irrigation Requirement 32, 35, 98, 99, 119, 290
Net long wave radiation 284, 305, 310, 312
Net radiation 275, 285, 302, 309, 312, 317
Non-soluble alkali soil 435

O

One dimensional flow without recharge 362
Open channel 6, 57, 58, 63, 66, 80, 85, 135, 136138, 141, 142, 144, 153, 156, 348, 491
Open ditch 36, 46, 495, 496, 504, 542
Operational efficiency 95, 106, 384
Orifice 69
Osmotic potential 207, 208, 440
Osmotic pressure 436, 437, 455

P
Pan evaporimeter method 293
Parallel deep ditch system 494
Parallel ditch system 493, 494
Parts per million 434
Pavlovskey's formula 146
Peak period consumptive use 119, 274, 275
Penman method 302, 305, 306
Perched aquifer 339, 381
Perennial 135, 451
Permanent 28, 34, 40, 109, 135, 177, 196, 384, 394, 481, 524
Permeability 30, 198, 201, 239, 338, 343, 344, 345, 346, 349, 350, 351, 352, 353, 354, 356, 368, 370, 377, 383, 401, 403, 433, 442, 443, 444, 492, 554, 562
Permeameter 349, 352
Philip's Infiltration Equation 216
Pitot tube 66
Placement of seed 456
Plant Parameter 113
Poiseuille's law 237
Ponding method 416, 430
Porosity 3, 48, 102, 104, 198, 200, 201, 204, 214, 231, 254, 262, 338, 339, 343, 344, 345, 348, 351, 535, 557, 574, 576, 578, 579, 585
Power canal 135
Prefabricated concrete sections 585
Pressure plate apparatus 229, 233, 264
Pressure potential 208
Priestly-Taylor estimate 308
Priority of Water Courses in Lining 413
Project irrigation efficiency 93
Psychrometric constant 279, 286, 302, 309

Q
Quality diagram 443

R
Radiation 114, 275, 278, 283, 284, 285, 286, 295, 297, 299, 300, 302, 305, 306, 308, 310, 312, 314, 316, 317, 322, 324, 328
Radiation method 297
Random ditch 493
Random or zigzag system 582
Rational formula 498, 499, 500, 507, 514
Red soils 194
Reference crop evapotranspiration 270, 308, 310, 312
Reference surface 277, 278, 279
Regime Approach 163
Relative density 178, 202
Relative humidity 281, 283, 298, 302, 306, 314, 320
Relative short wave radiation 284
Relative sun shines duration 284
Residual soils 193
Resistance to ET 275
Richard's equation 241, 263
Roof tiles 399, 411
Rough method of farmers 118
Runoff relationship 522
Ryves formula 502

S
Saline soil 435
Saline-alkali soil 435
Saline-sodic 435
Salinity classification 443
Salinity hazard 439
Salt Affected Soils 431, 437
Salt concentration 85, 86, 89, 435, 436, 437, 440, 448, 574
Salt Problems 46, 433
Sand 10, 29, 34, 41, 58, 82, 112, 118, 132, 143, 176, 192, 193, 194, 195, 196, 199, 214, 239, 344, 385, 386, 387, 392, 394, 395, 399, 401, 403, 451, 43, 584, 286, 287
Saturated bulk density 201204, 205
Saturation extract 435, 436, 448
Scheduling irrigation 109, 111, 120, 130, 131
Schilfgaarde (1963) formula 576, 579
Schilfgaarde (1964) formula 576
Seasonal consumptive use 274, 291, 292
Semi-confined 339, 346
Sharp crested weir 74
Shortcrete lining 394
Silt loam 41, 45, 143, 194, 196, 499, 508, 509, 510, 587
Sinking fund method 406
Slope of saturation vapor pressure 282
Sodicity hazard 440
Sodium adsorption ratio 435, 437, 461
Sodium calcium activity ratio 440
Soil and water amendments 462
Soil cement 396
Soil extract 435
Soil heat flux 275, 285, 309, 317, 323, 335
Soil Moisture Characteristics 228, 229, 230, 231, 232, 233, 234, 235, 242, 254

Soil moisture characteristics curve 231, 232, 233, 234, 235, 242. 254
Soil moisture content 35, 105, 109, 110, 112, 113, 119, 123, 124, 202, 204, 205, 214, 228, 234, 235, 248, 248, 263, 278, 291
Soil Parameter 113, 130
Soil structure 196, 197, 198, 199, 201, 262, 461, 490,
Soil Texture 33, 41, 194, 195, 198, 201, 228, 262, 291, 441, 461, 499
Soil Water Uptake by Plants 261
Soil water zone 337, 338
Soil-water potential 131, 207
Soil-Water Relationship 199
Solar or short wave radiation 284
Sorptivity model 217
Sources of excess water 489
Specific retention 344
Specific storage 381
Specific storage 341, 342, 343
Specific surface 344, 345, 436
Specific yield 344, 353, 354, 355
Steady State Sub-surface Drainage Design 536
Stomatal resistance 114, 277, 278
Stone and concrete block lining 395
Storage coefficient 342, 343, 375, 377
Straight-line method 406, 427
Strange's Table's 503
Stratification 350, 351
Stress in aquifer 340
Sunken screen pan evaporimeter 293
Surface drainage System 491, 497, 498, 535, 536, 588
Surface inlet 495
Surface tension 209, 210, 212, 236, 262, 344, 347

T

Tensiometer 112113, 124
Tension apparatus 229, 230
Thornthwaite method 300
Timing of irrigation 456
Toxicity 433, 457, 464, 465
Tracer method 55, 84
Transmissibility 346, 350, 368, 369, 371, 372, 375
Transpiration 261, 270
True SAR 435, 440
Types of lining 384, 385

U

Ultimate irrigation potential 470, 479, 486
Unconfined aquifer 338, 339, 340, 341, 354, 355, 358, 359, 360, 362, 363, 365, 367, 368, 370
Unconfined flow 358, 359, 360
Unconfined flow with recharge 359
Unsteady flow in confined aquifer 375
USWB Class A Pan 293

V

Vapor pressure 280, 281, 282, 283, 286, 313, 318, 322, 333
Velocity-area Method 57
Venturi meter 66, 68
Viscosity 211, 212, 214, 237, 262, 345, 348, 349, 351, 352
V-notch weir 74, 77, 78
Void ratio 200, 201, 202, 203, 205, 262, 344
Volcanic soils 193
Volumetric method 55, 57
Volumetric water content 202, 248

W

Water application efficiency 21, 32, 39, 42, 94, 95, 97, 106, 123
Water Bearing Formation 337
Water budget equation 113
Water conveyance efficiency 93, 94, 97, 101
Water courses 3, 93, 136, 187, 413, 426
Water distribution efficiency 103
Water meter 63
Water Movement in Soils 236
Water requirement and availability 473
Water Resources 471, 470, 473, 480, 481, 486
Water Resources of the World 480, 481
Water storage efficiency 84, 95, 97, 98, 106
Water use efficiency 12, 34, 93, 95, 96, 97, 100, 106, 130
Weight of soil 201, 205
Weighted curve number 526
Weir crest 73, 74, 77, 80
Weir pond 73, 74
Weir scale or gauge 74
Weirs 69, 73, 74, 77, 79, 82, 84
Wetted perimeter 21, 142, 144, 147, 156, 157, 163, 183, 405, 408, 413, 417, 418, 419, 420, 421, 423, 424, 427, 504, 515